17th International Conference on Microscopy of Semiconducting Materials 2011

Journal of Physics: Conference Series Volume 326

Cambridge, UK
4-7 April 2011

ISBN: 978-1-61839-296-1
ISSN: 1742-6588

Printed from e-media with permission by:

Curran Associates, Inc.
57 Morehouse Lane
Red Hook, NY 12571

Some format issues inherent in the e-media version may also appear in this print version.

Printed by Curran Associates, Inc. (2011)

For permission requests, please contact the Institute of Physics
at the address below.

Institute of Physics
Dirac House, Temple Back
Bristol BS1 6BE UK

Phone: 44 1 17 929 7481
Fax: 44 1 17 920 0979

techtracking@iop.org

Additional copies of this publication are available from:

Curran Associates, Inc.
57 Morehouse Lane
Red Hook, NY 12571 USA
Phone: 845-758-0400
Fax: 845-758-2634
Email: curran@proceedings.com
Web: www.proceedings.com

TABLE OF CONTENTS

INVITED PAPERS

SESSION A: EPITAXY OF WIDE BAND-GAP NITRIDES

SESSION B: STRUCTURAL CHARACTERISATION OF POLY- AND NANOCRYSTALLINE MATERIALS

SESSION C: STRAINED LAYER CHARACTERISATION

SESSION D: ANALYTICAL TRANSMISSION ELECTRON MICROSCOPY AND ATOM PROBE MICROSCOPY

SESSION E: QUANTUM WIRES AND QUANTUM DOTS

SESSION F: STUDIES OF PROCESSED SILICON AND RELATED MATERIALS

SESSION G: DEVICE STUDIES

SESSION H: ADVANCED SCANNING ION AND ELECTRON BEAM APPLICATIONS

SESSION J: SCANNING PROBE APPLICATIONS

PREFACE

This volume contains invited and contributed papers from the 17th international conference on 'Microscopy of Semiconducting Materials' held at Churchill College, University of Cambridge, on 4–7 April 2011. The meeting was organised under the auspices of the Institute of Physics and supported by the Royal Microscopical Society as well as the Materials Research Society of the USA.

This conference series deals with recent advances in semiconductor studies carried out by all forms of microscopy, with an emphasis on electron microscopy and related techniques with high spatial resolution. This time the meeting was attended by 131 delegates from 25 countries world-wide, a record in terms of internationality.

As semiconductor devices shrink further new routes of device processing and characterisation need to be developed, and, for the latter, methods that offer sub-nanometre spatial resolution are particularly valuable. The various forms of imaging, diffraction and spectroscopy available in modern microscopes are powerful tools for studying the microstructure, the electronic structure, the chemistry and also electric fields in semiconducting materials. Recent advances in instrumentation, from lens aberration correction in both TEM and STEM instruments, to the development of a wide range of scanning probe techniques, as well as new methods of signal quantification have been presented at this conference. Two examples of topics at this meeting that have attracted a number of interesting studies were:

- the correlation of microstructural, optical and chemical information at atomic resolution with nanometre-scale resolved maps of the local electrical fields in (In,Al)GaN based semiconductors and
- tomographic approaches to characterise ensembles of nanowires and stacks of processed layers in devices

Figure 1. Opening lecture by Professor Sir Colin J Humphreys.

Each manuscript submitted for publication in this proceedings volume has been independently reviewed and revised where necessary. In the end, 61 manuscripts were accepted for publication. The Editors are very grateful to the following colleagues for their rapid and careful reviewing of manuscripts: M Albrecht, J S Barnard, R Beanland, G A Botton, D Cooper, L Clement, A J Craven, A G Cullis, N Daneu, V Grillo, E Grünbaum, A Gustafsson, P-H Jouneau, O L Krivanek, D Larson, M Luysberg, S I Molina, F A Ponce, A Rosenauer, F M Ross, I M Ross, J-L Rouviere, and Z L Wang.

Prizes for student presentations at the conference were awarded to Mr Alberto Eljarrat, Universitat de Barcelona, and Mr Thibaud Denneulin, CEA-LETI Grenoble, both of whom are presenting their results also as contributions to this proceedings volume.

Entertainment during the conference dinner was provided by the college's string quartet, and after the meal we held the now traditional friendly competition between teams of microscopists; this year teams were asked to construct models of scientific instruments using only cardboard, paper, sticky tape and glue. By unanimous vote, the team who presented the 'Atomic Food Microscope (AFM)' prototype won first prize – the deft 'dipping mode' of the cardboard cantilever (see photo) was a highlight!

Figure 2. The prize winning 'Atomic Food Microscope (AFM)' in action!

The organisers are very grateful to the following companies who contributed to the success of the meeting by presenting trade stands during an exhibition on the evening of 5 April: Agar Scientific, Bruker, CamScan, FEI, Gatan, Hitachi High Technologies, HREM Research, IOP Publishing, JEOL, Leica Microsystems and MICOS.

Finally, we would like to thank the staff of the Institute of Physics for their expert assistance in planning and organising this conference and in particular Claire Garland for her dedicated professional support and her always joyful approach to any issues that arose.

October 2011
T Walther
P A Midgley

PEER REVIEW STATEMENT

All papers published in this volume of *Journal of Physics: Conference Series* have been peer reviewed through processes administered by the proceedings Editors. Reviews were conducted by expert referees to the professional and scientific standards expected of a proceedings journal published by IOP Publishing.

Microstructure and polarization fields in nitride semiconductors

F A Ponce
Department of Physics, Arizona State University, Tempe, AZ 85287-1504, USA

E-mail: ponce@asu.edu

Abstract: The nitride semiconductors are widely used in high efficiency light emitting devices and are currently being considered for photovoltaic applications. The reduced symmetry in the wurtzite structure compared to cubic semiconductors results in the growth of large densities of crystalline defects and in the presence of strong spontaneous and piezoelectric polarization effects. A correlation between the microstructure and the polarization fields can be achieved with electron holography in the transmission electron microscope. Electron holograms thus obtained can provide energy band profiles with sub-nanometer spatial resolution. The phase of the electron beam is sensitive to the electrostatic potential, and a direct measurement of the latter can be achieved by making the electron beam signal that traverses the specimen interfere with a reference electron beam that travels through vacuum. This technique has been quite useful in probing the fields and charges at dislocations and at interfaces in semiconductors, and it is particularly useful to determine the piezoelectric effects in group III nitride semiconductor heterostructures. A review of applications to InGaN and AlGaN based heterostructures is presented in this paper.

Key words: Electron energy band profiles, piezoelectric fields, nitride semiconductors, electron holography

1. Introduction

In the past two decades, the nitride semiconductors have acquired pronounced relevance due to their robustness and their range of band gaps[1]. The bond between nitrogen and the group III elements is strong; the cohesive energies of AlN, GaN, and InN are 11.67, 9.06, and 7.97 eV, respectively.[2] The corresponding band gap energies of 6.28, 3.41, and 0.7 eV span the electromagnetic spectrum from the infrared into the ultraviolet, including the full visible spectrum.[3] The growth of these materials is challenging due to the high temperatures required for synthesis and the absence of suitable substrates. High densities of dislocations are necessary in order to accommodate the lattice mismatch and the thermal expansion mismatch within the AlGaInN alloy system.[4] The lattice displacement is associated with those dislocations (Burgers vectors) are large compared with the interatomic bond length.[5] Heterojunction lattice mismatch often results in pseudomorphic thin film structures with large strain fields.[6-9] The absence of a center of symmetry in the hexagonal wurtzite structure results in significant spontaneous polarization. In addition, piezoelectric polarization results from the strain due to dislocations and strained heterointerfaces. The electrostatic fields due to spontaneous and piezoelectric polarization are of the order of several MeV/cm,[7] and are typically highly localized at the nanometer scale, and have a strong effect on carrier transport and carrier recombination.

Profiles with nanometer resolution of the electrostatic potential and charge distribution in solids can be obtained by electron holography in the TEM. This technique has been used to visualize the potential energy profile across p-n junctions in silicon devices used in microelectronics.[10] Electron holography has been particularly useful to measure the potential and charge density variations in the nitride semiconductors. Electron holography was first used by Cherns in 1999 to measure the piezoelectric field in InGaN quantum wells.[11] In this paper, a brief introduction is first given to the electron holography technique. This is followed by examples of applications to determine the nature of electrostatic potential variations in the vicinity of dislocations in GaN, at InGaN/GaN quantum wells, and at two-dimensional electron and hole gas heterostructures.

2. Electron holography in the transmission electron microscope

The use of electron holography has increased significantly in the last decade, particularly in the direct measurement of the electrostatic fields and charges in nitride semiconductor materials. In conventional TEM techniques, like those used for diffraction contrast analysis, only the intensity of the electron beam is recorded, while the phase information is lost. Electron holography is an interferometric technique which provides information about spatial distribution of the phase shift, $\Delta\theta$. In order to obtain the phase information, a highly coherent electron beam is obtained by using a field emission gun. A sample is prepared to electron transparency, but not too thin to avoid surface related effects such as strain relaxation and screening due to the surface.

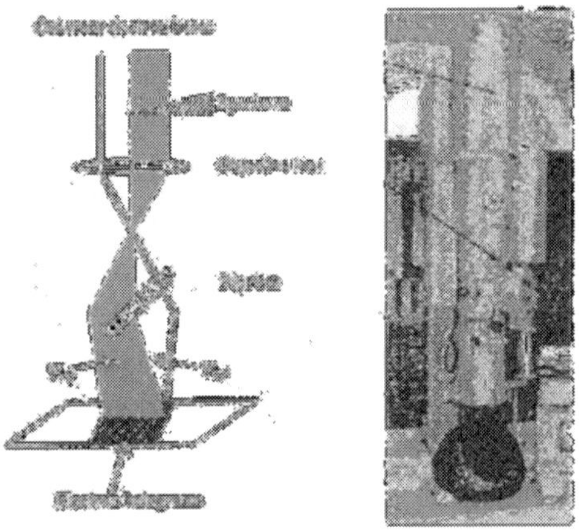

Figure 1. Schematic diagram of the electron beam path for electron holography in a transmission electron microscope.

The phase of the electron beam is modulated by spatial variations of the electrostatic potential of the specimen. Figure 1 shows the configuration used for electron holography. The sample is introduced into the electron beam path covering about half of the beam diameter. The other half of the beam travels through vacuum and it is used as a reference beam. The electrons that traverse the sample experience a change in velocity, in part due to the variation of the reference potential, and gain in phase compared to the electrons traveling through vacuum. The phase change is directly proportional to the electrostatic potential as we will see later. The phase information can be retrieved from the wave interference pattern.[12,13]

Journal of Physics: Conference Series **326** (2011) 012001 doi:10.1088/1742-6596/326/1/012001

Electron holography is an electron-interference technique that makes the recovery of the phase information possible. A conducting filament that acts as an optical biprism is used to produce an interference pattern. The biprism is aligned manually so that its image matches the boundary of the specimen at the image plane. When a positive bias is applied to the biprism, the beams traveling through vacuum (reference wave) and the sample (object wave) are made to overlap at the image plane, and their overlap is recorded. The geometric configuration around the biprism is depicted in Fig. 2(a). The biprism filament is along the *y*-direction, and the beam propagates along the *z*-direction. The biprism is at a given potential, while the two plates are grounded. In some microscopes these two plates are the edges of the selected area aperture.

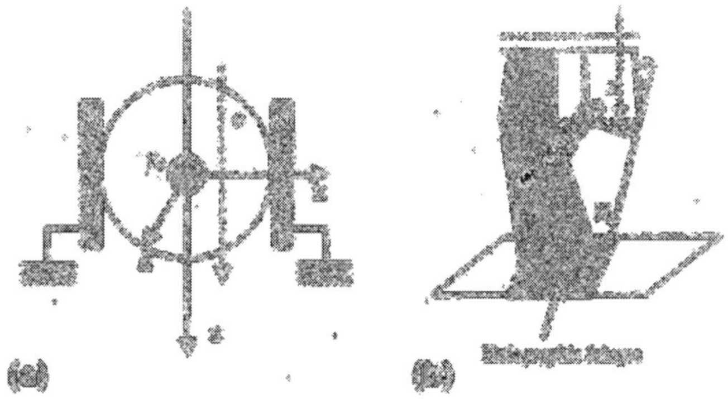

Figure 2. Holographic image formation. (a) configuration around the biprism. (b) Interference between two plane waves.[13]

As already indicated, the specimen under investigation is positioned to cover only half of the object plane, in that way one beam passes through the specimen (referred to as the object wave $\psi_{obj} = A_o\, e^{i\theta_o}$) and the other beam passes through the vacuum (referred to as the vacuum wave $\psi_{vac} = A_v\, e^{i\theta_v}$). A bi-prism, which resides in the plane of the selected area aperture, deflects the two electron beams in order to cause their interference. Applying a positive bias to the filament of the bi-prism, the object wave and vacuum wave converge and yield an interference pattern in the overlapping region. The intensity distribution can be represented by: [13]

$$
\begin{aligned}
I &= \left| A_v e^{i(\theta_v - \Delta)} + A_o e^{i(\theta_o + \Delta)} \right|^2 \\
&= A_v^2 + A_o^2 + 2A_v A_o \cos(\theta_o - \theta_v - 2\Delta)
\end{aligned}
$$

where the first two terms represent the background intensity and the third terms represent the superimposed interference fringes. The extra phase shift Δ is due to the biprism deflection, which is dependent on the applied field and the geometric configuration of the biprism, but is not a property of specimen. The information that we want to obtain is the phase shift θ_o - θ_v which is due to the modulation by the crystal potential *V*. The crystal electrostatic potential *V* is typically in the order of 10 eV which is much less than the electron beam energy, so that the difference of momentum between object and vacuum waves is proportional to *V*. Considering relativistic effects, the relationship between phase shift θ_o - θ_v and crystal potential *V* is:

$$
\theta_o - \theta_v = \int \frac{p_{obj} - p_{vac}}{\hbar} dz = \frac{eV(E - E_0)}{E(E + 2E_0)} \frac{2\pi}{\lambda}
$$

where E is the kinetic energy of incident electron beam, E_o is the rest energy of electron, t is specimen thickness and λ is the wavelength of electron beam. With a fixed accelerating voltage in the TEM, this expression can also be written as:

$$\theta_o - \theta_v = C_E V t,$$

where C_E is a constant. Therefore, the crystal electrostatic potential profile can be obtained from the phase shift information that is extracted from the electron hologram.

The phase reconstruction process[12] starts by taking the Fourier transform of the intensity equation shown above. Three delta functions appear, involving convolutions represented by the symbol $\otimes$.

$$FT(I) = \delta(q) \otimes FT(A_v^2 + A_v^2) + \delta(q - q_\Delta) \otimes FT\left(A_v A_o e^{i(\theta_v - \theta_o)}\right) + \delta(q + q_\Delta) \otimes FT\left(A_v A_o e^{-i(\theta_v - \theta_o)}\right)$$

The background intensity is transferred into an autocorrelation function and appears to be in the center, while the Fourier transform of the co-sinusoidal interference fringes provides two sidebands, which are away from the center by the carrier spatial frequency, q_Δ. The next step is to shift one of the sidebands to the center, and take the inverse Fourier transform. A complex object image with a modulus $A_v A_o$ and phase θ_o - θ_v is obtained. In addition, if we remove the specimen and take the hologram, a complex image with a modulus A_v^2 and zero phase shift can be obtained. The damping of amplitude of the electron beam across the specimen is due to the inelastic scattering process in the material and can be expressed as

$$\frac{A_o}{A_v} = e^{-t/2\lambda_i}$$

where λ_i here is the mean free path for inelastic scattering of electrons in the crystal. Therefore, the thickness profile of a specimen can be obtained from the amplitude image, while the electrostatic potential profile of the specimen can be obtained from the phase image. Furthermore, in a one-dimensional potential profile, the internal fields are proportional to the slope of the electrostatic potential. The corresponding electrostatic charges are found using Poisson's equation, and are related to the curvature of the potential profile.

3. Dislocations in GaN

As mentioned before, thin epitaxial films with optimized electronic properties have threading dislocation densities ranging from 10^8 to 10^{10} cm^{-2}.[4] They play an important role in accommodating epitaxial growth under adverse conditions associated with lattice mismatch and thermal mismatch with the substrate and at InAlGaN compound heterojunctions. Threading dislocations have been found to be of the edge, screw, and mixed types.[14] The Burgers vectors of the dislocations are large with respect to the lattice parameters, which often leads to the formation of coreless dislocations.[15] Dislocations are frequently associated to non-radiative recombination centers.[16] The electrostatic charges at dislocations lines were first studied using electron holography by Cherns[17] and Cai.[18] They found that dislocations have specific charges depending on the dislocation type.

4. Polarization fields in InGaN/GaN

In the absence of a center of symmetry, the hexagonal wurtzite structure exhibits noticeable spontaneous and piezoelectric polarization effects.[7] Polarization fields up to a few MeV/cm in the nitride semiconductors were first determined by applying a reverse bias to the p-n junction,[8] and later by electron holography.[11] The strong internal field can lead to positive and detrimental effects in light emitting diodes (LEDs). Reduced radiative recombination efficiency can result from the spatial separation of electrons and holes. Spatially separated by the Stark effect, their transition matrix elements and the radiative recombination rate can be significantly decreased. In addition, the internal fields facilitate the escape of carriers and reduce the carrier capture cross section by quantum wells.

Another detrimental effect is instability of the emission wavelength for devices operating under high injection conditions. The polarization field is screened and the emission wavelength is blue-shifted. The magnitude of this shift could be up to 100nm for large polarization cases, such as for InGaN/GaN green LEDs. Therefore, it is important to find ways to prevent polarization effects in the active region. Non-polar and semi-polar growth have been proposed and their fabrication has been successfully demonstrated. However, the in-plane polarization could still be a problem for the non-polar growth of quantum wells, as we will see later.

On the other hand, strong polarization fields could play a positive role in the conductivity of p-region and n-region in LEDs. In n-type AlGaN/GaN heterostructures, the positive polarization sheet charges at the interface will promote the formation of two-dimensional electron gas (2-DEG) with a sheet carrier concentration as high as 10^{13} cm^{-2}. Similarly in p-type AlGaN/GaN heterostructures, the negative polarization sheet charges at the interface could help with the formation of two-dimensional hole gas (2-DHG). By substituting the conventional n-GaN or p-GaN layer in LEDs with the proper AlGaN/GaN multi-heterostructure, the lateral conductivity of LEDs can be enhanced, and an improved current spreading can be expected.

In the following sections, we discuss the theoretical aspect of the polarization field in nitrides, and then the measurement of polarization effects by electron holography, including the mapping of internal electrostatic potential in the InGaN/GaN quantum wells and evidence of the formation of 2-DEG and 2-DHG in n-type and p-type AlGaN/GaN heterostructures, respectively.

5. Electrostatic fields and charges in light emitting diodes emitting in the green region of the visible spectrum

Various aspects of polarization fields in InGaN/GaN quantum wells have been reported[6-9] and measured.[19-21] In particular, there is much interest in the polarization field and charge distribution in InGaN/GaN multi-quantum wells in the active region of green LEDs. An important feature of electron holography is that it can provide a mapping of the electrostatic potential distribution at an atomic scale, while other techniques usually give average values. An example of a green LED structure is shown in Fig. 3.[22] This layer structure was grown by metal organic chemical vapor deposition on a c-plane sapphire substrate, the active region containing five InGaN/GaN quantum well periods, emitting light at ~2.3 eV. Cross- section samples were prepared for TEM using standard mechanical polishing and argon-ion milling techniques. Figure 3(a) is a high-resolution image of the active region. The InGaN QW and GaN barriers are uniform with a thickness of 4.6 and 9.3 nm, respectively. The interfaces are flat and atomically abrupt.

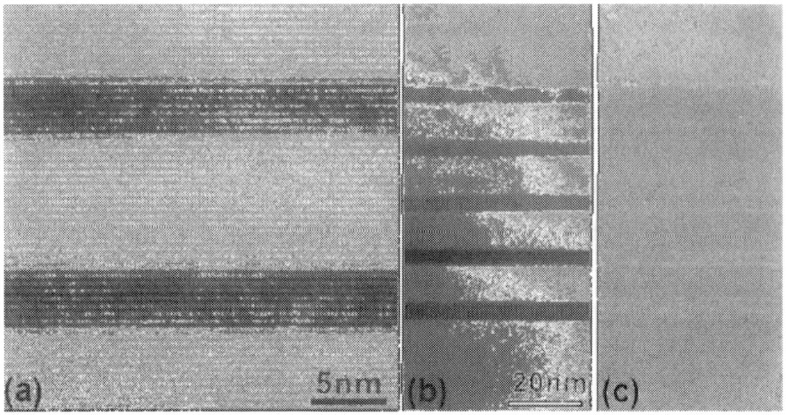

Figure 3. Active region of a green LED. (a) High resolution TEM of the quantum well region. (b) Electrostatic potential profile showing the p-n junction and quantum well region. (c) Charge density distribution due to polarization effects.[22]

Electron holograms of the active region were obtained with the sample aligned to minimize diffraction conditions while keeping close to an orientation where the electron beam was parallel to the interfaces. The holograms were digitally recorded using a charge coupled device camera. The amplitude and phase values were deconvoluted from the electron hologram; the results are shown in Figs. 3(b) and 3(c). The resulting spatial profile of the electrostatic potential and charge distribution are plotted in Fig. 4. The electrostatic potential distribution due to the p-n junction is clearly observed in Fig. 4(a) with a shift of about 3 eV in the conduction band edge. Also observed is the potential profile due to the quantum wells. It is noticed that the field and the polarization charge density in the quantum wells vary with distance to the p-n junction. The sheet polarization charge densities of the first three InGaN quantum wells have a similar value of ~ 0.018 C/m^2, corresponding to a 20% indium composition, for a non-screening case, and corresponding to piezoelectric fields of ~1.95 MV/cm. The sheet polarization charge densities in the fourth and fifth periods are estimated to be 0.007 and 0.0005 C/m^2, respectively. This corresponds to a significant screening of the fourth and fifth QWs.

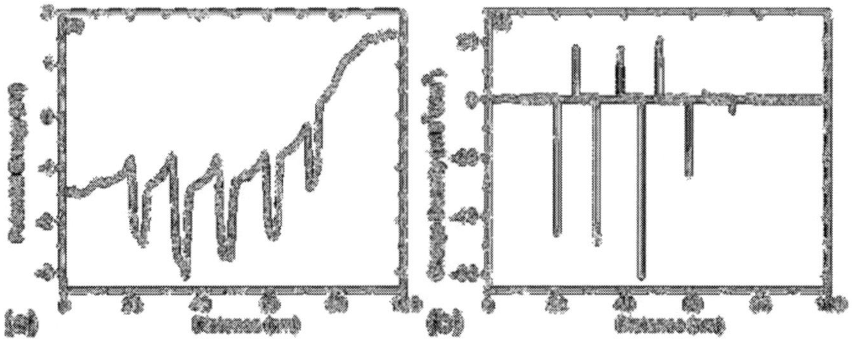

Figure 4. Active region of a green LED. (a) Electrostatic potential profile showing the p-n junction and quantum well region. (b) Charge density distribution due to polarization effects.[22]

The observation of inhomogeneous polarization fields within the multi-quantum well is consistent with a depth-resolved study using cathodoluminescence,[22] in which a low energy peak appears when the accelerating voltage is increased. The decay of piezoelectric fields in the fourth and fifth QWs could be due to strain relaxation. It is possible that the first three periods remain fully strained while the fourth and fifth periods undergo plastic relaxation. Another possibility involves the migration of hydrogen ions from the p-type region into the top two QWs.

6. Two-dimensional electron gas (2DEG) thin film structures

Strong polarization fields can result in two-dimensional carrier confinement, also known as two-dimensional electron gas (2DEG), in AlGaN/AlN/GaN heterostructures. The high electron conductivity of these thin film structures can be used to provide low resistant current paths to n- and p-type layers of the LED.[23-26] Fig. 5 shows part of a n-type AlGaN/AlN/GaN superlattice, exhibiting 2-D electron gas at the AlN/GaN interface. The AlGaN/AlN/GaN superlattice was epitaxially grown by MOCVD on an n-doped 10^{18} cm^{-3}, 3μm GaN layer on a c-plane sapphire substrate. The SL consists of 11 periods; each period starts with a 0.5 nm AlN layer, followed by a 12 nm AlGaN barrier layer with an Al concentration that is nominally graded linearly from 30% to 0%, and ends with a 28 nm GaN layer. The AlN layer is used to avoid interface alloy scattering effects and to improve the lateral mobility of the 2-DEG. Using a very thin AlN layer prevents significant reduction of the vertical conductivity. The AlGaN layer has an n-type doping level of 2×10^{19} cm^{-3}, and the n-type GaN in the SL is doped at 2×10^{18} cm^{-3}. In order to avoid impurity scattering, doping of the AlGaN layer was delayed by 4 s during growth, i.e., the first nanometer of the AlGaN is undoped; and the doping of the GaN layer was stopped 10 nm before starting the growth of the next AlN/AlGaN barrier. The characteristics of the full superlattice are observed in Fig. 5(a). A higher magnification TEM image showing the two top periods of the SL structure is shown in Fig. 5(b). The AlN layer appears with the brightest contrast, followed by the graded AlGaN layer with fading brightness and the GaN layer with relatively darker contrast.

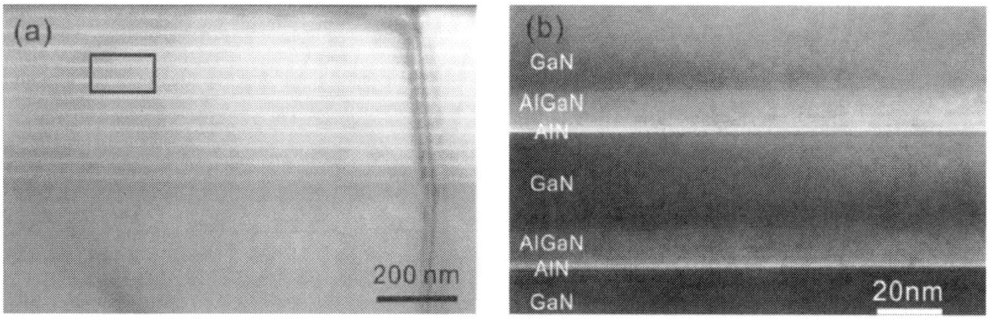

Figure 5. AlGaN/AlN/GaN n-type superlattice. (a) Bright field TEM image showing the eleven periods. (b) Higher magnification TEM image showing the top two periods.[24]

The electrostatic potential distribution obtained from the reconstructed phase and amplitude images are shown in Fig. 6(a). A profile with higher spatial resolution is shown in Fig. 6(b). An energy dip close to the GaN/AlN interface represents the accumulation of the 2-DEG. The curvature of the potential around the 2-DEG region is negative as expected. These results are consistent with calculated energy band diagrams.[25]

Journal of Physics: Conference Series **326** (2011) 012001 doi:10.1088/1742-6596/326/1/012001

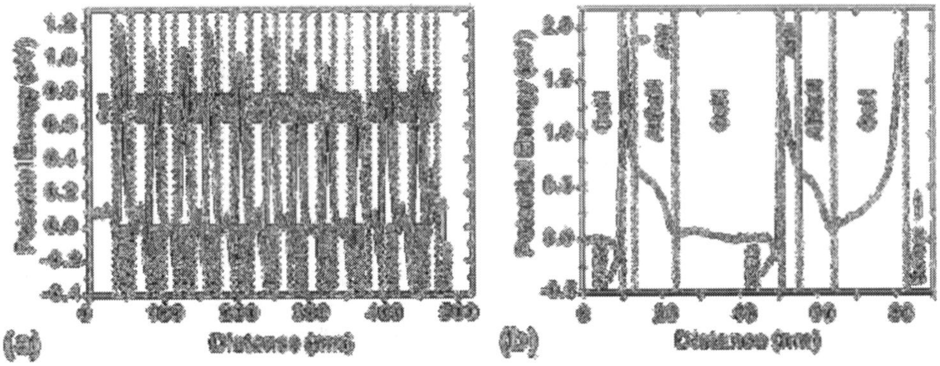

Figure 6. Potential energy profiles of AlGaN/AlN/GaN n-type superlattice showing the features of the 2DEG in the complete superlattice (a) and in the top two periods (b). [24]

7. Two-dimensional electron gas (2DHG) structures

If the AlGaN/GaN multi-heterostructure is made p-type, the band diagram would be different from the n-type case due to the alignment of Fermi-level. In order to achieve the accumulation of a 2-DHG, the growth sequence of GaN/AlN/AlGaN should be adjusted and the device structure should be optimized. We have studied a p-type AlGaN/GaN multi-heterostructure, which is designed for the purpose of high conductivity with the accumulation of 2-DHG. [25-26]

The AlGaN/AlN/GaN heterostructure in Fig. 7 was epitaxially grown by metal-organic chemical vapor deposition on a nominally un-doped 3-μm thick GaN layer on a c-plane sapphire substrate. The 2-DHG heterostructure consists of 6 periods; each period starts with a 0.5 nm AlN layer, followed by a 5 nm nominally un-doped GaN layer, and ending with a 12 nm AlGaN barrier layer with an aluminum concentration that is nominally graded linearly from 0% to 10%. The thin AlN layer is used to avoid interface alloy scattering effects and to improve the lateral mobility of the 2-DHG. In order to optimize the vertical conductivity, a very thin AlN layer is used. The initial portion of the AlGaN layer has a Mg doping level of ~1×10^{19} cm^{-3}, but the Mg flux is turned off for the last few nanometers. The nominally un-doped GaN shows, in secondary ion mass spectroscopy, a Mg background concentration in the range of $1\text{-}5\times10^{18}$ cm^{-3}, due to memory effects.

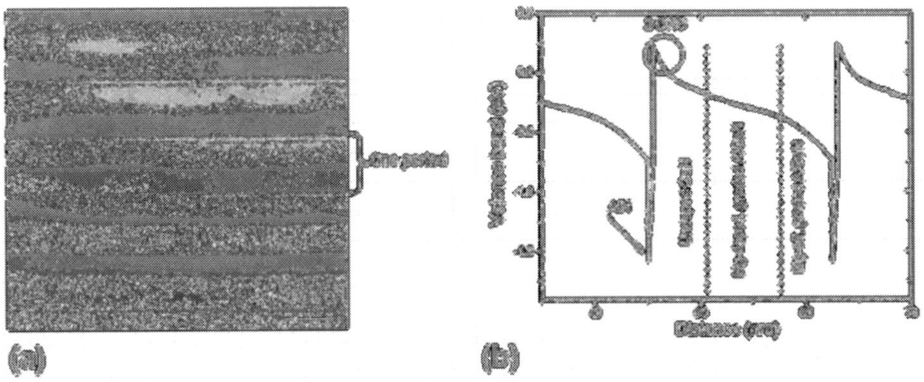

Figure 7. AlGaN/AlN/GaN superlattice structure. (a) TEM image of the superlattice. Brighter regions indicate high aluminum content. (b) Sketch of valence band profile of one period. The 2-DHG accumulation region is circled. [26]

Fig. 8 shows the electrostatic potential profile for 3 superlattice periods. The locations of the interface are determined by a careful match between diffraction contrast and phase image. The formation of a 2-DHG is revealed by a positive curvature in the GaN immediately to the right of the AlN layer in the potential energy profile. The two-dimensional hole gas (2-DHG) at the interface cancels the spontaneous polarization charges, and no discontinuity in the potential profile is observed. But the depletion of holes produces a positive curvature in the profile, in contrast to a negative curvature in the 2-DEG case.

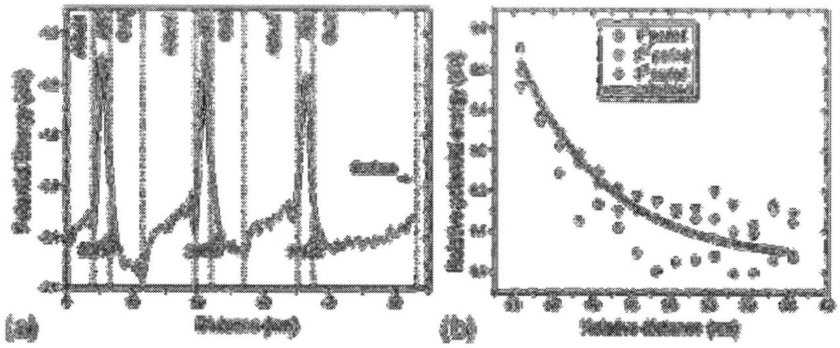

Figure 8. Potential profile of an AlGaN/AlN/GaN superlattice doped p-type, exhibiting a 2-DHG. No discontinuity is observed at the AlN/GaN interface. [26]

Generally, in semiconductor structures of this type, we find four major types of charges that are responsible for the variations of the electrostatic potential. (1) Fixed charges like those due to spontaneous and piezoelectric polarization, (2) ionized impurities, (3) interface fixed charges (band offset), and (4) free carriers (in our case holes) that redistribute under the effect of the other charges. For the 2-DHG case, the major contribution to the potential variation is from the holes confined on a nanoscale of length. In our case, the experimental and calculated potential distributions both show a noticeable curvature in the vicinity (~1nm) of the interface, but become relatively flat away from the interface, which indicate the 2DHG accumulation happens over only a few atomic monolayers of the interface. [26]

8. Conclusion

Defects and interfaces in the nitride semiconductors are closely related to the electronic and optical properties via the induced polarization fields. A detailed understanding of the electrostatic potential properties of semiconductor structures is possible at near the atomic scale using electron holography in the TEM. A brief introduction to the technique has been presented in this paper, and some applications to the study of electrostatic fields associated with dislocations, quantum wells, and 2DEG and 2DHG heterostructures of nitride semiconductors have been discussed. This type of studies is essential in order to achieve high efficiency light emitting devices. Similar studies should also be useful for the development of photovoltaic devices based on these materials.

References

[1] Ponce FA and Bour D P 1997 *Nature* **386**, 351-9.
[2] Stampfl C and Van de Walle C G 1999 *Phys. Rev. B* **59**, 5521-35.
[3] Wu J, Walukiewicz W, Yu K M, Ager III J W, Haller E E, Lu H, Schaff W J, Saito Y, and Nanishi Y 2001 *Appl. Phys. Lett.* **80** 3967-9.
[4] Lester S D, Ponce FA, Craford M G, and Steigerwald D A 1995 *Appl. Phys. Lett.* 66, 1249-51.

[5] Ponce FA, Cherns D, Young W T, and Steeds J W 1996 *Appl. Phys. Lett.* **69**, 770-2.
[6] Look D C and Sizelove J R 1999 *Phys. Rev. Lett.* **82**, 1237-40.
[7] Bernardini F, Fiorentini V, and Vanderbilt D 1997 *Phys. Rev. B* **56**, R10024-28.
[8] Takeuchi T, Wetzel C, Yamaguchi S, Sakai H, Amano H, Akasaki I, Kaneko Y, Nakagawa S, Yamaoka Y, and Yamada H 1998 *Appl. Phys. Lett.* **73**, 1691-3.
[9] Wei Q Y, Li T, Wu Z H, and Ponce FA 2010 *Phys. Stat. Sol. A* **207**, 2226-32.
[10] McCartney M R, Smith D J, Hull R, Bean J C, Voelkl E, and Frost B 1994 *Appl. Phys. Lett.* **65**, 2603-5.
[11] Cherns D, Barnard J, and Ponce FA 1999 *Solid State Comm.* **111**, 281-5.
[12] McCartney M R and M. Gaidardziska-Josifovska M 1994 *Ultramicroscopy* **53**, 283-8.
[13] Cai J and Ponce FA 2002 *J. Appl. Phys.* **91**, 9856-62.
[14] Ponce FA, Cherns D, Young W T, and Steeds J W 1996 *Appl. Phys. Lett.* **69**, 770-2.
[15] Cherns D, Young W T, Steeds J W, Ponce FA, and Nakamura S 1998 *Phil. Mag. A* **77**, 273-86.
[16] Ponce FA, Bour D P, Goetz W, and Wright P J 1996 *Appl. Phys. Lett.* **68**, 57-9.
[17] Cherns D and Jiao C G 2001 *Phys. Rev. Lett.* **87**, 205504-7.
[18] Cai J and Ponce F A 2002 *Phys. Stat. Sol. A* **192**, 407-11.
[19] McCartney M R, Ponce FA, Cai J, Bour D P 2000 *Appl. Phys. Lett.* **76**, 3055-7.
[20] Cai J, Ponce FA, Tanaka S, Omiya H, and Nakagawa Y 2001 *Phys. Stat. Sol. A* **188**, 833-7.
[21] Stevens M, Bell A, McCartney M R, Ponce FA, Marui H, and Tanaka S 2004 *Appl. Phys. Lett.* **85**, 4651-3.
[22] Wu Z H, Fischer A M, Ponce FA, Lee W, Ryou J H, Yoo D, and Dupuis R D 2007 *Appl. Phys. Lett.* **91**, 041915.
[23] Wu Z H, Stevens M, Ponce FA, Lee W, Ryou J H, Yoo D, and Dupuis R D 2007 *Appl. Phys. Lett.* **90**, 032101.
[24] Wu Z H, Ponce FA, Hertkorn J, and Scholz F 2007 *Appl. Phys. Lett.* **91**, 142121.
[25] Hertkorn J, Thapa S B, Wunderer T, Scholz F, Wu Z H, Wei Q Y, Ponce FA, Moram M A, Humphreys C J, Vierheilig C, and Schwarz UT 2009 *J. Appl. Phys.* **106**, 013720.
[26] Wei Q Y, Wu Z H, Sun K W, Ponce FA, Hertkorn J, and Scholz F 2009 *Appl. Phys. Exp.* **2**, 121001.

Impact of high-temperature metalorganic vapor phase epitaxial growth of AlGaN-based UV-A, UV-B and UV-C quantum wells on the improvement of their internal quantum efficiency

H Amano

Department of Electronics and Computer Science, Akasaki Research Center, Nagoya University, Furo-cho, Chikusa-ku, Nagoya 464-8603, Japan

E-mail: amano@nuee.nagoya-u.ac.jp

Abstract. High-temperature metalorganic vapor phase epitaxial (MOVPE) growth leads to the realization of high-quality AlN on a c-plane sapphire substrate and the successful growth of low-threading-dislocation-density (TDD) AlN and AlGaN by the grooved template technique. The ABC model in Shockley-Read-Hall (SRH) statistics was used to investigate the non-radiative and radiative processes of AlGaN-based quantum wells (QWs) having different TDDs. The internal quantum efficiency (IQE) was found to be equally dependent on TDD regardless of the emission wavelength. A similar dependence is confirmed in the relationship between TDD and nonradiative recombination coefficient A in UV-A, UV-B and UV-C QWs and in violet, blue and blue-green QWs.

1. Introduction

Considerable efforts have been carried out to improve the performance of AlGaN-based UV-A, UV-B and UV-C LEDs by many researchers. Nevertheless, their external quantum efficiency (EQE) is much lower than those of InGaN-based blue LEDs. Clarifying the factors that limit their EQE is necessary in improving their performance. In general, there are three factors that limit the efficiency of LEDs: (1) IQE, (2) carrier injection efficiency and (3) light extraction efficiency. Among these, IQE is dependent on the radiative recombination rate of each material and the rate of nonradiative recombination. Theoretically, the radiative recombination coefficient of free excitons in AlN is larger than that of GaN or InN [1]. Therefore, AlGaN-based QWs should be more radiative or should have a higher IQE than InGaN-based QWs. However, there is no report showing that the IQEs of AlGaN-based QWs are comparable to or even higher than those of InGaN-based QWs. Thus, a systematic study was conducted to confirm that the IQEs of UV-A, UV-B and UV-C QWs are comparable to those of violet, blue or blue-green QWs.

If UV-A, UV-B and UV-C LED structures are grown on a GaN substrate or a GaN template, crack generation caused by the grown-in tensile stress is a fatal problem that prevents the fabrication of the LEDs. Therefore, the growth of AlGaN-based UV-A, UV-B and UV-C LED structures on an AlN substrate or an AlN template is inevitable. If AlGaN is grown on AlN, the generation of mismatch-induced defects such as misfit dislocations is another serious problem. Therefore, reduction of the

TDD and clarification of the TDD dependence of IQE are the first steps toward improving the performance of UV-A, UV-B and UV-C LEDs.

Recently, we have developed a custom-built MOVPE system that can operate with a growth temperature of as high as 2,073 K, enabling the growth of AlN and AlGaN having an atomically flat surface with a growth rate higher than 6 μm/h on a c-plane sapphire substrate. This MOVPE system is also effective for growing a low-residual-impurity AlN underlying layer and a high-Al-content AlGaN layer; thus, the effect of carbon-related nonradiative recombination can be neglected [2]. AlN on a c-plane sapphire substrate with a TDD of as low as 3×10^{8} cm^{-2} can be successfully grown.

By using this high-temperature MOVPE system, the TDD in AlN and AlGaN on an AlN template can be controlled by a grooved template technique [3]. In this paper, the effects of the high-temperature MOVPE growth of AlN on a c-plane sapphire substrate and the growth on a grooved template on the reduction of the TDD in an AlGaN are shown.

Next, sets of QWs with emission wavelengths of 230, 250, 300 and 350 nm having different TDDs were grown, and the IQE of each sample was characterized by ABC model in SRH statistics [4]. As references, the TDD dependence of the IQEs of violet, blue and blue-green QWs grown on a free-standing GaN substrate or on a Si substrate was also compared.

2. Experiments

AlN of 2 μm thickness was grown at 1,400℃ on a c-plane sapphire substrate using a custom-built MOVPE system having a face down configuration with a water-cooled gas injector [2]. Figure 1 shows a schematic of the MOVPE reactor. Surface temperature was measured by using a pyrometer and monitored from the bottom of the flow channel. The carrier gas was hydrogen. Growth was conducted at 100 Torr. The molar fraction of trimethyl-aluminum (TMAl) to ammonia (NH_3) was 116.

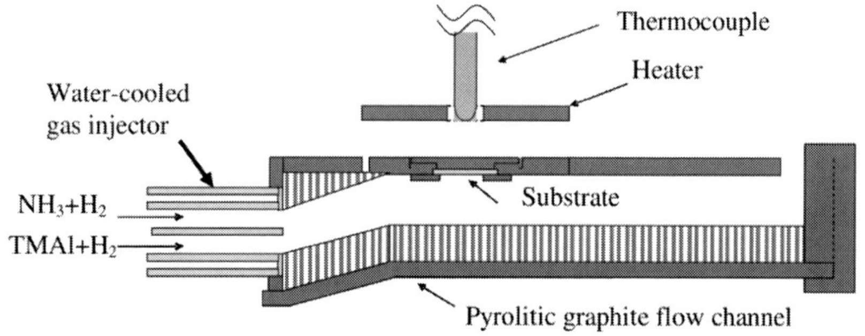

Figure 1. Schematic of the high-temperature MOVPE system.

After the growth of AlN, a periodic trench pattern was formed along <10-10> of AlN by dry etching. The groove depth and width were 2 and 3 μm, respectively. Then, again, samples were mounted on the susceptor in the reactor and AlGaN and QWs were successively grown. AlGaN and QWs were also grown on planar AlN.

The TDD of each sample was measured using the ω-mode X-ray rocking curve (XRC) obtained from c-plane and m-plane reflections. The full width at half maximum (FWHM) of the c-plane XRC is related to the density of dislocations having a screw component, and that of the m-plane XRC is related to the density of dislocations having an edge component. In every sample, the FWHM of the m-plane XRC is wider than that of the c-plane XRC, which means that the density of the edge-type dislocation is higher than that of the screw-type dislocation [5]. In some samples, the density of the edge-type dislocation was also confirmed by plan-view transmission electron microscopy (TEM) observation. The discrepancy of the calculated TDD obtained from XRC measurement and that obtained by counting using TEM images was less than 15%. The dark spot density determined by cathodoluminescence was also confirmed to coincide with the density of dislocations having edge components.

The IQEs of the UV/DUV MQWs with different TDDs were characterized by measuring excitation-density-dependent photoluminescence (PL) using an ArF excimer laser with a wavelength of 193 nm as the excitation source, which is known as SRH analysis [6-9]. Neutral density filters were used to control the excitation density. All the measurements were performed at room temperature.

When the generation rate (G) is equal to the total recombination rate (R), and if the excitation carrier density is sufficiently low for Auger recombination to be negligible, $G=R$ and PL integrated intensity are given using the following formulas.

$$G = R = An + Bn^2 \tag{1}$$

$$I_{PL} = \eta Bn^2 \tag{2}$$

Here, n is the excitation carrier density, A is the nonradiative recombination constant, B is the radiative recombination constant and η is a constant determined on the basis of the PL collection efficiency. In this study, we fix B as 0.2×10^{-10} cm^3/s [10]. This value was obtained by measuring of the radiative lifetime of a thin GaN layer on a c-plane sapphire substrate, although several values have also been reported. In the case of free exciton recombination, B is larger than those for free-electron and free-hole recombination [1]. Differences in the B value cause an error in the absolute IQE value. It is currently unclear which value of B should be used. However, the relative trend of the TDD dependence of IQE can be discussed regardless of the chosen value of B.

By combining the above two formulas, G can be rewritten as

$$G = \frac{A}{\sqrt{B\eta}}\sqrt{I_{PL}} + \frac{1}{\eta} I_{PL} \tag{3}$$

Then, nonradiative coefficient A is deduced by curve fitting and the IQE can be expressed as

$$IQE = \frac{Bn^2}{An + Bn^2} \tag{4}$$

3. Results and discussion

Figure 2 shows the atomic force microscopy (AFM) images of AlN grown at (a) 800℃, (b) 1,000℃ and (c) 1,400℃. The effect of the high temperature growth on the improvement of the surface flatness is evident. Root mean square surface roughnesses of the samples are 1.79, 0.47 and 0.098 nm, respectively.

Figures 3-5 show the AFM image, cross-sectional TEM image and plan-view TEM image of the AlN grown on a trench-patterned AlN template. The total thickness is approximately 20 µm. The coalescence position cannot be identified from the AFM image. The plan view TEM image shows that the TDD is less than 10^7 cm^{-2}.

(a)
(b)
(c)

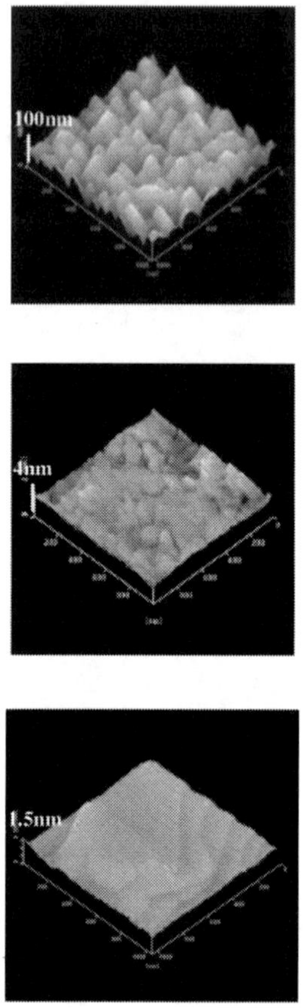

Figure 2. 1 μm^2 AFM images of AlN grown at (a) 800℃, (b) 1,000℃, and (c) 1,400℃.

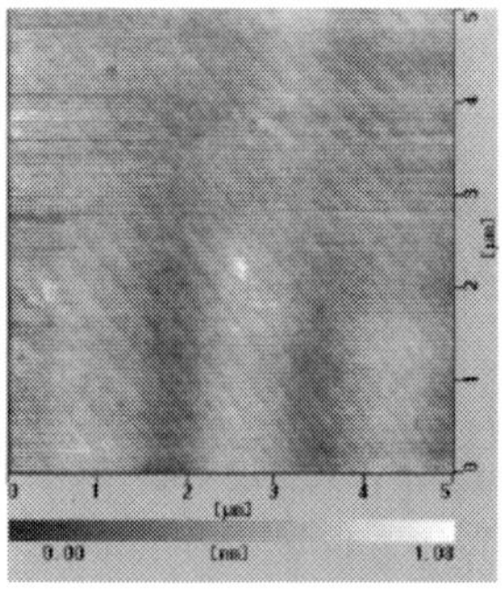

Figure 3. 5 µm × 5 µm AFM image of AlN grown on trench-patterned AlN template.

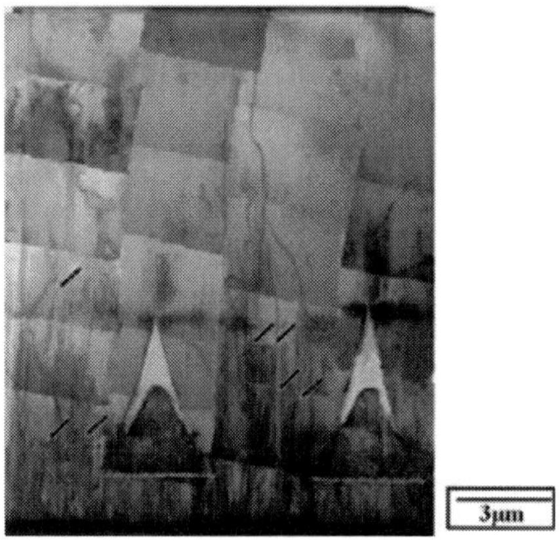

Figure 4. Cross-sectional TEM image of AlN grown on trench-patterned AlN template.

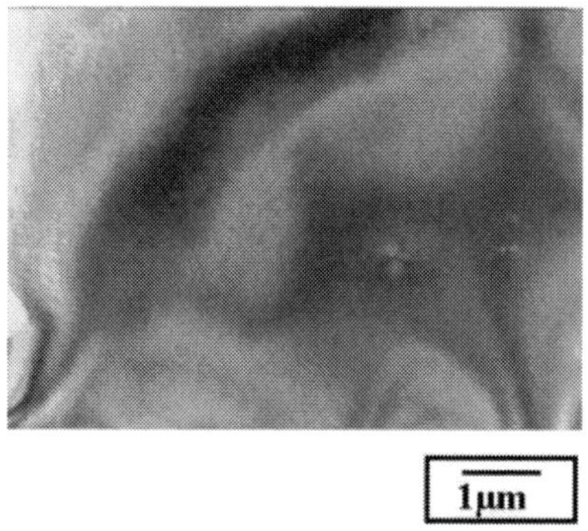

Figure 5. Plan-view TEM image of AlN grown on trench-patterned AlN template.

Figure 6 summarizes the compositional dependence of the TDD in AlGaN grown on a planar AlN template (broken line) and on a grooved AlN template (solid line). Thus, we can control TDD in AlGaN over the entire compositional range.

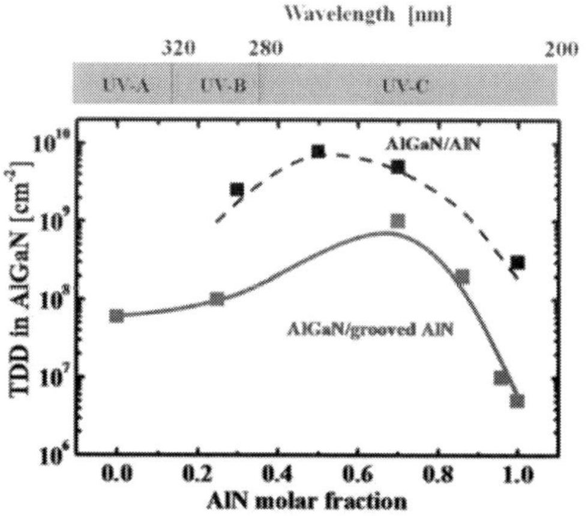

Figure 6. Compositional dependence of the TDD in AlGaN grown on planar AlN template (broken line) and on grooved AlN template (solid line).

Figures 7 and 8 show an example of the relationship between generation rate G and photoluminescence (PL) intensity (I_{PL}) (Figure 7) and IQE as a function of excitation carrier density n (Figure 8) for QWs emitting 230 nm UV-C having different TDDs.

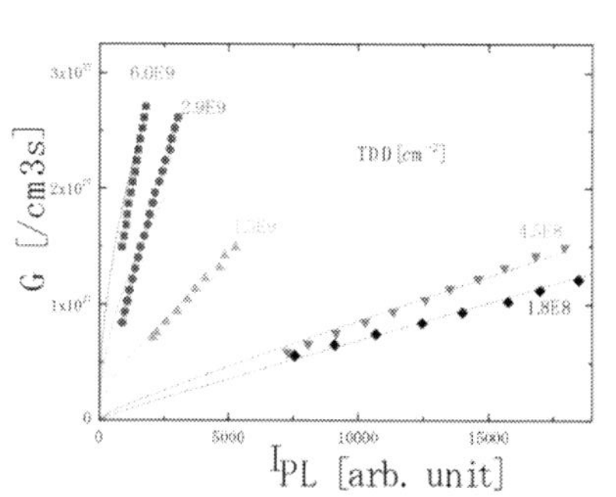

Figure 7. Experimental results and fitting curves of the G and I_{PL} for 230 nm QWs having different TDDs. Nonradiative coefficient A is deduced from this fitting curve using formula (3).

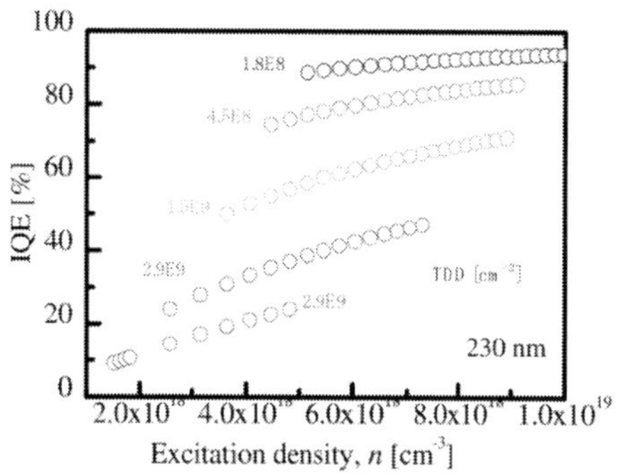

Figure 8. Calculated IQE using formula (4) as a function of n for samples shown in Fig.7.

The same procedure was performed for every sample to deduce the nonradiative recombination coefficient A. Figure 9 summarizes the TDD dependences of A for the 230, 250, 300 and 350 nm QWs. Those of 420 to 495 nm c-plane QWs grown on free-standing GaN having very low dislocation density and semipolar QWs on patterned Si substrate are plotted for comparison. From the results, it is clear that A is linearly dependent on TDD, A≒0.05×TDD+8.1E6 [s^{-1}], and independent of the emission wavelength. Again, it should be emphasized that this A coefficient is deduced assuming the B coefficient to be 0.2×10^{-10} cm^3/s. In this assumption, a nonradiative coefficient A of approximately 10^7 s^{-1} seems to be the limit of accuracy.

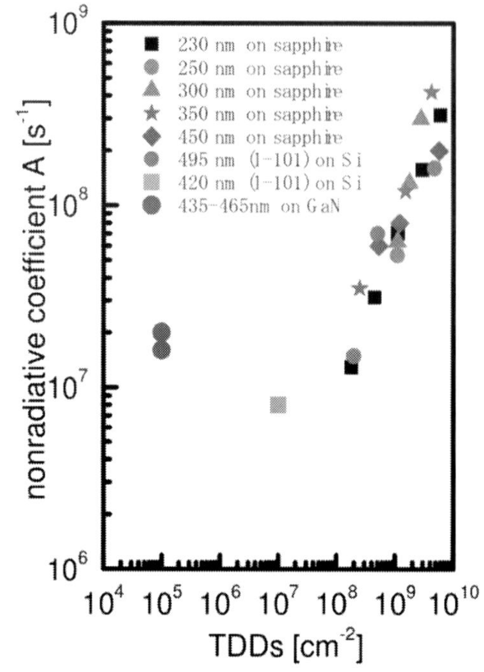

Figure 9. TDD dependence of the nonradiative coefficient A for QWs emitting different wavelengths from 230 to 495nm.

4. Summary

The impact of the high-temperature MOVPE of AlN and AlGaN and the effect of TDD on the IQEs of UV-A, UV-B and UV-C quantum wells were confirmed for the first time. There is still an ambiguity of the absolute value of the internal quantum efficiency when ABC model in Shockley-Read-Hall statistics is used because of the uncertainty of the radiative recombination coefficient B. However, it is evident that there is a very similar trend for AlGaN-based UV-A, UV-B, UV-C and InGaN-based violet, blue or blue-green QWs as a function of TDD. A simple conclusion can be obtained, that is, high-internal-quantum-efficiency UV QWs should be realized by reducing the threading dislocation density.

Acknowledgements: I would like to acknowledge Professors I Akasaki and S Kamiyama and Drs. T Takeuchi and M Iwaya, Meijo University, Professors Y Mori and Y Kitaoka, Osaka University, and Drs. M Yamaguti and Y Honda, Nagoya University, for fruitful discussions throughout this work. The majority of this work was carried out by the students of Meijo University and Nagoya University. In particular, I am very grateful to the efforts of Dr. M Imura, Mr. K Takeda, Mr. K Ban, Mr. J Yamamoto, Meijo University and Mr. T Tanikawa, Nagoya University. This work was supported by NEDO "Fundamental Technology Development of Next-generation Lighting of High-efficiency and High-quality Lighting" project.

References

[1] Dmitriev A and Oruzheinikov A, 1999 *J. Appl. Phys.* **86** 3241
[2] Imura M et al., 2008 *J. Crystal Growth* **310** 2308
[3] Nakano K et al., 2006 *phys. stat. sol.(a)* **203** 1632
[4] Ban K at al., 2011 *Appl. Phys. Exp.* **4** 052101
[5] Metzger T et al., 1998 *Philos. Mag.* **A77** 1013
[6 Shockley W and Read Jr. W T, 1952 *Phys. Rev.* **87** 835
[7] Ryu H-Y, Kim H-S and Shim J-I, 2009 *Appl. Phys. Lett.* **95** 081114
[8] Shen YC et al., 2007 *Appl. Phys. Lett.* **91** 141101
[9] Dai Q et al., 2009 *Appl. Phys. Lett.* **94** 111109
[10] Im JS, Moritz A, Steuber F, Harle V, Scholz F and Hangleiter A, 1997 *Appl. Phys. Lett.* **70** 631

Inversion boundary induced grain growth in ZnO ceramics: from atomic-scale investigations to microstructural engineering

Nina Daneu,* Slavko Bernik and Aleksander Rečnik

Department for Nanostructured Materials, 'Jožef Stefan' Institute, Jamova cesta 39, 1000 Ljubljana, Slovenia

*e-mail: nina.daneu@ijs.si

Abstract. In semiconducting materials special boundaries play the key role in crystal growth. They introduce an abrupt structural and chemical anisotropy, which is readily reflected in an unusual microstructure evolution, whereas their local structure affects the physical properties of semiconducting materials. These effects, however, can be exploited to tailor the electronic and optical properties of the materials, as demonstrated in this review. The presented topic fits in the field of preparatory stage of phase transformations, manifested through evolution of chemically induced structural faults. In the noncentrosymmetric structure of ZnO, inversion boundaries (IBs) are the most common type of planar faults that can be triggered by the addition of specific spinel-forming dopants (Sb_2O_3, SnO_2, TiO_2). In addition to conventional HRTEM techniques several new methods were developed to resolve crystallography and atomic-scale chemistry of IBs. The absolute orientation of the polar c-axes on both sides of the IB was determined by a novel quantitative microdiffraction method, providing a reliable identification of crystal polarity in noncentrosymmetric crystals. To determine sub-monolayer quantities of dopants on the IB, we developed a special technique of analytical electron microscopy using concentric electron probe (CEP) in EDS or EELS mode, providing more accurate and precise results than any other available technique. Knowing the local crystal chemistry of IBs we were able to design experiments to identify their formation mechanism. IBs nucleate in the early stage of grain growth as a dopant-rich topotaxial 2D reaction product on Zn-terminated surfaces of ZnO grains. Soon after their nucleation, ZnO is epitaxially grown on the inherent 2D phase in an inverted orientation, which effectively starts to dictate anisotropic growth of the infected crystallite. In very short time the grains with IBs dominate the entire microstructure in ZnO ceramics via IB-induced exaggerated grain growth mechanism. This phenomenon was used to design physical properties of ZnO-based varistor ceramics, whereas the bottom-up approach demonstrated here provides the basic tool for microstructural engineering of functional materials in virtually any system that is prone to formation of special boundaries.

1. ZnO-based varistor ceramics

Varistors are electronic components with highly nonlinear dependency between current (*I*) and voltage (*U*), therefore they are widely used for surge protection of electronic circuits and devices:

$$I = K \cdot U^{\alpha} \qquad \text{(Eq. 1)}$$

The coefficient α is a measure of the varistor's nonlinearity and its value in a commercial varistor is in the range from 30 to 80 [1]. A typical varistor *U-I* curve is shown in Figure 1a.

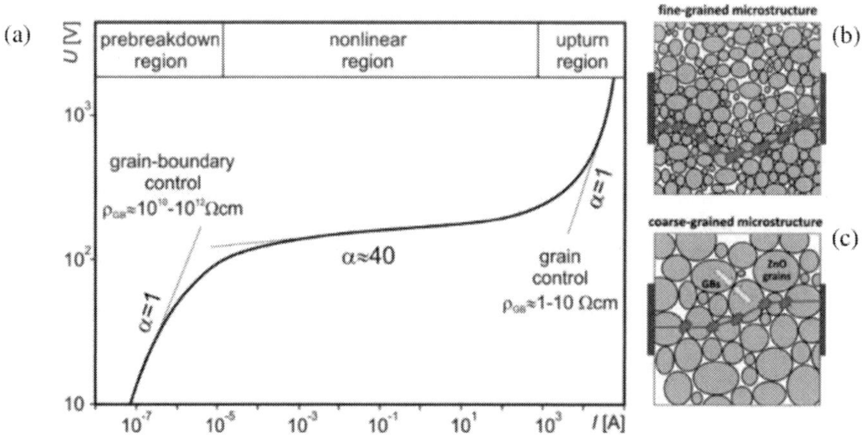

Figure 1: (a) Typical varistor curve showing the characteristic *U-I* dependence. Each ZnO-ZnO GB presents a charge barrier of ~3.5 V. The total breakdown voltage of the varistor depends on the number of active GBs between the two electrodes: (b) fine-grained microstructure for high-voltage applications and (c) coarse-grained microstructure for low-voltage applications.

The basic function of the protection device is that it acts as a resistor until the nominal *breakdown voltage* is reached, above this threshold, however, it suddenly (in nanoseconds) becomes orders of magnitude more conductive than the electronic device it protects. This quick response enables the surge voltage to be transferred through the varistor and prevents damage to the electronic device. Nowadays the vast majority of varistors is based on semiconducting zinc oxide (ZnO) ceramics sintered with various metal oxide additives in order to enhance the varistor's electrical properties. An ideal varistor is composed of highly conductive uniformly sized ZnO grains, separated by highly non-conductive nonlinear varistor-type grain boundaries (GBs), which produce the so-called *varistor effect*. In a varistor, an active ZnO-ZnO GB acts as a single switching device (diode) with a switch (or breakdown) voltage of about 3.5 V [1]. The most common additive that produces thin resistive GB layers in ZnO varistors is bismuth oxide (Bi_2O_3). During sintering the Bi-rich liquid phase distributes evenly along the GBs, where it forms a thin charge-barrier layer after solidification [2]. High GB resistivity defines the varistor's performance in the pre-breakdown region, where it conducts only a negligible *leakage* current. This unwanted effect mainly depends on the amount of secondary phases and non-resistive GBs [3]. The overall breakdown voltage of the varistor depends on the effective number of active GBs between the electrodes. As soon as this nominal voltage is surpassed the conductivity of ZnO grains becomes essential. To increase the conductivity of ZnO grains transition metal oxides (Mn, Ni, Cr, etc.) are added [4]. One of the major parameters for production of varistors for diverse voltage applications is therefore the number of active GBs. This can be simply controlled by the size of ZnO grains. In principle, varistors with smaller average grain size are implemented for high-voltage applications, whereas those with larger average grain size are used for low-voltage applications (Figs. 1b and 1c). ZnO grain size is controlled by a group of dopants (*i.e.* Sb_2O_3, SnO_2, TiO_2, *etc.*), which are called grain-growth modifiers due to their influence on ZnO microstructure development [4-8].

2. The role of dopants in ZnO-based varistor ceramics

Grain size control in ZnO varistor ceramics is quite challenging because commercial varistor compositions contain many interdependent dopants, which were empirically added to achieve the required electrical properties. Besides their influence on electrical properties, each dopant also influences ZnO grain growth. The grain growth is therefore not simply influenced by the initial grain size of ZnO and the additives and their recrystallization within the specific processing regime, but also by the formation of their reaction products in the various stages of sintering [9]. To make things even more complicated, different molar ratios of the dopants, for example, that between Bi_2O_3 and Sb_2O_3, lead to diverse reaction pathways during sintering, which all in their own way influence ZnO grain growth.

In order to understand the influence of individual additives their role in ZnO grain growth must be investigated. Sintering of pure ZnO is controlled by solid-state diffusion [10]. In the presence of Bi_2O_3, sintering of ZnO becomes liquid-phase assisted. Bi_2O_3 melts at 740°C (ZnO-Bi_2O_3 eutectic) enhancing the densification of the sample, whereas the ZnO grain growth rate depends on the amount of Bi_2O_3 addition. At very low additions (<0.05 mol%), ZnO grains that are in contact with Bi_2O_3 grow exaggeratedly due to partial wetting [2]. At moderate additions of Bi_2O_3 (0.05-1 mol%), the liquid phase present at every GB enhances the diffusion process and hence ZnO grains grow rapidly according to a solution-precipitation phase-boundary reaction mechanism [10]. Higher additions of Bi_2O_3 (>1 mol%) however, cause diffusion paths through Bi_2O_3 liquid phase to become longer and consequently ZnO grain growth is slowed down [11]. Typical additions of Bi_2O_3 in varistor compositions are about 1 mol%. In order to more efficiently control the ZnO grain size, other dopants are added to the binary ZnO-Bi_2O_3 varistor composition. The most common grain growth controlling additives for ZnO-based varistor ceramics are Sb_2O_3, SnO_2 and TiO_2 [4-8]. Empirically, Sb_2O_3 would be added when small ZnO grain size is desired, while the addition of TiO_2 generally results in large-grained microstructure.

Typical microstructures of high- and low-voltage ZnO-based varistor ceramics are shown in Figure 2. In spite of an obvious difference between the final microstructures, there has been a lot of dispute in what way these additives actually influence the ZnO grain growth. Popular explanations rely on the so-called particle pinning mechanism, based on Zener's drag phenomenon, which should act as an inhibiting force working against the ZnO grain growth [12]. This explanation was supported by the fact that adding higher concentrations of grain growth controlling additives (typically Sb_2O_3) resulted in the formation of more spinel grains [5,6]. As with the increasing concentration of spinel grains also the ZnO grain size becomes smaller, the spinels were consequently held responsible for inhibited ZnO grain growth. While this phenomenological correlation generally appears to be true, it fails to explain exaggerated grain growth in TiO_2-doped varistor ceramics [7], where despite an abundant presence of spinel grains some ZnO grains grow swiftly, disregarding any (spinel) particle pinning force that might be acting on the course of rapidly expanding ZnO grains. In addition to spinels, the grain growth controlling additives also cause the formation of IBs in ZnO grains. Their possible influence on ZnO grain growth was ignored until first TEM studies showed that spinel-forming dopants are present on IBs and made a clear connection that they are a consequence of the addition of grain growth controlling additives [13,14]. Soon IBs were suspected to have certain influence on grain growth, although their actual role was not truly understood: in TiO_2-doped varistor ceramics, for example, IBs were held responsible for exaggerated grain growth [7], while in Sb_2O_3-doped ceramics they were deemed to have an inhibiting effect on ZnO grain growth [5,6]. In our investigations we focused on evaluation of these effects on microstructure development in ZnO-based varistor ceramics.

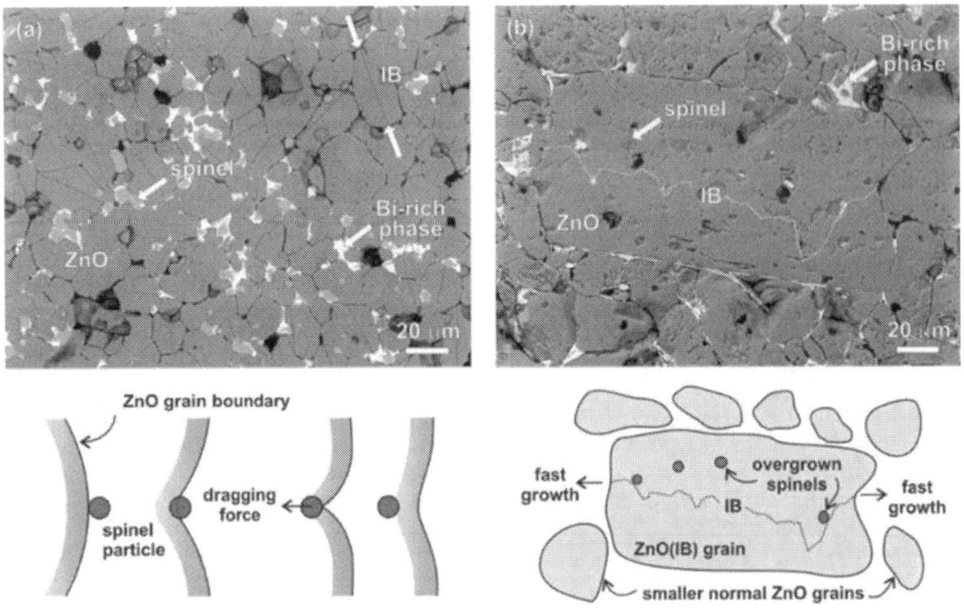

Figure 2: Typical microstructures of ZnO-based varistor ceramics: (**a**) high-voltage varistor ceramics with fine-grained microstructure - grain growth is inhibited due to (spinel) particle pinning mechanism (dopants: Sb_2O_3 and SnO_2). (**b**) Low-voltage varistor ceramics with coarse-grained microstructure - exaggerated grain growth is caused by the presence of IBs in some ZnO grains (TiO_2-doping). Spinel particles are overgrown.

3. Microstructure development in SnO_2-doped ZnO-Bi_2O_3 ceramics

To understand the role of grain growth modifying additives in the ZnO grain growth, basic ZnO-Bi_2O_3 varistor composition was doped with SnO_2 [8]. While the addition of SnO_2 to varistor ceramics is not favourable for the electric properties of the ZnO-based varistor, because it increases the leakage current [15], it turned out to be a perfect dopant for the grain growth mechanisms studies. Surprisingly we observed that adding very small amounts of SnO_2 to the basic varistor composition resulted in unexpected coarse-grained microstructure. The addition of even the lowest amounts of SnO_2 triggers formation of spinel particles and if the particle pinning mechanism was controlling the grain growth, the size of ZnO grains should be smaller than that in the dopant-free microstructure even at the lowest additions of the spinel forming dopant. But this was not the case. ZnO grains sintered with 0.1 mol% of SnO_2 were almost twice as large as the ZnO grains sintered without the spinel forming dopant. With the increasing concentration of SnO_2 the ZnO grain size was gradually reduced so that between 0.5-2.0 mol% of SnO_2 addition the ZnO grain size reached the grain size of undoped ZnO! This unusual behaviour shed a serious doubt on the popular explanation of grain growth inhibition by the presence of spinel particles. In all microstructures spinel grains were regularly present, but they did not present any obstacle for ZnO grain growth. Only at very high additions of SnO_2 the ZnO grain size became much smaller than that in normal ZnO without SnO_2 (Figure 3). We could confirm that with the increasing SnO_2 addition ZnO grain size was decreased, but the large initial increase of ZnO grain size at low SnO_2 additions could not be explained by any known grain growth mechanism.

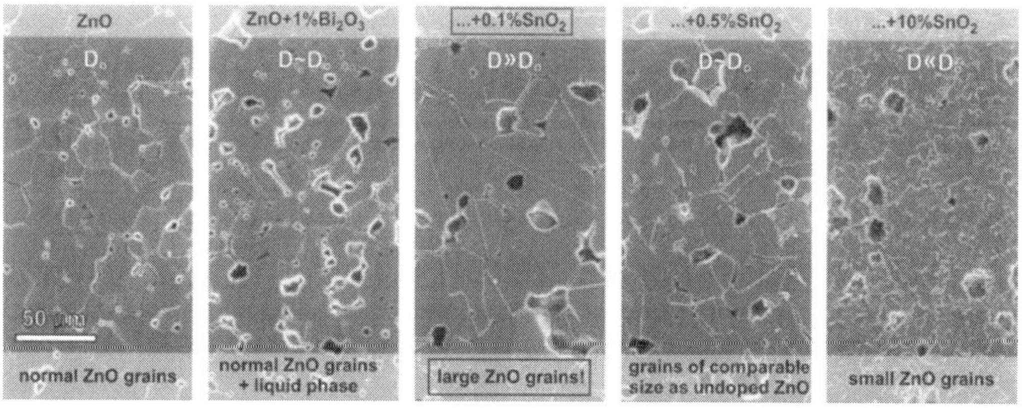

Figure 3. A sequence of microstructures starting from undoped ZnO, ZnO doped with 1 mol% of Bi_2O_3, and those with different additions of SnO_2. The grain sizes of undoped ZnO (D_0) and that with Bi_2O_3 addition are comparable ($D \sim D_0$). As soon as SnO_2 is added to this composition the grain size is surprisingly increased by a factor of 2 ($D \gg D_0$). With increasing SnO_2 addition ZnO slowly reaches the average grain size of undoped ZnO (D_0) and only at above 10% addition of SnO_2 the grain size is reduced by about a factor of 2 if compared with normal ZnO grain size ($D \ll D_0$).

To explain this unusual increase in ZnO grain size with small additions of SnO_2 we investigated the early stages of ZnO grain growth [8]. Along with spinel particles the addition of spinel forming dopant triggers the formation of IBs in ZnO grains. The influence of these planar defects on microstructure development was not studied before. At the beginning of grain growth IBs form in some ZnO grains. These grains are *nuclei* for anisotropic grain growth. In the samples with larger SnO_2 additions, more nuclei are formed, whereas in the samples with lower SnO_2 additions the number of nuclei is comparably smaller. As the growth proceeds these grains grow faster than normal ZnO grains. The growth of such crystallite is controlled by the growth of an inherent planar fault. This grain growth mechanism was called *IB-induced grain growth*. Following this mechanism grains with IBs grow preferentially in the direction parallel to the inherent IB plane and may reach unusually high aspect ratios. Resulting grains become highly anisotropic. When the longer axes of elongated grains reach lengths equal to the mean free distance between the grains with IBs, their anisotropic growth is stopped. If we have more nuclei the collisions occur sooner, and the resulting grains are smaller. After this stage another growth mechanism prevails, causing the elongated grains to thicken. This stage is to be compared with seed growth where according to the Ostwald's ripening Law grains with larger curvature radii grow at the expense of the surrounding smaller grains. ZnO grains with IBs grow exaggeratedly at the expense of normal ZnO grains until they entirely dominate the microstructure. The final microstructure is thus composed of only ZnO grains with IBs. The average size of the grains in the final microstructure equals the mean free distance between the IB nuclei, and is inversely proportional to the amount of spinel forming dopant added to the basic ZnO composition. Now the sudden increase of ZnO grain size on low additions of spinel forming dopant can be explained. Low addition of SnO_2 produces only a small number of IB nuclei and the resulting microstructure would be coarse-grained, whereas larger amount of SnO_2 would produce a large number of IB nuclei and result in fine-grained microstructure. Higher levels of SnO_2 addition also lead to increasing amount of secondary phases, however, their influence on ZnO grain growth is subordinated to the role of IBs [8].

In order to fully understand the nature of IBs and their true influence on ZnO grain growth, we performed a complete top-down chemical and structural analysis of these planar faults.

4. Atomic structure of inversion boundaries

The atomic structure and the chemistry of IBs in Sb_2O_3-doped ZnO were investigated by quantitative TEM techniques. Convergent beam electron diffraction (CBED) and high-resolution TEM (HRTEM) methods were used to determine the orientation of the polar *c*-axis on both sides of the inversion boundary and the translation state between the inverted ZnO domains. Quantitative energy-dispersive X-ray spectroscopy (EDS) combined with HRTEM analysis allowed us to determine the exact amount and the arrangement of Sb in the boundary layer. IBs were found to be head-to-head oriented, and the IB plane consists of a highly ordered $SbZn_2$ monolayer where the cations occupy octahedral interstices of the structure. In the octahedral boundary layer, Zn and Sb atoms constitute a honeycomb superstructure with a three-fold (*3m*) in-plane symmetry [16]. Both, the inversion as well as the composition of IBs have long reaching influence on physical properties and grain growth of ZnO.

4.1. Crystallography of IBs in ZnO

The orientation of the polar *c*-axis across the IB plane is considered the most important feature that must be determined to characterize the translational state of an interface in non-centrosymmetric crystals. As shown in our early study on Sn-rich IBs in ZnO simple chemical etching of polished microstructures including ZnO grains with IBs indicates an inversion of the polar c-axis across the IB plane by the reversal of triangular etch-pits in adjacent crystal domains [8]. In order to prove the inversion on the structural level we developed a new technique, which allows determination of polar axes in non-centrosymmetric crystals by a simple electron diffraction method in TEM [17,18]. The so-called micro-diffraction technique is based on the fact that due to the breakdown of the Friedel's Law the scattering factors for **g** and **-g** reflections are different in non-centrosymmetric structures. Based on semi-kinematic interactions between zero-order Laue zone reflections we showed that the absolute orientation of the *c*-axis could be obtained already from very thin crystals. The difference can be observed under convergent beam electron diffraction (CBED) conditions, using a small condenser aperture to avoid overlapping of the diffraction disks. The crystal polarity can then be simply deduced from the intensity of the diffraction disks. By comparing the computed and experimental micro-diffraction patterns we demonstrated that the polar *c*-axes of the two crystal domains are oriented towards the interfacial plane to form IBs in head-to-head (→|←) configuration (Figure 4).

The fact that IBs in Sb_2O_3-doped ZnO are head-to-head oriented is not only important for understanding the crystallography of these planar faults, but has far reaching implications related to electronic charge transport through the polycrystalline material. Namely, if a perfect ZnO crystal starts with an O layer and terminates with a Zn layer, then the electroneutrality of the crystal is retained. By the inversion, Zn atoms in the inverted crystal domain are shifted from *type-I* to *type-II* tetrahedral sites across the IB plane. This translation produces a new O-terminating layer on the other end of the crystal that was Zn-terminated prior to inversion, while the charge surplus is balanced in the IB layer. Duplication of O-terminating planes should markedly increase the current-voltage barrier at the ZnO grain boundaries. This alone would, in effect, improve the non-linear varistor characteristics. This aspect of surface-charge transport caused by the inversion of ZnO crystals [19] has not yet been studied in the scientific literature. These two effects, (i) the increased current-voltage barrier at the surface of the ZnO grains caused by a head-to-head inversion and, (ii) the internal barrier caused by the presence of IB layer, should not be present in ZnO ceramics without IBs. A supporting evidence for such effect can be seen in empirical α-values of varistor ceramics with or without Sb_2O_3 doping. Namely, comparable varistor compositions with Sb_2O_3 addition display twice as high α-value compared to crystals of that composition without this dopant [4]. Now that we know that Sb_2O_3 triggers formation of IBs in every ZnO grain this increase in α-value can be explained by the above mechanism.

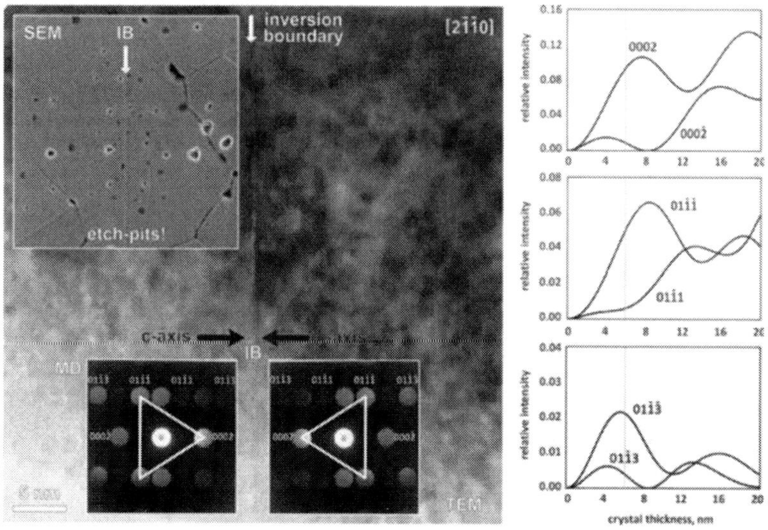

Figure 4. Inversion of crystal domains is revealed already by etching, where the orientation of the etch-pits is reversed across the IB (SEM). Micro-diffraction patterns are recorded on both sides of the basal-plane IB in ZnO crystal oriented along the $[2\bar{1}\bar{1}0]$ zone axis (TEM). Pairs of reflections not related to the mirror plane show different intensities and the patterns on both sides of the IB are inverted. Calculated intensities of these reflection pairs indicate a head-to-head (→|←) orientation of the adjacent ZnO domains across the IB.

Translational state of coherent basal-plane IBs was investigated by quantitative HRTEM. The reversal of the c-axis and an apparent lattice dilatation across IBs allowed us to distinguish these faults from basal-plane SFs, which are rather common in ZnO. HRTEM images in $[2\bar{1}\bar{1}0]$ projection revealed a rigid-body shift of $\frac{1}{3}\cdot[01\bar{1}0]$, whereas in the $[01\bar{1}0]$ projection no shift was observed, resulting in total displacement vector of $R_{IB} = \frac{1}{3}\cdot[01\bar{1}0] - \Delta\cdot[0001]$. This shift is identical to a displacement observed on basal-plane SFs, which is produced in a wurtzite lattice by a 180° rotation of (0002) layer into (faulted) cubic stacking. The SF shift is followed by inversion operation where Zn^{2+} ions are shifted from *type-I* to *type-II* tetrahedral sites across the fault plane to produce an IB. The basal plane displacement depends on the type of the dopant and its oxidation state that due to local repulsion forces produce a rigid block relaxation Δ. In Sb-rich IBs the $\Delta = 0.102$ was measured by quantitative HRTEM analysis [16]. Another very prominent feature that could be observed from HRTEM images is dopant ordering in the IB-layer implying the occupancy of either 1/3 or 2/3 of Sb atoms in the IB plane (Figure 6 [16]).

4.2. Chemical composition of IBs in ZnO

The exact chemical composition of IBs is one of the most important parameters for determination of the interface structure. The very first analyses of the IB plane involved a spatial-difference technique [13]. This method is based on acquiring one EDS/EELS spectrum on the interface of interest and another in bulk crystal. The amount of the dopant on the interface is obtained by the difference between the two spectra. Spatial difference EDS analysis of Sb-rich IBs in ZnO [13] indicated that 1.0 ± 0.5 monolayer of Sb is uniformly distributed in the IB plane. With spatial difference technique it is fairly impossible to accurately determine the specimen thickness, which leads to high systematic errors of such measurements. This technique therefore does not provide a sufficient accuracy that would allow a distinction between the two possible Sb arrangements (1/3 or 2/3) in the IB plane. To solve such problems a new analytical method for measuring small amounts of a dopants at interfaces was designed.

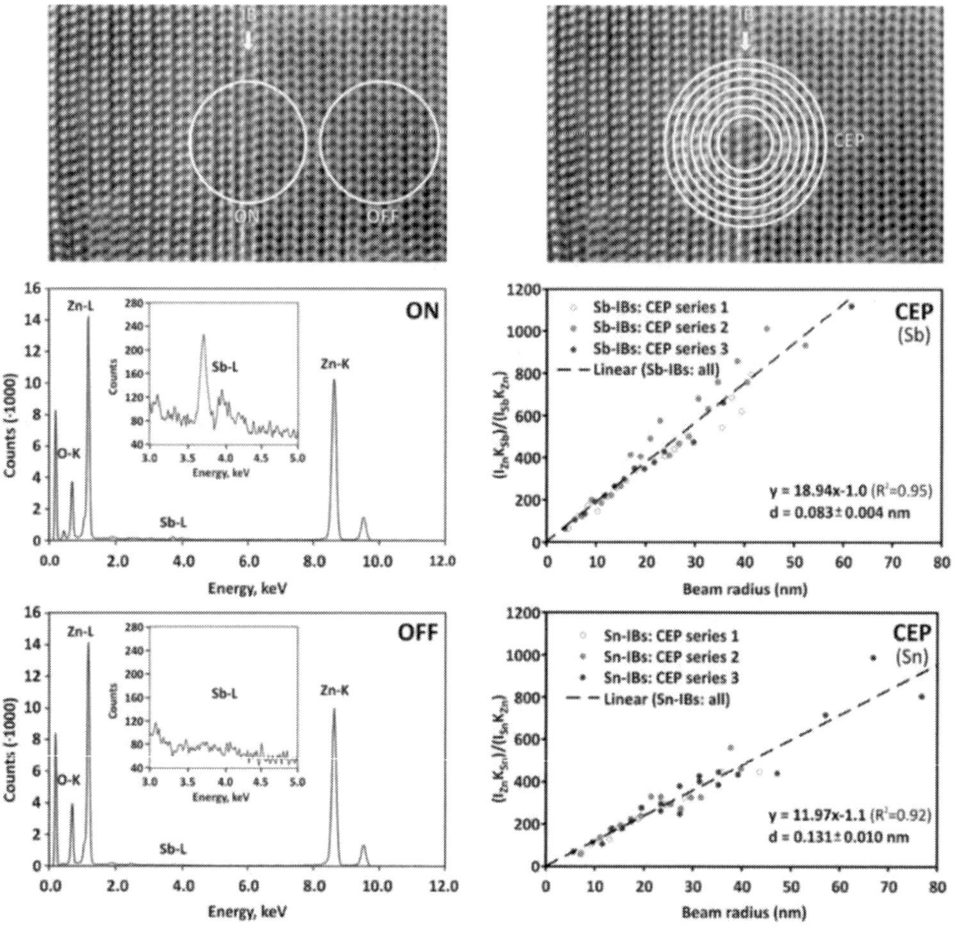

Figure 5. Schematic presentation of the spatial difference technique (left): one spectrum is recorded ON and another OFF the boundary. The amount of the dopant is estimated from the difference between the two spectra. The new CEP method (right) and results on Sb-rich and Sn-rich IBs in ZnO (below) [15,20,21].

The *concentric electron probe* (CEP) method is based on acquiring a sequence of EDS or EELS spectra with different beam diameters and the boundary always placed at the centre. Because the centre of analysis is identical for all measurements sample thickness variations have no influence on the accuracy of our measurements. The dependency between matrix (I_M) and dopant (I_D) intensity ratio *vs.* the beam diameter (r) can be described by a simple geometrical approximation [16,20,21]. One of the conditions that must be met if we want to obtain accurate results is that the dopant element is present only in the interface or has a very low solid solubility (x) in matrix phase. If the electron beam is wide and the solid solubility is high, then a considerable amount of the dopant signal stems from the bulk rather than from the boundary, and the resulting over-estimate of the effective boundary width can be very large. This effect would lead to a strong deviation from the linear dependence and to asymptotic convergence of the function $I_M/I_D=f(r)$ towards a constant value (x) for large beam sizes. In our case the dopant is present only in the IB plane. For very low solid solubility ($\lim x \rightarrow 0$) the I_M/I_D intensity ratio is linearly dependent on r, where I_M is proportional to the concentration of the matrix element in cylindrical volume (without the interface plane) excited by the incident electron beam of selected radius (r), whereas I_D of the dopant element stems only from the interface [16]:

$$\frac{I_M \cdot c_D}{I_D \cdot c_M} = \frac{\pi}{2d} r - 1 \qquad \text{(Eq. 2)}$$

where d is the arbitrary width of the interface occupied by the dopant element. To calculate the actual occupancy of the dopant, this value must be considered in terms of structural density for the specific crystallographic plane. IBs in ZnO lie in the basal planes of the wurtzite structure, which are separated at a spacing of $c/2$=0.2605 nm. In order to obtain the fraction of the interface occupied by the dopant element the arbitrary width d must be divided by this value for ZnO.

Our HRTEM analysis [16] showed that Sb is perfectly ordered in the IB layer and occupies either 1/3 or 2/3 of the IB octahedral sites. The results of our quantitative CEP/EDS analysis of three different IBs showed that Sb occupies 0.326 ± 0.040 of the IB-layer, which is close to 1/3 of a monolayer. High-angle annular dark-field scanning transmission electron microscopy (HAADF-STEM) studies of Sb-rich IBs confirmed the observed occupancy and a hexagonal ordering of Sb and Zn atoms in the IB plane [23]. Using complementary imaging and analytical techniques we could estimate the actual accuracy and precision of CEP method and quantitatively compare it with the spatial difference technique. In analytical methods only precision (statistical error) can be estimated, but the most important parameter accuracy is never discussed because of the lack of independent measurements or theory that would provide a reliable estimate of the measured quantity. Based on complementary structural studies [16] we know that in the case of Sb-rich IBs antimony occupies 1/3 of the boundary monolayer (highly ordered hexagonal arrangement). Hence the actual accuracy and precision of both methods can be estimated. The accuracy of the spatial difference measurement [13] was poor ±300%, whereas the accuracy of CEP measurements [16] approaches ±2%, which is two orders of magnitude better than any other analytical method! Most of the improvement is due to elimination of systematic errors by the CEP method. Precision, on the other hand, can be improved by careful experimental work and was ±50% for spatial difference measurement [13] and ±12% for the CEP method. More detailed theoretical background and the capabilities of the CEP method are given elsewhere [20-22]. The atomic fractions of Sn-rich and Fe-rich IBs in ZnO showed occupancies of 0.504 ± 0.038 [20-22] and 1.018 ± 0.076 [21], respectively. Both analytical methods are schematically illustrated in Figure 5.

4.3. Atomic arrangement in the IB layer

The final questions related to the structure of IBs are related to crystal chemistry and local charge balance at the atomic scale. HRTEM studies and CEP measurements on IBs with different dopants indicated that the oxidation state of particular dopant dictates the atomic arrangement in the IB plane.

According to Pauling's electroneutrality principle for ionic crystals [24] every Zn^{2+} in tetrahedral interstices of the ZnO structure obtains a charge of ½ e^- from each of the four closest O^{2-} ions and the average charge per tetrahedral site in the wurtzite structure is +II. Every cation located in the IB octahedral interstice, however, has six neighbouring O^{2-}, which produces an average charge per octahedral site of +III [16]. To satisfy the local charge balance according to Pauling's rule the expected in-plane composition of Sb-rich IBs would be $SbZn_2$, where one Sb^{5+} alternated with two Zn^{2+} cations would result in requested average oxidation state of +III for octahedrally coordinated cation in the IB plane. This is consistent with quantitative structural and chemical analysis of Sb-rich IBs [16].

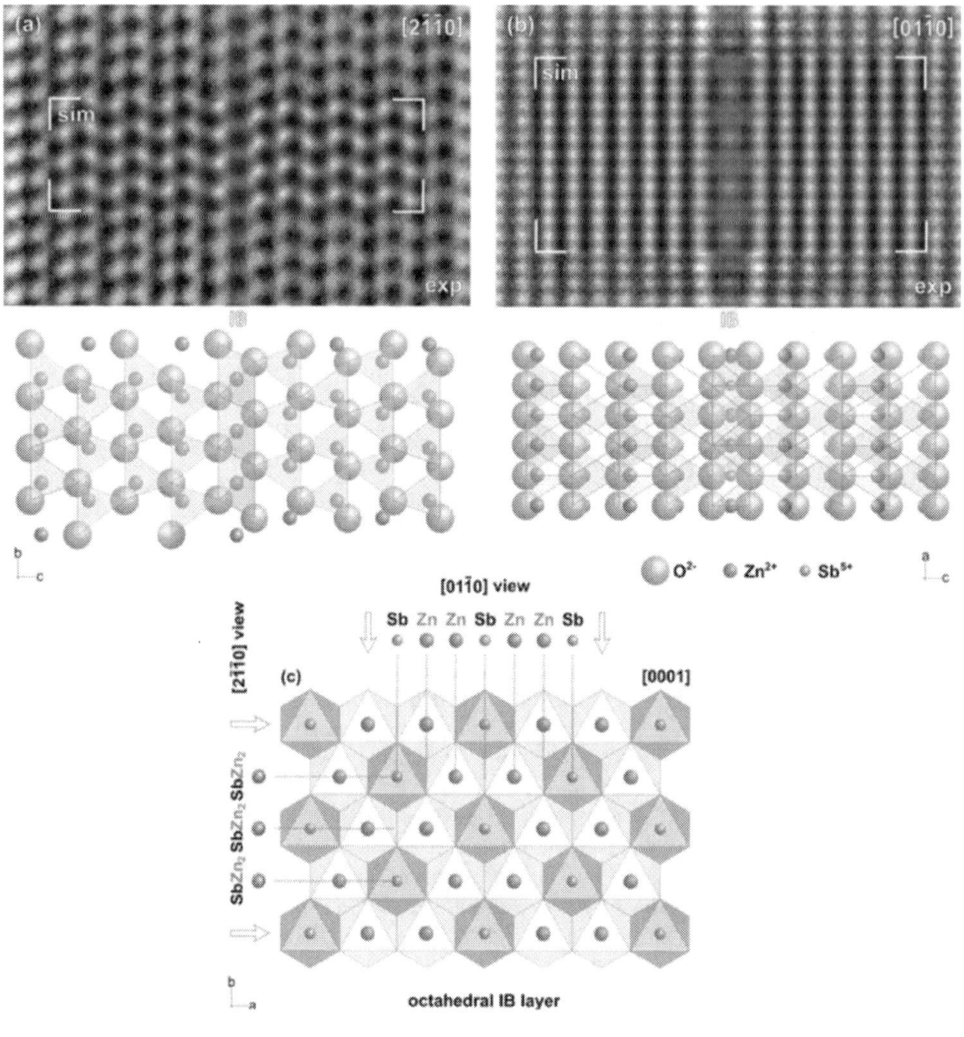

Figure 6. Atomic model of Sb-rich IBs in ZnO reconstructed from (**a**) [$2\bar{1}\bar{1}0$] and (**b**) [$01\bar{1}0$] projections of the interface. In [$2\bar{1}\bar{1}0$] view the atomic columns in the IB plane have equal intensity, which implies that their composition is identical. In [$01\bar{1}0$] view, however, an interesting *abb* in-plane periodicity is revealed, suggesting that the atomic columns have two different atomic densities. Based on these contrast features, CEP analysis of IBs and Pauling's principle of electroneutrality in ionic crystals the actual IB composition can be resolved [15].

Figure 6 shows HRTEM images of Sb-rich IB in ZnO in two low-index crystallographic orientations where the IB plane is viewed edge-on. From these two projections an atomic arrangement of the IB plane was reconstructed. Sb and Zn atoms in the IB plane form a highly ordered octahedral layer with a three-fold in-plane symmetry in the ratio of Sb:Zn=1:2 [16]. In this analogy we predicted that IB-layers with other pentavalent dopants (M^{5+}) should have similar atomic arrangement with an IB-layer composition of MZn_2 (1:2), tetravalent dopants (M^{4+}) will have an IB-layer composition of MZn (1:1), and trivalent dopants (M^{3+}) would form a full IB-layer M (1:0) of the corresponding dopant element. Investigations of Sn(IV)-rich and Fe(III)-rich IBs in ZnO have supported this prediction [21].

5. Nucleation and growth of inversion boundaries

Top-down structural analysis of IBs has clearly indicated that the main driving force for the inversion is the preservation of the local charge balance upon incorporation of IB-forming dopants. In the next step we have designed experiments to study the nucleation and growth of IBs in different systems. Experiments using ZnO single crystals embedded into ZnO powder with the addition of SnO_2, Sb_2O_3 and In_2O_3 [19] showed that, depending on the oxidation state of the dopants, we can identify two competing mechanisms of IB nucleation: (i) *internal diffusion*, and (ii) *surface reaction* and growth. The first mechanism is characteristic for +III dopants and is controlled by Zn-vacancy diffusion mechanism, whereas the second mechanism holds for all IB-forming dopants and is controlled by chemisorption of the dopants on Zn-deficient (0001) surfaces. Grain growth studies of ZnO ceramics have indicated that IBs are growth faults that control the growth of ZnO grains. Grain growth experiments showed that IBs form prior to the spinel phase in a particular phase system and grains with IBs grow exaggeratedly at the expense of normal ZnO grains until they dominate the microstructure [8].

5.1. Internal diffusion and surface reaction mechanisms of IB formation

To resolve the mechanism of IB formation we have to understand the principles of dopant diffusion into the ZnO structure and thermodynamics of chemical reactions that take place on the surface of growing ZnO grains. The diffusion of isovalent dopants into the ZnO structure can be explained by zinc self-diffusion and its evaporation at higher temperatures [25]. Evaporation of Zn atoms from the (0001)-Zn surfaces produces clusters of Zn-vacancies, which are more stable than isolated vacancies [29]. The easiest path for diffusion of Zn-vacancies and vacancy clusters are the close-packed (0001)-Zn planes, while Zn^{2+} ions were predicted to move more rapidly by jumping through a series of octahedral sites lined along the close-packed Zn-layers. At elevated temperatures the mobility and evaporation of Zn atoms near the surface of ZnO crystals becomes significant and the near-surface concentration of Zn-vacancies is increased. The cation diffusion mechanism would apply for any +II dopant that forms a solid solution with ZnO without changing the basic ionic configuration of the wurtzite structure [25]. Local charge deficiency, caused by the formation of Zn vacancies at elevated temperatures, is therefore compensated for by the incorporation of the dopant elements.

Zn deficiency may also be the main driving force for the migration of +III ions into close-packed basal planes of ZnO. On their path the +III ions would preferentially occupy the larger octahedral sites rather than the tetrahedral sites. In order to reduce the induced surplus of charge Zn^{2+} ions from the adjacent wurtzite layer are shifted from the *type-I* to the *type-II* tetrahedral sites, producing the head-to-head inversion boundary [16]. This generates a charge deficiency in the next Zn layer and triggers a progressive relaxation (inversion) of the complete ZnO domain. The charge balance of the entire ZnO crystal is achieved when the charge-relaxation front exits the crystal. Following the cation diffusion mechanism, accompanied by an inversion of the adjacent ZnO domain, the +III ions may propagate along parallel basal planes deep into the ZnO crystal, producing multiple parallel IBs on their way. Such cases are known in ZnO-Fe_2O_3 [26,27], ZnO-Ga_2O_3 [28] and ZnO-In_2O_3 [19,27] systems.

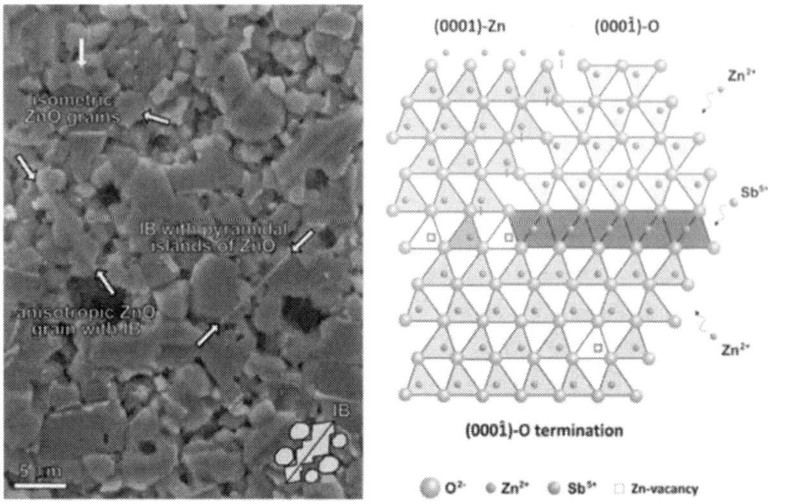

Figure 7. Initial stages of IB nucleation at 780°C in the ZnO-SnO_2 system (left). The temperature of IB formation is lower than the spinel-phase formation temperature. The illustration (on right) shows the basic principle of IB nucleation following the surface reaction mechanism that applies for +IV and +V dopants. To maintain the local charge balance the adjacent ZnO domain grows in inverted orientation with respect to the IB plane.

The dopants that have a higher oxidation state than +III follow a different nucleation mechanism because the surplus of charge brought by these dopants cannot be simply compensated by the internal diffusion. Instead these dopants use the weakness of Zn-vacancy clusters on the (0001)-Zn surfaces [29]. The octahedral sites located at the centres of these vacancy clusters would be very favourable for chemisorption of the IB-forming dopants. This process leads to formation of a stable (two-dimensional) surface compound. To avoid unfavourable face-sharing with neighbouring tetrahedral sites Zn^{2+} ions jump to their closest octahedral interstices, forming a mixed octahedral layer together with the IB-forming dopant. As soon as the octahedral interstices are occupied, the wurtzite structure on top of the IB layer starts to crystallise in an inverted orientation with respect to the base crystal, as illustrated in Figure 7. The main reasons for crystallisation in the inverted orientation are preserving the local charge balance, avoiding octahedral-tetrahedral face-sharing, and minimising the short-range cation-cation repulsion forces. To satisfy the local charge balance, the dopant ions are ordered in the IB layer in an appropriate ratio with the Zn^{2+} ions to satisfy the average oxidation state per IB interstitial site of +III that fulfils the charge balance for head-to-head IBs in ZnO [16].

5.2. Stability of the IB-layer and its influence on ZnO grain growth

Let us take a look on the thermodynamic reasons for IB nucleation. The necessity of the presence of IB-forming dopants to generate an IB implies that the resulting 2D structure embedded in the hosting crystal is more stable than the reactants alone (ZnO and the dopants). Nucleation of IB by either internal diffusion or surface nucleation causes local chemical reaction driven by the preservation of the local charge balance. The formation of such special boundaries, like IBs in ZnO, is therefore chemically triggered and takes place in the preparatory stage of binary-phase formation [30]. In general, the activation energy for the formation of special boundaries in host crystals is lower than the energy for the formation of the full binary phase. By low-temperature sintering of SnO_2-doped ZnO it was demonstrated that IBs nucleate below the formation temperature of the $ZnSn_2O_4$ spinel [19]. Because of the intimate structural relationship with the host ZnO structure the crystals containing IBs compete for the same building material as the crystals without such defects. Grain-growth studies in various binary systems showed that the composite grains with special boundaries are more successful in growth than the normal grains, implying that the formation energy of composite grains is significantly lower than that of the host crystal alone. An initial stage of IB formation and nucleation mechanism by surface reaction and growth is illustrated in Figure 7.

After their nucleation, IBs grow via solid-state reaction taking place at the IB front. The formation of IB structure is an efficient way of lowering the system energy by incorporating the impurity atoms. In the initial stages of growth grains with IBs develop extremely anisotropic morphology. During their growth, islands of ZnO are topotaxially deposited on both sides of the IB-lamella. While the driving force for growth of such composite crystal with the direction of the inherent IB plane is the incorporation of the IB-forming dopant, the main reason for their thickening normal to the intrinsic IB layer in the second growth stage is the relatively large surface area of the $(000\bar{1})$ terminations. In spite of the fact that the growth rate in $[000\bar{1}]$ direction is the slowest in ZnO [31], composite ZnO grains always have much larger curvature radii than normal ZnO grains, and they are always more successful in competition for building material (ZnO), regardless of the size of the normal grains [8]. The high stability of the composite grains and their ability to consume all the available ZnO has dramatic effects on microstructure development. Once an IB is formed this grain can never regress to a normal ZnO grain under regular conditions. The microstructures of fully evolved ZnO grains with IBs present a thermodynamically stable equilibrium state and typically show no further recrystallization [32].

5.3. Grain growth kinetics in ZnO ceramics in the presence of IBs

Time lapse grain growth experiments at different sintering temperatures revealed four distinct stages of microstructure evolution on doping with IB-forming additives were identified [32]: (i) nucleation of the IB on the surface of the primary ZnO grains, (ii) formation of composite ZnO(IB) grains followed by exaggerated anisotropic growth, (iii) thickening of the composite ZnO(IB) grains until all normal ZnO grains are consumed and (iv) reaching fully evolved equilibrium microstructure where all grains contain IBs. During nucleation the concentration of IBs is proportional to the amount of the added IB-forming dopant and the efficiency of the dopant distribution through the microstructure, *i.e.*, Sb_2O_3 with a higher mobility produces a higher initial number of ZnO(IB) grains than the less mobile SnO_2 or TiO_2 at equivalent temperature and dopant concentration [33]. In the next stage many IBs that have formed on the surface of ZnO grains are dissolved, while some remain and start to grow anisotropically. The material needed for the growth of the ZnO(IB) grains is supplied from normal ZnO grains, which are dissolved in the growth process. The grain growth in this stage is entirely controlled by the *IB-induced grain-growth* mechanism until ZnO(IB) grains reach the mean free path between the ZnO(IB) nuclei. The mean free path for ZnO(IB) growth depends on the initial concentration of nuclei (Figure 8). The grain growth during this stage is very rapid and the apparent energy for the ZnO grain growth is the lowest during the entire growth process [32]. After impingement of the ZnO(IB) grains growth is controlled by the Ostwald's ripening mechanism, where the existing ZnO(IB) grains fill up the whole microstructure at the expense of the remaining normal ZnO grains. In this growth stage the grain-growth rate is slowed down while the apparent energy for ZnO growth is increased. Close to equilibrium only local recrystallization processes take place. During this final stage equilibrium grain boundaries are formed due to relative surface energies in particular crystallographic directions. During this stage the number of grains and their size does not change with sintering time [32].

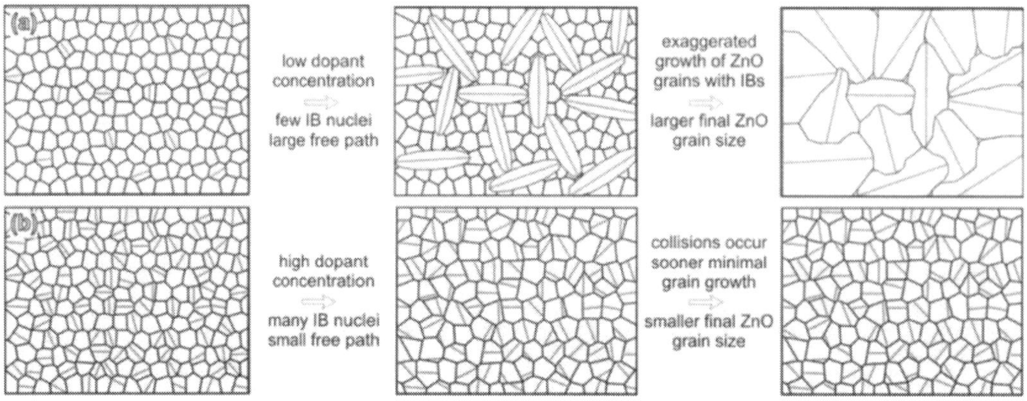

Figure 8. (**a**) Microstructure development in the ZnO sample doped with 1 mol% Bi_2O_3 and 0.1 mol% SnO_2 after 15, 120 and 240 min of sintering. (**b**) Schematic presentation of the IB-induced grain growth mechanism in ZnO sample with high addition of the IB-forming dopant.

6. Microstructural engineering of varistor ceramics

The knowledge about the atomic structure of IBs, local charge balance, the stability of the IB structure and the mechanisms of IB nucleation and growth can serve us to design experiments for optimization of the varistor's physical properties. The main technological challenges would be to improve the electrical properties, design components for any desired voltage and reduce the production cost by lowering the amount of additives, shortening sintering time/temperature, *etc*. Can this be done?

In principle, we have demonstrated that lowering the amount of IB-forming dopant leads to the formation of coarse-grained microstructure. Such microstructure is desired in low-voltage applications. Sb_2O_3 is generally known as the dopant to produce fine-grained varistor ceramics, however, via the *IB-induced grain growth* mechanism it should be possible to obtain coarse-grained microstructures even with Sb_2O_3 if we start with fewer IB-nuclei. In our diffusion doping experiment we have shown that microstructure development in Sb_2O_3-doped ZnO is controlled entirely by the exaggerated growth of ZnO grains containing IBs, whereas the presence of spinel grains has absolutely no effect on the final grain size [32]. According to the proposed *IB-induced grain growth* mechanism grains with IBs (nuclei) grow anisotropically and exaggeratedly along the direction of the inherent IB plane and during growth they consume surrounding normal ZnO grains without IBs. The final grain size simply depends on the quantity of Sb_2O_3 addition. At higher Sb_2O_3 concentrations, more nuclei form in the beginning of the grain growth process, the growing grains collide sooner and the final grain size is smaller. On the contrary, lower Sb_2O_3 additions lead to larger grain size. Our calculations showed that the additions of <30 ppm of Sb_2O_3 would produce microstructures of ZnO grains larger than the grain size of undoped ZnO, while higher additions of Sb_2O_3 would lead to fine-grained microstructures [33].

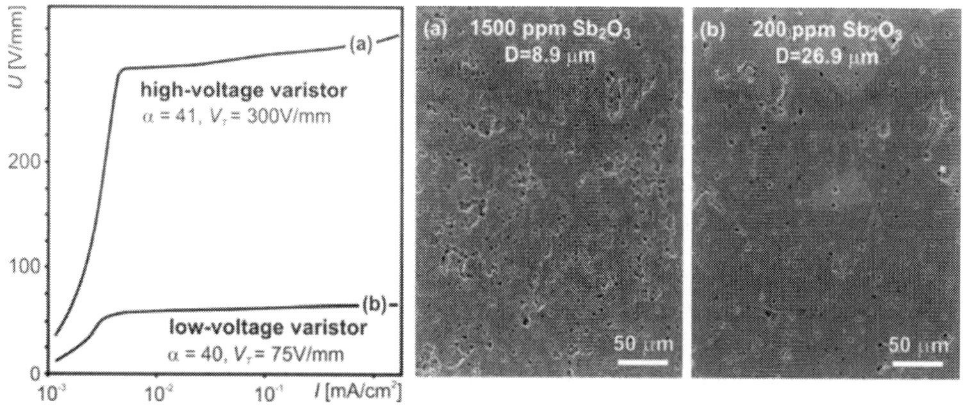

Figure 9. Electrical field vs. current density (*E-J*) curves for high- and low-voltage Sb_2O_3-doped ZnO-based varistor ceramics with optimized compositions with corresponding microstructures (**a**) fine-grained (**b**) coarse-grained.

These findings were implemented in optimization of ZnO-based varistor ceramics. First, addition of Sb_2O_3 was varied to study microstructure development in binary ZnO-Sb_2O_3 system [34]. It was found that additions of less than 75 (mol) ppm Sb_2O_3 lead to the formation of ZnO grains larger than that in the undoped ZnO ceramics. This value is about twice as high as the amount of Sb_2O_3 determined in the diffusion doping (in vapour phase) experiment [33]. Possible reasons for the loss of Sb are locally increased concentrations of Sb_2O_3 causing formation of nanosized $Zn_7Sb_2O_{12}$ spinel particles and evaporation of Sb_2O_3 at higher temperatures. After optimizing the processing parameters for the binary ZnO-Sb_2O_3 system other dopants were selectively added in order to produce the varistor effect (Bi_2O_3) and increase the conductivity of ZnO grains (Co_3O_4, Mn_3O_4, NiO, Cr_2O_3). With the addition of other dopants the total amount of Sb_2O_3 increases due to solid solubility of antimony in Bi-rich liquid phase and minor formation of $Bi_3Zn_2Sb_3O_{14}$ pyrochlore phase. To obtain the coarse grained microstructure up to 200 ppm of Sb_2O_3 are needed in complete varistor composition (Figure 9). Electrical measurements of this new generation of ZnO-based varistors show very promising nonlinearity coefficients ($\alpha \approx 40$) and low leakage current. Fundamental understanding of *IB-induced grain growth* in ZnO provided a solid foundation for completely new approach to tailor the microstructures of either low, medium or high-voltage varistor ceramics at radically lower amount of dopants, while gaining improved electrical properties at lower cost. Consequently varistor ceramics with the reduced amount of unwanted secondary phases at the GBs and increased effective contact area among the ZnO grains can be produced, providing improved energy absorption capability of varistors for any nominal voltage. Savings on raw material per varistor unit also give producers a considerable cost-advantage towards their competition at higher quality, and due to highly reduced concentration of heavy metal contaminants, also an ecologically advantageous product.

References

[1] Clarke DR 1999 *J. Am. Ceram. Soc.* **82** 485
[2] Wong J 1974 *J. Am. Ceram. Soc.* **57** 357
[3] Olsson E, Dunlop G and Osterlund R 1993 *J. Am. Ceram. Soc.* **76** 65
[4] Matsuoka M 1971 *Jpn. J. Appl. Phys.* **10** 736
[5] Kim J, Kimura T and Yamaguchi T 1989 *J.Mater.Sci.* **24** 2581
[6] Senda T and Bradt RC 1991 *J. Am. Ceram. Soc.* **74** 1296
[7] Makovec D, Kolar D and Trontelj M 1993 *Mat.Res.Bull.* **28** 803
[8] Daneu N, Rečnik A, Bernik S and Kolar D 2000 *J. Am. Ceram. Soc.* **83** 3165
[9] Inada M 1980 *Jpn. J. Appl. Phys.* **19** 409
[10] Senda T and Bradt RC 1990 *J. Am. Ceram. Soc.* **73** 106
[11] Dey D and Bradt RC 1992 *J. Am. Ceram. Soc.* **75** 2529
[12] Nunes SI and Bradt RC 1995 *J. Am. Ceram. Soc.* **78** 2469
[13] Bruley J, Bremer U and Kraševec V 1992 *J. Am. Ceram. Soc.* **75** 3127
[14] Makovec D and Trontelj M 1994 *J. Am. Ceram. Soc.* **77** 1202
[15] Bernik S and Daneu N 2001 *J. Eur. Ceram Soc.* **21** 1879
[16] Rečnik A, Daneu N, Walther T and Mader W 2001 *J. Am. Ceram. Soc.* **84** 2657
[17] Tafto J and Spence JCH 1982 *J. Appl. Cryst.* **15**, 60
[18] Mader W and Rečnik A 1998 *Phys. Status Solidi* A **166** 381
[19] Rečnik A, Daneu N and Bernik S 2007 *J. Eur. Ceram Soc.* **27** 1999
[20] Walther T, Daneu N and Rečnik A 2004 *Interface Science* **12** 267
[21] Walther T, Rečnik A and Daneu N 2006 *Microchim. Acta* **155** 313
[22] Walther T, Rečnik A and Daneu N 2005 Proc. 14th Conf. Microsc. Semicond. Materials, Oxford, *Springer Proc.in Physics* **107** 199
[23] Yamazaki T, Nakanishi N, Rečnik A, Kawasaki M, Watanabe K, Čeh M, Shiojiri M 2004 *Ultramicroscopy* **98** 305
[24] Pauling L 1940 *Cornell University Press* The Nature of the Chemical Bond
[25] Tomlins GW, Routbort JL and Mason TO 2000 *J.Appl.Phys.* **87** 117
[26] Wolf F, Freitag BH and Mader W 2007 *Micron* **38** 549
[27] Hörlin T, Svensson G and Olson E 1998 *J. Mat. Chem.* **8** 2465
[28] Barf J, Walther T and Mader W 2004 *Interface Science* **12** 213
[29] Dulub O, Diebold U and Kresse G 2003 *Phys.Rev.B* **68** 545409
[30] Rečnik A, Čeh M and Kolar D 2001 *J. Eur. Ceram. Soc.* **21** 2117
[31] Lee JS and Wiederhorn SM *J. Am. Ceram. Soc.* 2004 **87** 1319
[32] Daneu N, Rečnik A and Bernik S 2011 *J. Am. Ceram. Soc.* **94** 1619
[33] Daneu N, Rečnik A and Bernik S 2003 *J. Am. Ceram. Soc.* **86** 1379
[34] Bernik S, Bernard J, Daneu N and Rečnik A 2007 *J. Am. Ceram. Soc.* **90** 3239

Assessment of lattice strain, rotation and dislocation content using electron back-scatter diffraction

Angus J Wilkinson

Department of Materials, University of Oxford, Parks Road, Oxford OX1 3PH, U.K.

angus.wilkinson@materials.ox.ac.uk

Abstract. Cross-correlation based analysis methods have been developed for electron back scatter diffraction (EBSD) patterns that improve the angular sensitivity to ~10^{-4} rads. This enables EBSD to be used to study the much smaller misorientations and even local elastic strain fields that are typical in semiconducting materials. Mapping of the lattice rotations and elastic strain variations provides sufficient detail for quantitative analysis of the threading dislocation density through the Nye tensor. The analysis will be briefly described and applications given to GaN and Si/SiGe based systems. Measurements of tilt, twist and elastic strain variations in GaN layers on basal plane sapphire will be reported and compared to results for some epitaxial lateral over grown (ELOG) GaN samples. The effects of misfit interfacial dislocations on the spatial distribution of the full strain and rotation tensors in a partially relaxed SiGe layer will also be shown.

1. Introduction

Over the last two decades electron back scatter diffraction (EBSD) has become an established tool for microstructural characterization of structural materials, particularly metallics [1, 2]. Although EBSD has been used to identify substrate – epilayer orientation relationships [3, 4] and grain boundary types in multicrystalline Si [e.g. 5, 6] its uptake for characterization of functional and semiconducting materials has been much more limited than for metallics. To a significant extent this can be attributed to the angular resolution limit of ~0.5°(~10^{-2} rads) on lattice misorientations. Although acceptable for metallics which are most often polycrystalline and with relatively high lattice defect density, to date EBSD measurements have been too insensitive for semiconductors which are often single crystal and with relatively low lattice defect content.

Early attempts at using cross-correlation-based analysis by Troost et al [7] and Wilkinson [8-10] have recently seen significant development and improvement by Wilkinson, Meaden and Dingley [11, 12] to provide the sensitivity required for characterisation of many semiconductor samples. The approach relies on the fact that elastic strains and small rotations cause slight shifts in features, such as zone axes, within the EBSD patterns that can be measured to high precision using cross correlation techniques. The method will be described in more detail below. EBSD is now emerging as a laboratory based technique that, in some ways, is competitive with many large scale centralised synchrothon facilities for high spatial resolution strain mapping.

The cross-correlation-based analysis of EBSD patterns has been applied to unrelaxed SiGe blanket films on Si in planview [8, 10] and elastic relaxation caused by sectioning [12] and by growth on mesas of finite width [13-15]. Vaudin et al [16] have used the method to study stresses generated by wedge indents in Si. They found excellent agreement between EBSD and micro-Raman measurements and assessed the stress sensitivity to be ~10 MPa. Very recent work by Tomita et al [17] have also shown good agreement between EBSD and micro-Raman measurements for strained Si on insulator and SiN/Si systems. Villert et al [13] have made measurements across the depth of a bent Si sample finding good agreement with theory. Excellent spatial resolution of the technique is evident in studies on cross-sections of AlGaN/GaN heterostructures by Ishido et al [18] which includes a linescan across a 5 nm AlN/25 nm GaN multilayer in which the 50 repeats can be distinguished. Alternative approaches for using EBSD to analyze strain in semiconductors include the use of pattern blurring to demark regions of high elastic strain gradients [19], and a modified three parameter Hough transform [20].

2. High Accuracy EBSD Methodology

EBSD systems comprise a ~40 mm diameter scintillator screen mounted in front of a highly tilted (~70°) sample and viewed with a low light level CCD camera. In most cases, including in our systems, the CCD and screen are coupled using a fast lens, though more direct coupling with a fibre optic bundle is used in some cases. The camera is mounted outside the SEM vacuum chamber and connected to a PC via a 'firewire' interface. TSL OIM software was used to control the SEM beam position and acquire the EBSD patterns at the full (~1k by 1k by 12 bit deep) resolution of the camera. The patterns are recorded to hard disc for subsequent batchwise analysis for strain variation using CrossCourt software from BLG productions [21].

The following describes the conceptual basis of strain analysis using EBSD. Changes in elastic strain and lattice rotations cause small shifts in the positions of zone axes and other features in the EBSD patterns. The magnitude and direction of these shifts vary with position across the pattern in a systematic way that has been related to the size and nature of the strain and rotation state within the diffracting volume [10-12]. One pattern from the series to be analyzed is chosen as a reference pattern. Ideally the reference pattern should come from a location on the sample for which the strain is known, or in many cases is known to be zero. A set of square sub-regions are defined across the patterns. The shifts relative to positions in the reference pattern of diffraction features in each of these sub-regions is then determined using cross-correlation analysis for each of the test patterns. The cross-correlation functions are calculated via the Fourier domain, with a weighting function used to prevent issues with aliasing. Within the Fourier domain band pass filtering is applied to remove high frequency noise, and low frequency background intensity variations. Details of the significance of these image analysis procedures on the resulting pattern shift measurements are given in [12]. The pattern shifts in 256 by 256 pixel sub-regions can be measured to better than ±0.05 pixels though of course pattern quality and signal to noise influence this noise floor. The magnitude of the cross-correlation function peaks are also recorded. These are normalized to give unity for an autocorrelation and a geometric mean over all sub-regions on a pattern is calculated and used to assess the quality of the pattern matching. Typically for deformed metallics peak heights below ~0.3 are considered unsatisfactory and the corresponding data are discarded as unreliable. For most semiconductor applications the peak heights are generally considerably higher due to the lower lattice defect content.

To distinguish the effects of strains and tilts it is necessary to determine the shifts in the positions at multiple sub-regions of the EBSD patterns. We generally use at least 20 sub-regions combined with either least square error, or iterative robust fitting methods to obtain the strain and rotation tensors that best fit the measured pattern shifts. Infinitesimal deformation theory is used in the analysis. The quality of this best fit solution can be assessed by calculating the (weighted) mean angular error in the shifts actually measured at each sub-region of the pattern compared to shifts expected on the basis of the best fit solution for strain and rotation. If the measured strains and rotations are larger than the mean angular error there is good confidence in the measurements, but if they are smaller then they are below the noise limit and should be disregarded.

Variations in the lattice rotations about three orthogonal axes and the three shear strains are found directly, however the analysis is insensitive to hydrostatic dilatation and as a result only the differences between pairs of normal strains are determined. To separate these normal strains we can assume that the sample surface (normal along x_3) is traction free so that we can write

$$\sigma_{33} = 0 = C_{13}\varepsilon_{11} + C_{23}\varepsilon_{22} + C_{33}\varepsilon_{33} + C_{34}\varepsilon_{23} + C_{35}\varepsilon_{31} + C_{36}\varepsilon_{12} \qquad \text{(equation 1)}$$

in which C_{ij} and ε_{23}, ε_{31}, ε_{12} are the known single crystal elastic constants and elastic shear strains in the sample reference frame. This equation along with measurements of the differences (ε_{11}-ε_{33}) and (ε_{22}-ε_{33}) allows all 6 strain terms to be isolated. This traction free surface condition is taken to be acceptable in most instances in the structural metallics, however, for measurements very near abrupt interfaces in semiconductors it is possible for the stress (and/or elastic strain) field to vary significantly within the sampling depth of EBSD (a few tens of nanometers). In any such instances the separation of the normal strains should be considered in more detail. The elastic constants often simplify when measurements are made on a sample with surface normal along a symmetry axes.

3. Applications

3.1. Epitaxially Lateral Over Grown GaN

Epitaxial lateral overgrowth (ELOG) [22] and pendeoepitaxy [23] have been shown to give a significant reduction in the density of threading dislocations in GaN layers grown on sapphire and SiC substrates. In the laterally grown wing regions dislocation densities of 10^6 cm^{-2} can be achieved

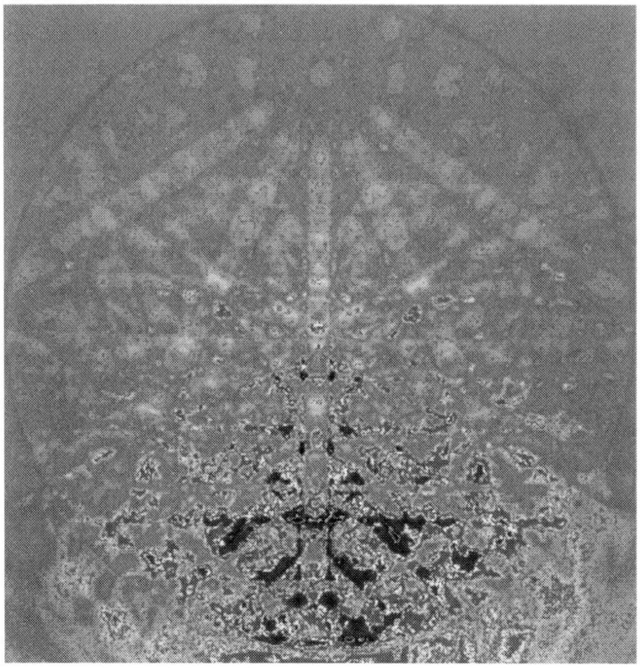

Figure 1. Example EBSD pattern from GaN grown directly above the central seed region of an ELOG sample.

compared to 10^{10} cm^{-2} in the seed regions. In ELOG GaN the crystal quality (defect density), crystal orientation, stress and strain thus all vary with position relative to the underlying mask structure. EBSD measurements were conducted on an ELOG GaN sample grown by Dr Ian Watson and co-workers at the Institute of Photonics, University of Strathclyde [24]. The sample studied (STR129/3) consisted of a (0001) sapphire substrate on which a 2.5 μm planar GaN seed layer was grown. Over the seed layer a 4 μm thick ELOG GaN layer was grown through a mask of 9 μm wide 200 nm thick SiO_2 stripes repeated with a period of 16 μm. The sample studied had been grown to a point just beyond coalescence. An example pattern from the seed region is shown in figure 1, and preliminary measurements on the sample were reported in [25]. The measurements were taken on a JEOL 6500F, operated at 20 keV and an estimated probe current of ~3 nA with sample tilted 60° from normal incidence. A line scan was made on the top surface of the ELOG GaN, using a step size of 80 nm and spanning ~3 repeats of the structure. A reference pattern was selected within the central seed region and used within CrossCourt to conduct the cross-correlation measurements at 50 256 by 256 pixel sub-regions and to make the subsequent analysis for lattice rotations and strain variations. Figure 2 shows some of the outputs from the analysis. Two data quality parameters described in section 2 above are routinely used to assess the data. The first of these is the geometric mean of the cross-correlation peak heights, which remains above 0.9 throughout but shows lowest values near the wing-wing coalescence boundaries.

The second data quality parameter is the mean angular error which is significantly below 10^{-4} rads throughout the dataset but shows distinct peaks above the wing-wing coalescence boundaries, and lower values of 2-3x10^{-5} rads above the seed regions. Visual inspection of the patterns indicates a very subtle loss of acuity near the wing-wing coalescence boundaries though the effect is very minor and possibly subjective and steered by knowledge of the trend shown in figure 2a.

Selected components of the lattice rotation and strain variations are shown in figure 2b. The terms not shown all had only minor variations along the line scan, with the exception of the out of plane normal strain ε_{33} which varied in the opposite sense to ε_{11} and to a lesser magnitude as expected from equation 1. These near zero results are all expected from the traction free surface boundary condition, and the plane strain like conditions expected from the long extent of the stripe structure along the x_2 axis.

There is good reproducibility in the results from the ~three periods of the mask pattern. The tilts w_{31} about the x_2 axis parallel to the stripe direction show marked changes at the seed-wing interface and wing-wing coalescence boundary (figure 2b). At seed-wing interfaces tilts of ~0.006 rad (~0.3°) occur, the sense of the tilt being opposite for the wings grown from either side of a given seed region. As a consequence the tilt changes by ~0.012 rads (~0.6°) at wing-wing coalescence boundaries. The change in tilt occurs more abruptly at the wing-wing coalescence boundary than at the seed-wing interface. There is also a significantly smaller and unexpected variation of lattice rotations w_{12} about the surface normal (i. e. twists) which again show a gradual increase in size across the wing regions and a more abrupt change at the wing-wing coalescence boundaries.

The strain variations are all relative to the unknown strain state at the reference point. The 'in plane' strain ε_{11} perpendicular to the stripes varies considerably and becomes progressively less compressive as we move from the centre of a seed region toward the wing regions. However near the wing-wing coalescence boundary there is a very pronounced but localized increase in the 'in plane' compression.

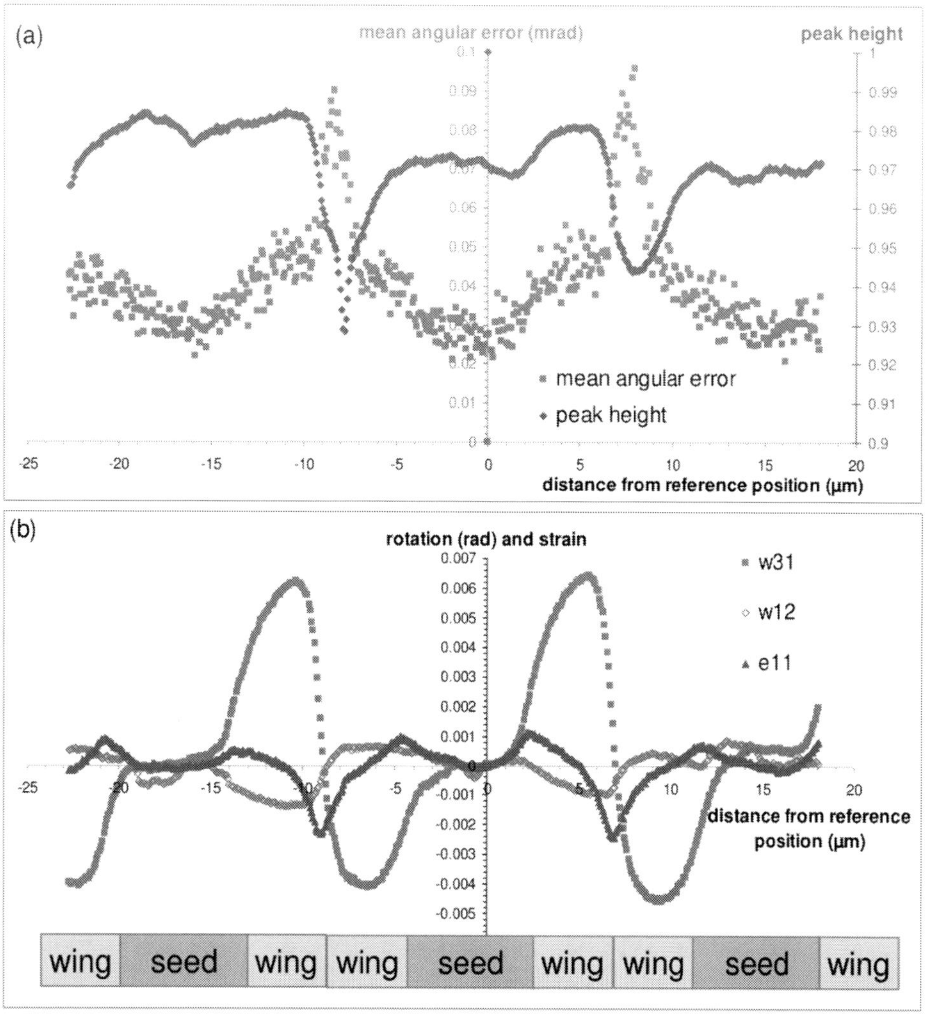

Figure 2. Measurements for EBSD line scan across top surface of an ELOG GaN sample. (a) data quality parameters, (b) selected components of lattice rotation and strain variation.

3.2. GaN Buffer Layer on Sapphire

The second example application is to the analysis of tilt and twist mosaics in a samples of 900 nm thick GaN film grown on (0001) oriented sapphire single crystals by metalorganic vapour phase epitaxy (MOVPE). The sample was one of a set grown by Prof Peter Parbrook and co-workers while he was at the EPSRC National Centre for III-V Technologies, University of Sheffield.

EBSD analysis was performed on a JEOL JSM 6500F using a beam energy of 15 keV, and current of approximately 10 nA. The sample was tilted toward the EBSD detector by 70° from normal incidence about the [1010]GaN axis. A 6 μm by 6 μm area was mapped using a square grid with a 100 nm step size. Pattern shifts were again measured at 50 sub-regions across each pattern and were used to determine a least square error best fit strain and rotation tensor. A pattern toward the top left hand corner of the map was used as a reference pattern. The mean of each strain and rotation component was determined over the map and these values then subtracted so that the maps shown in figure 3 show the deviations from the mean value. The mean peak height map (figure 3a) shows some reduction from the top to bottom which corresponds to a loss of pattern contrast due to a build-up of carbonaceous contamination on the sample surface during the scan. The mean angular error (figure 3b) gives an average value of 5×10^{-5} rads for the entire dataset, and shows a slight decrease in the data quality towards the lower part of the map due to a build-up of surface contamination during the scan. The axes system used to describe the strains and rotations are given in figure 3c. The twist rotations w_{12} about the surface normal x_3 are clearly larger than either of the tilt rotations w_{23} and w_{31} about the two in plane axes x_1 and x_2. Elastic strain variations are generally smaller than the lattice rotations, though this difference is not nearly as marked as for deformed metals [26, 27]. In this crystal orientation the two out of plane shear strains ε_{23} and ε_{31} simply scale linearly with the corresponding stress terms σ_{23} and σ_{31}, which should go to zero at the traction-free surface. The results do indeed show that variations in ε_{23} and ε_{31} are noticeably smaller than the other strain components in accord with the expected boundary condition. Figure 4 gives histograms showing the distribution of different rotation components about the mean value. Comparison is made to rotations measured in a similar way in a near perfect GaAs single crystal which effectively show the noise level on the measurements. For the GaAs data the histogram shows the distribution averaged over the three rotation components. These 'pseudo-rocking curves' show quantitatively the greater spread in the twist mosaic compared to the tilt mosaics that was visually noticeable in the maps (figure 3). The distributions of elastic strains around the mean value are seen to be significantly smaller than for the rotation and only just above the noise level indicated by data from the GaAs sample. A greater spread for the twist mosaic than for the tilt was also seen in the Hough-transformed based EBSD measurements by Rao *et al* [28] though in the current analysis the angles are an order of magnitude smaller indicating a lower defect density in the GaN film.

Nye [29] gave an elegant geometrical construction allowing the lattice curvature to be linked to the density of geometrically necessary dislocation (GND) content within the crystal. Given the densities, line and Burger's vectors of a set of dislocations the lattice curvature of the crystal can be calculated directly and unambiguously, however, the reverse process of determining the dislocation content from the curvatures leads to ambiguities in most cases due to there being too many possible dislocation types. However, a lower bound GND density can be established which supports the measured curvatures and minimizes some other term such as the total dislocation line length or energy (an L1 minimization). Kroner [30] extended this analysis to include the elastic strain gradients that Nye had assumed to be small (which we have found to be the case in deformed metals). In the GaN sample although the elastic strain variations are smaller than the rotations the difference is not so great that the strain field can be ignored in the analysis. When elastic strain gradients are included in the analysis only three of the nine terms in the Nye dislocation tensor can be found and these relate to dislocation types that thread through the sample surface [31]. For this GaN sample we assume that the three measured lattice curvatures (rotation and elastic strain gradients) are accommodated by combinations of screw dislocations along [0001], and three possible edge dislocation types with [0001] line direction and ⅓<1120> Burgers vectors. Even with this small and simple set of possible dislocation types there is ambiguity in solving for the dislocation densities which arises from the non-independence of the three edge dislocations. Of the many solutions that support the measured curvature we select one that also minimizes the total line energy in which we account for the differing magnitudes of the Burgers vectors and differentiate between screw and edge types. Maps showing the density distributions of the four dislocation types are given in figure 5. As expected the larger twist mosaic propagates through the GND density analysis to result in a greater density of <a> type edge dislocations.

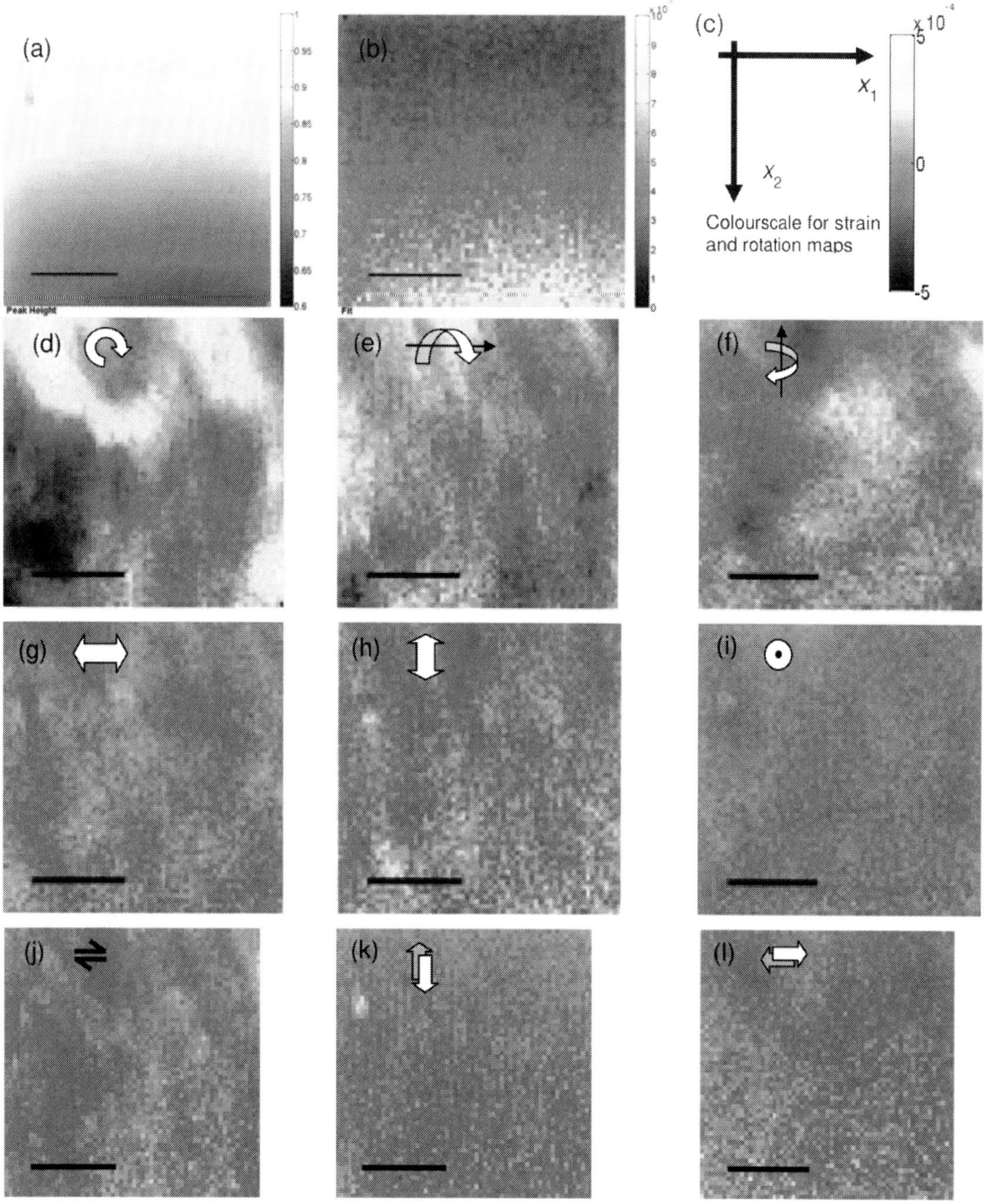

Figure 3. High accuracy EBSD maps of a 900 nm thick GaN layer grown on sapphire. (a) cross-correlation peak height, (b) mean angular error (rads), (c) axis system and colour scale for strains and rotations, (d), (e), (f) rotations about the x_3, x_1, and x_2 axes respectively, (g), to (l) variations from the mean of elastic strains ε_{11}, ε_{22}, ε_{33}, ε_{12}, ε_{23}, and ε_{31} respectively. The scalebars are 2 μm long and the step size was 100 nm.

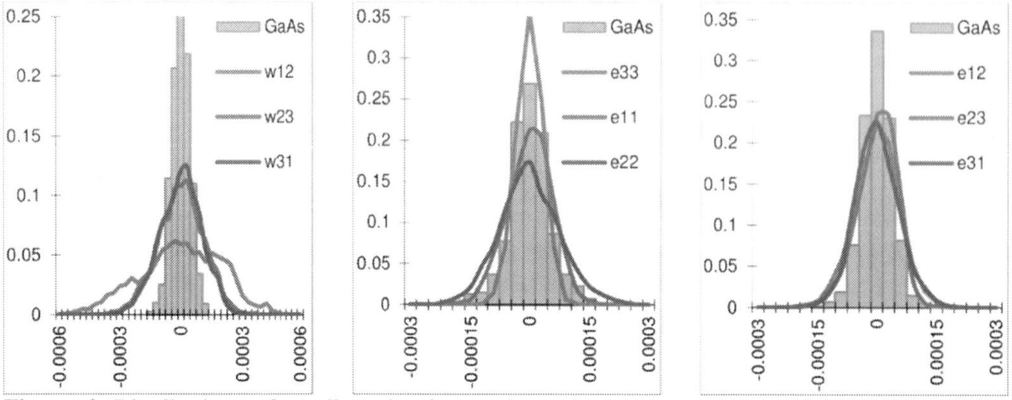

Figure 4. Distributions of (a) tilt and twist rotations, (b) normal strains, and (c) shear strains in a 900 nm thick GaN layer on sapphire. Solid bars show similar distributions for a GaAs single crystal averaged over the selected rotation or strain components.

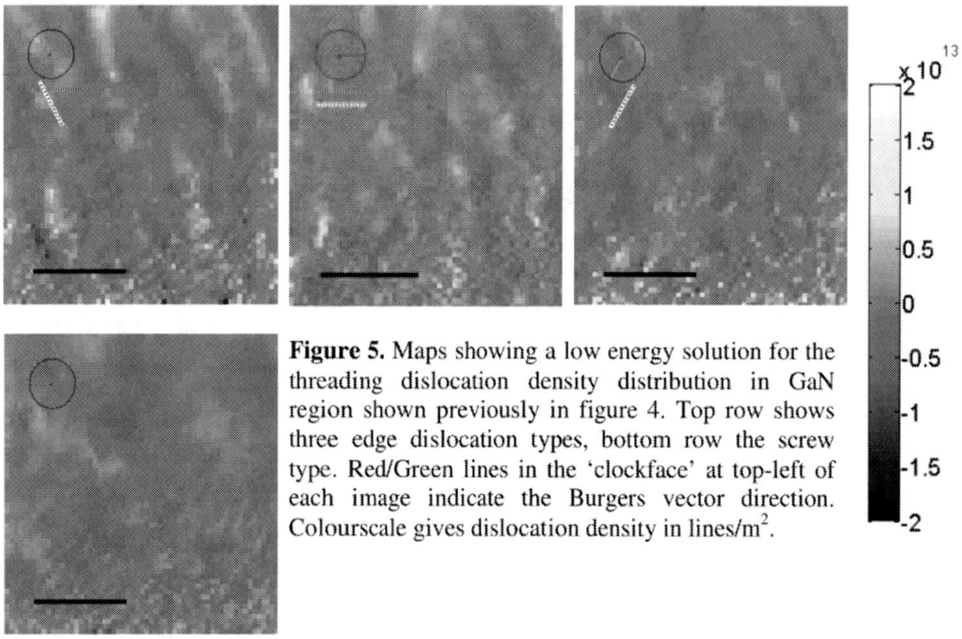

Figure 5. Maps showing a low energy solution for the threading dislocation density distribution in GaN region shown previously in figure 4. Top row shows three edge dislocation types, bottom row the screw type. Red/Green lines in the 'clockface' at top-left of each image indicate the Burgers vector direction. Colourscale gives dislocation density in lines/m^2.

3.3. SiGe/Si – plastic relaxation

The final illustration to be given here is from a series of SiGe/Si samples supplied a considerable time ago by the B. T. Research Laboratories (Martlesham Heath) by Tuppen, Gibbings and Hockly [32]. The particular sample studied consisted of a 1.2 μm thick $Si_{1-x}Ge_x$ alloy grown by molecular beam epitaxy on a Si (001) substrate. The nominal Ge content was 10%, for which the epilayer thickness is well above the critical value required for strain relaxation by interfacial misfit dislocations [32]. SEM conditions employed were 15 keV beam energy, a beam current of ~3 nA, with a relatively long exposure time for the EBSD pattern acquisition of approximately 1 second. An 80 μm by 80 μm map was made with a 0.5 μm step size. The data was collected on a LaB_6 SEM and a slight scan instability can be seen in the maps but does not detract from the discussion and striking comparison with the GaN data. The cross-correlation analysis was undertaken in an off-line, batchwise, automated process using CrossCourt 3 software [21], using 32 sub-regions for the analysis. A reference pattern was selected at the centre of the map. Within the mapped area the mean angular error was found to have an average value of 1.1×10^{-4} rads, with 5% of points having values less than 0.9×10^{-4} rads, and 95% being below 1.3×10^{-4} rads.

As the strain and rotation at the reference point are not known the results are presented in figure 6 as maps showing the deviation of the individual strain and rotation components from their (unknown) mean values. X-ray diffraction methods could be used to establish the mean strain values. The map of lattice rotations (w_{12}) about the surface normal (figure 6a) show two orthogonal sets of features running along the near vertical [110] and near horizontal [110] axes. The rotations about the two in-plane axes are larger and only show the features running near parallel with the rotation axis, i.e. vertical features for the w_{31} component and horizontal for the w_{23} component. The features are clearly generated by the two orthogonal sets of misfit dislocation lines generated by the partial relaxation of the large misfit stress initially present in the epilayer. The edge component of these dislocations generate rotations about their line direction, so that each set of dislocations is only present in either the w_{31} map or the w_{23} map, but not present in the other. Rotations about the surface normal are generated by the screw components of any dislocation with line direction parallel to the surface plane and so both sets of dislocations appear in the w_{12} map. The screw components of the dislocations also cause a (tensor) shear strain ε_{12} in the surface plane (figure 6g), but do not contribute to the normal strains. Again the effects of the two sets of dislocations are separated in the maps of the two in plane normal strains (ε_{11} and ε_{22}) with the dislocations not affecting the strain along their line directions. As the edge components of the interfacial dislocations relieve the misfit strain there is a concomitant relaxation of the tetragonal distortion which is seen as a change in the out of plane strain ε_{33} which is affected by both sets of dislocations.

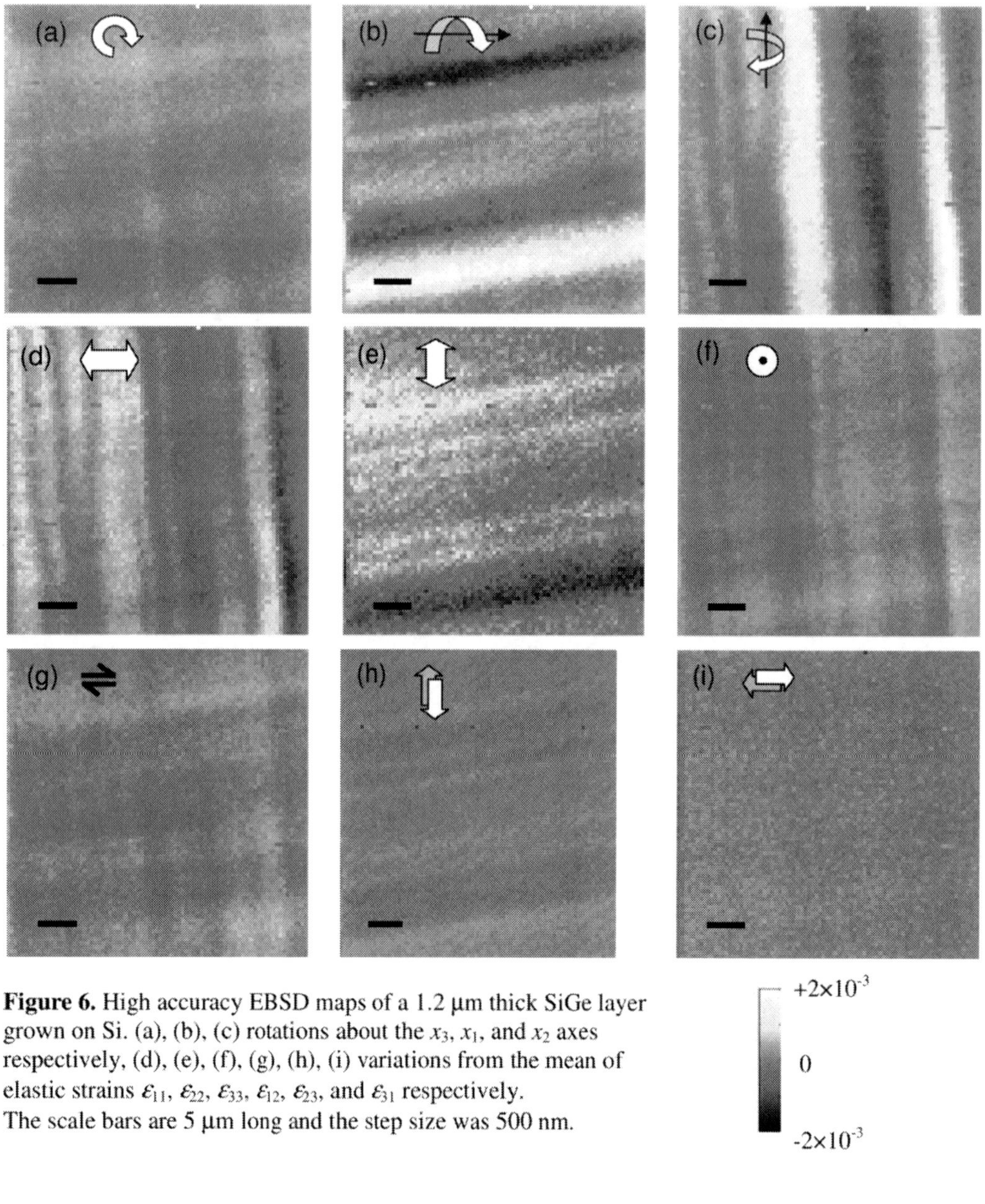

Figure 6. High accuracy EBSD maps of a 1.2 μm thick SiGe layer grown on Si. (a), (b), (c) rotations about the x_3, x_1, and x_2 axes respectively, (d), (e), (f), (g), (h), (i) variations from the mean of elastic strains ε_{11}, ε_{22}, ε_{33}, ε_{12}, ε_{23}, and ε_{31} respectively.
The scale bars are 5 μm long and the step size was 500 nm.

4. Summary

The development of cross-correlation based analysis of EBSD patterns has generated a step change in the sensitivity of the method and dramatically increases its scope for characterising many semiconductor samples. The residual elastic strains and lattice rotations can be measured and mapped to approximately $\pm 10^{-4}$ (rads). The method is now very straight forward to use and data sets of the size presented here take up to an hour or two to collect and a similar time to analyse to the standard used here. Quicker analysis is possible by using fewer sub-regions for cross-correlation which can be useful if a faster preliminary look at the data is required. All 9 components of the displacement gradient tensor (strain and rotations) are provided with minimal experimental complexity, so that the fullest representation of the strain state is generated, from which standard tensor analysis can be used to provide, for example, principal strains, or maximum resolved shear stresses (though for stresses single crystal elastic constant are required). A lower bound estimate of the dislocation density is also possible using the Nye/Kroner framework.

The main challenges for further development of the method for applications to semiconductors are (i) spatial resolution, and (ii) absolute (as opposed to relative) strain measurement. Spatial resolution will continue to be a major concern in applying EBSD to some device structures. Although in some instances the goals seem unattainable, in others there is already sufficient spatial resolution available. Improvements in spatial resolution are likely to come from improved detectors allowing operation at lower beam energies, and further into the future with energy filtering/discrimination. The need to have a reference pattern from a region of known strain is an issue receiving considerable attention at the current date [33]. It seems likely that dynamical diffraction simulations of EBSD patterns [34, 35] may have sufficient fidelity to the real patterns to allow their use as reference patterns (simple kinematic simulations are not sufficient). The challenge has now become an experimental one of determining the detector geometry and distortions with sufficient accuracy so as to avoid introducing 'phantom strains' through small artifacts in the cross-correlation measurement of shifts. This challenge is likely to be met in the next year or so and will give a further step change in the scope of problems that can be tackled with EBSD.

References

[1] Dingley D 2004 *Journal of Microscopy* **213** 214
[2] Wilkinson A J and Hirsch P B 1997 *Micron* **28** 279
[3] Hart R, Midgley P A, Wilkinson A and Schwarzacher W 1995 *Appl. Phys. Lett.* **67** 1316
[4] Ji Y A, Wang L, Zhang Y Q, Wei B, Wang J H, Cheng Y L and Suo H L *Ultramicroscopy* **111** 314
[5] Chen J, Sekiguchi T, Xie R, Ahmet P, Chikyo T, Yang D, Ito S and Yin F 2005 *Scr. Mater.* **52** 1211
[6] Gallien B, Duffar T, Lay S and Robaut F *J. Cryst. Growth* **318** 208
[7] Troost K Z, Vandersluis P and Gravesteijn D J 1993 *Appl. Phys. Lett.* **62** 1110
[8] Wilkinson A J 1996 *Ultramicroscopy* **62** 237
[9] Wilkinson A J 1997 *Mater. Sci. Technol.* **13** 79
[10] Wilkinson A J 2000 *J. Electron Microsc.* **49** 299
[11] Wilkinson A J, Meaden G and Dingley D J 2006 *Mater. Sci. Technol.* **22** 1271
[12] Wilkinson A J, Meaden G and Dingley D J 2006 *Ultramicroscopy* **106** 307
[13] Villert S, Maurice C, Wyon C and Fortunier R 2009 *Journal of Microscopy-Oxford* **233** 290
[14] Wilkinson A J 2006 *Appl. Phys. Lett.* **89** 3
[15] Wilkinson A J, Meaden G and Dingley D J 2009 *9th International Workshop on Beam Injection Assessment of Microstructure in Semiconductors* **45** 285
[16] Vaudin M D, Gerbig Y B, Stranick S J and Cook R F 2008 *Appl. Phys. Lett.* **93**

[17] Tomita M, Kosemura D, Takei M, Nagata K, Akamatsu H and Ogura A 2011 *Japanese Journal of Applied Physics* **50** 010111
[18] Ishido T, Matsuo H, Katayama T, Ueda T, Inoue K and Ueda D 2007 *IEICE Electron. Express* **4** 775
[19] Keller R R, Roshko A, Geiss R H, Bertness K A and Quinn T P 2004 *Microelectron. Eng.* **75** 96
[20] Maurice C and Fortunier R 2008 *Journal of Microscopy-Oxford* **230** 520
[21] BLG Productions Ltd. (www.blgproductions.co.uk)
[22] Wang H M, Chen C Q, Gong Z, Zhang J P, Gaevski M, Su M, Yang J W and Khan M A 2004 *Appl. Phys. Lett.* **84** 499
[23] Linthicum K, Gehrke T, Thomson D, Carlson E, Rajagopal P, Smith T, Batchelor D and Davis R 1999 *Appl. Phys. Lett.* **75** 196
[24] Watson I M, Liu C, Kim K S, Kim H S, Deatcher C J, Girkin J M, Dawson M D, Edwards P R, Trager-Cowan C and Martin R W 2001 *Phys. Status Solidi A-Appl. Res.* **188** 743
[25] Trager-Cowan C, Sweeney F, Trimby P W, Day A P, Gholinia A, Schmidt N H, Parbrook P J, Wilkinson A J and Watson I M 2007 *Phys. Rev. B* **75** 8
[26] Britton T B, Liang H, Dunne F P E and Wilkinson A J 2010 *Proc. R. Soc. A-Math. Phys. Eng. Sci.* **466** 695
[27] Karamched P S and Wilkinson A J 2010 *Acta Materialia* **59** 263
[28] Rao D V S, Beanland R, Kappers M J, Zhu D and Humphreys C J 2009 *16th Int. Conf. Microscopy of Semiconducting Materials, J Phys. Conf. Series* **209** 012022
[29] Nye J F 1953 *Acta Metallurgica* **1** 153
[30] Kroner E 1958 *Continuum theory of dislocations and self stresses* Springer-Verlag
[31] Wilkinson A J and Randman D 2010 *Philosophical Magazine* **90** 1159
[32] Tuppen C G, Gibbings C J and Hockly M 1989 *Journal of Crystal Growth* **94** 392.
[33] Britton T B, Maurice C, Fortunier R, Driver J H, Day A P, Meaden G, Dingley D J, Mingard K and Wilkinson A J 2010 *Ultramicroscopy* **110** 1443
[34] Trager-Cowan C, Sweeney F, Winkelmann A, Wilkinson A J, Trimby P W, Day A P, Gholinia A, Schmidt N H, Parbrook P J and Watson I M 2006 *Materials Science and Technology* **22** 1352
[35] Winkelmann A, Trager-Cowan C, Sweeney F, Day A P and Parbrook P 2007 *Ultramicroscopy* **107** 414

Aberration-corrected scanning transmission electron microscopy of semiconductors

OL Krivanek, N Dellby and MF Murfitt

Nion Co., 1102 8th St., Kirkland, WA 98033, USA

E-mail: krivanek@nion.com

Abstract. The scanning transmission electron microscope (STEM) has been able to image individual heavy atoms in a light matrix for some time. It is now able to do much more: it can resolve individual atoms as light as boron in monolayer materials; image atomic columns as light as hydrogen, identify the chemical type of individual isolated atoms from the intensity of their annular dark field (ADF) image and by electron energy loss spectroscopy (EELS); and map elemental composition at atomic resolution by EELS and energy-dispersive X-ray spectroscopy (EDXS). It can even map electronic states, also by EELS, at atomic resolution. The instrumentation developments that have made this level of performance possible are reviewed, and examples of applications to semiconductors and oxides are shown.

1. Introduction

The performance of scanning transmission electron microscopes has grown very considerably in the last decade. In 2002, a sub-Å electron probe was formed for the first time in an electron microscope [1]. Now there are more than 100 STEMs in the world capable of forming such probes, and probe sizes of the order of 0.5 Å have been reached [2-4]. In 2004, a single atom in a crystalline solid was identified spectroscopically for the first time [5]; now elemental maps with single atomic resolution and sometimes also single-atom sensitivity have been produced in several laboratories, using three types of signals: EELS, EDXS, and quantitative analysis of annular dark field (ADF) intensities. Here we review these developments from the perspective of what they are likely to mean for the microscopy of semiconductors.

2. Modern STEM instruments

The method employed by a STEM to analyze samples is very simple: form a small but intense probe of electrons, scan it over a thin sample, and record various signals that originate from the scattering of the incident electrons by the sample. Atomic resolution and sensitivity are possible because of two key facts: electrons can be focused into a probe that is smaller than one atom, and they interact rather strongly with matter, producing a detectable signal from every atom in a sample.

The STEM's ultimate performance is determined by many factors, but three of them are predominant:

i) the electron source, which determines the current that can be packed into a small electron probe,

ii) the quality of the STEM's optics, which determines how small the probe can ultimately be, and

Published under licence by IOP Publishing Ltd

iii) the quality and variety of the STEM's detectors, which determine what kinds of signals can be collected and with what efficiency.

Important progress has been made in the last decade on all these fronts. In the area of electron sources, the W(310) cold field emission (CFE) gun, whose theoretical brightness is about 10x higher than that of Schottky guns [6], has been improved to give reduced brightnesses of $3x10^8$ A/(m^2 sr V) [4], which amounts to a "coherent probe current" [7] of 0.43 nA. This is about 3x larger than the values normally measured for W CFE sources, and there are indications that even higher values are possible. In practical terms, it means that a current of 1 nA can now be packed into a 0.14 nm, 200 keV electron probe [4], a setting that is liable to produce small holes rather quickly in many types of samples. But the increased brightness is very useful when the accelerating energy is reduced to avoid knock-on damage, and it is very welcome indeed when the electron source is equipped with a monochromator, which reduces the probe current substantially.

In the area of electron optics, aberration correctors have improved the ultimate attainable probe size by about 3x, to about 50 pm at 200-300 keV primary energy, and about 100 pm at 60 keV. These advances have been covered in many research articles and several volumes, e.g. [8]. Monochromators capable of reducing the energy spread of the incident beam to about 40 meV have been built [9], and new designs that promise to lower the resolution to about 10 meV have appeared [10]. These developments promise to endow EELS with fundamentally new capabilities.

In the area of detectors, an annular bright field detector has allowed the detection of hydrogen columns in YH_2 [11], and a reconfigurable segmented annular detector has been introduced [12]. The collection efficiency of the EELS signal has been increased to >90% through aberration-corrected post-sample optics [13], and the collection efficiency of energy-dispersive X-ray (EDXS) detectors has been improved about 3x relative to previous detector arrangements, to ~7% (0.9 sr) [14].

Other advances have been realized in the areas of stability, with some STEMs now able to keep the electron probe stable to better than 10 pm r.m.s. relative to the sample [7], low primary energy performance [15, 16], and in the quality of the vacuum at the sample, with pressures in the 10^{-8} Pa regime now regularly attained in some STEM instruments [4]. The improved vacuum is making it possible to examine samples longer without running into sample contamination or etching [17].

Acting together, these advances have allowed a wide range of new applications in many areas of materials science. To give just one example, a monolayer of BN containing substitutional impurities has been analyzed atom-by-atom, and the position and chemical identity of every single atom in a significant part of the sample has been determined [18]. The sample was of course rather special: its monolayer nature made atomic identification by measuring ADF intensities possible, and its high resistance to radiation damage (at a primary energy of 60 keV) allowed a high irradiation dose that produced an image with the high signal-to-noise ratio needed for the identification. But the work nevertheless shows that the ultimate goal of resolving and identifying every atom in a solid has now been reached in certain samples, and that the new capabilities are bound to produce many new results.

3. Applications to semiconductors

Investigations of semiconductor materials that are carried out in electron microscopes can be divided into three broad classes: a) investigating the structure of the material and of its defects, b) mapping its composition and c) investigating its electronic properties. The STEM can make major contributions in all the three categories, as shown by the examples below.

Figure 1 shows a high-angle annular dark field (HAADF) image of a PbS quantum dot in polycrystalline ZnS. The key unknown was the size and shape of the quantum dots. Lead being heavier than the other elements present made the PbS dots considerably brighter than the ZnS grains. Just one PbS dot is present in the shown field of view, indicated by the double arrow. It is faceted, and about 2.5 x 4 nm in size. Individual Pb atoms can be seen outside of the dot, both on the ZnS matrix and outside it, on the sample support. These may be an artefact due to ion milling displacing atoms rather than removing them, but they are interesting because they show that single lead atoms were readily visible even though the matrix was not particularly light. Another notable feature of the image is that despite the fact that several ZnS grains show lattice fringes and are thus diffracting, the incoherent nature of the HAADF image (e.g. [8]) has prevented diffraction effects from introducing major contrast artefacts into the image and allowed the PbS grain to remain clearly visible.

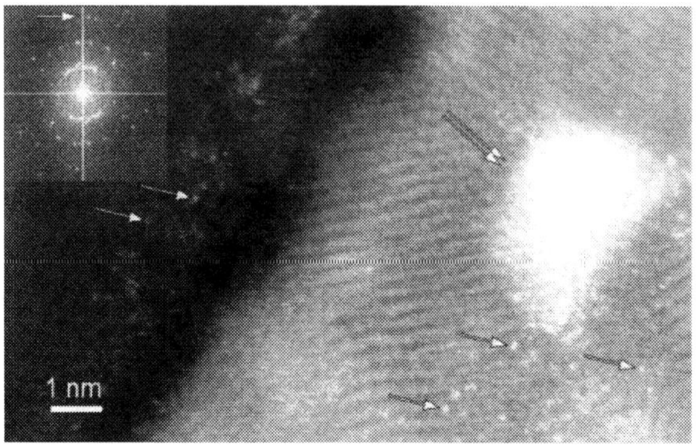

Fig. 1. HAADF image of PbS quantum dot, indicated by two long parallel arrows, in a ZnS matrix. Many lead atoms are visible both within the ZnS matrix and outside; some are marked by single arrows. Fourier transform insert (of a larger image area) shows that spatial frequencies up to 1 $Å^{-1}$ (arrowed) have been transferred into the image. Nion UltraSTEM200, 200 keV. Sample courtesy Profs F.B. Prinz, U. of Vienna and R. Bosch, Stanford U.

Figure 2 illustrates elemental identification using atomic intensities in ADF images of monolayer graphene with one N and several Si impurities. The nitrogen atom gave an image intensity that was 1.3x higher than the average C atom intensity (after correcting for image contributions due to probe tails [18]) and the nitrogen atoms gave 4x the intensity of carbon. The illumination and collection angles for the image were the same as those used for imaging monolayer BN by us previously [18], and the power law relating the intensity I_Z of the atomic image to the atom's atomic number Z was also the same: $I_Z \alpha Z^{1.64}$. The example shows that the ADF signal can be used for accurate elemental mapping, provided that the contribution of each atom to the images of its neighbors is carefully accounted for [18].

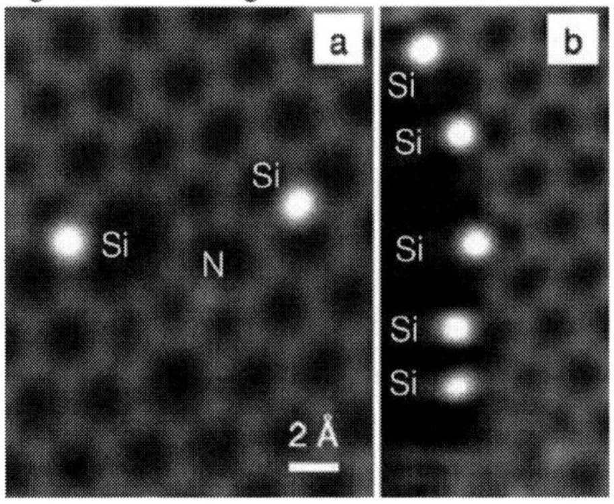

Fig. 2. a) ADF images of N and Si atoms substituting for C at and near topological defects in graphene, b) Si decorating a graphene edge. Nion UltraSTEM100, 60 keV. Courtesy Matt Chisholm, ORNL and Venna Krishnan and Gerd Duscher, U. of Tennessee.

Figure 3 shows EELS elemental mapping of high concentration elements. The sample was GaAs, viewed in the [011] projection, in which Ga and As atoms separate into different atomic columns.

Ga and As $L_{2,3}$ edges, at 1115 and 1323 eV energy loss respectively, were used for the mapping. They were considerably weaker than edges corresponding to energy losses smaller than 1 keV, but even so, largely because of the efficient coupling of the EELS signal into the spectrometer, the mapping could be carried out at 20 ms per pixel with only 100 pA of beam current, i.e. in 82 s total acquisition time for the 64x64 pixel map. Principal component analysis (PCA) was used to minimize the statistical noise.

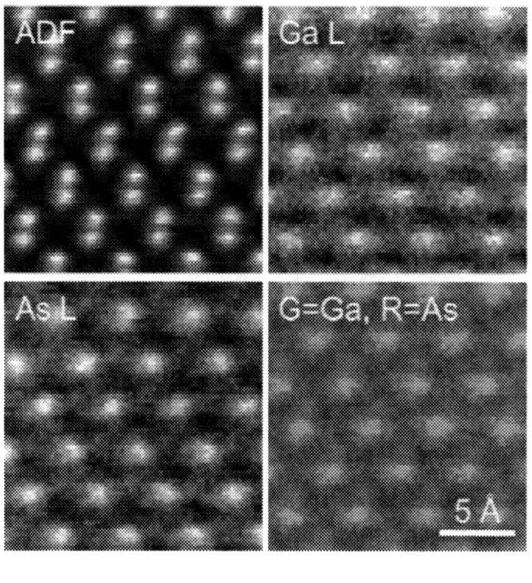

Fig. 3. Elemental mapping of GaAs. Nion UltraSTEM100 Gatan Enfina EELS, 100 keV. Courtesy Maria Varela, ORNL and Sergio Molina, U. Cadiz, Spain. (colour online)

EELS delocalization is not a major concern for energy-losses greater than 1 keV when either the illumination or the collection angles (or both) are large [7]. This was the case here, with both angles >30 mr. It is therefore not surprising that the atomic peaks in the EELS map are only slightly less sharp than the ADF ones.

Figure 4 shows elemental mapping of La atoms in a TiO_2 matrix [19]. The sample was grown on an $LaAlO_3$ substrate, and this allowed La to penetrate into low-angle grain boundaries in the TiO_2. An ADF image is shown along with Ti and La elemental maps, made using the Ti $L_{2,3}$ and La $M_{4,5}$ edges, at 455 and 832 eV respectively. A profile integrated vertically through the combined EELS map (upper right) shows that was there was essentially no La dissolved in the matrix on the left side of the boundary, but that there was La enrichment on the right side of the atomically sharp boundary.

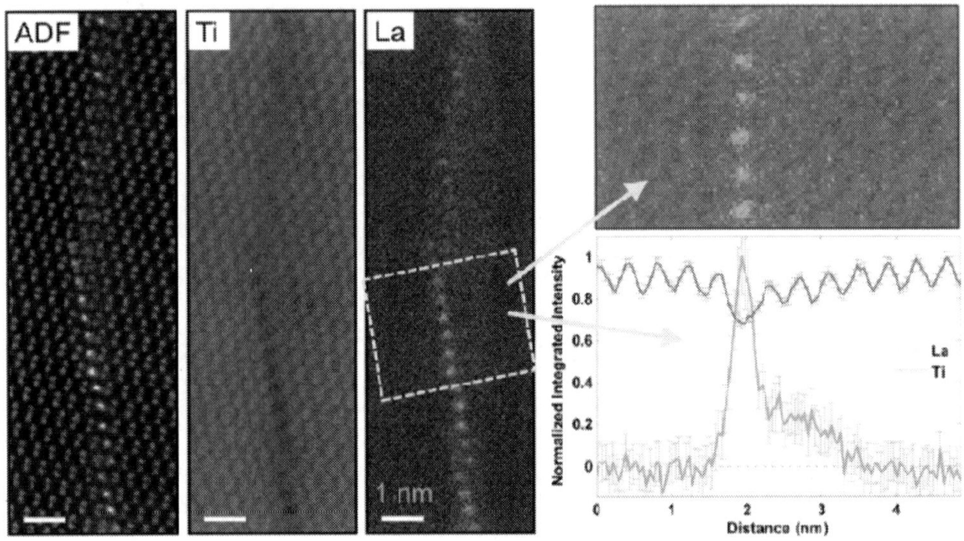

Fig. 4. EELS mapping of La segregation at a low-angle grain boundary in TiO_2 grown on $LaAlO_3$. Nion UltraSTEM100, Gatan Enfina EELS, 100 keV, 100x444 pixel spectrum-image. Courtesy Lena Kourkoutis and David Muller, Cornell U. (colour online)

Figure 5 shows a case of EELS elemental mapping used for structure-determination purposes. The sample was $LaMnO_3$ and elemental maps were obtained at 60 keV primary energy, using the O-K, Mn-$L_{2,3}$ and La-$M_{4,5}$ edges respectively. The ADF image (not shown) was dominated by the signal from the heavy La (Z=57) and Mn (Z=25) columns, and the part of the

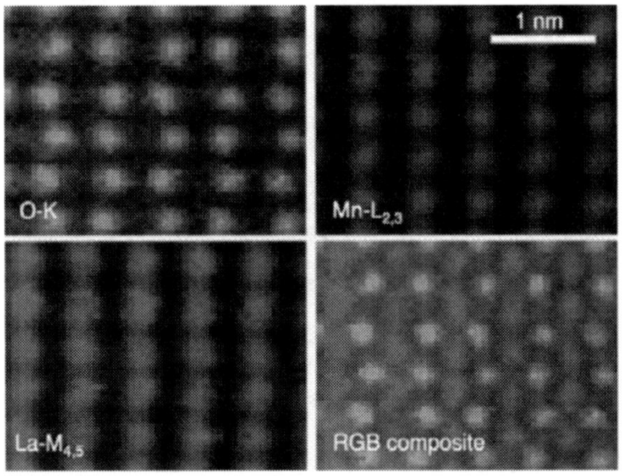

Fig. 5. EELS imaging of oxygen octahedral rotations in LaMnO3. Nion Ultra-STEM100, Gatan Enfina EELS, 60 keV. Courtesy Maria Varela and Stephen Pennycook, ORNL. (colour online)

signal that was produced by the O atoms was not at all clear. The elemental map, however, makes it obvious that the oxygen columns are not laid out in a regular rectangular pattern. Their arrangement corresponds to rotated oxygen octahedra, and it would not be visible without being able to image the oxygen atoms separately from the other atoms present. It needs to be noted, however, that there were also oxygen atoms present in the La columns, and that these were not imaged along with the oxygen-only columns seen in the oxygen image [20]. This was due to the fact that the heavy La atoms rescattered the O EELS signal to high angles, outside the EELS entrance aperture. Detailed image simulations have reproduced the experimental image accurately.

Figure 6 shows the mapping of atomic bonding in $EuTiO_3$ grown on $DyScO_3$ [19]. The figure first shows a "conventional" EELS elemental map of Eu, obtained using the $M_{4,5}$ edge at 1131 eV (a), and a portion of a HAADF image recorded simultaneously (b). It then shows that it is possible to identify two distinct components of the Eu edge, which are shifted by about 2 eV (c). The final part shows a 2D colour map of where the different components identified in the spectrum came from (d), demonstrating that the higher energy loss component occurred only at the interface, in a single atomic layer. It most likely corresponds to a layer of Eu 3+ at the interface, with the bulk $EuTiO_3$ layer containing Eu 2+.

Changes in chemical bonding have been mapped by EELS before [e.g. 21], but not with atomic resolution. The fact that chemical bonding can now be mapped column-by-column shows just how far EELS has managed to progress in recent years.

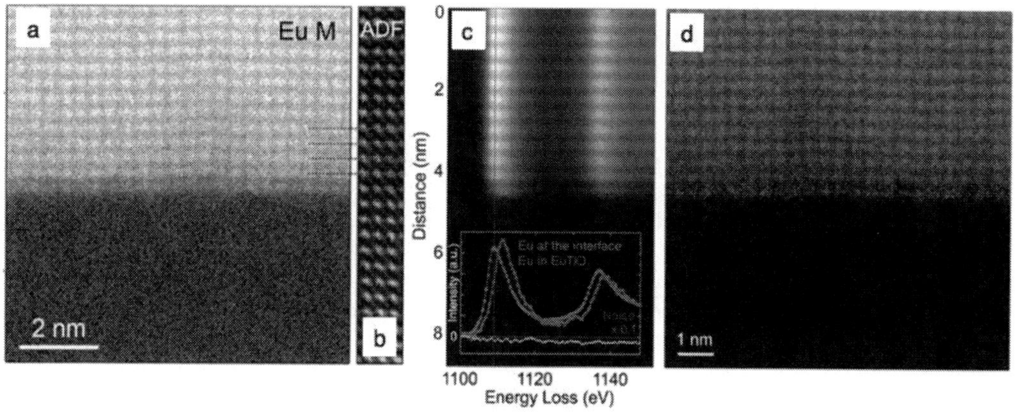

Fig. 6. Mapping bonding changes at the $EuTiO_3/DyScO_3$ interface through spectroscopic imaging. a) Eu elemental map showing reduced Eu concentration at the interface, b) part of the simultaneously recorded HAADF image, c) evolution of the horizontally averaged Eu-$M_{4,5}$ edge fine structure across the interface and the three components, d) three-component fit to the full spectrum-image. Nion Ultra-STEM100, Gatan Enfina EELS, 100 keV. Courtesy Lena Kourkoutis and David Muller, Cornell U. (colour online)

Apart from Figure 3, the examples presented above do not show classical semiconductors. However, their implication is clear: it is now readily possible to explore elemental distributions on the atomic scale, using EELS mapping in the STEM. The major limitations of the technique are that a high electron exposure, of the order of 10^8-10^{10} electrons/nm^2 is needed, but must not lead to the sample undergoing major changes, that the sample must be reasonably thin (typically less than 20 nm), and that not all elements have EELS edges that are good for EELS mapping. Nevertheless, it is easy to see that impurities concentrating at dislocations and other defects in III-V and other semiconductors should be readily identifiable, that it should be possible to analyze the electronic properties of a wide variety of electronic interfaces, and that the strong signal and high efficiency of EELS may ultimately yield elemental maps of dopants with single atom sensitivity.

Elemental mapping using the EDXS signal [14] promises to be nearly as powerful as EELS mapping and will enjoy some advantages, such as being able to examine thicker samples, and having spectra that have better peak-to-background ratios and are simpler to interpret. However, the signal strength and the collection efficiency will be poorer than for EELS, and probably not sufficient to map single atoms reliably [22]. Moreover, with the EDXS energy resolution typically >100 eV, examination of changes in the bonding or in the electronic states of a material will not be possible.

Elemental mapping by quantitative analysis of ADF intensities, which can be done with electron doses about 100x smaller than EELS and EDXS mapping, will probably become the technique of choice for mapping elements in monolayers. In thicker samples, its interpretation will be less clear-cut, but even so it is likely to be a useful technique to fall back on when the sample is radiation-sensitive and the large incident dose typically needed for EELS and EDXS mapping cannot be used.

4. Conclusion

Aberration correction scored two major successes in the 1990s: it corrected the vision of the initially defective Hubble space telescope [23], and it corrected the imaging of electron microscopes. The corrected Hubble telescope opened fundamentally new vistas on the large-scale universe, and aberration-corrected electron microscopes are opening new vistas on the world of atoms. Never before have we been able to image and analyze non-periodic matter with the levels of precision, sensitivity and flexibility that are now possible. It is an exciting time to be working on applying the new instruments to materials problems of major practical significance, such as those found in the semiconductor field, and to be pushing the power of the techniques even further.

References

[1] Batson PE, Dellby N and Krivanek OL 2002 Nature **418** 617.
[2] Erni R Rossell MD Kisielowski C and Dahmen U 2009 Phys. Rev. Lett. **102** 096101
[3] Sawada H et al 2009 Journal of Electron Microscopy **58** 357
[4] Dellby N et al 2011 Europ J Appl Phys in press
[5] Varela M et al 2004 Phys Rev Lett **92** 095502
[6] Schwind GA, Magera G and Swanson LW 2006 J Vac Sci Technol **24** 2897
[7] Krivanek OL et al 2011 *in: Scanning Transmission Electron Microscopy: Imaging and Analysis* (Pennycook and Nellist eds, Springer-Verlag, Berlin and New York, 2011) 615
[8] Pennycook SJ and Nellist PD (eds) 2011 *Scanning Transmission Electron Microscopy: Imaging and Analysis* (Springer-Verlag, Berlin and New York)
[9] Essers E et al 2010 Ultramicroscopy **110** 971
[10] Krivanek OL et al 2009 Phil. Trans. R. Soc. **A 367** 3683
[11] Ishikawa R et al 2011 Nature Materials **10** 278
[12] Shibata N et al 2010 Microscopy Microanalysis **16** (Suppl 2) 124
[13] Krivanek OL et al 2008 Ultramicroscopy **108** 179
[14] Schlossmacher P et al 2010 Microscopy Today **18** 14
[15] Krivanek OL et al 2010 Ultramicroscopy **110** 935
[16] Sasaki T et al 2010 Journal of Electron Microscopy **59** S7
[17] Meyer J 2011 private communication
[18] Krivanek OL et al 2010 Nature **464** 571
[19] Kourkoutis LF 2010 Proceedings IMC17 (Rio de Janeiro) 354
[20] Varela M et al 2011 *in: Scanning Transmission Electron Microscopy: Imaging and Analysis* (Pennycook and Nellist eds, Springer-Verlag, Berlin and New York, 2011) 429
[21] Kurata H Isoda S and Kobayashi T 1996 Journal of Electron Microscopy **45** 317
[22] Watanabe M 2011 *in: Scanning Transmission Electron Microscopy: Imaging and Analysis* (Pennycook and Nellist eds, Springer-Verlag, Berlin and New York, 2011) 291
[23] http://hubble.nasa.gov/missions/sm1.php

Strain, composition and disorder in ADF imaging of semi-conductors

V Grillo[1,2,*], K Mueller[3], K Volz[4], F Glas[5], T Grieb[3] and A Rosenauer[3]

[1] S3-NANO, CNR Via Campi 213A, I-41125 Modena, Italy

[2] CNR-IMEM Institute Parco Area delle Scienze37/A, I-43010 Parma, Italy

[3] Institut für Festkörperphysik, Universität Bremen, Otto-Hahn-Allee 1, D-28359 Bremen, Germany

[4] Materials Science Center and Faculty of Physics, Philipps University Marburg, Germany

[5]Laboratoire de Photonique et de Nanostructures, CNRS, Route de Nozay, 91460 Marcoussis, France

* Corresponding author: vincenzo.grillo@unimore.it

Abstract: The effect of strain, composition and disorder in ADF images is systematically studied as a function of detection angle in order to understand the main contrast mechanisms. We demonstrate that the complex phenomenology in ADF images can be accounted for by accurate simulations and modelling. The advantage of an accurate modelling on the image interpretation will be demonstrated in the case of dislocations, chemical analysis of InGaP/GaAs and most noticeably the measurement of both In and N content in quaternary InGaAsN.

1. Introduction

Whereas the initial hopes on scanning transmission electron microscopy (STEM) with a high-angle annular dark-field (HAADF) detector were for an easily interpretable chemically sensitive technique it is becoming now clear that many effects determine the contrast with this technique, especially when imaging along high symmetry zone axes [1]. We mention in particular here strain and static disorder effects [2-4]. On one hand these contrast effects may complicate compositional analysis but on the other hand new interesting information can be gained from more detailed experiments. The most recent microscopy studies have demonstrated a high interest in STEM based techniques. It is becoming clear that new and more reliable information can be gained with different kinds of detector settings beyond the "classical" HAADF setting. This is the case of LAADF (low angle annular dark field) [5,6] and even ABF (annular bright field) that pushes the interest of analysis in the region of low angle scattering [7,8].

We aim to demonstrate in this work that new insight into materials properties can be gained by the comparison of imaging with different camera lengths. We will show that it is possible to investigate the presence of surface strain relaxation, gauge the static displacement, evaluate quantitatively the composition in quaternary alloys and perform defect analysis. A qualifying point of a successful analysis of the different contrast mechanisms is the comparison of imaging with different experimental conditions that permit to distinguish the different contrast contributions.

We will concentrate in particular on the analysis of the angular distribution of scattering as a function of different structural properties of the material, namely composition, static disorder and long range strain. In order to understand the possible phenomena and contrast mechanisms we used both approximate modelling and rigorous multislice simulation with frozen lattice algorithms [9,10]. The latter has been implemented using an updated version of the STEM_CELL parallel computing algorithm [11,12].

The first part of this work will be dedicated to developing a kinematical model of diffuse scattering and study how it relates to chemical effects, static displacement and strain. In the second part a few experimental examples of applications are shown. These examples will regard dislocations in GaN, the effect of alloy disorder on the contrast between InGaP and GaAs and the composition in InGaAsN quaternary alloy. These examples have been chosen to highlight the role of each single contrast mechanism.

2. Theory

Diffuse scattering arises from the displacement of atoms from the periodic ideal position. This may occur because of

1) thermal agitation,
2) the local atomic static displacement (SD) to reach equilibrium configuration. This typically occurs on a very local scale as a form of disorder in alloys or because of point defects.
3) long range strains field are imposed on the analyzed structure.

Since diffuse scattering is dominant in HAADF and very important in medium and low angle dark field it is relevant to pinpoint the mechanism of diffuse scattering production.

A first common characteristic of diffuse scattering, especially at medium-high angle, is the incoherent character, i.e. the diffuse scattering produced by an atomic column is to a good level of approximation the sum in amplitude without phasing of the contributions from all atoms. These waves can then undergo additional elastic scattering that are at the origin of, for example, Kikuchi lines or additional diffuse scattering.

If this diffuse multiple scattering is neglected and the elastic rearrangement is azimuthally averaged in the integration over the detector it is possible to introduce a kinematical approximation typically used for high angle scattering that consider the HAADF image as the convolution of a probe and an object function [13]. We refer in particular to the notation of Voyles [14] that considers a depth (hereafter z) dependent probe $J(z)$ which can be identified with the forward propagating part of the electron wave that has undergone only elastic scattering

$$I_{HAADF} = \int J(z) \otimes O(\vec{r}, z) dz \tag{1}$$

$\vec{r}$ here is the coordinate vector in the plane (x,y) orthogonal to electron propagation and O is then the object function that contains the scattering from single atoms (at position $\vec{r}_i$) and that can be written to some extent as

$$O(\mathrm{r}) = \sum_{\mathrm{i=atoms}} \delta(\vec{\mathrm{r}} - \vec{\mathrm{r}}_{\mathrm{i}})\sigma \tag{2}$$

For the cross section σ of the diffusely scattering single atom a common expression that is valid for scattering angles θ much larger than the typically low order Bragg angles is [15]

$$\sigma \propto \int_{\theta\,\min}^{\theta\,\max} f^2(\theta)\left[1 - e^{-2M(\theta^2/\lambda^2)}\right] d^2\theta \tag{3}$$

This expression has been introduced to describe the thermal diffuse scattering in the Einstein model. It contains the single atom scattering factor f that produces at high angles the well known Z^a dependence at high angles. However it is customary to extend this to the case of static displacement, although not being perfectly applicable. An effective Debye Waller factor is therefore introduced [4] as

$$M = M_{\mathrm{TDS}} + M_{\mathrm{Stat}} \tag{4}$$

M_{Stat} accounts for the average static displacement. This simple quasi-thermal modelling permits to give a qualitative account of the characteristics of static displacement, namely that it produces an increase of low angle scattering. This is graphically represented in figure 1a showing the average cross-section according to eq. 3 for pure GaAs and for GaAs after the introduction of a SD of the same magnitude as what would be produced by 3% of N. This small amount of N has a relatively low chemical effect (here neglected) but produces a large disorder due to the strong difference in covalent/ionic radius between As and N [16]. A similar effect was already found on B-doped Si [17].

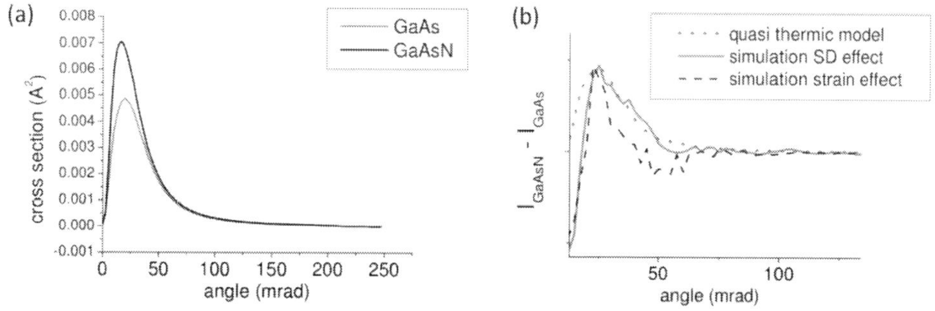

Figure 1. a) Comparison of the cross-section for GaAs and GaAsN as a function of the scattering angle for a beam voltage of 200kV: the difference is more pronounced at low angle. b) Angular dependence of the Huang scattering (see text). The solid line refers to rigorous simulations for the $GaAs_{1-y}N_y$ (y=3%)/GaAs case. The dotted line is a quasi-thermal modelling. The dashed line refers to the effect of surface strain (structure as in ref. [2]) and has been obtained by comparison between strained and unstrained GaAs. All profiles have been smoothed by three point lateral averaging to reduce the fluctuation inherent to the limited numbers of frozen lattice configurations.

According to our valence force field (VFF) simulations the additional non-thermal displacement v_{SD} for 3% N in GaAsN is 0.0087nm for group III element and 0.0051 nm for group V elements. The most important finding is that the difference between the two cross-sections concentrates in the low angle regime. Following a nomenclature typical of X-rays we will call this additional static diffuse contribution "Huang scattering" [18] to distinguish it from the thermal contribution. Another characteristic that emerges from this model is that diffuse scattering tends to saturate for very large disorder since for very large M (which would make the crystal similar to an amorphous alloy) the cross-section would tend to a constant.

In Figure 1b the angular distribution of Huang scattering has been plotted according to rigorous simulations (solid line) and according to the quasi-thermal model (dotted line). The details of the rigorous simulation are described in the experimental and simulation section below. The distribution in the quasi-thermal model has been translated laterally by a constant value to fit experimental data. The best fit is obtained for a translation by 6 mrad and accounts for the convergence of the impinging beam (11.9 mrad). Both solid and dotted profiles are obtained as the difference between the GaAsN and GaAs case. In spite of some differences the two profiles are quite similar. The additional dashed line refers to the case of long-range strain that will be considered below. All profiles are normalized to their maxima.

Previous work [4] has demonstrated that a correction of the cross-section as above is not sufficient to account for experimental phenomenology as, in fact, SD produces important effects on the forward propagating wave. In particular in zone axis conditions the most relevant part of J in equation (1) is the 1s Bloch wave component of J. The 1s state is responsible for the localization of most of the probe current on the atomic column (channelling) and for this reason its excitation is more directly bound to the generation of diffuse scattering.

Simulations have demonstrated that the introduction of SD produces a faster damping of the 1s excitation and therefore the J factor in eq. 1 for the diffuse scattering [3]. This phenomenon goes under the name of dechanneling and since it affects only the term J in eq. 1 it produces a similar effect at both low and high angle. However it is interesting to note that at low angle SD has two competitive effects: dechanneling reduces the intensity (through the decrease of J) while Huang scattering increases it (through the increase of σ). This produces a non monotonic dependence of the LAADF intensity on the size of SD [4]. Conversely, SD through dechanneling produces always a decrease of intensity in HAADF (even at very high detection angles). A way to summarize the above effects is to remark that SD translates diffuse scattering from high angle to low angle.

If the above discussion permits to define a framework for the thermal and static local effects this does not allow to treat the case of long-range strain. The use of an effective Debye Waller factor with average displacement in one direction v_x equal to a component or the module of elastic displacement vector $\vec{u}$, for example, is not satisfactory since $\vec{u}$ depends on which part of the crystal is used as a reference and can be different from 0 even at distances far from an extended defect.

It has been highlighted that long-range strain and local static displacement have a different effect on the propagation of the forward wave function J [19]. We will try to highlight also some characteristics of the additional Huang scattering produced by the strain.

In fact, the dashed curve in figure 1b is the plot of the angular distribution of the Huang scattering produced by the surface relaxation at the barriers of a strained QW (as in ref. [2]). As in the precedent cases the Huang scattering has been deduced as the difference between the strained and unstrained case for a specimen thickness of 40nm. This deduction is based on the fact that the other effect, namely dechanneling, is only a multiplicative factor in the whole angular distribution. This does not mean that strain related dechanneling is negligible: Walther and Humphreys [20] for example have found strain effects extending to detection range out of 200 mrad where Huang scattering is certainly negligible. Conversely, what we want to highlight here is that the simulation shows that the angular distribution of strain related Huang scattering is quite different, namely more oriented in the forward direction with respect to the case of SD. This can be an effect of the more localized nature of the SD.

A qualitative but effective description of the spatial location of this Huang scattering contribution comes from a recent simplified model that describes the forward propagating wave J in terms of its sole 1s component [21]. This approach permits to calculate a qualitative image able to reproduce the main experimental features and to clarify that the important quantities for ADF contrast are the surface tilt of atomic columns that determines the initial excitation of the 1s state and the curvature of the atomic planes, i.e. a variation of the column inclination. When such a variation occurs the states at the two parts of the crystal need to be matched. With increasing curvature the overlap between the two states (and in particular the 1s state) is reduced but an additional diffuse scattering is observed. This has been summarized in the equation system in ref. [2] and is ere extended to the 2D case with the addition of the Huang scattering contribution I_{Huang} which is just proportional to the curvature Γ through an adjustable factor H.

$$\Gamma^2 = \left(\frac{\partial^2 u_x}{\partial z^2}\right)^2 + \left(\frac{\partial^2 u_y}{\partial z^2}\right)^2 \tag{5}$$

$$\begin{cases} \dfrac{d^2\Phi_{1s}(\rho,z)}{dz^2} = \left[-2\mu_{1s}\dfrac{d\Phi_{1s}(\rho,z)}{dz} - 2\left(\dfrac{\Gamma^2}{\sigma_\theta^{\,2}}\right)\Phi_{1s}(\rho,z)\right] & (6) \\ \dfrac{dI_{Huang}(\rho,z)}{dz} = 2H\left(\dfrac{\Gamma^2}{\sigma_\theta^{\,2}}\right)\Phi_{1s}(\rho,z) & (7) \\ I = \int(\Phi_{1s}+C)\left[\sum_{i=atoms}\sigma_i\delta(z-z_i)\right]dz + I_{Huang} & (8) \end{cases}$$

Here Φ_{1s} is the 1s excitation factor, μ_{1s} is the 1s absorption factor, $\sigma_\theta$is the size of 1s in reciprocal space, C is an adjustable parameter to account for the non-1s states contribution. The non trivial initial condition for the 1s state is

$$\Phi_{1s}(z=0) \approx \frac{1}{\Omega}\left[1 - \frac{1}{\sigma_\theta^2}\left(\theta_t - \left.\frac{\partial u_x}{\partial z}\right|_{z=0}\right)^2 - \frac{1}{\sigma_\theta^2}\left(\left.\frac{\partial u_y}{\partial z}\right|_{z=0}\right)^2\right] \tag{9}$$

where Ω is the area of the unit cell and the specimen tilt θ_t has been set for simplicity along x direction.

3. Experimental and simulations

Three kinds of sample have been analyzed for this work. The first is nominally undoped hexagonal GaN grown at 1150°C on a GaN buffer grown at 560°C on (0001)Al_2O_3 substrates in an AIXTRON MOVPE system equipped with a rotating substrate [22]. This sample has been prepared in plan-view to observe dislocations end-on. Conventional two-beam analysis has been also conducted. As second sample we used an $In_xGa_{1-x}P$/GaAs ([x]=51%) structure grown by MOCVD at 700 °C. The InGaP was capped with other layers obtaining the following layer sequence along [100]: GaAs substrate/GaAs buffer /InGaP /GaAs QW/AlGaAs cap. As third sample we used a 4 period multiple quantum well of InGaAsN/GaAs. The nominal compositions are [x]=20% In and [y]=2.4 % N. The structure has been gown by MOCVD on GaAs. The last two samples have been prepared in cross section along the <110> direction. For (S)TEM experiments the InGaAsN has been prepared in cross-section by FIB. As for GaN and InGaP samples the specimen preparation has been performed by mechanical polishing and ion milling using a GATAN DUO MILL.

The STEM observation for these two samples has been performed using a JEOL 2200 FS operated at 200keV and with an objective lens spherical aberration Cs=0.5 mm implying an optimal convergence of 11.9 mrad. In the case of InGaP the specimen thickness has been measured by energy-loss thickness mapping and convergent beam imaging. The experiments on InGaAsN samples have been performed on a Titan 80/300 TEM/STEM operated at 300 keV with an objective lens spherical aberration Cs=1.2 mm implying an optimal convergence of 9 mrad. The detector response in this case has been calibrated for different camera lengths according to the well established procedure. This has permitted to obtain a direct evaluation of sample thickness along the zone axis in the GaAs region [23].

As for simulations, rigorous calculations have been performed based on STEM_CELL software, a parallel computer simulation which is largely based on Kirkland's routines [10]. The input cells have been calculated by VFF or, in the case of surface strain relaxation (see fig. 1b), by decorating Finite Elements calculation with atomic positions as described in ref. [2]. The detection in the diffraction plane has been divided into circular rings (every segment about 2.5 or 5mrad wide). The results for intermediate thickness have been stored every 4 or 8 slices up to the maximum thickness of 200nm. The slices were 0.2 nm thick. The advantage of the segmentation of the detector in small angular regions is that this permits to introduce after the simulation the efficiency of the detector and different camera lengths. The post processing of data has been performed within the STEM_CELL user interface [12]. An optimal sampling [24] has been used for a single GaAs <110> cell of 12 x 8 pixels and 24 x 16 for alloys where 4 unit cells were simulated to reduce compositional fluctuations. The number of frozen lattice runs was set to 20. The slices have been sampled on 1024 x 1024 pixels while thermal Debye Waller factors were taken from [25]. For simplified simulation the solution of equation (8) has been calculated within STEM_CELL graphic user interface that implements both an Euler and a Runge Kutta solver for the equation.

4. Results and discussion

4.1 Long-range strain

Grillo et al. [5] have studied the HAADF and LAADF contrast from dislocations oriented in a plane orthogonal to the electron propagation direction. In this contribution we will demonstrate the case of ADF contrast from end-on screw dislocation in GaN. The theme has been addressed by different authors and worth mentioning is in particular the work of Cosgriff et al. who carefully studied the effect of surface relaxation by accurate multislice simulations [26].

The first interesting result is that using eq. (6-8) we can reproduce a similar contrast. The strain field of a screw dislocation has been calculated with the classic analytical formula accounting for surface relaxation [26]. The result of simulation for exact on-axis conditions is visible in figure 2a showing both the simulated profile and the expected 2D image. The simulation produces a result qualitatively similar to the rigorous simulation within 1 minute calculation time on a PC. It is, in particular, possible to notice the dip at 1 nm from the core and the slow rise of the intensity with increasing distance. As for the absolute value of the contrast in the dip eq. (6-8) are not able to produce a quantitative simulation as they depend on the parameter choice for the simulation and on the specimen thickness. However, using a standard choice [2] of the main parameters a good agreement is obtained with rigorous simulations.

Using the same equations it is possible to extend the simulations to the case of a tilted sample. In particular figure 2b is obtained with a tilt of 10 mrad. It shows a very peculiar dipole contrast oriented along the tilt direction with dipole sign that depends on the orientation of the Burgers vector. This shows some similarity with the contrast in 2-beam dark field, however, the dipole orientation is not related to the Bragg vector direction (in this case orthogonal to the TEM foil plane) but depends only on the actual sample mistilt.

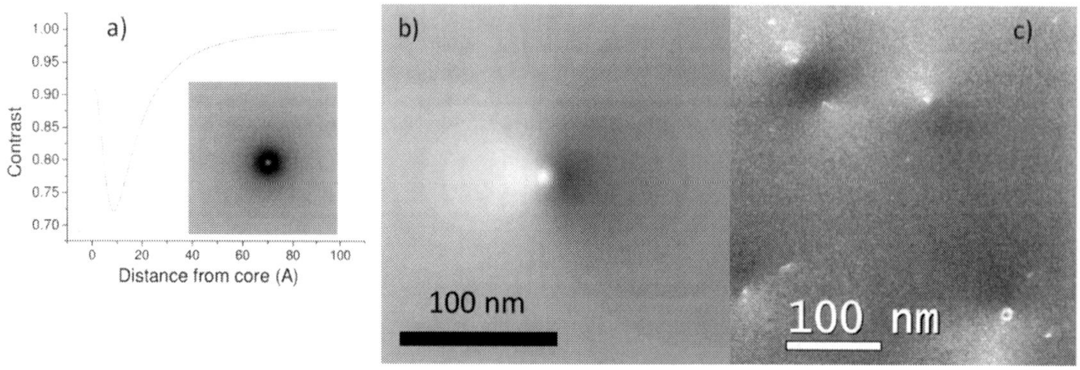

Figure 2. a) Simulation according to eq. (8) of the dependence of the intensity on the distance from the core for a screw dislocation. Specimen thickness was set to 20 nm. The inset shows a 2D rendering of the dislocation. b) simulation of the ADF contrast of a screw dislocation for a mistilt of 10 mrad. c) experimental ADF image (46 – 128 mrad) of screw dislocations.

This has been checked by experiments on a GaN sample prepared in plan-view. Figure 2c shows a typical ADF image with a detection acceptance semi-angle of 46-128mrad. The image shows a few screw dislocations with a bright core and the predicted dipole-like contrast. In particular, in the top left part of this figure two dislocations with opposite dipole orientations, and therefore opposite dipole signs, are visible. However, except for the sign all dipole have similar orientation. Moreover, the apparent dipole orientation could effectively be changed in the experiment by appropriate tilting. The dependence on detection conditions has also been checked against simulations (here not shown) and it was verified that for low detection angles the contrast of the core was strongly increased.

4.2 Static displacements in alloys: InGaP/GaAs

The demonstration of the importance of the SD in HAADF images has already been presented in [3]. We want to study here, by a comparison between simulations and experiments, the more complicated effect in the middle–low angular regime. To this aim we use the $In_xGa_{1-x}P$/GaAs system. When the In molar fraction approaches [x]=51% the two lattice parameters are lattice matched. Moreover, the expected HAADF intensity on the basis of Z^a considerations in the two systems is quite similar. With a=1.7 the expected intensity ratio I_{InGaP}/I_{GaAs} is 0.896. In this situation static displacement is expected to be dominant since chemical differences are relatively low. The GaAs is a binary system and therefore presents no static displacement while InGaP is an alloy with composition close to 50% where SD is maximum. According to VFF calculation the static displacements in InGaP are 0.0044 nm on group-III elements and 0.0086 nm on group-V elements.

A proof of the importance of SD in this system is found in the experiment of figure 3a for different camera lengths on a sample with a rapid and non-linear thickness increase in the horizontal direction. For comparison simulations for a wedge shaped sample are also reported. In this example a direct quantitative comparison with experiments is complicated due to the selective milling of the InGaP layer that can be easily observed in the experimental image. We will perform only a qualitative analysis of the contrast inversion. Simulations demonstrate that a slightly (i.e. 5%) thinner InGaP layer only shifts the point of contrast inversion to larger thickness (in the case of the range 48-128mrad the shift would amount to about 50nm). For detection within 30-84 mrad and at very low thickness it is certainly possible to observe that the GaAs is effectively slightly brighter than the InGaP. However, InGaP takes over quite quickly, at a thickness of about 40 nm (according to EELS measurements). The approximate thickness of contrast inversion is indicated by a line in figure 3a.

At small camera length (larger detection angle) the InGaP/GaAs contrast inversion occurs at very thick sample regions. Eventually, at very small camera length (in this case detection angles of 170-470 mrad) no inversion is observed. The Huang scattering due to SD is at the origin of the additional InGaP intensity in the 34-80 mrad detector that builds up within a typical channelling length of some tens of nm. As discussed in the theory section the Huang scattering is concentrated in the region between 30-50 mrad. It is therefore natural that for a detector starting from 30 mrad this contribution is very important while for detection beyond 46 mrad (second row) this contribution is lower.

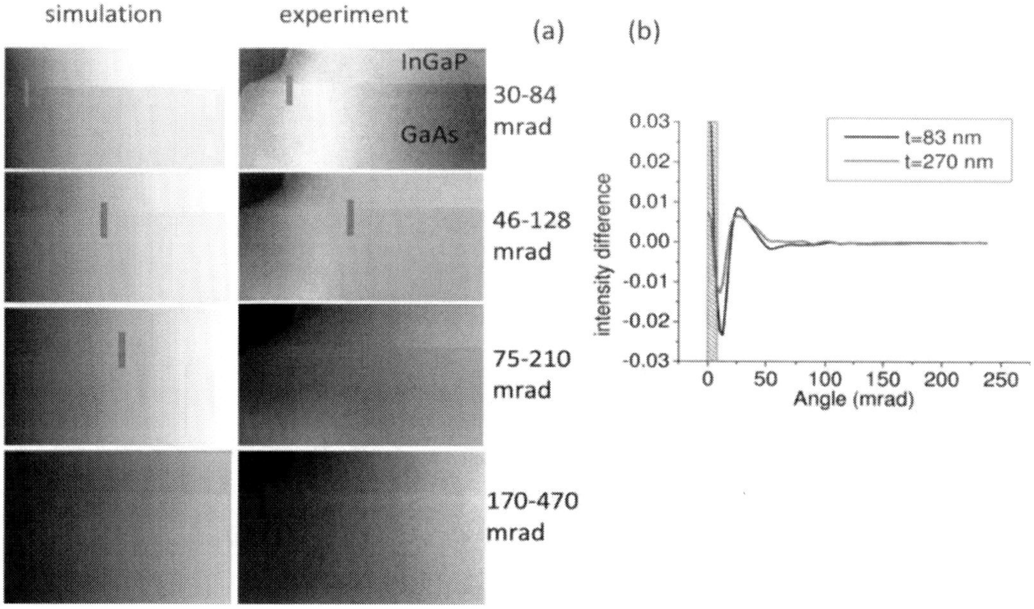

Figure 3. a) Qualitative comparison of simulation and experiment for the interface InGaP/GaAs for different angular detection ranges. The thickness increases in both cases from left to right. The contrast inversion, indicated by the vertical bars, is quite sudden for LAADF in the range 30-84 mrad but becomes slower for increasing detection angle. At very high angle (170-470 mrad) no inversion is observed. The maximum thickness in the simulation is 250nm while it is roughly 350 nm in experiments. Faint striations in the simulations are graphical artefacts due to the discretisation of the grey scale on 255 levels. b) Simulation of the Huang scattering angular distribution (calculated as the difference between InGaP and GaAs) for two thickness values.

The simulated distribution of the Huang scattering for this specific system for two different thicknesses is shown in figure 3b that confirms this conjecture. To evaluate the Huang scattering distribution, as in the case of GaAsN, we used the fact that the chemical contribution is very similar, therefore we just calculated the difference in the radial scattering distribution.

It is in particular interesting to notice that the radial distribution of the Huang scattering in the diffraction plane is slightly broader in the case of thicker sample. This is due to multiple scattering of the Huang scattered electrons [27]. The multiple scattering of diffuse waves brings the additional Huang contribution to larger angles and this explains the fact that even with larger detection angle an inversion of the InGaP/GaAs level occurs. It is worth mentioning that this phenomenology can be explored only with the frozen lattice approach since other approximate methods cannot account for multiple scattering.

4.3 Chemical analysis of quaternary alloys

Having in mind the above effects it is now the turn to tackle the most complex case, the InGaAsN system. This is a complicated system since both In and N content can vary and need to be determined independently. The experiments reported here regard the case of InGaAsN QW on GaAs. The presence of GaAs barriers is useful to have a safe reference.

The large interest in this system is due to the fact that the introduction of In has a real "chemical effect" that should be well visible in HAADF. Conversely, N is a very light element and should be visible mainly through the static displacement it introduces in the lattice. A convincing proof of this fact is visible in simulations in figgure 1b showing that practically no effect should be visible in the low angle regime. It is therefore reasonable to believe that a comparison between low angle and high angle ADF experiments should permit to separate In and N contribution.

A systematic experimental study on the effect of detection angles on the chemical contrast has been published for ternary InGaAs/GaAs and binary SiGe/Si systems and used to derive the In or Ge composition [28, 29]. We will treat here the information at different angles as independent in order to separate the In and N contribution.-In order to study quantitatively the above phenomenon simulations have been performed for 26 sets of In and N compositions. The resulting dependence on [x] and [y] of LAADF ($26<\theta<161$ mrad) and HAADF ($50<\theta<301$ mrad) has been plotted in figures 4a and 4b. The data have been with fitted to a 2D polynomial to obtain smooth interpolation.

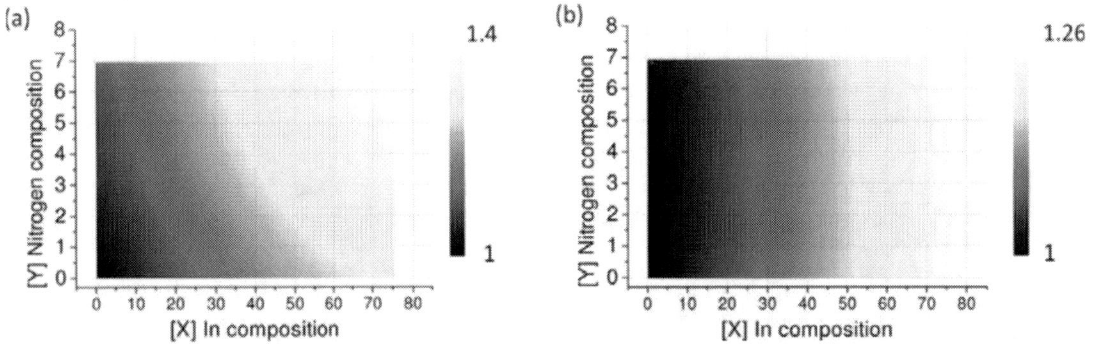

Figure 4. Outcome of simulations for LAADF (a) and HAADF (b) imaging of $In_xGa_{1-x}As_{1-y}N_y$. The simulations refer to a detection range of 26-161 mrad and 50-300 mrad respectively. The intensity levels are normalized to GaAs.

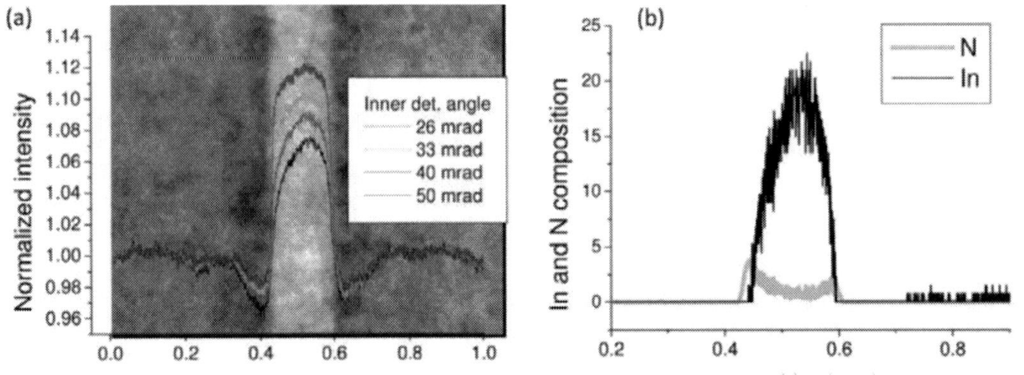

Figure 5. a) Experimental profiles of the InGaAsN quantum well for different detection range. The underlying image is an experimental image for the detector range 50-300 mrad. b) Compositional profiles as deduced from figure 5a and comparison with simulations in figure 4a. The intensity at interfaces is strongly affected by surface strain artefacts.

The simulations refer to a thickness of 88 nm, relevant for comparison with experiments. All intensities have been normalized to the GaAs level. The simulations confirm our original guess that low angle ADF is much more sensitive than HAADF to the N content; in fact, in HAADF the intensity remains the same for any N content up to at least 7% (on Y-axis); it increases monotonically for increasing In content (on X-axis). The situation is completely different for low angles where N has an important effect. The method to extract the In and N content is based on a point to point comparison of normalized LAADF and HAADF intensity between simulation and experiment.

Figure 5a shows an experimental image under HAADF conditions (50-300 mrad); superimposed are normalized profiles for different detector conditions. The inner angle is directly indicated while the outer is about 6 times larger. Both the image and the profiles show the characteristic dip in the barrier and the smooth profile that are typical artefacts of surface strain relaxation [2]. It is clear that this kind of effect reduces the reliability of the composition map.

The rise of the intensity in the centre of the QW for decreasing detection angle is conversely the effect of an increase of Huang scattering due to SD which is in larger part contributed by N and to a minor extent by In. A reduced detection angle produces an increase of the intensity also in the barriers where both In and N are absent. This is clearly due to the Huang scattering produced by the long-range strain according to equation (1). Figure 5b is an attempt to reconstruct In and N profiles from the profiles obtained with inner detection angles of 26 and 50 mrad. The profiles are clearly affected by the artefacts related to the strain and additional Huang scattering in the regions close to the interface. The most reliable region remains the centre of the QW where the atomic column tilt due to surface relaxation should be 0. Here the best fit between simulations and experiments is for $[x]=19\%$ and $[y]=1\%$, which is to be compared with nominal values of 20.5% and 2.4%. This result can be considered as an important progress toward quantitative compositional evaluation of quaternary alloys. The relative error on In evaluation is low as the quantitative analysis is related to the well established high-angle scattering [3, 28]. The larger relative error on the apparent content of N that is more related to low-angle scattering. Therefore, a more accurate description of LAADF scattering is necessary [5].

Finally, figure 5a permits to outline a possible strategy to remove the strain effects from the composition profile at least in simple cases. In fact we notice that the intensity dip due to dechanneling can be partially compensated by Huang scattering and the compensation increases with decreasing detection angle. Unfortunately, it can be estimated that a satisfactory compensation would occur for very low detection angle possibly close to annular bright field conditions where eventually other effects take place. Alternatively, it is possible to extrapolate the Huang scattering at very low angle from data in the allowed range. The result is just a linear combination of HAADF and LAADF profiles. In figure 6 we found that a combination of the profiles at 50 and 26 mrad in the form of $I_{50}+1.5(I_{26}-I_{50})$ shows a good compensation of strain effects. This profile can be considered a good qualitative demonstration of the abruptness of the interface which could not be deduced from figure 5a alone, but no separate composition information on In or N can be deduced from this plot.

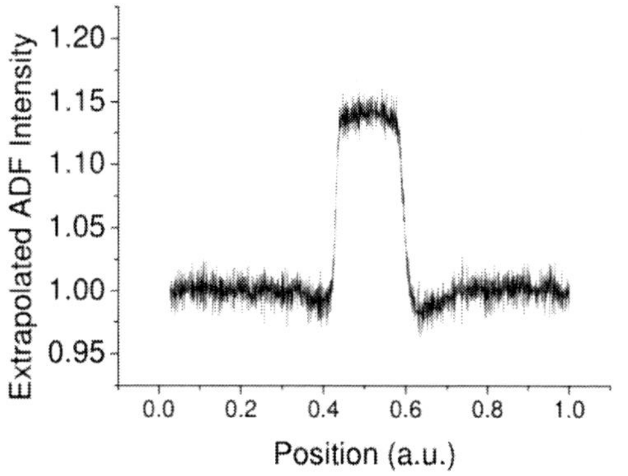

Figure 6. Combinatorial profile obtained as $I_{50}+1.5(I_{26}-I_{50})$. This linear combination of profiles permits to minimize the strain effect.

5. Conclusions

We have analyzed systematically for screw dislocations, InGaP and InGaAsN the contrast mechanism in ADF imaging under different detection conditions. In the examples we demonstrated that it is possible to account for the strain contrast in dislocations and the contrast inversion as a function of static displacement, however, the most interesting application of a multiple detection condition approach is in the case of InGaAsN where the In and N compositions can be independently estimated. In spite of the demonstration of principle, the uncertainty on N composition and the problem with surface relaxation indicate that the research has to be oriented toward a more detailed description of the low angle scattering. In conclusion, it is becoming evident that advanced experiments with multiple detection ranges and simulations can help solving the complex interplay among composition, strain and static disorder.

Acknowledgements
The author (VG) would like to thank Dr. Cesare Frigeri and dott. F. Rossi for contribution and discussion. Part of the computation has been performed under the HPC-EUROPA2 project (project number: 228398) with the support of the European Commission - Capacities Area - Research Infrastructures.

References

[1] Nellist P D and Pennycook S J 1999 *Ultramicroscopy* **78** 111
[2] Grillo V, 2009 *Ultramicroscopy* **109** 1453
[3] Grillo V, Carlino E and Glas F 2008 *Physical Review B* **77** 054103
[4] Yu Z, Muller D A and Silcox J, 2004 *Journal of Applied Physics* **95** 3362
[5] Grillo V and Rossi F 2011 *J. Crys. Growth* **318** 1151–1156
[6] Frigeri C, Shakhmin AA, Vinokurov D A and Zamoryanskaya M V 2011 *Physica status solidi (c)* **8** 1269
[7] Huang R et al 2011 *Angewandte Chemie Int .ed.* **50**, 1
[8] Ishikawa R, Okunishi E,Sawada H, Kondo Y, Hosokawa F and Abe E 2011 *Nature Materials* **10**, 278
[9] Xu P, Loane R F and Silcox J., 1991 Ultramicroscopy **38**, 127
[10] Kirkland E J 1998 *Advanced Computing in Electron Microscopy*, (New York: Plenum Press)
[11] Carlino E, Grillo V, and Palazzari P 2007 Proc. Microscopy of Semiconducting Materials (edited by A. Cullis and P. Midgley) *Springer Proc. Phys.* **120**, 177
[12] http://tem-s3.nano.cnr.it/software.htm
[13] Browning N D and Pennycook S J 1996 *J. Phys. D: Appl. Phys.* **29**, 1779–1798
[14] Voyles P M, Grazul J L and Muller D A 2003 *Ultramicroscopy* **96**, 251–273
[15] Pennycook S J and Jesson D E, 1991 *Ultramicroscopy* **37**, 14
[16] Wu X, Robertson, MD, Gupta JA and Baribeau J-M, 2008 *J. Phys.: Condens. Matter* **20**, 075215
[17] Hillyard S and Silcox J 1995 *Ultramicroscopy* **58**, 6
[18] Wang ZL 1994 *Ultramicroscopy* **53**, 73
[19] Plamann T and Hytch MJ 1999 *Ultramicroscopy* **78**, 153
[20] Walther T and Humphreys CJ 1997 Proc. EMAG 97 *Inst. Phys. Conf. Ser.* **153**, pp 303-306
[21] Broeckx J, Ob de Beeck M and Van Dyck D 1995 *Ultramicroscopy* **60,** 71
[22] Salviati G et al 2001 *Institute of Physics Conference Series* **169**, pp 251–254
[23] Rosenauer A et al. 2009 *Ultramicroscopy*, **109**, 1171
[24] Dwyer C 2010 *Ultramicroscopy* **110**, 195–198
[25] Schowalter M, Rosenauer A, Titantah J T and Lamoen D 2009 *Acta Cryst. A* **65**, 5
[26] Cosgriff EC et al. 2010 *Philosophical Magazine* **1478**, 4361
[27] Treacy MMJ and Gibson JM 1993 *Ultramicroscopy* **52**, 31
[28] Walther T, Qiu Y and Cullis AG 2010 *Journal of Physics: Conference Series* **241**, 012068
[29] Walther T 2006 *J. of Microscopy* **221**, 137

Aberration-corrected STEM and EELS of semiconducting nanostructures

K Cui[1,2], S Hosseini Vajargah[1], S Y Woo[1], M Couillard[1], S Lazar[3], R N Kleiman[4], D A Thompson[4] and G A Botton[1,*]

[1] Department of Materials Science and Engineering, McMaster University 1280 Main Street West, Hamilton, Ontario, L8S4L7, Canada
[2] Department of Electrical and Computer Engineering, McGill University, Montreal, Quebec, H2A3A7, Canada
[3] FEI Company, Eindhoven, The Netherlands
[4] Department of Engineering Physics and Centre for Emerging Devices Technology, McMaster University, 1280 Main Street West, Hamilton, Ontario, L8S4L7, Canada

E-mail: gbotton@mcmaster.ca

Abstract. We review some applications of aberration–corrected electron microscopy for the detailed characterization of semiconducting nanostructures using a combination of high-angle annular dark-field scanning transmission electron microscopy and electron energy loss spectroscopy. The study of self-assembled quantum wires shows that it is possible to determine the composition of the nanostructures with better than 1 nm resolution down to the atomic level while the contrast in the high-angle annular dark-field images is used to determine the presence of wetting layers separating quantum wires and the strain field arising from the local compositional changes. The local measurements of energy loss spectra demonstrate the shift of plasmon peaks consistent with the changes in lattice parameters. High-angle annular dark-field images are also used to study the contrast in GaSb thin films deposited and study the presence of anti-phase domain boundaries. These examples show that aberration-corrected microscopy combined with electron energy loss spectroscopy provide not only enhanced resolution but also increased sensitivity to atomic site compositional changes.

1. Introduction

Group III-V semiconductor nanostructures such as self-assembled nanowires have potential uses in optoelectronic applications such as long wavelength lasers [1] and broad band emitters [2], and have been demonstrated to emit at room temperature in the telecommunication range [3]. Also of great promise for optoelectronic applications such as photodetectors and photovoltaics is the use of III-V compounds grown on Si. While it is clear that in the latter case, the greater range of optical properties of III-V compounds gives much enhanced flexibility in design and tuneability of the optoelectronic properties, there is still significant difficulty in growing high-quality films exhibiting low defect densities, due to the intrinsic lattice parameter mismatch between the films and substrate. Although conventional electron microscopy provides excellent insight on the nature of the defects, defect density, and allows to visualize both chemical/compositional contrast, it is now apparent that the new imaging techniques and compositional analysis methods can provide complementary and essential information that can be used to gain further insight on the very local composition, strain, and local changes in polarity information that could only be obtained with great effort with conventional techniques due to the strong dynamical scattering. The recent development of electron microscopes with aberration correctors has opened the field for such studies by providing enhanced stability of the microscopes, improved resolution and spectroscopic sensitivity.

Here we present some examples of studies related to the characterization of InAs quantum wires grown on quaternary alloys lattice matched to InP. These materials have shown photoluminescence emission at room temperature and tuneability of emission with changes in quantum wire size and growth conditions. We show that the improved resolution of the aberration corrected instruments, together with atomic resolved electron energy spectroscopy mapping, provide key information about the presence of wetting layers connecting the quantum wires as well as the detection energy shifts in the plasmon resonances. We demonstrate that these changes are very localized to the quantum wire structures. We also highlight recent results related to the study of growth mechanisms and defects of GaSb thin films deposited on Si where high-resolution imaging provides key evidence of local polarity change complementing convergent beam electron diffraction information as well as direct visualization of the anti-phase boundaries highlighting the transition between two out-of-phase domains.

2. Experimental methods

Conventional contrast imaging was carried out on a Philips CM12 microscope operated at 120keV. Ultrahigh-resolution microscopy was carried out with a FEI Titan 80-300 "cubed" microscope equipped with a high-brightness source, an electron monochromator, a CEOS hexapole corrector for the probe-forming lens and a similar corrector for the image-forming lens. The system is also fitted with a Gatan 866 energy filter for electron energy loss spectroscopy. This system is also located in an ultrastable environment with demonstrated temperature stability of better than 0.1 °C, low electromagnetic fields (<0.05 mG) and low vibrations [4]. Quantitative elemental maps of the In composition were also obtained with a field emission scanning transmission electron microscope (JEOL 2010F) fitted with a Gatan Tridiem Energy Filter. Samples for conventional imaging were prepared with the conventional mechanical polishing method, followed by dimpling and ion milling to electron transparency (Gatan PIPS system equipped with a liquid N_2 cold stage, 4 keV argon ions at an incident angle of ±4°). For ultrahigh resolution electron microscopy, samples were prepared by the tripod method to electron transparency using an Allied Tech Multiprep unit. Samples are then cleaned using a Technoorg Linda Gentle-mill using 300-500 eV ions to remove the residues of polishing contamination from the samples. This last step, as well as the mechanical polishing, has been shown to provide excellent quality samples suitable for the aberration-corrected microscopy work shown here.

3. Self-assembled quantum wires

Self-assembled InAs quantum wires were grown by gas source molecular beam epitaxy on the surface of InGaAlAs which is used as a confinement/capping layer for the electrons (figure 1). This quaternary alloy is also lattice matched to the InP substrate so that the In content is 0.53 of the group III elements. These wires have been shown to grow along the $[0\bar{1}1]$ direction of the InGaAlAs surface as a consequence of the anisotropy of the surface reconstruction and nature of surface steps, as demonstrated from AFM imaging [5]. With the deposition on the InGaAlAs of 4-5 monolayer of InAs, quantum wires structures are formed as a result of strain relaxation due to the 3.2% lattice mismatch. As a consequence, the effects of strain, composition and curvature of the quantum wires are intimately linked with each other and ultimately control the chemical potential for group III elements of the surface front during subsequent MBE overgrowth [6]. Due to the complexity of the structures, the effect of strain and the coupling of this strain with the composition also affect the image contrast. Therefore, several imaging methods and chemical analysis tools have been used to characterize these materials. In the following sub-sections we present imaging with conventional contrast transmission electron microscopy, elemental mapping with EELS as well as aberration-corrected imaging so as to elucidate some of the structural features observed in these materials. Previous work has used this chemical information, combined with finite elements calculations, so as to determine the stress state in the wires and the surrounding areas [6].

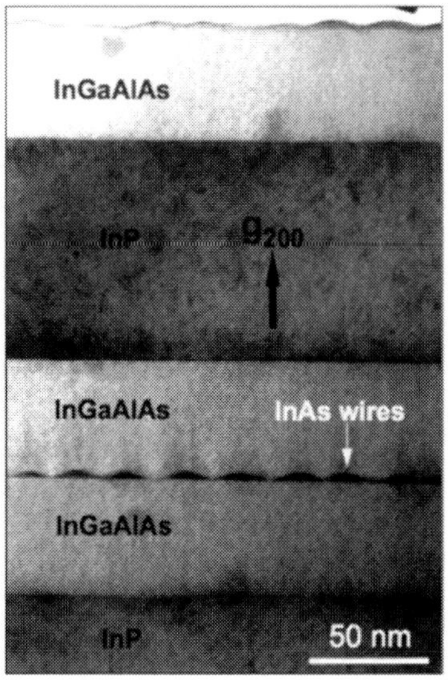

Figure 1. General overview of the wire structure with the confinement layers. The cross-section shows the InAs wires enclosed in the InGaAlAs confinement layers, the InP substrate and capping layer. An additional InGaAlAs layer is used as a reference to generate an uncapped array of InAs wires at the very top of the surface used for atomic force microscopy analysis.

3.1. Conventional diffraction contrast imaging

Conventional imaging shows the complexity of the structures and the interplay between the strain and composition in TEM images. By selecting the chemically sensitive ***g***=[200] reflection in two-beam conditions (figure 2a), the contrast between the wires and the surrounding quaternary InGaAlAs confinement layers is highlighted. Alternatively, the selection of the ***[022]*** reflection highlights the strain contrast and the strong strain interaction between the wires (figure 2b). In the case of multiple stacks of wires used for laser applications, it is also clearly possible, when using the chemically sensitive ***g***=[200] reflection to see the change in the alignment between the subsequent layers of quantum wires. With this imaging method, we have also detected dark bands above the wires and connecting the layer stacks as shown in figure 3 and [7]. Since there is potentially ambiguity in the chemically sensitive images due to possible residual strain contrast, elemental mapping with EELS was used to assess such effect.

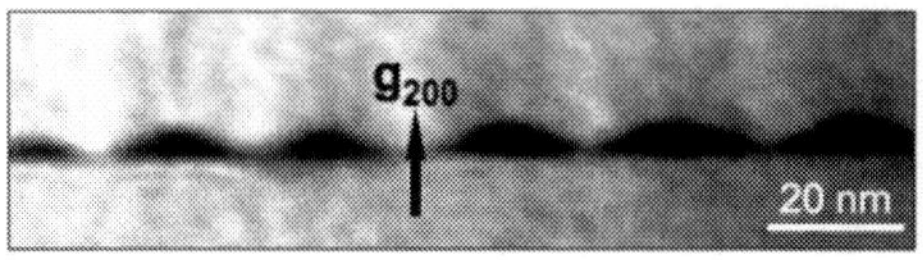

Figure 2a. Conventional cross-sectional TEM imaging of quantum wires. Chemically sensitive contrast image using the ***g***=200 reflection

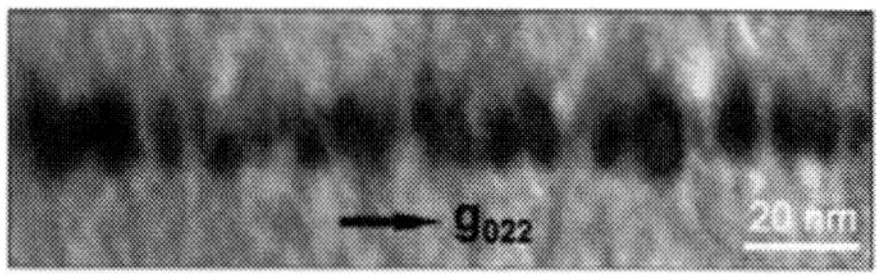

Figure 2b. Conventional cross-sectional TEM imaging of quantum wires. Strain sensitive contrast image using the ***g***=022 reflection.

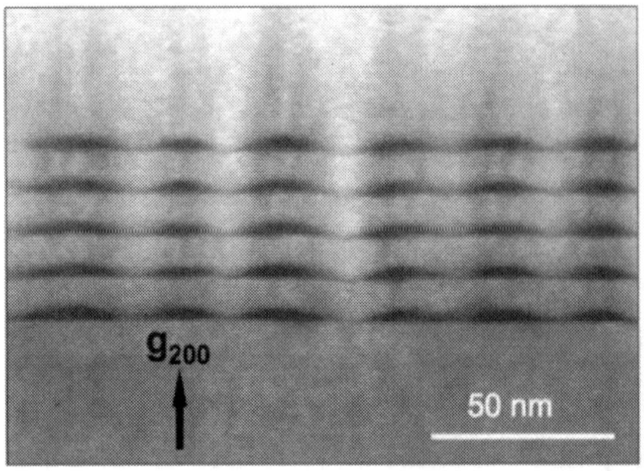

Figure 3. Multiple stacks of wires. Bright-field TEM image of multiple stacks of wires obtained with the ***g=200*** reflection.

3.2. Chemical analysis with EELS and EDXS

In order to identify without any ambiguity the origins of the stripe-like contrast observed in the conventional TEM images as seen in figure 3, we used EELS spectrum imaging and the detection of the In $M_{4,5}$ edge in the spectra. Using an energy window from 450 eV to 520 eV energy loss and the power-law background extrapolation, we can extract the In signal and this can be quantified based on the reference InGaAlAs layer (with In content of 0.53) assumed to be of known composition based on X-ray diffraction measurements. The resulting In maps thus provide quantitative composition with a resolution confirmed to be in the 1 nm range when using a non-aberration-corrected field emission TEM [5]. Elemental maps show that In-rich stripes are detected not only on single layer structures [5] but also between multiple stacks of wires (figure 4). Elemental maps of a row of wires (figure 5), suggest relatively uniform composition from wire to wire but, at the resolution of the conventional scanning transmission electron microscopy (STEM) instrumentation, the data does not provide clear evidence on the possible presence of a wetting layer, connecting adjacent wires. This wetting layer is an important factor that determines whether the growth mechanisms follows the Stranski-Krastanov (island formation on wetting layer) or Volmer-Weber (island without wetting layer), In addition, a wetting layer has the potential of affecting the electron confinement and the interactions between the quantum wires and thus any simulations on the optical properties. Therefore, additional high-resolution mapping in combination with high-angle annular dark-field imaging are required.

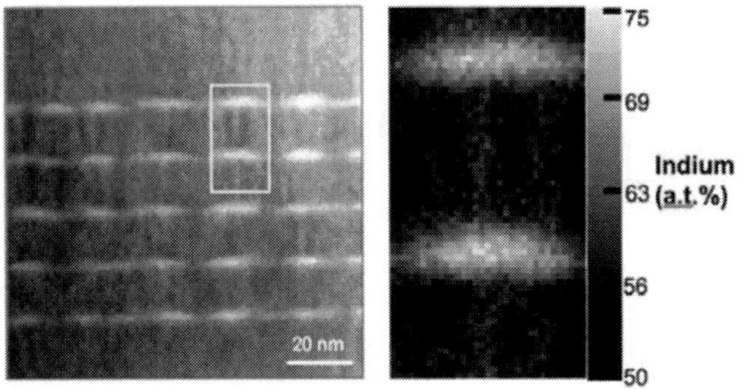

Figure 4. EELS SI map with multiple wires. Left: HAADF image and boxed area of interest; Right; In $M_{4,5}$ elemental map and quantitative gray scale lookup table.

Journal of Physics: Conference Series **326** (2011) 012007 doi:10.1088/1742-6596/326/1/012007

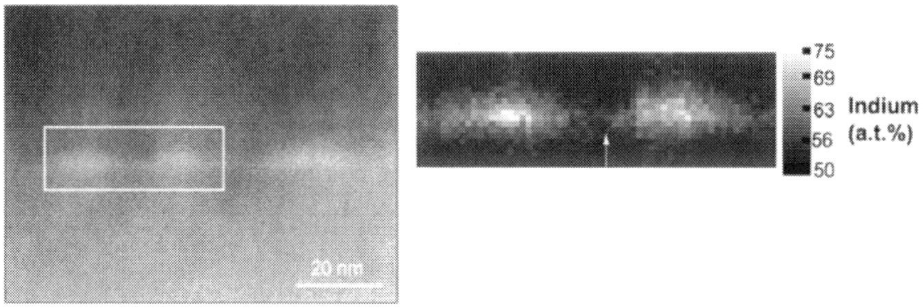

Figure 5. EELS SI map with multiple wires with the possible wetting layer.

In order to ascertain the best possible resolution achievable in elemental maps for the elements of interest here, we have used a reference InGaAlAs material as a test case. Atomic resolution mapping has, for example, been demonstrated in oxides with a relatively large lattice parameter [8] but no atomic resolved mapping on III-V alloys has been reported in the open literature so far. With aberration correction and a probe current approaching 100 pA, elemental maps of In and As demonstrate that atomic resolved mapping is indeed possible and that the peak intensity of the In or As maps is specific to the column corresponding to the related high-angle annular dark field (HAADF) intensify (figure 6). For example, the In map has peaks precisely corresponding to the brightest dumbbell pair coinciding with the In column while the As map has peaks precisely coinciding to the lower intensity dumbbell in the HAADF image. These results demonstrate that, in the conditions used for mapping and in particular for a collection semi-angle larger than 80 mrad, the signal is localized to the atomic column channelling the electron intensity. Since the projected spacing between the In and As columns is in the order of 1.6 Å, this demonstrates a spatial resolution sufficient to distinguish the two atomic columns.

Although the detection efficiency of a standard Si(Li) energy dispersive X-ray spectrometer is significantly lower than what can be achieved with EELS and the newest silicon drift detectors [9] by a factor of about 9 (based exclusively on the solid angle), our measurements show that atomic resolved signals can also be detected in an EDXS map (figure 7).

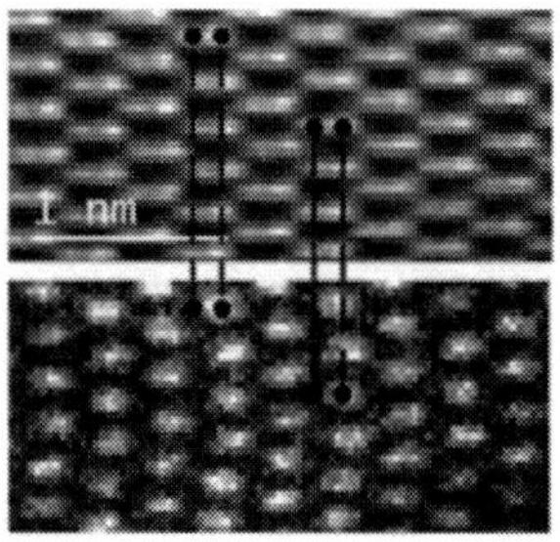

Figure 6a. Atomic resolved EELS maps, for In ($M_{4,5}$ edge) at 450-520 eV loss. Top: HAADF signal, Bottom: In $M_{4,5}$ map. Image quantified following multivariate statistical analysis and noise removal [10]. The thickness of the sample, relative to the total inelastic mean free path at 200keV, (t/λ), is approximately 0.2.

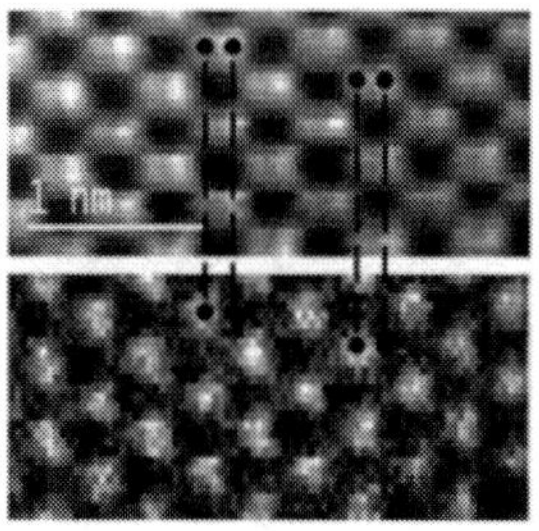

Figure 6b. Atomic resolved EELS map for As ($L_{2,3}$ edge) at 1300-1400 eV loss. Top: HAADF signal, Bottom: As $L_{2,3}$ map. Image quantified following multivariate statistical analysis and noise removal [10]. The thickness of the sample, relative to the total inelastic mean free path at 200keV, (t/λ), is approximately 0.2.

The raw EDXS data based on the In L and As K peak signals mapped in figure 7 shows in fact a regular pattern centred at the atomic columns. These results show that, in spite of the known effects of beam broadening, the contribution of channelling of electrons on the atomic columns and the highly focused electron beam, significant intensity changes are detected between the different atomic positions, particularly for the In signal which is most intense and has lower noise levels. Further work would be necessary to ascertain whether, for a sharp interface, these results are fully consistent with atomic resolution elemental mapping as demonstrated in EELS experiments [8].

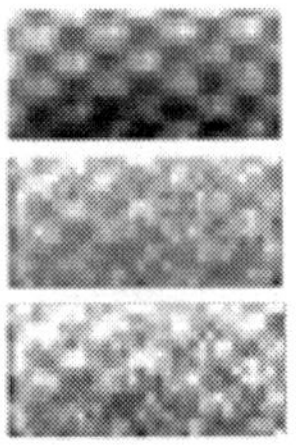

Figure 7. Elemental maps obtained with a Si(Li) drift (0.13 sr solid angle) detector on the aberration-corrected STEM. Top image is the annular dark-field (ADF) signal recorded simultaneously as the EDXS signal. Middle image is the As signal and bottom image is the In. (dwell time 50 msec/pixel). Maximum counts for the In map is 25, for the As map 19. 32x15 pixel map (2.1x1.1nm field of view). The thickness of the sample, relative to the total inelastic mean free path at 200keV, (t/λ), is approximately 0.2.

Because of the demonstration of atomic resolution in EELS maps for In and As, we have focused our attention to the characterization of the wetting layer separating two quantum wires studied previously above at lower resolution (figure 5). The two dimensional EELS map extracted over the area between two adjacent quantum wires shows that the EELS intensity in the In column increases over 3 atomic planes (figure 8) thus confirming that there is a residual wetting layer separating two wires. The implication of the presence of this wetting layer is important in terms of both the growth mechanism and also for the electronic coupling and interaction between the wires. As the nature of a potential well between the wires is intrinsically affected by the local composition, the presence of this wetting layer suggests an anisotropy in the confinement of the wires with a lateral potential well somewhat reduced between adjacent wires. The effect of this feature on the optical properties has been discussed recently [11].

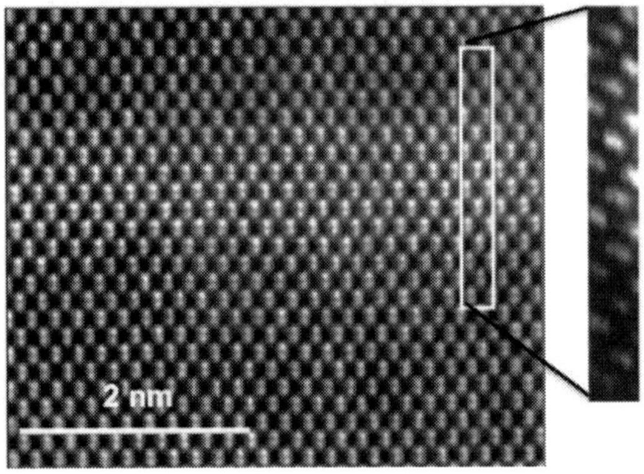

Figure 8. HAADF image of the wetting layer (left) and atomic resolved EELS map of In (right) obtained from the highlighted area surrounded by a white frame. The In map was extracted from the spectrum image following background subtraction of the $M_{4,5}$ edge. Data was processed with multivariate statistical analysis to reduce some of the noise contribution [10].

3.3. Aberration-corrected imaging

The EELS elemental mapping of the wetting layer suggests that there is an increased In concentration within a few atomic planes from the confinement layer. With the enhanced HAADF resolution of a probe-corrected STEM, it is also possible to complement the chemical information from EELS with the intensities of HAADF images (figure 9). From an intensity profile, it is apparent that the intensities of the group III column across the wetting layer increases significantly while the intensity of the group V column is only slightly increased. The background intensity between the dumbbell columns also shows increased intensity, due to the tails of the probe, contributing to an overall increase of the

HAADF signal. With these profiles, it is very clear that the wetting layer consists of 3 to 4 atomic layers with increased In concentration. By combining the EELS maps with the HAADF images, unambiguous confirmation of the presence of the wetting layer is thus possible. The measurement of the localized strain using the geometric phase analysis (GPA) from the HAADF images is also feasible in a very stable instrument and microscope environment. This technique was initially developed for phase contrast HRTEM imaging [12] and demonstrated recently on HAADF images of the GaSb/Si interface [13]. The HAADF image of two stacked wires shows the strong contrast between the confinement layer and the InAs wires while the analysis of strain with the GPA using the [111] and [$\bar{1}$11] reflections shows the local strain in the growth direction extending almost exclusively to the wires themselves, at least within the noise level of the observed in the confinement layer (figure 10).

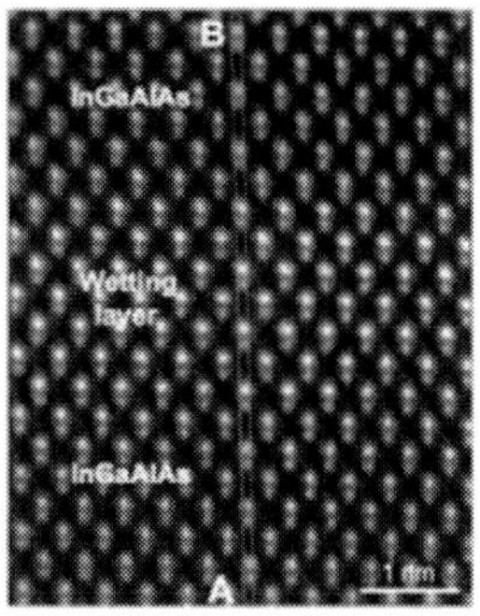

Figure 9a. HAADF image of wetting layer. The vertical thin line indicates the area selected for the intensity profile shown in figure 9b.

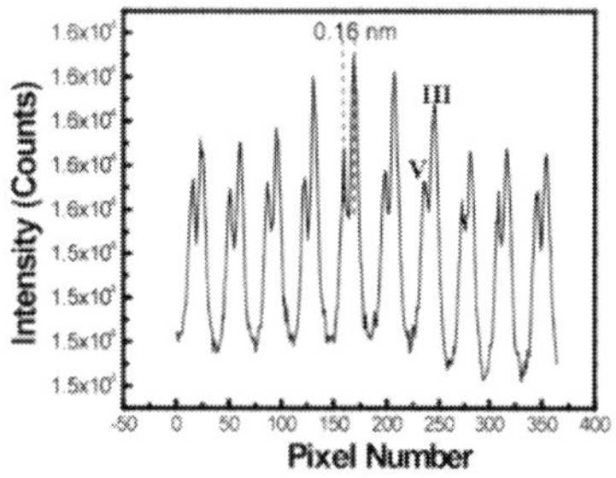

Figure 9b. HAADF intensity profile from the same area identified in figure 9a from point A (pixel "0") to point B.

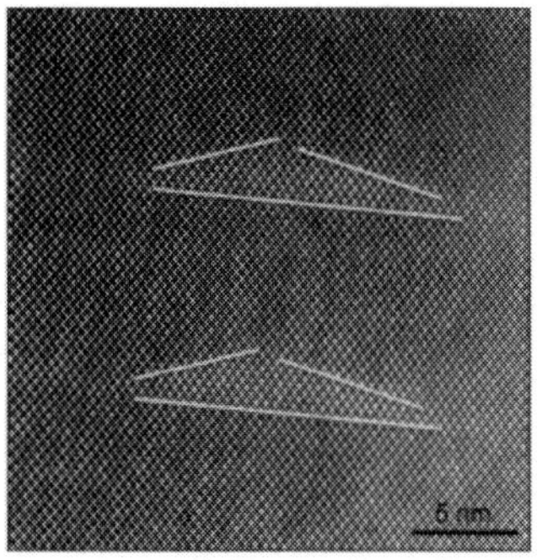

Figure 10a. HAADF image of two stacked quantum wires. Surrounded by the straight lines are the "boundaries" of the dots as a reference comparison in figure 10b. The selected frame is the same as for figure 10b.

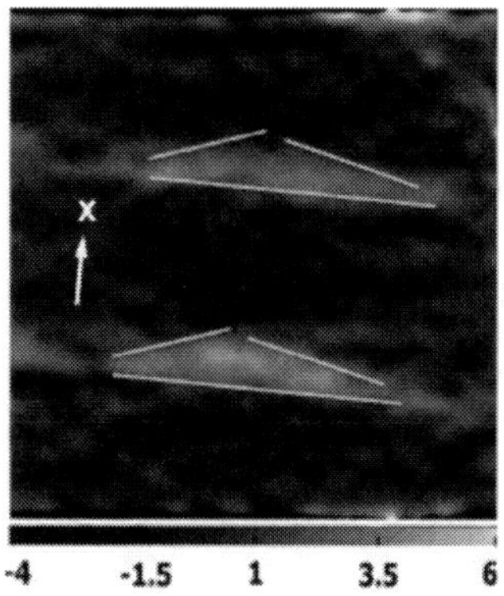

Figure 10b. Strain image ε_{xx} obtained from the geometric phase analysis of the HAADF image shown in figure 10a. The straight lines identify exactly the same area shown in figure 10a and delimit the boundaries with increased HAADF signal. The gray scale bar at the bottom of the figure is the strain in % units.

3.4. Low-loss mapping on quantum wires

Measurements of the plasmon energies have been used extensively as an effective way to measure indirectly the composition of samples in alloys [14] and metal hydrides [15]. In the field of semiconductors, plasmon imaging has been used to distinguish Si and SiOx based on the strong contrast between the two phases. [16]. More recently, low energy losses have been used to image surface plasmon polaritons in metallic nanowires in the infra-red optical regime and in the visible range [17, 18]. In III-V compounds, bulk plasmon shifts can potentially reveal changes in valence electron density resulting from changes in the composition and volume of the unit cell due to the strain. Two-dimensional maps acquired on the cross-sectional samples reveal very significant changes in the bulk plasmon position and shifts that can be quantified (figure 11). These changes are consistent with the increase in the local lattice parameter as demonstrated from the GPA measurements and further work at lower energy losses is in progress to detect any possible interaction between the valence electrons between stacked quantum wires.

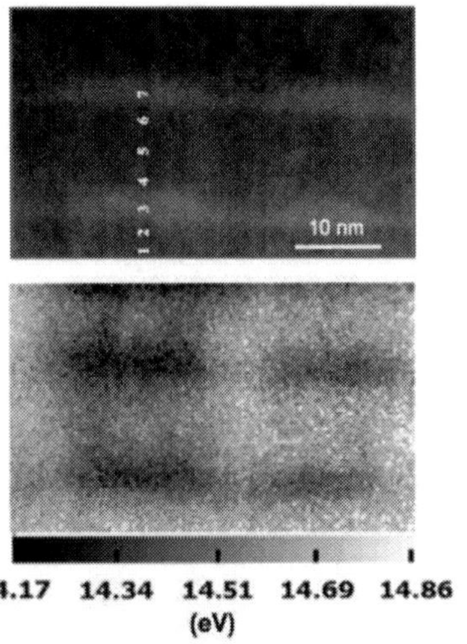

Figure 11a. HAADF STEM image (top) of stacked quantum wires and plasmon shifts map (bottom) of exactly the same area. Gray scale bar shows the energy shift of the plasmon peak position.

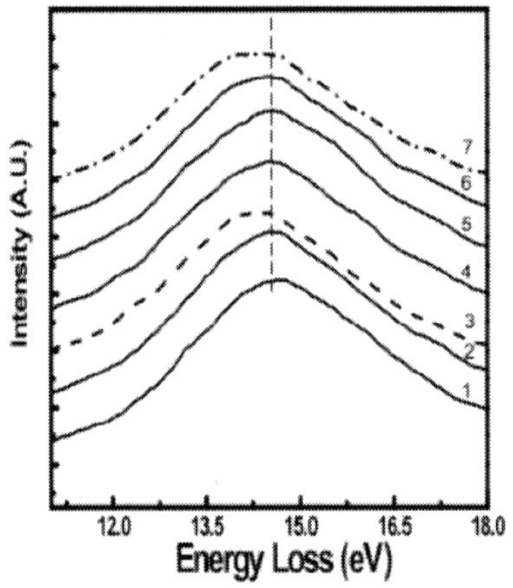

Figure 11b. Low-loss spectra extracted from the spectrum image dataset used to map the peak positions extracted in figure 11 (a). Spectra taken from confinement layers (spectra 1, 2, 4,5 and 6) and quantum wires (spectra 3 and 7). The vertical dashed line depicts the peak positions of spectra taken from the capping layers as a reference.

4. GaSb on Si layers

In the highly lattice mismatched system of GaSb films grown on Si by molecular beam epitaxy, the nature of the defects is still a subject of significant debate. Even in films grown with an AlSb buffer layer, arrays of dislocations have been observed with conventional imaging [19], high-resolution TEM [20] and HAADF [13] but there have not been conclusive observations of the predicted anti-phase boundaries (APBs). Even though APBs are expected due to the likely presence of quarter-unit cell surface steps [21], direct and unambiguous evidence of APB in films has not been demonstrated so far in spite of the importance of such defects on the electronic properties. By growing GaSb films and a strained-layer-superlattice (SLS) of AlSb/GaSb to prevent further propagation of threading dislocations and possibly other extended defects, we have thus investigated whether anti-phase domains (APDs) are indeed present and their structure at the atomic level with the ultimate purpose of understanding the interaction between defects and the potential electrical activity of the range of defects present in the films.

4.1. Conventional diffraction contrast

Using a combination of diffraction contrast imaging, and convergent beam electron diffraction (CBED) with conventional TEM we have identified adjacent APDs separated by boundaries (figure 12). The superlattice reflection dark-field (DF) images show that some of the APBs intersect the SLS and propagate to the surface while some self-annihilate (figure 12). Misfit dislocations, threading dislocations and microtwins are also generated at the Si interface (figure 13) even when an AlSb buffer layer is present. Previous work using cross-sectional TEM and EDXS mapping has shown that the AlSb buffer layer is present as islands [13] and that a network of misfit Lomer dislocations is present. CBED measurements have confirmed that domains separated by the defects shown in figure 12 are indeed of opposite polarity [22].

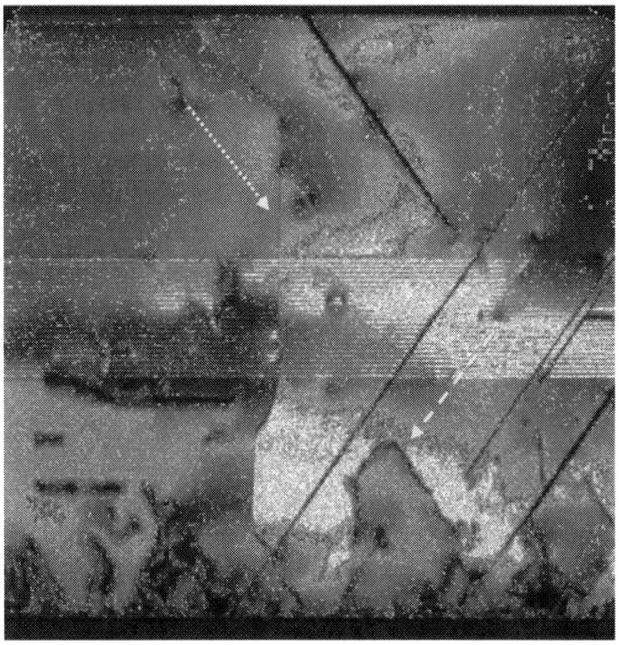

Figure 12. Conventional TEM DF image of GaSb film grown on Si showing both propagating (dotted arrow) and self-annihilated (dashed arrow) APBs. Chemically sensitive contrast image using the $g=00\bar{2}$ superlattice reflection. The dotted arrow indicates the APB separating two anti-phase domains on the left and right of the boundary. The dark features running obliquely are microtwins running along the {111} planes.

The SLS also allow clear visualization of the effect of microtwins on the growth of the films with strong contrast enhanced using the chemically sensitive reflections. Already from the diffraction contrast image, it is possible to identify the tilting of the SLS within the microtwin (figure 13) and the fact that the microtwins do not necessarily appear to form right at the interface with the substrate. Due to the finite thickness of the foils, however, this effect does not consider the propagation of the microtwins in the third dimension as discussed further in [22]. These observations are discussed in the following section.

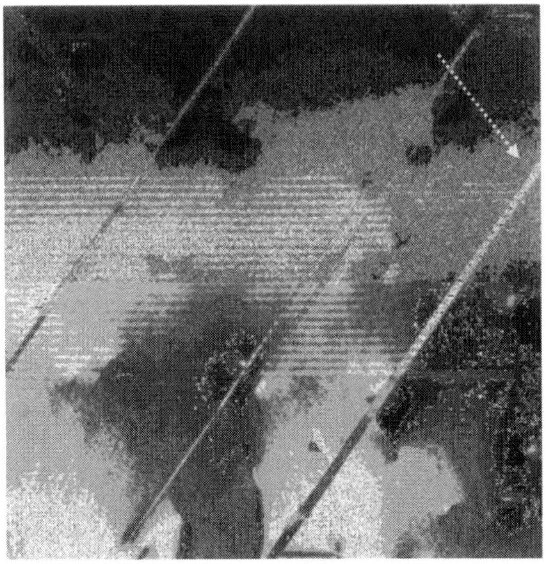

Figure 13. Conventional TEM DF image of the AlSb/GaSb strained-layer-superlattice and various microtwins intersecting the layers using the $\boldsymbol{g}=\bar{2}20$ fundamental reflection. Pointed by the dotted arrow is evidence of a growth front change from planar to stepped due to the microtwin.

4.2. Aberration corrected microscopy

Imaging with higher resolution and sensitivity to the atomic contrast demonstrates the detection of local deviations from a planar surface due to the presence of the microtwins (figure 14). HAADF imaging also shows the formation of atomically sharp interfaces between the strained layers, with the transition between the AlSb and GaSb just occurring over one single atomic plane (figure 15). HAADF images also demonstrate the detection of lone Ga atoms at the $\sum=9$ tilt boundary resulting from intersection of two coherent twin boundaries. Defects are marked by the array of continuous arrows in figure 15.

Figure 14. HAADF image of the strained-layer-superlattice intersecting a microtwin. The growth front locally changes from a planar interface. The bright layer is the GaSb component of the SLS while the dark layer is the AlSb component of the SLS.

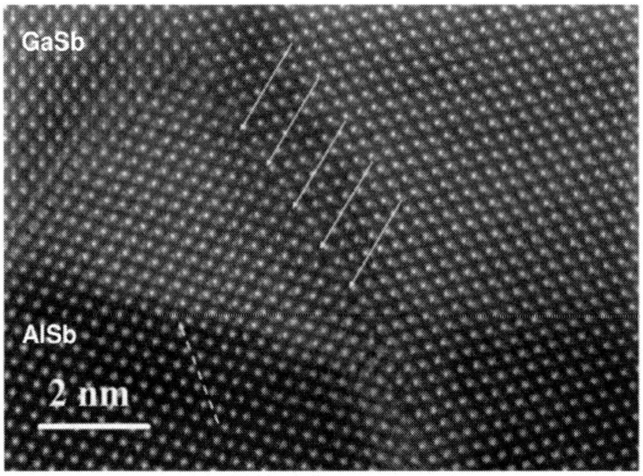

Figure 15. HAADF image of a defect in the strained-layer-superlattice due to the propagation of a twin from the interface between GaSb and Si. Also visible are missing dumbbell pairs pointed to by the continuous arrows and an atomically sharp interface between the AlSb and GaSb layers. In the AlSb layer, only the Sb column is visible in the dumbbell pair while, on GaSb, both the Ga column (of weaker intensity) and the Sb columns are visible.

Also very clearly visible, based on the intensity of the reversal of the dumbbell pairs in the SLS, are APBs that appear to be about 4-5 atomic planes wide (figure 16). The contrast of the APBs in the AlSb clearly suggest that the Sb column also changes polarity over the thickness of the TEM sample and that the intrinsic width is thus most likely less than the apparent projected dimension. The region where the polarity changes within the APB in the AlSb component of the SLS thus appears as a bright band due to the channelling of electrons on the Sb atoms, now present on both atomic sites over the thickness of the thin film rather than just one of the two sites.

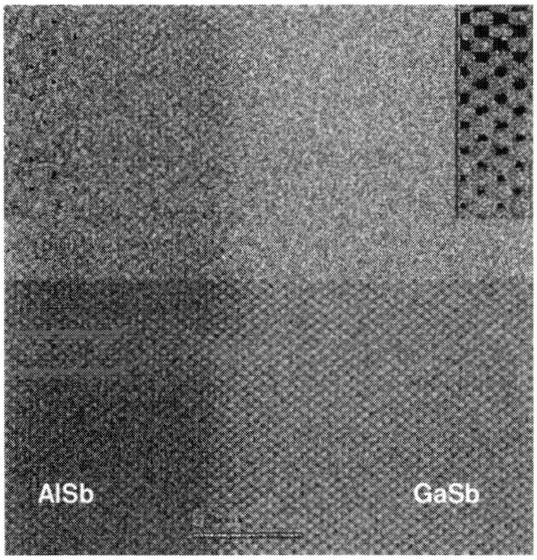

Figure 16. HAADF image of an APB detected in the AlSb/GaSb strained-layer-superlattice as delimited by the two lines. The AlSb layer is identified by the lower intensity at the right of the image (growth direction is from left to right) and the GaSb is on the right of the image, HAADF simulations of AlSb and GaSb images are show as inserts left and right respectively. In the simulations and in the experiments, only the Sb column is visible in AlSb while both the Sb and Ga columns are visible in GaSb. Within the APB in the AlSb layer, both columns are of significant intensity. For clarity of the APB visibility, the image is rotated with respect to the growth direction, which is now from left to right.

At the interface with the Si substrate we have also observed APBs and interactions of these with other defects. Further work detailing the origins of the APBs at the interface with the substrate will be presented elsewhere [23]. These results with high-resolution HAADF clearly show that aberration correction not only provides improved resolution but also enhanced detection of anti-phase boundaries and potentially electrically active lone atomic sites.

5. Conclusions

We have demonstrated the application of conventional and aberration-corrected electron microscopy for the detailed study of semiconducting nanostructures. The application of these two techniques has been shown in the context of the study of structure, strain and composition of InAs quantum wires and in the detailed investigation of GaSb films deposited on Si. Chemical mapping with atomic resolution has been successfully used to determine unambiguously the chemical nature of a wetting layer present between InAs wires while plasmon mapping has been used to identify changes in the local electron density. High-resolution high angle annular dark-field imaging has provided insight on the distribution of strain in quantum wires and the detection of defects in GaSb films grown on Si. The sensitivity to atomic contrast in HAADF imaging has also been exploited to provide insight on the nature of anti-phase boundaries and the interaction between these and strained-layer-superlattices.

Acknowledgements

This work was partially supported by the Ontario Centres for Excellence, the Discovery Grants program of NSERC and ARISE Technologies. The electron microscopy was carried out at the Canadian Centre for Electron Microscopy, a national facility supported by NSERC and McMaster University.

References

[1] Reithmaier J P, Eisenstein G, and Forchel A 2007 *Proc. IEEE* **95**, 1779
[2] Lin Z C, Lu C Y and Lee C P 2006 *Semicond. Sci. Technol.* **21**, 1221
[3] Grundmann M 2002 *Nano-Optoelectronics* (New York: Springer)
[4] Botton G A 2008 *Int. J. Nanotechnology* **5** 1082
[5] Cui K, Robertson M D, Robinson B J, Andrei C M, Thompson D A and Botton G A 2009 *J. Appl. Phys.* **105** 094313
[6] Cui K, Robinson B J, Thompson D A and Botton G A 2010 *J. Appl. Phys.* **108** 034321
[7] Cui K, Robinson B J, Thompson D A and Botton G A 2010 *J. Cryst. Growth* **312** 2637
[8] Botton G A, Lazar S and Dwyer C 2010 *Ultramicroscopy* **110** 926
[9] Schlossmacher P, Klenov D O, Freitag B and von Harrach H S 2010 *Microscopy Today* **18** 14
[10] Dudeck K, Couillard, M, Lazar S, Dwyer C, and Botton G A 2011 *Micron, in press*
[11] Cui K, Robinson B J, Thompson D A and Botton G A 2011 *J. Appl. Phys.* **109** 124311
[12] Hÿtch M J, Snoeck E and Kilaas R 1998 *Ultramicroscopy* **74** 131
[13] Hosseini Vajargah S, Couillard M, Cui K, Ghanad Tavakoli S, Robinson B, Kleiman R N, Preston J S and Botton G A 2011 *Appl. Phys. Lett.* **98** 082113
[14] Hunt J A 1995 *Microbeam Analysis* (New York: VCH Publishers) p 215
[15] Woo O T and Carpenter G J C 1992 Microsc. Microanal. **3** 35
[16] Carpenter G J C 2004 *Microsc. Microanal.* **10** 435
[17] Rossouw D, Couillard M, Vickery J, Kumacheva E and Botton G A 2011 *Nano. Lett.* **11** 1499
[18] Nelayah J, Kociak M, Stéphan O, García de Abajo F J, Tencé M, Henrard L, Taverna D, Pastoriza-Santos I, Liz-Marzán L M and Colliex C 2007 *Nat. Physics.* **3** 348
[19] Akahane K, Yamamoto N, Gozu S, Ueta A and Ohtani N 2006 *Thin Solid Films* **515** 748
[20] Kim Y H, Lee J Y, Noh Y G, Kim M D, Cho S M, Kwon Y J and Oh J E 2006 *Appl. Phys. Lett.* **88** 241907
[21] Kroemer H 1987 *J. of Cryst. Growth* **81** 193
[22] Woo S Y, Hosseini Vajargah S, Ghanad-Tavakoli S, Kleiman R N and Botton G A, to be submitted.
[23] Hosseini Vajargah S, Kleiman R N, Preston J S and Botton G A, to be submitted.

Microscopy needs for next generation devices characterization in the semiconductor industry

L Clement[1], C Borowiak[1], R Galand[1], K Lepinay[1], F Lorut[1], R Pantel[1], G Servanton[1], R Thomas[1], P Vannier[1] and N Bicais[1]

[1] STMicroelectronics, 850 rue Jean Monnet, F-38926 Crolles, France

Email : laurent-renaud.clement@st.com

Abstract. In this paper we present the different imaging based techniques used in the semiconductor industry to support both manufacturing and R&D platforms at STMicroelectronics. Focus is on fully processed devices characterization from large structure (3DI, Imager sensors) to advanced MOS technologies (28-20 nm). Classical SEM and TEM (mainly EFTEM) based techniques are now commonly used to characterize each step of the semiconductor devices' process flow in terms of morphology and chemical analysis. However to address specific issues, dedicated imaging techniques are currently being investigated. With the "High-*k* Metal Gate" stack involved in the more advanced MOS devices (28-20 nm), new challenges occur and therefore advanced characterization is mandatory. Some relevant examples are pointed out through (STEM) EELS and EDX experiments. Analysis of stressors mainly used to improve carrier mobility in next generation devices, is also presented with different approaches (NBD, CBED and Dark-field holography). Advanced STEM and AFM based techniques applied to characterize dopants and junction in MOS devices and also in more relaxed structure such as imager sensors is discussed too. Concerning back-end (interconnects) and 3D integration (3DI) issues, focus is on nano-characterization of defects by classical techniques (EFTEM, STEM EELS-EDX) and with dedicated ones still in development. To illustrate this topic some 3D FIB/SEM and E-beam tomography experiments are presented. Examples of microstructure and texture determination in poly-crystalline materials such as copper line by coupling SEM/EBSD and TEM techniques are also shown.

1. Introduction

Scanning and Transmission Electron Microscopy (SEM and TEM) coupled with Focused Ion Beam (FIB) are essential techniques to ensure the process development of future devices in the semiconductor industry. At STMicroelectronics Crolles, these tools are now commonly used to support both Manufacturing (200 mm and 300 mm *fabs*) and R&D platforms. In this industrial context with fast process evolution, large numbers of sites have to be analyzed with short turn-around-times (TAT) but new and innovative characterization techniques have also to be run up.

Nowadays, physical characterization faces some challenges: on the one hand, analysis of very small scale defects and features with the highest resolution (at nanometer and atomic scales) is required. Furthermore, in parallel with this miniaturization, new smart structures and complex materials are incorporated into devices. Also new transistor architectures that have appeared during the last years extended the classical "Moore" law towards the "More Moore" concept.

[1] To whom any correspondence should be addressed.

On the other hand, the functional diversification of semiconductor-based devices for derivative applications have also emerged, designed as "More than Moore" approach. In this case, more relaxed structures (like 70x70 µm Through Silicon Via (TSV), copper pillars, objects in 3D chips integration (3DI) scheme, imager sensors or MEMS…) have also to be characterized with dedicated techniques or tools.

2. Down-scaling of CMOS transistors in the "More Moore" approach

Following the "Moore law", device shrinking has continued for forty years from micrometric nominal dimension to about 40 nm. This was possible without great innovation of the CMOS architecture based on $Si/Si0_2$/poly-Si stack. Down-scaling of CMOS transistors for the 32/28 nm node and beyond now needs the introduction of new materials in the devices process' flow. For instance, concerning MOS gate, the implementation of a High-*k* (HK) dielectric with a metal gate (MG) stack is now required for advanced technology. Moreover, atomic mono-layers (known as capping layers in the "gate first" approach) of specific elements are used to tune the metal work function for N and PMOS transistor. Regarding backend (BE) flow, some impurities as aluminium (Al) or cobalt (Co) have to be added in copper (Cu) metal lines to limit electromigration phenomenon in interconnects and therefore ensure reliability of the chips.

So, new challenges have occurred and manufacturing processing have raised a need for development of advanced characterization techniques to address specific issues like HKMG, silicide, dopants profiling, stressors and interconnects characterization as described in the following paragraphs.

2.1. High-k Metal Gate (HKMG) and silicide characterization

The solution chosen for the 28 nm technology developed at the IBM Fishkill ISDA alliance (in which STMicroelectronics is a partner) is to incorporate: lanthanum (La) for NMOS and aluminium (Al) for PMOS at the metal (TiN)/High-*k* (HfSiON) interface. The quantitative measurement of the initial La and Al doses (about one monolayer) and their post anneal diffusion observation is crucial for the process development. These analyses thus require atomic resolution which fortunately is available in the TEM Cs corrected microscopes. Figure 1 presents HRSTEM images of a Si/SiON/HfSiON/La/TiN (gate) stack before and after 1050° C RTP annealing. The observations were carried out using the MINATEC's TITAN microscope integrating a Cs probe corrector (CEOS Company). The combination of bright field (BF) STEM image (blue), with dark field (DF) HAADF image (red), highlights the heavy and light elements distribution. It is shown that the annealing process induces diffusion and modification of the bottom SiON/HfSiON interface.

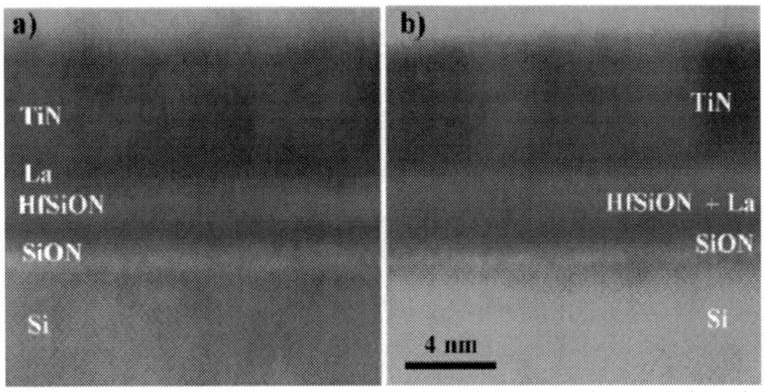

Figure 1. HRSTEM images combining HAADF signal (red) with Bright Field detector signal (blue): **a)** before annealing, **b)** after 1050°C RTP annealing. (TITAN 200 keV probe Cs corrected at CEA Grenoble-LETI MINATEC).

During the last years, analytical TEM developments (high brightness sources, SDD EDX detectors, fast EELS spectrometers) have drastically improved the STEM spectroscopy techniques. Fast chemical mapping with sub-nanometre resolution is now possible. STMicroelectronics have now a FEI TECNAI-Osiris on Crolles site, which integrates such technical advances: high brightness electron source (XFEG), four SDD windowless EDX detectors giving 0.9 srad collection angle (SuperX), and a Gatan FS1 Enfina EELS spectrometer [1]. Large EELS and EDX spectrum images can now be acquired in a short time using an electron probe with a small diameter (~0.3 nm) and a high beam current (1 nA). Figure 2 presents a STEM HAADF image of a cross section of a 28 nm SRAM bit cell cross section. A STEM-EELS map (top right) obtained using the Osiris TEM at 120 keV, shows the Hf, La and Si elemental distributions at the bottom of a NMOS gate. The lanthanum which was deposited on top of the HfSiON layer is now observed after annealing as distributed near the SiON/HfSiON bottom interface. This lanthanum diffusion creating a dipolar layer at the High-*k*/SiON interface is beneficial for the optimisation of the NMOS threshold voltage.

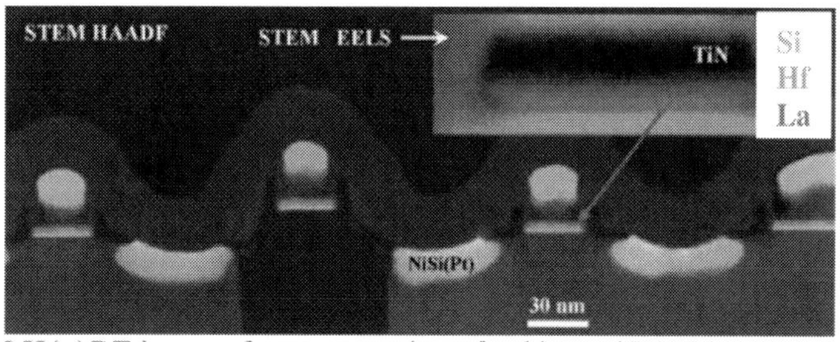

Figure 2. STEM HAADF image of a cross section of a 28 nm SRAM bit cell. At the right top a STEM-EELS map presents the Hf, La and Si distributions. (TEM Osiris at 120 keV).

As is shown in Figure 2, nickel silicide containing platinum (NiSiPt) is used for the 28 nm technology at gate, source and drain level. Past studies for 40 nm technology have shown anomalous nickel silicide growth [2] can induce electric failure. In order to diminish this effect for advanced technology, the platinum concentration inside the Ni target is increased (10 at % instead of 5 at %). However the effect of this higher Pt concentration on the silicide kinetics formation, the redistribution of dopants and platinum must be studied inside real devices. Figure 3 presents STEM-EDX maps of a 28 nm NMOS transistor showing in particular Ni, Pt and As distributions. The platinum is poorly incorporated inside the Ni silicide, whereas As atoms are rejected at external interfaces. This shows that the annealing steps are not efficient for the Pt incorporation inside the NiSi silicide. Figure 4 now shows STEM EDX maps of a laser annealed NiSi(Pt) layer. In this case, part of the platinum has been incorporated inside the silicide, particularly at grain boundaries. The arsenic is still rejected at the NiSiPt outsides interfaces. This confirms that laser annealing is promising for the 28-20 nm technologies, but other extensive studies will be needed in order to validate this assessment.

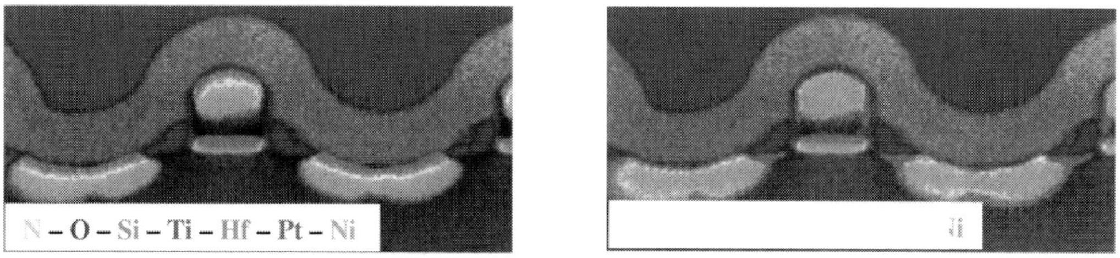

Figure 3. STEM EDX maps of a cross section of a 28 nm NMOS transistor showing mainly Ni, Pt and As distributions. The platinum seems to stay at the NiSi surface. (TEM Osiris at 120 keV).

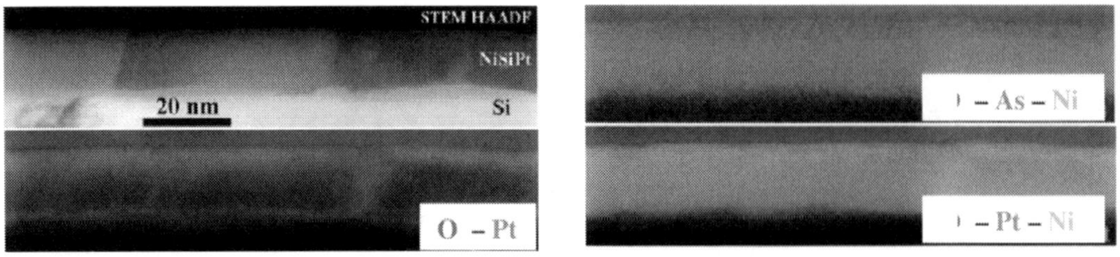

Figure 4. STEM EDX maps of a cross section of a NiSi(Pt) layer laser annealed. In this case platinum atoms incorporate better inside the NiSi(Pt), especially at grain boundaries.(TEM Osiris at 120 keV).

2.2 Dopants profiling at nanometer scale

Knowledge of the dopant (As, P, B) distribution in today's transistors junctions is necessary to improve their electrical properties. The detection and visualization of those impurities in nanoscale silicon devices has been highlighted by the ITRS[2] as a mandatory and difficult task for the development of future generation devices. Today, few characterization techniques are able to map dopant distribution both with a spatial resolution at the nanometer scale coupled with a high sensitivity.

Conventional transmission electron microscopy, despite atomic resolution, gives a too weak dopant contrast. However phase change contrast (revealed by holography technique) or inelastic interactions (measured in spectroscopy mode) can be exploited and provide alternative solutions. Using holography, it is possible to map the electrically active dopants [3]; however specific specimen preparation (parallel side milling, low amorphisation, no gallium contamination) using Focused Ion Beam (FIB) tool is mandatory to perform resolved electron holography experiments [4]. This technique is not used in our lab anymore since the specimen preparation and the minimum specimen thickness are not compatible with nowadays SRAM transistors size (< 100 nm).

Recent STEM, EELS and EDX spectroscopy developments (high brightness sources [5], new spectrometers, aberrations correctors…) place these techniques as candidates for sub-nanometer characterization of doped areas. However, sensitivity detection limit is obtained at the price of high local electron doses and silicon radiation damages at 200 keV become unavoidable. Those undesirable effects have been experimentally evaluated and it has been shown that they are significantly reduced if the electron probe acceleration is lowered to 120 keV or below [6]. During these studies, the STEM EELS technique, which is compatible with classical FIB lamella preparation, has been optimized on a TECNAI F20, for two-dimensional dopant maps from high density SRAM circuits. Using the Spectrum-Imaging approach [7], the first quantitative maps of arsenic dopant have been demonstrated from a 40 nm device gate length by recording more than 10 000 EEL spectra during an hour [1, 8]. Figure 5 shows that using this approach, arsenic dopant distribution from Source, Drain and Low Doped Drain (LDD) regions is revealed with a spatial resolution close to 2 nm and a dopant sensitivity in the low 10^{19} at.cm^{-3} (0.01%). This two-dimensional quantitative dopant map clearly reveals the presence of As inside the poly-silicon gate and local segregations at the gate edges are evidenced. Arsenic concentration profiles may be extracted across the LDD-Channel length to measure the chemical channel length. This kind of information is relevant for process engineering in order to optimize implantation conditions and, in consequence, maximize the devices electrical performances.

[2] www.itrs.net

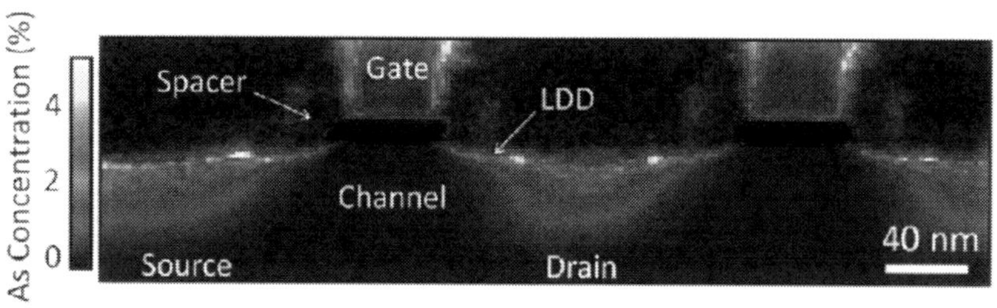

Figure 5. Two-dimensional quantitative arsenic dopant map acquired from a 40 nm gate length transistor using the STEM EEL spectrum imaging approach (Osiris TEM).

In the past EDX spectroscopy compared to EELS suffered from poor signal level detection. This has recently changed with the introduction of Silicon Drift Detectors (SDD). With the new S/TEM Osiris[3] [9], at STMicroelectronics Crolles, it is now possible to retrieve nanoscale chemical compositional analyses using STEM EDX. Figure 6 presents arsenic and phosphorous dopant maps acquired from a 40 nm transistor with a total of (800 x 400) pixels for a complete acquisition time of 20 minutes. Arsenic and phosphorus distributions from Source and Drain are clearly revealed using the STEM EDX approach with a high probe current (4 nA) and a probe size estimated lower than 1 nm. The dopant inside the gate seems to segregate at the poly-silicon gate edges. Following the example of STEM EELS, the STEM EDX approach leads to precise chemical channel length measurements inside "real" nanoscale silicon devices. Besides, it is interesting to note the presence of As and P at the top of nitrite spacers which played their role by protecting the LDD from the high energy implant of the Source and Drain regions.

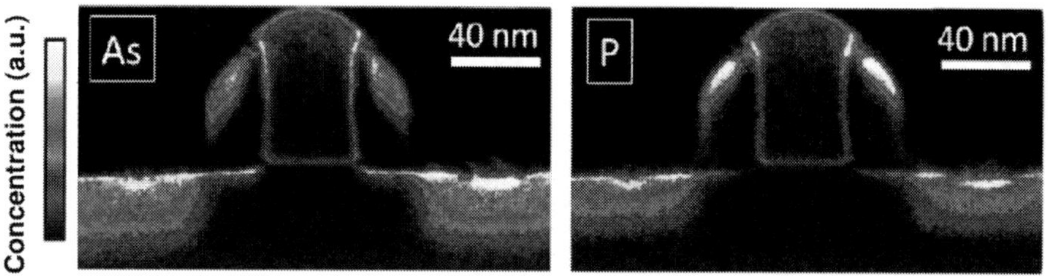

Figure 6. Two-dimensional (non quantitative) arsenic (left) and phosphorous (right) dopant maps acquired from a 40 nm gate length transistor using the STEM EDX spectrum imaging (Osiris TEM).

2.3 Stressors integration and characterization

To enhance carrier mobility and drive current in MOS transistors, several approaches based on strain engineering have been used in the semiconductor industry [10, 11, 12]. Unfortunately, till very recently, no physical characterization techniques were available to measure strain through the transistor channel with a sufficient sensitivity and spatial resolution. Considered for a long time as one of the most powerful techniques to measure lattice parameters [13, 14], Convergent Beam Electron Diffraction (CBED) technique suffers from some drawbacks largely described in the literature [15, 16, 17] leading to some difficulties to apply this technique in case of real devices analysis.

For a few years now, two TEM based techniques have emerged as very promising techniques for the semiconductor field: Nano Beam Diffraction (NBD) [18] and Dark-field holography (DF Holography) recently developed by the CEMES lab at Toulouse [19].

[3] The first worldwide S/TEM Osiris system has been installed at STMicroelectronics Crolles in April 2010.

The NBD technique uses a nearly parallel electron beam that can be obtained in a 2-condenser system by lowering the C2 aperture size down to about 10 μm. The corresponding semi-convergence angle classically obtained is 0.3 mrad with a probe size around 5-8 nm in a FEI Tecnai F20. Recent development on this technique using an FEI Titan with a 3-condenser system at CEA/Leti shows very promising results with strain sensitivity as low as $6x10^{-4}$ [20]. The main interest of this technique compared to CBED one is the device on-axis analysis capability. So, no sample tilt is required anymore and spatial resolution is then directly linked to probe size. Figure 7 shows NBD results obtained on a 45 nm PMOS transistor with SiGe source and drain stressors. Compressive strain through Si channel along x direction is clearly measured with the corresponding tensile strain along z direction.

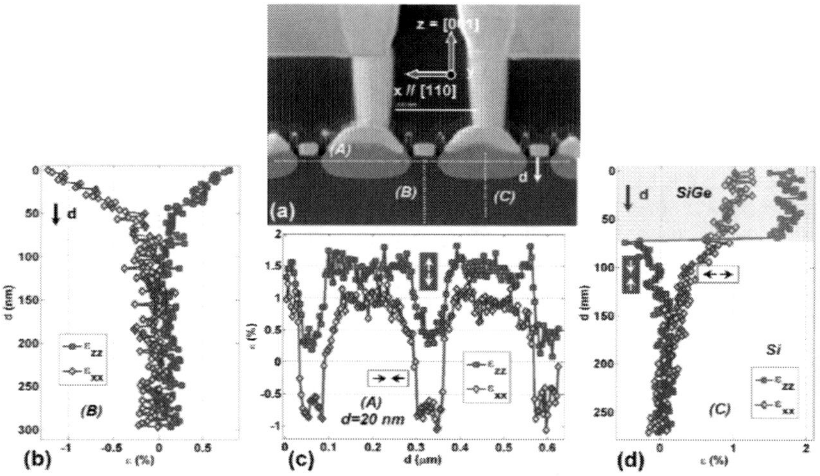

Figure 7. NBD results on a 45 nm metal gate PMOS transistor with SiGe source and drain. (a) STEM image of the investigated device. Strain components values along x and z directions along line scan (B) under the gate (b), along line scan (A) through source, channel and drain regions (b) and along line scan (C) just below one stressor (d).

However strain mapping remains difficult to obtain with this technique due to large and time consuming data processing. Furthermore, NBD technique is still tricky to apply on low stressed devices.

In this case, dark-field electron holography based on the analysis of a hologram obtained by the interference between one diffracted beam through the region of interest with another through an unstrained (reference) region, could be an alternative. The main advantage of this technique is the principal strain components mapping feasibility on real devices. STMicroelectronics is involved in the development of this technique for microelectronics field applications through the "HDStrain" ANR project carried out by the CEMES lab in Toulouse.

We have developed in particular dedicated sample preparation by coupling backside milling (to avoid curtaining effects) with low energy milling at 8 kV (to limit amorphous layers thickness) in dual beam system. The aim is then to obtain parallel sided lamella with no bending and uniform thickness through a large field of view to ensure high-quality dark-field holography experiments.

2.4 Interconnection issue

Concerning transistors interconnects (backend, BE) issue, one of a main failure modes in IC technology is electromigration (EM). This phenomenon is the migration of atoms under electron flow in metal lines leading to formation of voids (figure 8b) and thus to the loss of chip functionality. To characterize reliability of interconnects, dedicated EM structures (figure 8a) are tested in accelerated conditions (high temperature and current density) and monitored on resistance evolution with time. Some studies are currently performed to understand what drives EM (influence of copper microstructure?) and to find best process conditions to improve interconnect resistivity (surface treatment, impurities adding...).

The first relevant information after an EM test is the localization of the void(s) that need to be performed on very small features: 45 nm wide (W) and 100 nm thick (T) EM copper line geometry for the 28 nm node technology (figure 8a). In case of SEM imaging, interconnect is revealed by classical FIB cut whereas in case of TEM imaging, lamella preparation is required with thickness less than 45 nm in this case to avoid any 3D effects as illustrated in Figure 8b.

a) 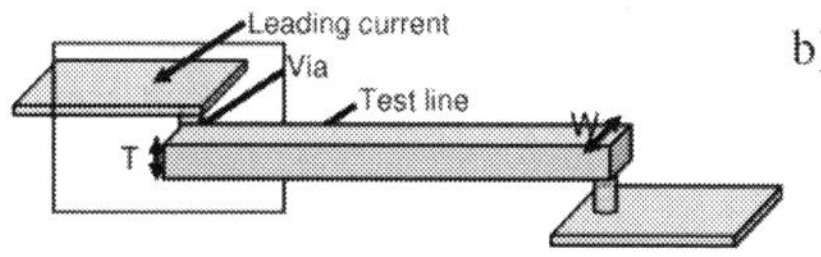

b) 

Figure 8. a) Electromigration test structure sample **b)**TEM image of copper interconnect with small void localized under the via.

Some experiments are actually performed [21] to show if local microstructure inside a metal line can influence or not electromigration mechanism. For this, electron back-scattered diffraction (EBSD) technique in an SEM is used to obtain texture and grain size information.
Figure 9 illustrates an example of global microstructure analysis for 140 nm wide copper lines where changing copper deposition process results in much higher (111) texture but also much smaller mean grain size. This explains why reliability performances decrease for this process variation.

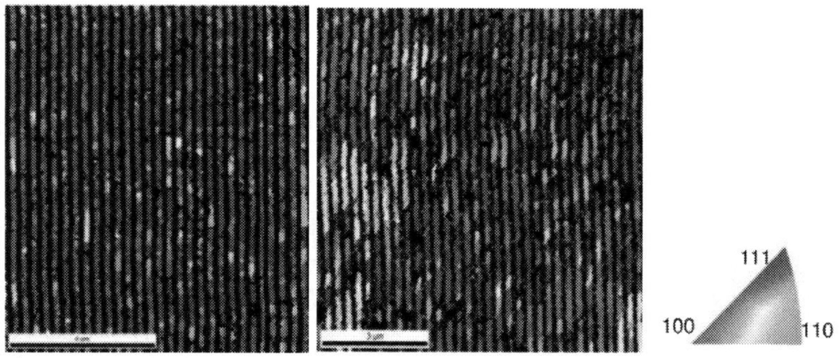

Figure 9. EBSD maps for two different copper deposition conditions (reference on the right)

Unfortunately, drift issue induced by charge effect is a major concern for EBSD experiments due to the presence of dielectric material separating copper lines. Thus a compromise must be made between mapping size and resolution to restrain this drift. Moreover, limits of EBSD technique begin to appear for these very small dimensions where grain size (mainly defined by line width) is comparable with SEM spot size in EBSD mode (with a 70° tilted stage). An alternative is to define process dedicated to EBSD analyses to limit sample preparation which can be complex (with top-down polishing) and take advantage of cleanliness of wafer going out of manufacturing.

To further understand EM phenomenon, we also need to access local information of void environment and ideally at the beginning of the void nucleation process step. In this case SEM-EBSD technique is not sufficient anymore and must be coupled with (S)TEM technique allowing a more precise void localisation (as illustrated in figure 10) [22,23]. Through this example we demonstrate the interest in combining several characterization techniques and get complementary information. The localization of a void on EBSD maps (thanks to STEM image) shows that its place corresponds to multiple boundaries with five grains common in void.

Figure 10. EBSD map and STEM image of tested interconnect in longitudinal cross section at the beginning of void growth. Red circles on EBSD map show possible void places. Red circle on STEM image highlights the real position of the void.

For such analysis, sample preparation has become a key parameter to ensure relevant results. As STEM imaging requires a lamella preparation, we have to take into account possible damage induced by FIB on crystallography. So an optimum FIB voltage and current must be found to make compatible sample preparation with EBSD analysis.

Finally, it seems obvious that for actual developed technology, EBSD performed in SEM is limited. We must consider TEM imaging for highly resolved microstructure characterization. A commercialized system known as ASTAR allows performing equivalent of EBSD with indexing by template matching in a TEM. This system combines advantages of EBSD (with the TEM spatial resolution) to determine orientation grain, misorientation and localize small defects like voids at the beginning of growth which is less sensitive to sample preparation, contrary to EBSD.

2.5 *Advanced morphology characterization : towards 3D visualization?*

With even smaller dimensions of devices being developed in the semiconductor industry, physical characterization faces a challenge to provide information about the structures being developed: a relevant TEM analysis relies on the ability to prepare TEM lamellae with a thickness less than the feature to be observed, to not get projection effects of overlapping structures. For advanced technology nodes, such a sample preparation becomes a critical operation to pinpoint the exact location of the device, while thinning down a lamella to a few tens of nanometers.

As it allows analyzing specific features within 'large' volumes without projection effects, Electron Tomography (ET) overcomes these hard points in the physical characterization process. In the ET process, images of the sample to characterize are recorded in HAADF STEM mode every 1°, along a large tilt range, [-90°; +90°]. The obtained tilt series is then processed using the FEI Inspect3D™ software (for image alignment and tilt axis definition), to finally reconstruct a three dimensional volume from the aligned series using either weighted or iterative back projection techniques. Volume rendering and slice extraction from this virtual volume is then performed using ResolveRT™ software; it is thus possible to isolate a specific feature within the virtual volume without projection effect.

As the sample containing the feature of interest is tilted to record the ET series, the effective thickness of a TEM lamella as seen by the incident beam would drastically increase with the tilt angle, with a result of poor information extracted from the recorded images. Therefore, a specific sample preparation dedicated to ET is preferred to conventional TEM lamellae [24]: a chunk with the feature of interest is prepared and placed on top a probe, and then milled to obtain a cylinder-shaped sample (to keep a constant effective thickness all along the tilt range). The probe is then used in a specific sample holder, compatible with such specimens and allowing a full tilt range without shadowing effect (no 'missing wedge' effect).

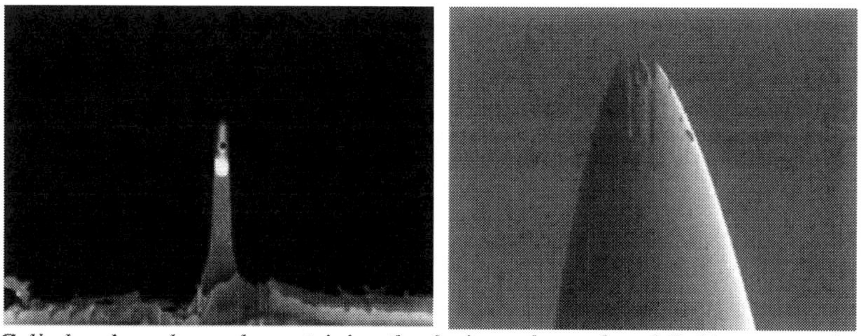

Figure 11. Cylinder-shaped sample containing the device to be analyzed (left) – mounted on top of a probe holder (right).

With materials in the devices to be analyzed presenting large variation in atomic number Z, images recorded in HAADF STEM mode exhibit a large contrast between high Z and low Z compounds. Therefore, the camera length, i.e. the sample-to-detector distance, needs to be tuned to get a correct ratio between high Z and low Z signals, with the constraint to restrict it to the lower acceptable value, to not get diffraction contrast(that is tilt-angle sensitive). From an experimental point of view, such a compromise may be difficult to achieve, with the result to get poor information from compounds of low Z in the reconstructed volume. As a consequence, ET appears to be more adapted to characterize back-end interconnected structures (with metallic compounds such as Cu, Ta, Ti, Al) rather than front-end structures, where a large variety of compounds - from Si, C, N to Hf, La, W, will alter the quality of the reconstructed volume) [25].

Figure 12 presents an ET study of a shared contact in a SRAM cell in C45 technology node. The shape of this tungsten plug can be easily resolved, as well as the NiSi silicide layer on top of the gate and under the plug. On the contrary, almost no contrast difference is observed between silicon (bulk), polysilicon (gate), SiCN (spacers), SiOC (dielectric). This example highlights the advantages of the technique to characterize metallic interconnects, and its limits where materials of low Z/high Z co-exist.

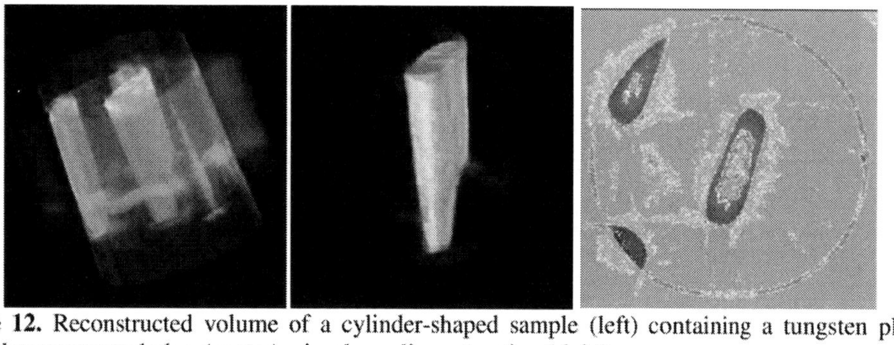

Figure 12. Reconstructed volume of a cylinder-shaped sample (left) containing a tungsten plug – isolated reconstructed plug (center) – in-plane slice extraction (right).

3. Dedicated physical characterization techniques for "More than Moore" approach

3.1. FIB/SEM 3D for 3DI application

For some devices to be analyzed, physical characterization through either surface or cross section observation of the structure may not be sufficient to provide relevant results. As a fact, the need for physical characterization all along the volume of the device pushes for alternative solutions, such as FIB/SEM 3D.

This technique relies on the use of a dual column system, with the consecutive use of SEM imaging and FIB milling (figure 13): the structure is peeled in multiple cross sections, using the FIB to guarantee a constant thickness of the slices milled. After every milling operation, a SEM image of the physical cross section is recorded. The final data stack, that consists of typically several hundreds of images, will be treated (images alignment, volume filtering) to obtain a reconstructed volume. This allows 3D visualization and cross sections in the 3 directions (one real and two reconstructed).

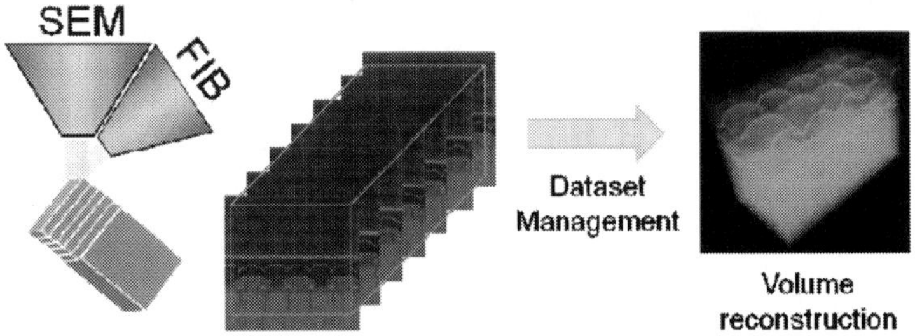

Figure 13. FIB/SEM 3D principle.

To illustrate the technique, an example of 3D integration device is presented figure 14. Reliability tests have been performed on a (2x15 µm) Through Silicon Via (TSV) with cycles of heating ramps. This particular TSV has failed the electrical test. The origin of the failure is assumed to be due to the presence of a void in the metal pad under the TSV. The goal of this analysis is therefore to confirm the presence of the void and characterize it (location, volume). Exact location of the void with respect to the TSV can be given as well as the surface within the pad and the volume of the void. These provide useful data about the failure mode of the TSV.

In addition to the 3D information provided, FIB/SEM 3D demonstrates the advantage of top-down virtual deprocesssing without any induced artifact; as a fact, such data would not have been obtained using step by step top-down deprocessing of this structure by mechanical polishing of the surface, as the void would have been filled during the polishing process and the structure would have been degraded.

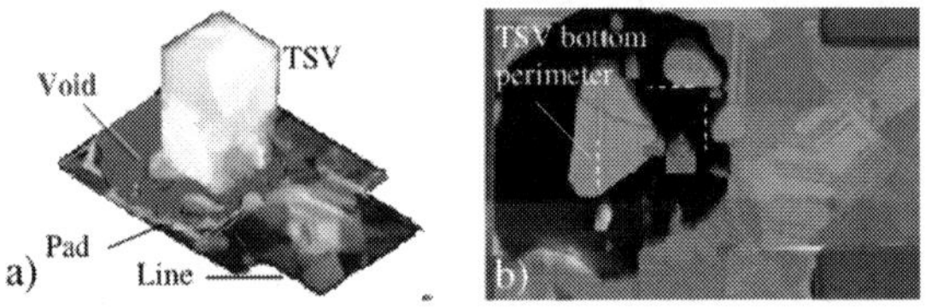

Figure 14. a) 3D reconstruction of a (2x15 µm) TSV and copper line b) Reconstructed planar section (TSV bottom) [26].

Devices with micrometer scale dimensions are well adapted for the FIB/SEM 3D technique. As the technique is based on SEM imaging, the technique faces the same limitations of conventional SEM imaging: the achievable resolution restricts the use of such a technique to a minimum voxel size of 2^3 nm^3. Furthermore, mechanical instabilities even make data acquisition at high magnification very difficult. Consequently, FIB/SEM 3D is better suited for devices in the 100 nm to 30 µm range, while other 3D techniques such as E-beam tomography or X-ray tomography fill the gap for devices of, respectively, lower or higher dimensions.

3.2. Dopant characterization in relaxed structures

Scanning Capacitance Microscopy (SCM) and Scanning Spreading Resistance Microscopy (SSRM) are two AFM based techniques able to characterize doping through respectively capacitance and resistance measurement. The detection limit of these techniques scales from 10^{15} at.cm^{-3} to 10^{20} at.cm^{-3}. The advantage of both SCM and SSRM technique is the very high dynamic range of the detection limit compared to TEM based techniques like EELS or EDX. However, they suffer from difficult sample preparation. Indeed, for such experiments, the sample surface must be cleaned with no damage, so the use of FIB (which induces amorphisation and ion implantation) is forbidden for cross-section analysis. That is why these techniques are currently dedicated to relaxed structures (~ μm) like imager sensors characterization (figure 15) for which sample preparation by simple cleaving is possible.

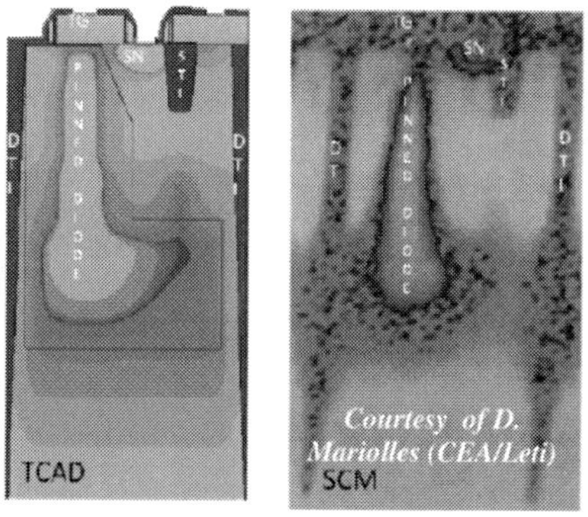

Figure 15. SCM map on vertical junction in a relaxed imager structure performed at CEA/Leti. Distance between two consecutive DTI (Deep Trench Isolation) corresponds to pixel size (~1.5 μm). TCAD simulation and SCM map are in very good agreement.

4. Sample preparation

In semiconductor industry, one of today's lab challenges is to provide in a volume mode and with shortest turn-around-times, high quality sample preparation methods allowing advanced physical characterizations of both very small and large scale features.

4.1. Sample preparation for small scale features

In case of small scale features, dual beam systems (FIB/SEM) are the preferred solutions serving failure analysis and process control applications (Figure 16). These tools are used very extensively. After localisation of the site of interest, cross-sections can be FIB-cut manually to obtain very precise single cross-sections or automatically with successive slices to provide 3-D information on defects or integrated features. They offer defect localisation within failing structures like memory cells and SEM characterizations capability without time consuming sample pre-preparation and without handling the sample after cutting.

Furthermore, if SEM resolution is insufficient to reveal the defect origin or the necessary process details, it is possible to prepare TEM samples using a controlled, easy to learn in-situ process and to efficiently continue the characterization with a high resolution TEM. FIB/SEM systems have revolutionized sample preparation for TEM. Several FIB based techniques exist to perform extraction of a TEM lamella: the conventional technique (using a sawed bar extracted from the wafer), the ex-situ and in-situ lift-out techniques. The in-situ lift-out technique is the privileged one providing lamella free from material re-deposition, compatible with plasma cleaning and if necessary, further lamella thinning. When the lamella has been carefully prepared, TEM imaging is the final step of the analysis, TEM results can be available within 5 to 6 hours.

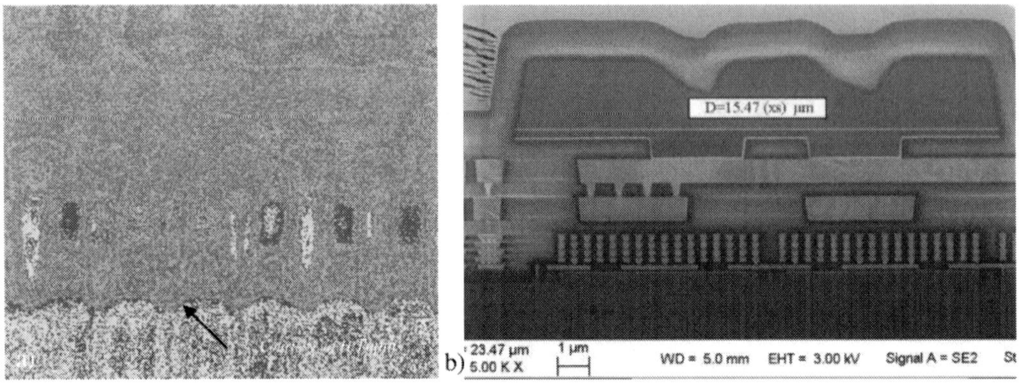

Figure 16. **a)** SEM view showing an open contact in 28nm cell memory **b)** general SEM view of metal stacks in 45nm products.

The continuous development of ion columns leads to sample preparation improvement. In fact, the use of low energy milling can now be used easily in dual beam tools to limit the amorphous layer thickness on both sides of the TEM lamella. Alternative methods for sample preparation are also under investigation like backside milling to produce very thin parallel-sided lamellae for specific application like holography in TEM.

4.2. Sample preparation for large scale features

For the larger structures, like 70 µm x 70 µm Through-Silicon-Vias (TSV's), bumps and copper pillars that are integrated in 3DI technologies or new packaging solutions, standard dual beams are not appropriate because the gallium based FIB milling rate (from 0.2 to 2.0 mm^3/s) does not allow to remove high material volumes.

The use of conventional mechanical polishing techniques and / or cross-polishers is more suitable. The mechanical polishing action can induce scratches and other artifacts that reduce image quality. That is the reason why cross-polishers (CP) are now commonly used in semiconductor lab's for larger scale features (figure 17). Use of broad argon ion beam eliminates this problem and can provide wide section fields of view (up to 2 mm), high quality sections free from artefacts for SEM imaging and EDX analysis if necessary (figure 18). The main drawback of methods using cross-polisher is that it requires several sample pre-preparation steps, essentially if the sample surface is not planar or protected:

1-sample sawing
2-capping with resist
3-mechanical polishing to approach the site of interest and
4- cross-polishing.

Therefore, the sample preparation is time-consuming (from 8 hrs to 12 hrs, depending on the section height). Another disadvantage of the method is that it allows neither successive sections nor SEM imaging to localize defects nor 3D-information. It is well appropriate for random construction analysis of large feature which does not require precise localisation (<15µm).

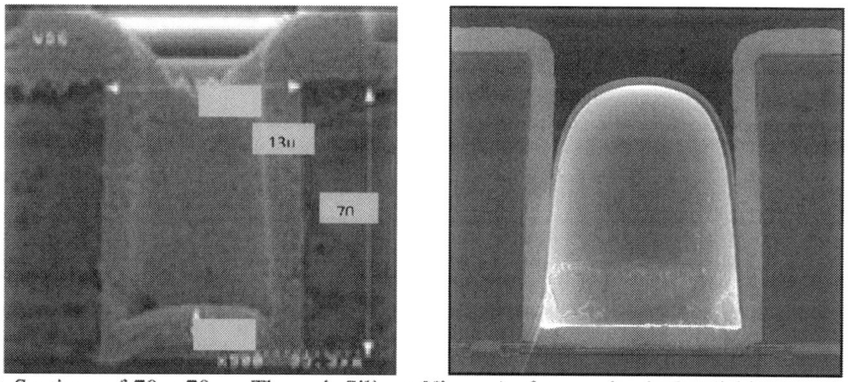

Figure 17. Sections of 70 x 70µm Through Silicon Vias : a) after mechanical polishing b) after final milling using a cross-polisher (high quality sample surface allowing high quality characterizations).

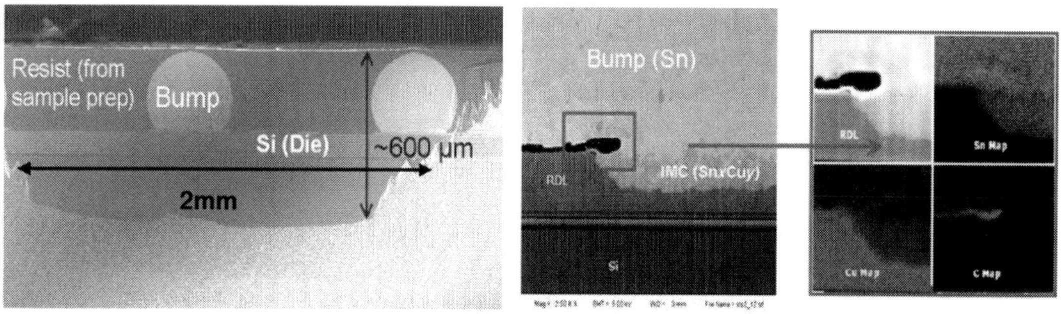

Figure 18: a) General SEM view of a cross-polisher section integrating a few bumps, b) zoom of bump base with EDX analysis of the RDL/bump interface.

To overcome these drawbacks (lack of precision, time consuming sample preparation, no capability to offer 3D information) semiconductor plants are very interested in new high current FIB milling that, recently, has been introduced to market. The FIB columns integrated in new systems are capable of high material removal rates at wider apertures (beam currents of 60-65nA). The SEM characterisation of a 80µm deep TSV as shown in figure 19b would take around 12hrs with standard dual-beams; with the new generation tools, the time is ~4 hrs.

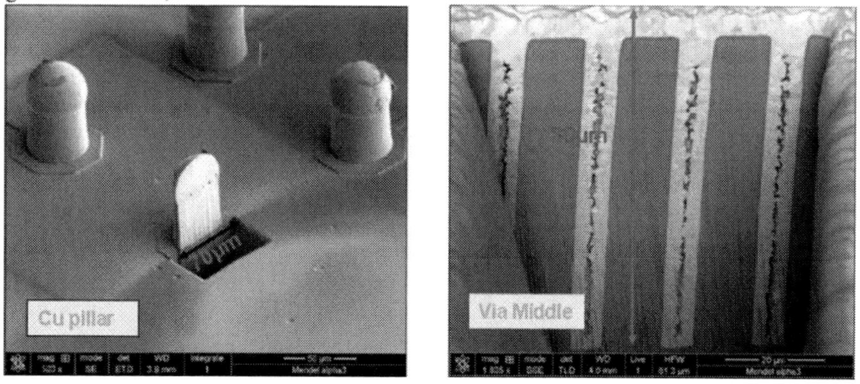

Figure 19. a) SEM views of Cu pillars, b) high aspect ratio vias cut with a high ion beam current (FEI Helios 450 dual beam).

5. Conclusion

In this paper an overview of the different SEM and TEM based techniques to address specific issues at STMicroelectronics have been shown. Some needs have been raised for the future both in term of high resolution (at atomic scale in TEM) to measure very thin layers but also in terms of detection sensitivity for chemical analysis. Stressors integration in future technology will also require advanced techniques to precisely measure strain in the transistor channel of real devices.

Sample preparation (for both SEM and TEM) has now become a key parameter to ensure physical characterization reliability leading to some developments on FIB/SEM systems for both small and large features to analyse. We have also shown that 3D analysis (with E-beam tomography) could be an alternative for the future to avoid 3D effects in TEM lamella for very small structures characterization.

References

[1] Pantel R, Servanton G, Delille D, Klenov DO 2010. *Proceedings from the 17th International Microscopy Congress* (IMC17), Rio de Janeiro, September 19-25
[2] Imbert B 2010. *Microelectronic Engineering* **87**:3 294-297
[3] Rau WD, Schwander P, Baumann FH, Hoppner W, Ourmazd A 1999. *Physical Review Letters* **82** 2614–2617
[4] Cooper D et al., 2007. *Appl. Phys. Lett.* **91** 143501
[5] Freitag B et al. 2008. *Proceedings of the 14th European Microscopy Congress*
[6] Servanton G, Pantel R, Juhel M, Bertin F 2010. *Proceedings of the 16th Conference Microscopy of Semiconducting Materials*, March 2009, *Journal of Physics: Conference Series* **209** 012044
[7] Colliex C, Tence M, Lefevre E, Mory C, Gu H, Bouchet D and Jeanguillaume C 1994 *Mikrochimica Acta* **114** 71-87
[8] Servanton G, Pantel R, Juhel M and Bertin F 2010. *Micron* **41** 118-122
[9] Von Harrach HS et al. 2010. *Microscopy and Microanalysis* **16**:S2 1312-1313
[10] Lee ML, Fitzgerald EA, Bulsara MT, Currie MT, Lochtefeld A 2005. *J. Appl. Phys.* **97** 011101
[11] Thompson SE, Sun GY, Choi YS, Nishida T 2006. *IEEE Trans. Electron. Dev.* **53** 1010–1020
[12] Payet F, Boeuf F, Ortolland C, Skotnicki T 2008. *IEE Transactions on Electron Devices* **55** 4
[13] Armigliato A, Balboni R, Carnevale GP, Pavia G, Piccolo D, Frabboni S, Benedetti A, Cullis AG 2003. *Appl. Phys. Lett.* **82** 2172
[14] Zhang P, Istratov AA, Weber ER, Kisielowski C, He H, Nelson C, Spence JCH 2006. *Appl. Phys. Lett.* **89** 161907
[15] Clement L, Pantel R, Kwakman LTTZ, Rouviere JL 2004. *Appl. Phys. Lett.* **85**:4 651-653
[16] Houdellier F, Roucau C, Clement L, Rouviere JL, Casanove MJ 2006. *Ultramicroscopy* **106** 951-959
[17] Clement L, Cacho F, Pantel R, Rouviere JL 2009. *Micron* **40**:8 886-893
[18] Usuda K, Mizuno T, Tezuka T, Sugiyama N, Moriyama Y, Nakaharai S, Takagi S 2004. *Appl. Surf. Sci.* **224** 113
[19] Hytch M, Houdellier F, Snoeck E 2008. *Nature* **453** 1086-1089
[20] Beche A, Rouviere JL, Clement L, Hartmann JM 2009. *Applied Physics Letters* **95** 123114
[21] Galand R, Haxaire K, Arnaud L, Petitprez E, Clement L, Waltz P and Wouters Y 2010 *Microelectronic Engineering* **88**:5 661-665
[22] Galand R, Clement L, Waltz P and Wouters Y 2010. *AIP conference proceedings* **1300** 47-54
[23] Galand R, Clement L, Waltz P and Wouters Y 2010. *Proceedings of SPIE* **7729** 77290B Scanning Microscopy
[24] Lorut F and Delille D 2008 *Microscopy and Microanalysis* **14**:S2 1002-1003
[25] Lorut F et al. 2008. *Proceedings from the 34th International Symposium for Testing and Failure Analysis* 499-504
[26] Frank T, Chappaz C, Leduc P, Arnaud L, Lorut F, Moreau S, Thuaire A, Farhane RE, Anghel L 2011 IITC

Quantitative Determination of Nanoscale Electronic Properties of Semiconductor Surfaces by Scanning Tunnelling Spectroscopy

R M Feenstra[1] **and S Gaan**

Department of Physics, Carnegie Mellon University, Pittsburgh, PA 15213, USA

Email: feenstra@cmu.edu

Abstract. Simulation of tunnelling spectra obtained from semiconductor surfaces permits quantitative evaluation of nanoscale electronic properties of the surface. Band offsets associated with quantum wells or quantum dots can thus be evaluated, as can be electronic properties associated with particular point defects within the material. An overview of the methods employed for the analysis is given, emphasizing the critical requirements of both the experiment and theory that must be fulfilled for a realistic determination of electronic properties.

1. Introduction

The scanning tunneling microscope (STM) enables direct imaging of structures on semiconductor surfaces with lateral resolutions on the order of 0.3 nm and vertical resolution of 0.01 nm or better [1]. It is well known however that the images actually reveal the electronic charge density (integrated over an energy window with width on the order of an eV on either side of the Fermi energy) [2,3]. Hence, deriving atomic structures and/or compositions is challenging. Comparison of the measured results to theoretical predictions, over a range of sample-tip bias voltages, has proven to be an effective means of determining structural models. Spectroscopic measurements, i.e. scanning tunneling spectroscopy (STS), provide important additional input for this type of analysis [4,5,6]. For semiconductor surfaces with complex reconstructions, the aforementioned methods provide a powerful means of determining the structure of the surfaces. (For *elemental* semiconductors, structures can sometimes be deduced directly from STM images, but for *compound* semiconductor the added complexity demands the use of theoretical predictions of structures [7]).

A somewhat different class of problems that can be investigated by STM/S is associated with surfaces for which the arrangement of surface atoms is quite simple and for which there is *not* a high density of surface states throughout the band gap, e.g. the nonreconstructed (110) surfaces of cleaved III-V semiconductors [8] or passivated surfaces such as H-terminated Si(001) [9]. In many cases, large well-ordered regions of such surfaces can be prepared, and then the presence of isolated defects, often intentionally introduced into the structure, can be studied. Also, heterostructures of e.g. III-V semiconductors can be cleaved and viewed in cross-section, thus revealing quantum wells, quantum dots, or other similar nanostructures [10].

In these sort of surfaces with relatively simple structure and low density of surface states, an important effect that occurs in STM/S is tip-induced band bending – the extension of the electric field

[1] To whom any correspondence should be addressed.

across the vacuum gap into the semiconductor as illustrated in figure 1. This tip-induced band bending can have dramatic effects on the spectra, not only producing shifts in the position of spectral lines or edges, but also greatly affecting the magnitude of the tunneling current or conductance over certain voltage ranges. In order to obtain accurate energy values from spectra (e.g. for spectral lines, band offsets, etc.) it is necessary to quantify the effects of tip-induced band bending on the spectra.

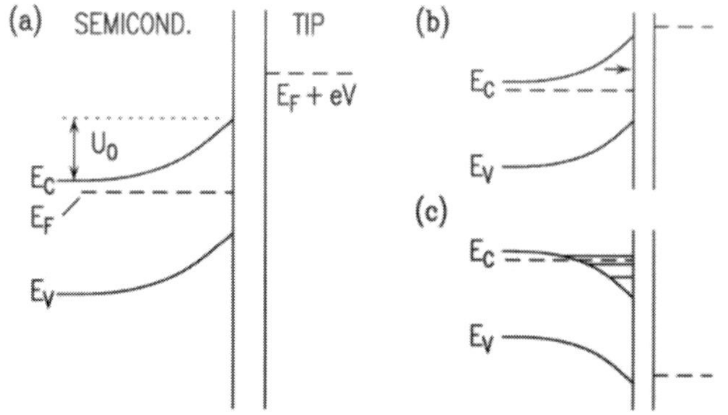

Figure 1. (a) Schematic diagram of energy bands with tip-induced band bending, showing the valence band maximum E_V and the conduction band minimum E_C. The sample Fermi level is denoted by E_F with the tip Fermi level at $E_F + eV$ where V is the applied sample-tip bias voltage. The band bending at the surface is denoted by U_0. Quantum effects within the semiconductor are illustrated in (b) and (c) for wavefunction tailing (arrow) through a depletion region and for localized accumulation state formation, respectively.

Early studies of semiconductor surfaces permitted computation of tip-induced band bending in a qualitative or semi-quantitative manner [8,11-13]. In the past decade a realistic method for obtaining the band bending in three dimensions (3D) [14] has been developed, and associated computer codes are available [15]. With those electrostatic potentials, computations of tunnel current are relatively straightforward within an effective mass (envelope function) treatment. Two methods have been developed for such computations, one employing a planar tunneling geometry [16] and the other treating a nonplanar tip with a plane wave expansion method [17]. The former yields a detailed spectrum of states with arbitrarily fine energy spacing, although energy levels are approximate since the planar geometry has been assumed (a proper nonplanar tip is used in computing the potential, but not for the current). The latter method permits a correct treatment of the tip geometry, but because of computational limitations in the plane wave expansion, only a small (≈10 nm on a side) region in the semiconductor around the point opposite the tip apex can be treated. Nevertheless, the results of the two methods are in good agreement for situations where the radial dependence of the potential is not too large (i.e. when no region of laterally localized potential exists in the semiconductor).

In this paper we provide an example of computed tunneling spectra for a semiconductor system, namely InAs quantum dots in GaAs. We demonstrate how, through fitting of the computed spectra to experiment, the tip-related parameters such as tip radius, tip-sample separation, and tip-sample contact potential (work function difference) can be determined. Furthermore, additional features in the observed spectra, discrete states in quantum dots in the present case, can be derived from the observed spectra. We also briefly discuss limitations in the computational methods, particularly those associated with non-equilibrium occupation of carriers in the semiconductor.

2. Experimental Methods

In addition to the usual requirements in any STM/S experiment of stable operation of the microscope and clean metallic probe tips, comparison of experiment to theory for detailed semiconductor spectroscopy has several additional experimental requirements. First, it is crucial that the spectra have high dynamic range, i.e. extending over 3 – 4 orders of magnitude at least, in order to permit a comprehensive comparison with computed results. As illustrated many years ago, band edges are in general *not* well defined for spectra that extend only over 1-2 orders of magnitude [8]. One convenient means of achieving this high dynamic range, and simultaneously acquiring the data in a short period of time (a few seconds), is to employ variable tip-sample spacing during the measurement [18]. The separation is varied as $s = s_0 + \Delta s$ with $\Delta s = a|V|$, using a value for a of about 0.1 nm/V. For comparison of theory to experiment, the data is normalized to constant separation using a multiplicative factor of $\exp(2\kappa\Delta s)$, where a voltage-averaged value of κ obtained from experiment is employed. Ideally the κ value would be close to a theoretically ideal value of $\approx$10 nm^{-1}, although in some cases lower κ values are found in the experiments. One possible reason for this nonideality are transport limitations in the current, further discussed in section 4. In any case, so long as the κ value does not deviate too much from an ideal one, then this method of normalization seems to be adequate. (A normalized conductance of the form $(dI/dV)/(\overline{I/V})$ is very convenient for *qualitative* display of experimental data [6,18], but this normalization itself can somewhat distort the data so that for detailed quantitative analysis we find the conductance at constant separation to be a better quantity).

A second important requirement for the experimental data is that spatially resolved spectra must in general be acquired, both at some typical location of the bare semiconductor surface and at specific locations near the region of interest on the surface. The former spectra are needed in order to enable determination of the tip-related parameters – tip radius, tip-sample separation, and tip-sample contact potential. With those, then spectra acquired near the region of interest can be used to determine values of additional parameters associated with that region. The simultaneous need for high dynamic range and spatially resolved spectra is a significant demand in STM/S work, since thermal drift in the instrument and/or unintentional changes in the tip structure can lead to incomplete data sets (use of a low temperature STM can aid in minimizing these instrumental difficulties).

3. Computational Methods

The geometry of the problem consists of a tip-vacuum-semiconductor junction. To obtain the electrostatic potential a finite-element method has been developed that can efficiently solve the 3D problem including possible occupation of electronic states on the surface or in the bulk of the semiconductor [12]. The solution is found iteratively, thus permitting evaluation of Poisson's equation including a possible high degree of nonlinearity in the charge density. In the absence of any electrostatic potential in the semiconductor, the charge density for a bulk band is denoted by $\rho(E_F)$ and for a surface state by $\sigma(E_F)$ where E_F is the Fermi energy (the temperature dependence of ρ, σ, and E_F is fully included in our computations but are not indicated here for ease of notation). In the presence of an electrostatic potential energy $U(x,y,z)$ (i.e. $-e$ times the electrostatic potential where e is the fundamental charge), the charge densities are given in a *semi-classical* approximation simply by $\rho(E_F - U(x,y,z))$ for a bulk band or $\sigma(E_F - U(x,y,0))$ for a surface band. The electrostatic potential energy is given by Poisson's equation

$$\nabla^2 U = \frac{\rho(E_F - U(x,y,z))}{e\varepsilon\varepsilon_0} \qquad (1)$$

where ε_0 is the permittivity of vacuum, ε is the relative permittivity (dielectric constant) of the semiconductor, and with $\rho = 0$ in the vacuum. The boundary condition at the surface is given by

$$\frac{\partial U}{\partial z}(x,y,0_+) = \varepsilon \frac{\partial U}{\partial z}(x,y,0_-) + \frac{e\sigma(E_F - U(x,y,0))}{\varepsilon_0} \tag{2}$$

with the semiconductor spanning the half-space $z<0$. For the finite-element evaluation, in the vacuum a generalization of prolate spheroidal coordinates (ξ,ϕ,η) are used so as to match the shape of the probe tip [19], and in the semiconductor cylindrical coordinates (r,ϕ,z) are used. In both cases the spacing between grid points is not constant but increases both with the radius r in the semiconductor (or with ξ in the vacuum) and with the depth z into the semiconductor. In this way, solutions out to very large (effectively infinite) values of those coordinates can be obtained, with $U \to 0$ at those large values. On the surface of the probe tip $U = eV + \Delta U$ where V is the sample-tip voltage and

$$\Delta U = \phi_m - \chi - (E_C - E_F) \tag{3}$$

is the contact potential (work function difference) between tip and sample, where ϕ_m is the tip work function, χ is the semiconductor electron affinity, and $E_C - E_F$ is the separation of conduction band (CB) edge deep inside the semiconductor from the Fermi energy.

The iterative method that has been developed permits solution of equations (1) and (2) for any specified set of bulk and/or surface bands. Also, solutions are possible not only for the semi-classical case described in those equations but also for a fully quantum mechanical treatment of particular states, as needed e.g. for a self-consistent treatment of accumulation or inversion situations in the semiconductor [20] (in which case the bulk charge density is expressed as *functional* of $U(x,y,z)$).

An example of a computed electrostatic potential is shown in figure 2, for the case of an InAs quantum dot (QD) embedded within GaAs [17]. The presence of the QD does not affect the potential (assuming equal dielectric constants for the InAs and GaAs), but it *does* affect the tunnel current by introducing electron and hole states within the GaAs band gap. The tip apex in figure 2 is positioned 4 nm from the center of a lens-shaped QD (shown in grey), and a strip of charge density associated with a surface step is located 6 nm on the other side of the QD (the physical step is not represented in the theory, only the charge density associated with the step). A constant state-density throughout the bandgap arising from the step is assumed, with value 2.5 nm^{-1}eV^{-1} and charge-neutrality level 0.25 eV above the VB maximum [17].

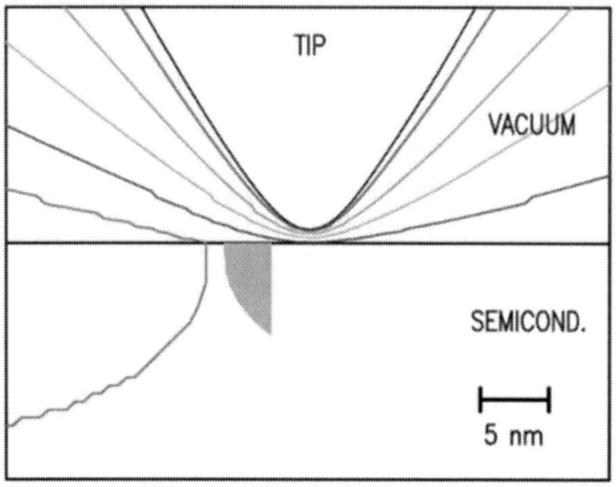

Figure 2. Electrostatic potential energy for a 2-nm radius-of-curvature probe-tip located 1.0 nm from a semiconductor surface. The sample-tip voltage is set at +1.0 V and the contact potential between tip and sample is -0.87 eV, so the electrostatic potential energy of the tip relative to a point deep inside the semiconductor is +0.13 eV. Contours are shown for potential energies (eV) of 0.155 (nearest to the tip: red), 0.232 (green), 0.310 (cyan), 0.388 (blue) and 0.465 (furthest from the tip: magenta). At distances further inside the semiconductor the potential falls gradually to 0 eV. (From Ref. [17]).

The above method for evaluating the electrostatic potential provides an exact solution. In contrast, evaluation of the tunnelling current requires some approximations. It should be noted that quantum effects within the semiconductor as illustrated in figures 1(b) and (c) can be very important in determining the current, so a treatment of the problem using e.g. only a transmission factor for the tunneling through the vacuum would be very unrealistic. It is necessary to compute the wavefunctions through both the vacuum and the semiconductor region where the potential is varying. From the wavefunctions, the current can be obtained using the Bardeen formalism [21], written for the case of a sharp tip by using the Tersoff-Hamann approximation [3,17],

$$I = \frac{8\pi\hbar e}{m} R^2 k_F \sum_{\mu} \left[f(E_\mu) - f(E_\mu - eV) \right] \left| \psi_\mu(0,0,s) \right|^2 \tag{4}$$

where m is the free-electron mass, R is the tip radius-of-curvature, k_F is its Fermi wavevector, E_μ is the energy of an eigenstate of the sample, $f(E_\mu)$ is the Fermi-Dirac occupation factor for that state, $f(E_\mu + eV)$ is the occupation factor for the corresponding state in the tip, and $\Psi_\mu(0,0,s)$ is the wavefunction of the sample state evaluated at the position of the tip apex: $x=0$, $y=0$, and $z=s$ where s is the tip-sample separation. The wavefunctions are evaluated within an effective mass (envelope function) approximation, writing Schrödinger's equation for a CB as

$$-\frac{\hbar^2}{2m_e}\nabla^2\psi_\mu + \left(E_C + U(x,y,z)\right)\psi_\mu = E_\mu \psi_\mu, \tag{5a}$$

for a valence band (VB) as

$$+\frac{\hbar^2}{2m_h}\nabla^2\psi_\mu + \left(E_V + U(x,y,z)\right)\psi_\mu = E_\mu \psi_\mu, \tag{5b}$$

for the probe tip as

$$-\frac{\hbar^2}{2m}\nabla^2\psi_\mu + \left(E_F + eV - W\right)\psi_\mu = E_\mu \psi_\mu, \tag{5c}$$

and for the vacuum as

$$-\frac{\hbar^2}{2m}\nabla^2\psi_\mu + \left(E_C + \chi - U(x,y,z)\right)\psi_\mu = E_\mu \psi_\mu \tag{5d}$$

where E_V and E_C are fixed locations of the band edges (i.e. corresponding to their values deep inside the semiconductor), m_e and m_h are effective masses, and W is the width of the metallic band in the semiconductor below its Fermi energy. In a typical computation, a single CB and three VBs (light-hole, heavy-hole, and split-off) are used. At the semiconductor surface, the wavefunctions are taken to be continuous and their derivatives with respect to z are related by

$$\frac{\partial\psi_\mu}{\partial z}(x,y,0_+) = \frac{m}{m_i}\frac{\partial\psi_\mu}{\partial z}(x,y,0_-). \tag{6}$$

where m_i equals m_e or m_h, as appropriate. At the surface of the probe tip, both the wavefunctions *and* their derivatives are continuous.

Two schemes have been employed for evaluating the wavefunctions. In the first, the tunnel current is written in a form appropriate for *planar* tunneling. It can be argued that this approximation is equivalent to a semi-classical treatment of the radial (and angular) components of the wavefunctions while maintaining a fully quantum-mechanical treatment of their z-component [16]. The z-components of the wavefunction are then obtained directly by numerical equation of the corresponding z-part of the Schrodinger equation. This approximation for the wavefunctions is valid so long as there are no regions of localized potential in the semiconductor and the probe tip is not too sharp (and actually it works quite well even for tip radii as small as 1 nm) [17].

Regions of localized potential are what occur e.g. around quantum dots, so that in such cases the planar geometry is not suitable and one must employ a type of plane wave expansion method: Plane waves in the z-direction are matched to decaying exponentials in the vacuum, with the wavefunctions thus obtained also incorporating the boundary condition of equation (6) [17]. Solution of the

Schrödinger equation is then accomplished with the usual eigenvalue method. Such methods are very computationally demanding, even for a modest number of plane waves (≈10) in each of the x, y, and z directions. Consequently, only a relatively small region of the semiconductor (≈10 nm in each direction) is sampled by the method. Nevertheless, this small region turns out to be sufficient for providing a reasonable description of a tunnelling spectrum, at least for temperatures that are not too low. Results using both of these methods, in the absence of a localized potential in the semiconductor, are in good agreement with each other [17].

Computations performed as described above yield self-consistency between the potential and the current for situations of semiconductor depletion, which is the usual case in most spectra. But, for situations of accumulation or depletion, the semi-classical charge density is not realistic. In those cases, a quantum-mechanical charge density must be formed from the wavefunctions, and the potential re-evaluated with this new charge density. A self-consistent solution is then obtained iteratively. This method provides a complete solution of the problem when planar evaluation of the current is used [20], but for the plane wave expansion method the new charge density is obtained only over a restricted region of the semiconductor. Some method to extend that charge density over the entire semiconductor would be needed, and a detailed formulation of that procedure has not been derived to date.

4. Comparison of theory to experiment

It should be emphasized that all of the results obtained from the computational methods described in section 3 depend entirely on the input parameters. The basic parameters for any STS tunnelling computation on a semiconductor surface are tip radius-of-curvature R, tip-sample separation s, tip-sample contact potential ΔU (i.e. work function difference), and an effective tunnelling area of the apex of the tip. This latter parameter could be deduced from the tip radius for a perfectly spherical tip apex, but we take it to be an independent parameter (which affects only the overall magnitude of a spectrum), to allow for a small protrusion on the tip apex. In general the values of these parameters are not something that are of interest *per se*, but rather, their values are needed only to obtain a spectrum that agrees overall with experiment. Then, in a typical study, there will be one or more additional parameters (such as band offsets) whose values *are* quantities of interest. Comparison between experiment and theory therefore reduces to a curve fitting process, often involving several spectra acquired/simulated at different points. With the four tip-related parameters, together with additional parameters (such as band offsets) associated with the point of interest or other surface phenomena (such a surface charge), many thousands of evaluations of the theoretical spectra are required. Evaluation of a single current at one sample-tip voltage requires a time ranging from minutes to hours depending on the complexity of the problem, and typically there might be 30 such evaluations needed for a spectrum (15 voltage points, and 2 evaluations at each in order to obtain the conductance dI/dV). Hence, significant computational resources (multiple processors) are required.

An example of a detailed comparison between experiment and theory is given in figure 3, showing spectra for the InAs QD already introduced in figure 2. A spectrum acquired 6 nm from the QD is shown in (a), revealing states associated with the bare GaAs surface. A spectrum containing features arising from the QD confined states is shown in (b), acquired 4 nm from the QD center (at distances nearer to the QD these confined-state features grow in size, but charging of the QD is a problem, so the analysis is restricted to only this QD spectrum for which the current through the QD states is relatively small) [17]. For the spectrum of the bare GaAs, this would normally be fit using four parameters, s, R, ΔU, and an overall magnitude for the spectrum. In the present case however, atomic steps existed on this cleavage surface (arising from strain in the heterostructure due to the QDs). Very little tip-induced band bending is apparent in the spectra of figure 3(a) (i.e. the apparent bandgap is close to the GaAs gap of 1.42 eV), revealing the influence of the states of the steps. These states are modelled using an energetically uniform state-density across the bandgap, with density σ and charge neutrality level E_N. Consequently, six parameters are used to fit the spectrum of the bare GaAs. This is a large number of parameters, so the model possibly overfits the experimental data, hence producing rather high correlation between the ΔU and E_N parameters and values for the R parameter that are somewhat smaller than expected on the basis of other investigations [16]. But importantly, the added features found with the GaAs bandgap region for the QD spectrum are not at all dependent on these parameter values of s, R, ΔU , σ , or E_N , so those parameters only serve to provide a "background" description of the bulk GaAs part of the spectra from which the QD features can then be separated.

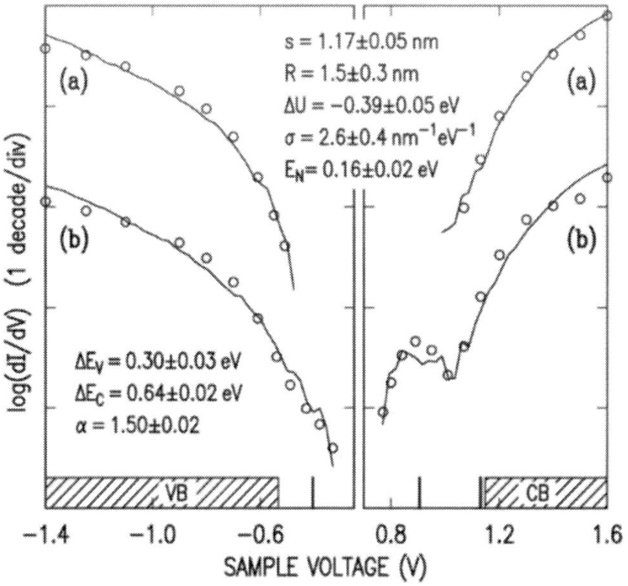

Figure 3. Tunnelling spectra, acquired (a) 6 nm from an InAs quantum dot (QD) and (b) 4 nm from the QD. Experimental results are shown by lines and theoretical fits by circles. The curves (a) and (b) are shifted from each other for ease of viewing. Values for the fitting parameters are listed. The location of states from the theory is shown at the bottom of the figure, giving the voltages at which the tip Fermi energy is aligned with QD states (for the VB, only light-hole states are shown) and with the hatched regions showing unconfined VB and CB states in the near-surface region. (From Ref. [17]).

To describe the QD, a minimum of two additional parameters are needed: the offsets between the InAs and GaAs CBs and VBs, ΔE_C and ΔE_V , respectively. With those two parameters the theory yields confined electron and hole states for the QD. However, it turns out that for the simple effective mass theory used here that the *density* of such states (i.e. number per unit energy) is significantly underestimated [22], so that even if the lowest energy state is correctly found in the theory then the next higher energy state would be incorrect. To account for this failing of the theory, one additional parameter was employed, a scale factor α by which the overall size of the QD was increased. The true physical size of the QD was precisely measured from STM images, including considerations of strain relaxation around the QD [17]. This size was then expanded by about 50% such that the density of electronic states (i.e. position of both the ground state and the first excited state for electrons) matched between experiment and theory. The final fit is shown in figure 3, where the values of all nine parameters (the eight listed there, plus one overall magnitude) is found by simultaneous fitting to both experimental spectra. In this way, the energies of the confined electron and hole states are determined, as schematically indicated in the lower part of the figure. These energies, in conjunction with the size, shape, and composition of the QD determined from the STM images [17], constitute a complete determination of the electronic and structural properties for this nanostructure. This information is useful since it can be used to distinguish between current *predictive* theoretical models for the confined state energies associated with QDs [17].

5. Summary

In summary, we have discussed in this work the application of computational methods to the quantitative analysis of tunnelling spectra from semiconductor surfaces. The electrostatic potential in the semiconductor, arising from the voltage difference between tip and sample as well as from occupation of electronic states on the surface or in the bulk of the semiconductor, can be obtained using an iterative finite-element method. Tunnel currents are then computed within the effective mass (envelope function) approximation by using either a semi-classical treatment of the lateral parts of the wavefunction or a completely quantum-mechanical treatment but one that is restricted to a relatively small region around the tunnel junction.

A number of limitations of the present theory exist, as discussed elsewhere [22]. One limitation in particular arises the assumption assumed that the tunneling process is the rate-limiting step in the transport of carriers from the probe-tip to the semiconductor, so that an "equilibrium" situation exists in which the occupation of carriers in the semiconductor is simply described by a constant Fermi energy. But in actuality, situations arise (at low temperature, or low doping, or for confined states, etc.) in which transport in the semiconductor itself is somewhat limited. In those cases, a much more complicated theory is needed to describe the distribution of carriers and resulting charge densities on the surface and in the bulk of the semiconductor. Such a theory could prove to be useful in enabling the determination of transport parameters for the carriers, even in nanoscale situations as occur in the STM geometry.

Acknowledgements
This work was supported by the U.S. National Science Foundation, grant DMR-0856240.

References

[1] Binnig G, Rohrer H, Gerber Ch and Weibel E 1983 *Phys. Rev. Lett.* **49**, 57
[2] Baratoff A 1984 *Physica B* **127B**, 143
[3] Tersoff J and Hamann DR 1985 *Phys. Rev. B* **31**, 805
[4] Becker RS, Golovchenko JA, Hamann DR and Swartzentruber BS 1985 *Phys. Rev. Lett.* **55**, 2032
[5] Hamers RJ, Tromp RM and Demuth JE 1986 *Phys. Rev. Lett.* **56**, 1972
[6] Stroscio JA, Feenstra RM and Fein AP 1986 *Phys. Rev. Lett.* **57**, 2579
[7] Smith AR, Feenstra RM, Greve DW, Neugebauer J and Northrup JE 1997 *Phys. Rev. Lett.* **79**, 3934
[8] Feenstra RM and Stroscio JA 1987 *J. Vac. Sci. Technol. B* **5**, 923
[9] Boland JJ 1990 *Phys. Rev. Lett.* **65**, 3325
[10] For review, see Feenstra RM 1994 *Semicond. Sci. and Technol.* **9**, 2157
[11] McEllistrem M, Haase G, Chen D and Hamers RJ 1993 *Phys. Rev. Lett.* **70**, 2471
[12] Dombrowski R, Steinebach Chr, Wittneven Chr, Morgenstern M and Wiesendanger R 1999 *Phys. Rev. B* **59**, 8043
[13] Jäger ND, Weber ER, Urban K, Krause-Rehberg R and Ebert Ph 2002 *Phys. Rev. B* **65**, 195318
[14] Feenstra RM 2003 *J. Vac. Sci. Technol. B* **21**, 2080
[15] Program SEMITIP, available at http://www.andrew.cmu.edu/user/feenstra/
[16] Feenstra RM, Dong Y, Semtsiv MP and Masselink WT 2007 *Nanotechnology* **18**, 044015
[17] Gaan S, He G, Feenstra RM, Walker J and Towe E 2010 *Appl. Phys. Lett.* **97**, 123110
Gaan S, He G, Feenstra RM, Walker J and Towe E 2010 *J. Appl. Phys.* **108**, 114315
[18] Mårtensson P and Feenstra RM 1989 *Phys. Rev. B* **39**, 7744
[19] Dong Y, Feenstra RM, Semtsiv MP and Masselink WT 2008 *J. Appl. Phys.* **103**, 073704
[20] Ishida N, Sueoka K and Feenstra RM 2009 *Phys. Rev. B* **80**, 075320
[21] Bardeen J 1961 *Phys. Rev. Lett.* **6**, 57
Duke CB *Tunneling in Solids* (Academic Press, New York, 1969), p. 217
[22] Wang LW, Williamson AJ, Zunger A, Jiang H and Singh J 2000 *Appl. Phys. Lett.* **76**, 339
[23] Feenstra RM 2009 *Surf. Sci.* **603**, 2841

Microscopy of nitride layers grown on diamond

B Pécz[1], L Tóth[1], Á Barna[1], G Tsiakatouras[2], AO Ajagunna[2], A Kovács[3] and A Georgakilas[2]

[1] Research Institute for Technical Physics and Materials Science, Hungarian Academy of Sciences, P.O.Box 49, H-1525 Budapest, Hungary
[2] IESL/FORTH and Physics Department/University Crete, PO Box 1527, 71110 Heraklion, Greece
[3] Center for Electron Nanoscopy, Technical University of Denmark, DK-2800 Kgs. Lyngby, Denmark

E-mail: pecz@mfa.kfki.hu

Abstract. GaN films were grown on three different orientation diamond substrates: (001), (110) and (111). In all three cases the hexagonal GaN is grown with (0001) direction parallel to the surface normal, but otherwise matching to diamond in a different way. The appropriate orientational relationships are determined by selected area electron diffraction. Besides threading dislocations a high number of inversion domains (ID) were formed in some GaN films. The preparation of the diamond surface and the growth conditions proved to affect significantly the formation of crystal defects such as threading dislocations and IDs. Single polarity GaN films with a low density of dislocations were achieved for the optimized growth conditions. The highest quality GaN layers were grown on AlN buffer in which two crystalline variants were nucleated, but one of them was overgrown already in the thickness of the buffer layer.

1. Introduction

The application area of GaN and related materials is very wide covering the optoelectronic devices, sensors and transistors as well. By now the high temperature stable high power devices reached power values above 10 W/mm. In the operation of them the most difficult technical problem is the heat dissipation during their operation. For that purpose substrates with excellent heat conductivity are needed. One of the best candidates is diamond despite the limited availability and also the limited knowledge on the growth of GaN on diamond.

Very few nitride growth experiments are known on diamond. Probably the first MOCVD (Metal Organic Chemical Vapour Deposition) growth experiment of Hageman et al. [1] was carried out on single crystalline diamond and resulted in poor quality GaN. However, overgrowing the GaN by HVPE (Hydride Vapour Phase Epitaxy) the quality improved substantially. Dreumel et al. grew also polycrystalline GaN on nanocrystalline diamond/Si template [2]. Recently the same group could grow oriented GaN crystals on polycrystalline diamond substrates [3].

This paper describes the microstructure of GaN grown on diamond single crystals by molecular beam epitaxy (MBE) with nitrogen RF plasma source [4]. The nitridation of diamond influenced the quality of the grown layers and that is discussed as well.

Journal of Physics: Conference Series **326** (2011) 012010 doi:10.1088/1742-6596/326/1/012010

2. Experimental

GaN layers with the thickness about 1 μm were grown on diamond single crystals by molecular beam epitaxy with nitrogen RF-plasma source. A thin AlN layer was initially grown as the nucleation layer on diamond. In some of the growth experiments the diamond surface was nitrided before initiating the III-nitrides growth. The temperature of nitridation was varied from 125°C to 750°C, while the duration time was varied from 1 minute to 60 minutes. The single crystalline diamond substrates were 5x5 mm^2 large and prepared by Element6 (www.e6.com) in three different orientations (100), (110) and (111).
The grown structures have been characterized by transmission electron microscopy (TEM) on cross sectional specimens prepared by Ar ion milling. The sample preparation process itself was long because of the cutting by a special saw in which a rotating Mo disc was fed with diamond paste from a small container placed below the disc. Bright field, dark field images and selected area diffraction patterns were taken in a Philips CM 20 microscope working at 200 keV, while high resolution images were taken in a JEOL 3010 electron microscope at 300 keV. Some of the high resolution work was done on an FEI Titan microscope on which also EDS (Energy Dispersive X-ray Spectroscopy) spectra were taken on line scans.

3. Results

The GaN film grown on (100) diamond is shown in Fig. 1. (Here we have to note, that the surface of the grown GaN layer was chemically etched to determine its polarity.) The cross section shows a few dislocations and a lot of inversion domains (ID). Fig. 2 shows the appropriate selected area diffraction pattern.

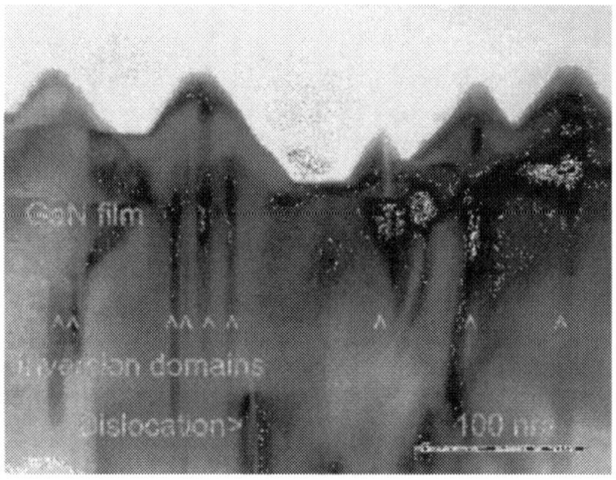

Fig.1: Bright field image showing many inversion domains in the GaN grown on (100) diamond. Note, that the surface of the layer was chemically etched.

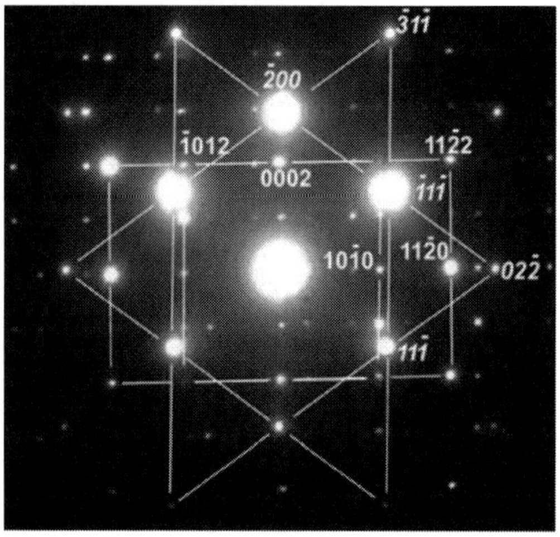

Fig. 2: Selected area electron diffraction pattern. The three digit indices mark the diamond reflections and the four digit indices mark the GaN reflections.

For (100) diamond the grown GaN is composed of two different domains having a common 0002 direction and matching like either (11$\underline{2}$0)GaN//(022)diamond, or (10$\underline{1}$0)GaN//(022)diamond.
In the case of (110) and (111) oriented diamond substrates single crystalline GaN was grown successfully. In all three cases the hexagonal GaN was grown with its (0001) direction parallel to the surface normal, but otherwise matching to diamond in a different way.
Further investigation of samples by high resolution microscopy revealed that the inversion domains were formed at the AlN/diamond interface (Fig.3). Otherwise the AlN buffer was homogeneous in thickness.

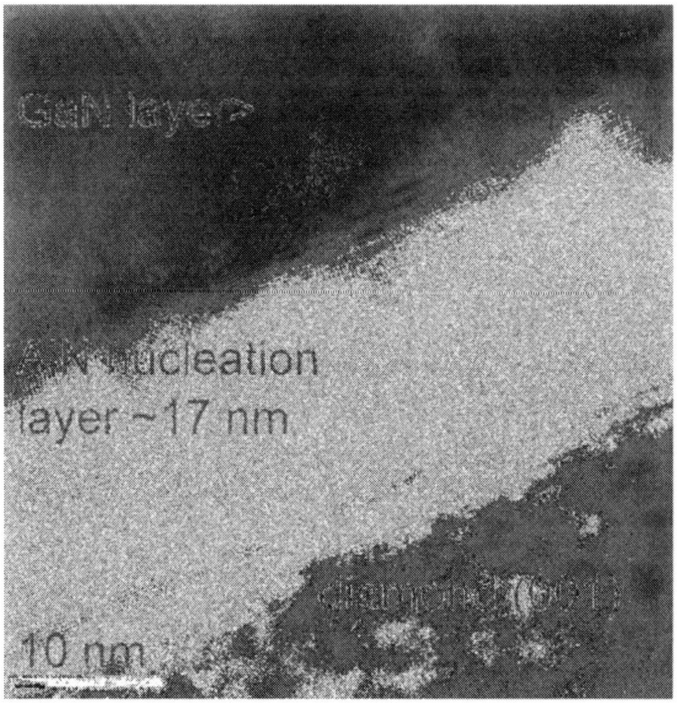

Fig. 3. HREM image taken at the AlN buffer region.

The image shows, that there are already IDs in the buffer layer as well. Consequently those were formed already on the surface of (001) diamond.

In the next set of experiments the surface preparation of diamond was modified and also thicker AlN buffer layers were grown. The next image (Fig. 4) shows the top region of a nitride structure, which was grown on (100) diamond nitrided prior to the AlN growth at 750°C for 30 minutes.

Fig. 4. Bright field image showing the surface region of nitride layers grown on diamond (100) after nitridation of the substrate.

The preparation of the diamond surface and the growth initiation conditions proved to affect significantly the formation of crystal defects such as threading dislocations and IDBs. Single polarity GaN films were achieved at the optimized growth conditions.

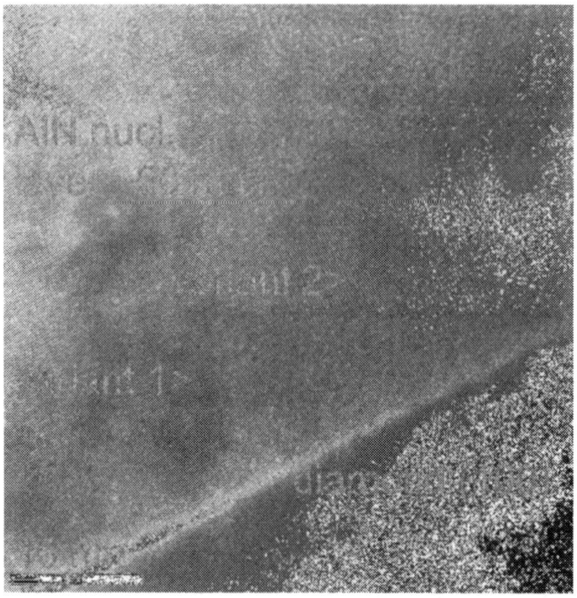

Fig. 5: High resolution image showing the interface of diamond single crystal substrate with the AlN nucleation layer.

It is well visible in Fig. 5 that the growth of AlN starts with two variants both epitaxially oriented to the (100) face of the substrate. One of the variants is locally overgrown by the other one in the first 12-15 nm of the nucleation layer. This was observed for the other investigated diamond substrate orientations as well. However, in this particular case of (100) diamond, the thick GaN was still composed of two variants rotated by 30 degree around their (0002) axis (like the case shown in Fig. 2). The interface looks atomically smooth and sharp, no intermediate layers could be observed. This observation together with the EDS line scans exclude the formation of C_3N_4 phase, which was observed by Kusunoki et al. [5], in their nitrogen implantation experiments. That may be attributed to the absence of energetic nitrogen ions in the nitrogen beam impinging the diamond surface.

4. Conclusions

In conclusion, it was shown that hexagonal GaN was grown successfully onto single crystalline diamond of all three orientations. The high density of inversion domains (and threading dislocations) was successfully reduced by the optimization of the growth conditions.

Acknowledgements

This work is supported by the European Commission in the framework of FP7 project MORGaN No. 214610. B. Pecz also thanks the support of OTKA project No. 75735.

References

[1] Hageman PR, Schermer JJ, Larsen PK, 2003 *Thin Solid Films* **143,** 9

[2] van Dreumel GWG, Buijnsters JG, Bohnen T, ter Meulen JJ, Hageman PR, van Enckevort WJP, Vlieg E, 2009 *Diamond and Related Materials* **18,** 1043

[3] van Dreumel GWG, Bohnen T, Buijnsters JG, van Enckevort WJP, ter Meulen JJ, Hageman PR, Vlieg E, 2010 *Diamond and Related Materials*, **19**, 437

[4] Iliopoulos E, Adikimenakis A, Dimakis E, Tsagaraki K, Konstantinidis G, Georgakilas A, *J. Cryst. Growth* **278**, 426 (2005)

[5] Kusunoki I, Sakai M, Igari Y, Ishidzuka S, Takami T, Takaoka T, Nishitani-Gamo M, Ando T, 2001 *Surface Science* **492**, 315

Defect investigation in $Al_{0.87}In_{0.13}N$/AlN/GaN heterostructures by scanning force microscopy methods

A Minj, D Cavalcoli and A Cavallini

Dipartimento di Fisica, Università di Bologna, Viale Berti Pichat 6/2, 40127 Bologna, Italy

albert.minj2@unibo.it

Abstract. AlInN/AlN/GaN heterostructures were characterized by atomic force microscopy (AFM) in semi-contact and conductive mode. These indium-related alloys contain threading dislocations (TDs) with a density around 10^8 ~10^9cm^{-2}, originating from the GaN (0001) substrate grown on sapphire. The TDs, with screw or mixed components, terminate at the surface of overgrown layers as V-defects. Using semi-contact AFM (phase-imaging) mode, we traced sites of indium segregation at the V-defects. These sites in V-defects were found to be highly conductive by current-AFM and could be a possible cause for the leakage current in Schottky diodes.

1. Introduction

Research on III-nitride semiconductors with a wide band gap varying between 0.63 eV (InN) and 6.2 eV (AlN) is achieving new heights due their potential wide range of applications in photonics and electronics. Heterojunctions based on III-Nitrides with high band-offsets have allowed Multiple Quantum wells (MQWs) (AlN/GaN or InN/GaN) based devices like Light Emitting Diodes (LEDs) and Laser Diodes (LD). They have shown promising results in terms of high emission efficiency, lifetime and have also realized red to UV/blue emission [1]. In the field of electronics, Heterojunction Field Effect Transistors (HFET) like High Electron Mobility Transistors (HEMT) based on AlGaN/GaN or AlInN/GaN have shown promising results in high frequency, high power operating-regime [2, 3] as they inherit properties like high electron saturation velocity and high breakdown voltage. However, these devices suffer from short lifetime and poor efficiency because of the presence of defects, poor surface/interfaces and phase-separation at inversion domains produced during their fabrication (growth) [4]. Due to availability of very few studies on their effect on optical and electrical properties, realization of high quality devices is hindered. Nanoscale investigation would be of considerable interest in determining their effect on the device electrical and optical properties [5]. Here, from the perspective of electronic devices (AlInN/GaN and AlInN/AlN/GaN HEMT structures), defects like threading dislocations (TDs) could be one of the deciding factors of device lifetime and efficiency. We have studied the electrical effects of these defects by tapping-mode AFM and conductive-AFM.

2. Methods

2.1. Current-Atomic Force Microscopy (contact mode)
Current AFM is operated in contact-mode with a constant bias voltage either applied to the AFM tip (with respect to the sample) or to the sample (with respect to the AFM tip). Here, as the AFM tip scans the surface, current is measured simultaneously. The image obtained providing details on spatial distribution of current, is called current-map. On a conductive sample where the AFM tip makes an ohmic contact with the surface, a current-map would give a one-to-one relation with the spatial distribution of resistivity as current is inversely proportional to the resistance. This is often called spreading resistance map (SSRM). However, when dealing with semiconductors, current-maps may not be proportional to spatial resistivity. Due to the inhomogeneity from phase-separation, clustering or defects, various factors like barrier height, band-gap, localized states and potential barrier would contribute to the current-map.

2.2. Phase-contrast image (semi-contact mode)
Nowadays, an AFM is often fitted with a cantilever oscillating at its resonant frequency, which is driven by piezoelectric crystal. During the scan, alteration in the oscillation amplitude (smaller than the free amplitude) from the surface features acts as a feedback signal to the electronic controller, which maintains the preset oscillation amplitude by adjusting the height of the scanner (piezoelectric crystal). This height is used to image the surface topography. Simultaneously alongside topography, the phase difference between cantilever oscillation and the driving signal applied to the piezoelectric crystal can also be monitored (phase-contrast image). This carries additional and complimentary information relating to certain surface properties including friction, adhesion and visco-elasticity, as they would affect the phase. One needs to be careful when associating phase-contrast with surface properties as above as it may be related to nanoscopic topographical artifacts, arising from enhanced adhesion due to an increased interacting area (e.g. at a grain boundary) or acting lateral forces at a step-feature. All the effects could be fundamentally described as an elastic/inelastic interaction of the AFM tip with the surface when operated at constant-amplitude [6, 7]. The surface energy of any material will vary spatially with a different chemical composition (originating from presence of oxide/aqueous layer, inhomogeneity and defects, not excluding contamination). In a real case situation, there will always be an inelastic interaction, and this loss of energy between the tip and surface can be summarized as the cause of phase-contrast. This type of imaging is a very powerful technique in providing surface properties.

2.3. Error mode image (semi-contact mode)
In semi-contact mode, during scanning the current value of the signal related to cantilever oscillation is the "error signal" of the feedback loop system. In this mode, feedback loop gain is adjusted in such a way that the feedback loop is able to follow the smooth features of the surface topography but not the abrupt features. Abrupt features in surface topography may occur from adsorbed nano-particles, structural defects or fabricated structures like trenches on the surface. So, abrupt features will appear with high contrast and smooth features with low contrast in error mode image. Therefore, error mode imaging is a useful method to enhance the contrast of very small (nm range) abrupt features.

3. Results and discussion

3.1. GaN
3 µm-thick GaN layers (000-1) were grown on sapphire at 1050 °C using LT (low temperature) GaN nucleation layer. This leads to the formation of semi-insulating (non-intentionally doped) n-type GaN. Semi-contact AFM topographical (fig. 1a) and phase-mode maps (fig. 1b) [8] show inverted pyramidal

Journal of Physics: Conference Series **326** (2011) 012011 doi:10.1088/1742-6596/326/1/012011

shaped facets that could also be distinguished in phase contrast mode. These features can be related to inverted pyramidal shapes (V-pits), as it was found that in n-type GaN substrates, threading dislocations open as N-terminated V-pits (V-defects) [9, 10]. From the corresponding phase-contrast image, it is seen that the facets of these pyramidal structures show a phase different from other regions

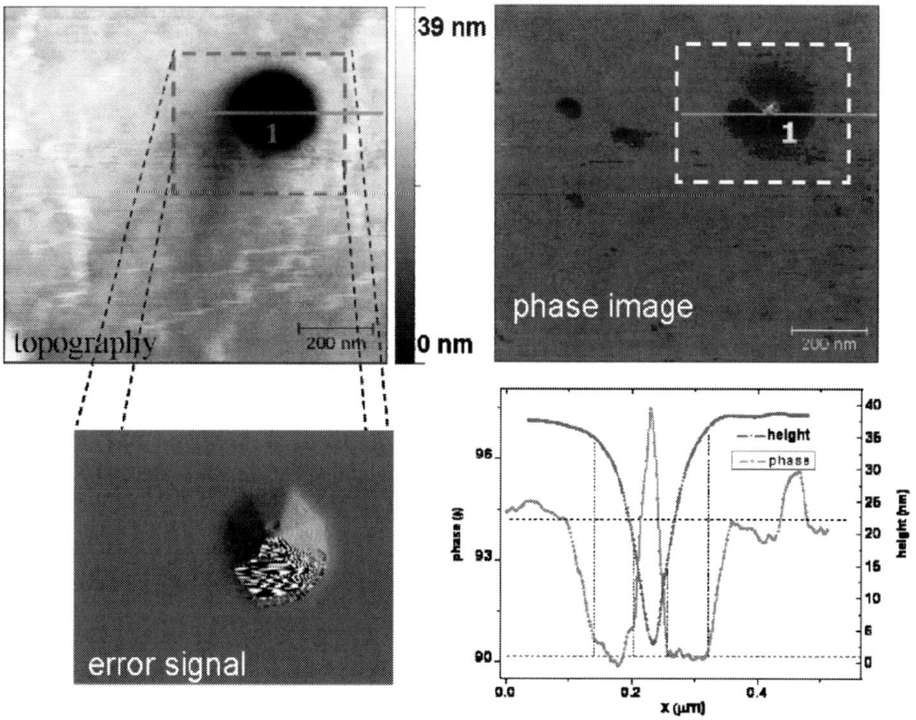

Figure 1. (a) Topography of GaN surface obtained in semi-contact mode. Pyramidal facet of TD (V-pit) is encircled. In the inset, its semi-contact error-signal image showing V-pits in GaN obtained at low feedback gain is shown and the corresponding phase-image (b) of the topography shown in (a). (c) Phase-profile and topography profile across the V-pit show a phase-difference between facets and GaN-surface.

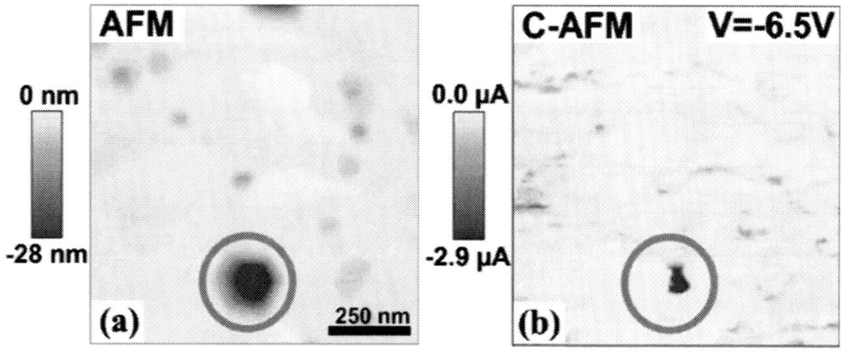

Figure 2. (a) topography of GaN surface obtained in contact mode. TD with screw-component is encircled (b) its corresponding current-map obtained at an applied bias of -6.5 volts to AFM probe. Reprinted with permission from [9]. Copyright [2008], American Institute of Physics.

(see figs.1c & 1d). This indicates that facets of TDs can be distinguished from their different surface energy. N-termination at the TD-facets due to surface reconstruction [9, 10] could explain the change in surface energy and hence, the contrast in phase-image.

It was already shown by Lochthofen et al. that N-terminated surfaces would also increase the electron affinity, which would lower the barrier locally [9]. In their current-map, here reproduced in fig. 2, it is seen that the TD facets show a relatively high current density compared to TD free surface (see figs. 2a & 2b). The same result has been also confirmed by the present analyses [11].

3.2. AlInN/AlN/GaN heterostructures

Alloys of AlN and InN ($Al_xIn_{1-x}N$) grown on c-plane GaN-buffer have played a very important role in further improving HEMT devices. The association of polarization fields, both spontaneous and piezoelectric, with AlInN and GaN allows the induction of polarization charges, which is compensated by formation of a two-dimensional electron gas (2DEG) [3]. For an indium concentration of 17%, AlInN is in lattice-match with GaN, allowing pseudomorphic-growth [12] and the presence of strong polarization fields in AlInN alone itself could cause high 2DEG density ($\sim10^{13}cm^{-2}$). However, the large difference in lattice-constants and thermal decomposition of AlN and InN prevent good quality growth of AlInN.

AlInN (33 nm)/AlN (1nm)/GaN (3 μm) heterostructures were grown on sapphire. The density of V-pits associated to TDs was found to be $\sim10^9cm^{-2}$. As shown in figure 3(a), two different features related to TDs can be distinguished, V-pits and channels. Channels are a series of V-pits situated in proximity and as a result appear continuous. It is known that indium preferentially segregates at the core of threading dislocations on the GaN surface [1, 13, 14]. As the AlInN layer grows, threading dislocations propagate towards the surface forming an In-rich or metallic indium nano-pipe extending from the AlInN/GaN interface to the AlInN surface. Interestingly, threading dislocations with screw-component terminate at the surface with inverted pyramidal structures consisting of 6-facets due to surface construction as a part of strain relaxation (called V-defects). Migration of In-adatoms is strain driven and as a result In-N bonds (weaker bonds) occupy low-coordinated sites like the edges of these pyramidal structures, while Al-N bonds (stronger bonds) occupy high-coordinated sites like the bulk and the surface of V-defects [1]. While fig. 3a shows a topography map of the sample surface, fig. 3b is a phase contrast image of the surface, the dark contrast appearing at the edges of TDs can be here noted. As in phase imaging the contrast could be related to compositional variations [6, 7], this dark contrast at the edges of TDs can be related to In, where In is expected to be localized.

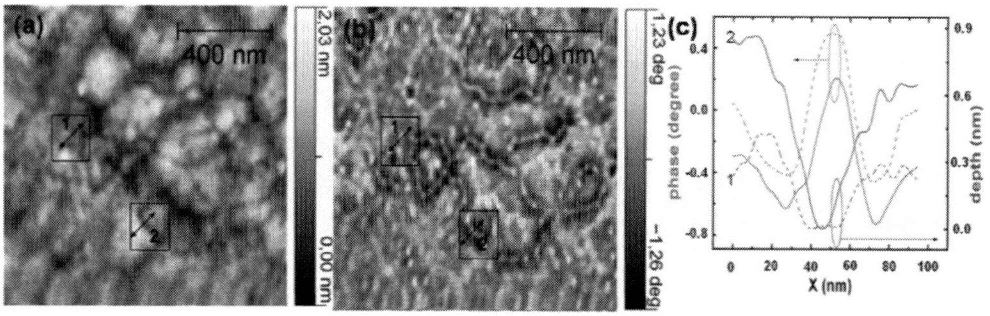

Figure 3. (a) Topography image of AlInN surface obtained in semi-contact mode. A pit is marked as 1 and channel is marked as 2. (b) Phase-contrast image of AlInN surface showing phase-contrast change around pits and threadlike features. There is dark contrast around the edges of threadlike features and pits whereas bright contrast in their interior. (c) Depth-profile and the corresponding phase-contrast profile across a channel and a pit marked in (a) and (b). Solid-line and dashed-line indicate channel and pit, respectively. Reprinted with permission from [15]. Copyright [2010], American Institute of Physics

One could also expect a localized increase or decrease of band-gap of AlInN due to the formation of Al-rich or In-rich regions, respectively (by Vegard's law). The indium segregated could also show a metallic behavior. Current-map obtained by applying 10 volts between the tip and the surface shows that the edges of V-defects, where dark-contrast in phase-image is also seen, shows a high current density compared to other regions (see fig 4a and 4c). The presence of high indium concentration, which would reduce band gap locally or would act as shunt between the surface and 2DEG could explain the high current density.

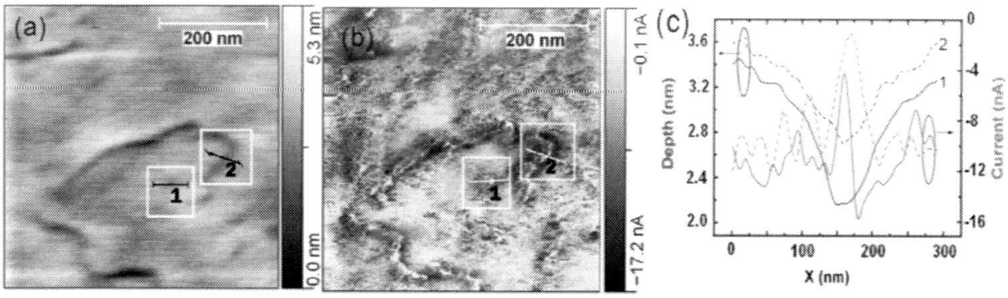

Figure 4. (a) Topography image of AlInN surface obtained in contact mode focusing on V-defects marked as 1 and channels marked as 2 in its insets. (b) Current-map of AlInN surface corresponding to (a) showing the presence of high conductivity along the edges of channel and pit. Bias voltage = 10 V. (c) Depth-profile and the corresponding current-profile across a channel and a pit marked in (a) and (b). Solid-line and dashed-line indicate channel and pit, respectively. Reprinted with permission from [15]. Copyright [2010], American Institute of Physics

4. Conclusion

From phase-contrast and C-AFM analyses of GaN and AlInN, compositional variation in V-defects and their effect on electrical properties have been inferred. Phase contrast image shows a different chemical composition on the facets of V-defects in GaN and at the edges of V-defects in AlInN. In these regions, C-AFM shows higher conductivity due to the increase of electron affinity in the case of GaN and due to the high indium content at the edges of the V-defects in the case of AlInN. The presence of such conductive channels and V-defects would facilitate high gate leakage current in AlInN/GaN heterostructure based HEMT devices.

Acknowledgement

H. Behmenburg, C. Giesen and M. Heuken of Aixtron AG, Herzogenrath, Germany, are gratefully acknowledged for providing the samples. This work was supported by the EU under the Project No. PITN-GA-2008 213238-RAINBOW.

References

[1] Kehagias Th, Dimitrakopulos GP, Kioseoglou J, Kirmse H, Giesen C, Heuken M, Georgakilas A, Neumann W, Karakostas T and Komninou P 2009 *Appl. Phys. Lett.* **95**, 071905
[2] Asif Khan M, Bhattarai AR, Kuznia JN and Olson DT 1993 *Appl. Phys. Lett.* **63**, 1214
[3] Tang H, Fang ZQ, Rolfe S, Bardwell JA and Raymon S 2010 *J. Appl. Phys.* **107**, 103701
[4] Kaneta A, Funato M, Narukawa Y, Mukai T and Kawakami Y 2006 *phys. stat. sol. (c)* **3**, No. 6, 1897
[5] Kalinin S and Gruverman A 2007 Scanning *Probe Microscopy: Electrical and Electromechanical Phenomena at the Nanoscale*, Springer Science and Business Media
[6] Garcia R, Magerle R and Perez R 2007 *Nature Mater.* **6**, 405

[7] Jung M and Choi JW 2010 *Ultramicroscopy* **110**, 670
[8] Magonov SN and Whango M-H. 1996 *Surface Analysis with STM and AFM. Experimental and Theoretical Aspects of Image Analysis.* VCH Weinheim.
[9] Lochthofen A, Mertin W, Bacher G, Hoeppel L, Bader S, Off J and Hahn B 2008 *Appl. Phys. Lett.* **93**, 022107
[10] Shiojiri M, Chuo CC, Hsu JT, Yang JR and Saijo H 2006 *J. Appl. Phys.* **99**, 073505
[11] Minj A, Cavalcoli D and Cavallini A 2011 *Physica B* (in press)
[12] Lorenz K, Franco N, Alves E, Watson IM, Martin RW and O'Donnell KP 2006 *Phys. Rev. Lett.* **97**, 085501
[13] Song J, Xu FJ, Yan XD, Lin F, Huang CC, You LP, Yu TJ, Wang XQ, Shen B, Wei K and Liu XY 2010 *Appl. Phys. Lett.* **97**, 232106
[14] Mouti A, Rouvière J-L, Cantoni M, Carlin J-F, Feltin E, Grandjean N and Stadelmann P 2011 *Phys. Rev. B* **83**, 195309
[15] Minj A, Cavalcoli D and Cavallini A 2010 *Appl. Phys. Lett.* **97**, 132114

Effects of the annealing temperature on the structural and electronic properties of MBE grown InGaN/GaN quantum wells

F Ivaldi[1], N A K Kaufmann[2], S Kret[1], A Dussaigne[2], B Kurowska[1], M Klepka[1], J Dąbrowski[1], P Dłużewski[1] and N Grandjean[2]

[1] Institute of Physics, Polish Academy of Sciences, Al. Lotników 32/46 02-668 Warsaw, Poland
[2] Institute of Condensed Matter Physics, Ecole Polytechnique Fédérale de Lausanne, CH-1015 Lausanne, Switzerland

ivaldi@ifpan.edu.pl

Abstract. A sample consisting of a 2.5nm thick $In_{0.25}Ga_{0.75}N$ quantum well grown by molecular beam epitaxy (MBE) was annealed for 2 hours in a hydride vapour phase epitaxy (HVPE) reactor under an NH_3 atmosphere at 880°C, 920°C and 975°C. High-resolution TEM and STEM combined with electron energy-loss spectroscopy (EELS) investigations on the local electronic properties have been carried out to correlate these with the structure evolution during annealing. Image processing techniques such as geometric phase analysis of HR-TEM images reveal significant fluctuations in the indium distribution inside the well on a nanometric scale. An increase of the emission characteristic has been observed for annealing at 920°C.

1. Introduction and experimental procedures

It is well known that the annealing process has great influence on a sample's emission characteristics, so that the full understanding of the effects of such process reveals crucial to improve III-N heterostructures [1].

A sample consisting of a 2.5nm thick $In_{0.25}Ga_{0.75}N$ quantum well grown by NH_3-MBE on a sapphire/GaN template was annealed for two hours in an HVPE reactor under NH_3 atmosphere at various temperatures ranging from 800 to 975°C. The photoluminescence (PL) and micro-PL characteristics of these heterostructures show a more uniform and red-shifted emission from the sample annealed at 920°C. To understand this phenomenon we investigated the structural changes in the specimen during the annealing process and determined the influence of such changes on the electronic properties.

Investigations have been performed on a FEI TITAN Cubed 80-300kV microscope equipped with Cs-corrector, monochromator and a 966 quantum GIF detector (0.2eV resolution) for EELS analysis and a JEOL 2000EX microscope equipped with a LaB_6 electron gun operating at 200kV.

Observations have been performed in the [11-20] and [1-100] direction on cross-sectional samples prepared by two different methods: by focused ion beam (FIB) using a Helios Nanolab FIB-SEM integrated system and by mechanical polishing and dimpling followed by ion milling. Electron transparency was obtained by PIPS ion milling with 4kV accelerated argon ions and LN_2 cooled stage. The final acceleration voltage used for surface cleaning has been 200V.

Local strain distribution and compositional information has been collected by processing HR-TEM images by geometric phase analysis (GPA), analyzing the tetragonal distortion in the heterostructure using the method described in [2]. Thin foil relaxation has been taken into account with the help of finite element (FEM) calculations [3]. Two beam condition technique has been applied to obtain images of 0002 lattice fringes tilting the sample by 3-4 deg around [0001] direction away from the exact zone axis. The incident electron beam was inclined to the optical axis of the microscope midway between the [0002] and [0000] beams. The application of Vegard's law allowed us to obtain detailed information on the indium distribution along the quantum wells.

2. Results and discussion

2.1. High resolution (S)TEM

Figure 1a shows a STEM low magnification cross-sectional image of the heterostructure prepared by FIB. It is evident that the presence of the mounds, formed due to kinetic roughening that takes place during NH_3-MBE growth [4], influences the morphology of the well, distorting its geometry evidently. From SEM images we could determine a periodicity of the mounds as ~500nm. These form plates in which the well runs parallel to the [0001] lattice planes and valleys in which the *c*-planes and the well are not parallel. In these regions a step growth of the quantum well can be observed. Atomic steps could be observed during STEM observations (fig.1b). V-pits or threading dislocations are preferentially to be found at the bottom of the valleys as figure 1a shows. Here the quantum well's width decreases due to the distortion caused by the dislocation.

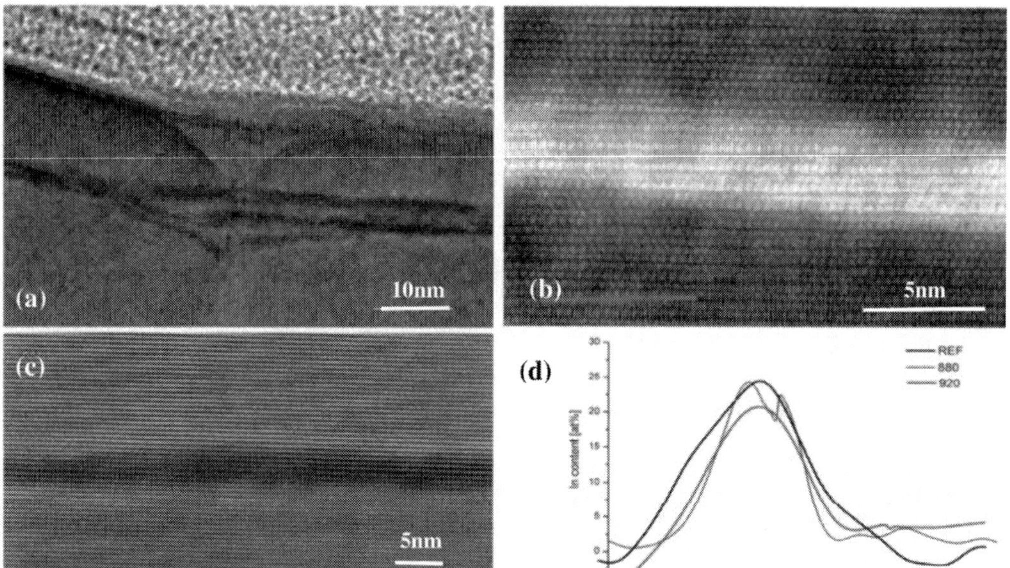

Figure 1. (a) V-pit dislocation at the bottom of an MBE mound; (b) STEM-HAADF image showing atomic steps at the InGaN/GaN interface at mound's slope, (c) Indium content map obtained by GPA superimposed on a TEM image of the reference sample; (d) Laterally averaged profiles of REF, TA880 and TA920 samples. The blue curve is characterized by two peaks presumably due to defects at the upper InGaN/GaN interface.

Geometric phase analysis of the reference sample (fig. 1c) allowed us the determination of the exact strain distribution and indium content along and across the wells. Our results show a uniform distribution and a concentration of 25-28at% of indium atoms in the wells. No phase separation or clustering has been observed. The average quantum well thickness could be determined to be 3nm. The originally 3nm thick quantum well reduces its width to 2.5nm after annealing at 880°C and to 2.2nm after annealing at 920°C. The indium concentration in the wells of these two samples has been determined to be respectively 23% and 22%. We observe a more homogeneous indium distribution in the well annealed at 880°C, which leads to a less broad PL spectrum. The indium concentration inside the well after the annealing process at 975°C could be determined to be 7-9%. We could observe an increase of the strain in the cap layer, which we suggest to attribute to vacancy activated indium diffusion towards the surface. Further investigations on the effects of the annealing on the vacancies in the cap layer will be carried out in future.

Surface diffusion effects have been observed. For increasing annealing temperatures the valleys and mounds observed in the as grown sample level and create a smoother surface. The thickness of the cap layer along the samples is then inhomogeneous.

From the lateral averaged profile of the strain maps calculated by geometric phase analysis shown in figure 1d, we see a clear increase of the strain inside the quantum well as well as a reduction of the full width at half maximum (FWHM). For the sample annealed at 975°C a reduction of the strain and a leveling of the averaged profile confirm the out-of-plane diffusion and desorption of indium atoms. Integration along the calculated profiles allowed us to confirm the decrease in the total indium content in the samples for increasing annealing temperatures.

2.2. Electronic properties

EELS investigations of the annealed samples have been performed on cross-sectional samples and correlated to theoretical calculations made using the Wien2k program. This simulation software is based on full augmented plane wave (FLAPW) methods for the calculations of the energy loss spectra of the chosen compounds [5]. Our calculations have been performed for $\boldsymbol{q}$ perpendicular to $\boldsymbol{c}$. We used for GaN lattice parameters of $a_{GaN} = 0.3189$nm and $c_{GaN} = 0.5185$nm. For the simulation of different indium contents we used lattice parameters derived by Vegard's law with $a_{InN} = 0.3548$nm and $c_{InN} = 0.5760$nm.

Through the simulation of various low loss spectra for different indium contents we calculated the dependence of the plasmon position on the atomic indium content. We notice a slight shift of the peak position for the annealed samples in comparison to the as-grown one. This is consistent with the difference in indium content revealed by GPA analysis. For 880°C and 920°C the differences in the plasmon peak position are minimal.

The measured bandgap for the sample annealed at 880°C has been determined through the acquisition and averaging of several spectra. Deconvolution of the zero loss peak from the acquired low loss spectrum and Kramers-Kronig analysis (KKA) have been applied to remove the contribution of surface losses to the acquired spectra [6]. Cerenkov effects could be neglected since the thickness of the investigated specimen fulfilled the requirements described in [7]. We measured in this way a bandgap of ~2.5eV. A narrower bandgap could be measured for the reference sample and the one annealed at 920°C (~2.3eV).

PL measurements showing a 55meV shift of the emission for the sample annealed at 880°C and an opposite shift for the one annealed at 920°C can be explained by the narrowing as well as by the different indium concentrations in the quantum wells and by the formation of clusters for higher temperatures.

In contrast to other groups that observed a relation between measured spectra and core-hole calculations, we find our measurements well correlated to ground state calculations of the low loss region of the investigated material [8].

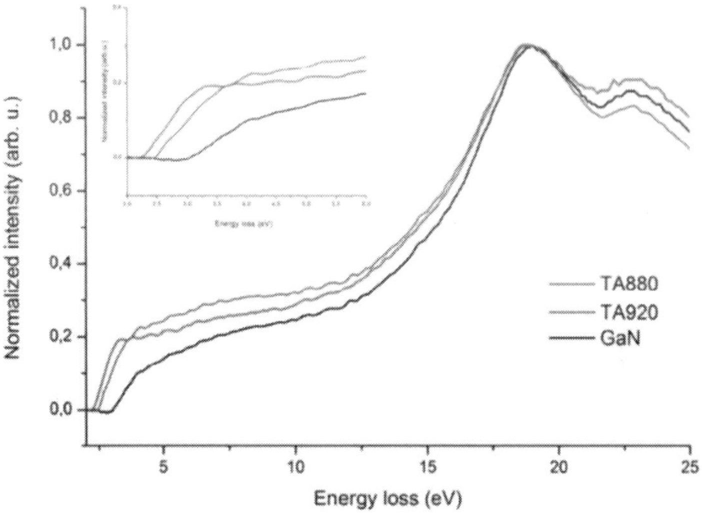

Figure 2. Low-loss spectra acquired for the annealed samples and a GaN reference. Insert: bandgap determined after KKA-analysis

3. Conclusions

We investigated the influence of different annealing temperatures on the structural properties of InGaN/GaN quantum wells with nominal 25% of indium content. We observed diffusion of In atoms to the sample surface and consequent desorption. A decrease of the total indium content in the samples for increasing annealing temperatures could be determined from laterally averaged profiles of strain maps. A more homogeneous indium distribution along the wells annealed at lower temperatures (<900°C) has been observed.

Electronic properties and the structure have been correlated through EELS measurements. A shift towards longer wavelength has been observed by measurement of the band-gap in the quantum well region for annealed samples at 920°C even though a slight decrease of the total indium content was observed in comparison to the as-grown sample. Indium diffusion and desorption is compensated by the narrowing of the well thickness, leading to a stronger QCSE and by indium clustering introducing size effects.

Acknowledgments

This work was supported by the EU project ITN RAINBOW, Contract N: PITN-GA-2008-213238. The TEM investigation has been supported by the Innovative Economy fund N: POIG.02.01.00-14-032/08.

References

[1] Tournie E et al. 2002 *Appl. Phys. Lett.* **80**, 4148
[2] Kret S et al. 2001 *Phys. Stat. Sol. (b)* **227**, 247
[3] Kret S et al. 2007 *Nanotech.* **19**, 465707
[4] Vezian S et al. 2004 *Physical Review B* **69**, 125329
[5] Blaha P et al. 1990 *Comput. Phys. Commun.* **59**, 399
[6] Stoeger-Pollach M. 2008 *Micron* **39,** 1092
[7] Gu L et al. 2007 *Phys. Rev. B* **75**, 195214
[8] Keast VJ et al. 2002 *Phys. Rev. B* **66**, 125319

TEM and XANES study of MOVPE grown InAlN layers with different indium content

S Kret[1], A Wolska[1], M T Klepka[1], A Letrouit[1], F Ivaldi[1], A Szczepańska[1], J-F Carlin[2], N A K Kaufmann[2] and N Grandjean[2]

[1] Institute of Physics, Polish Academy of Sciences, Al. Lotników 32/46 02-668 Warsaw, Poland
[2] Institute of Condensed Matter Physics, Ecole Polytechnique Fédérale de Lausanne, CH-1015 Lausanne, Switzerland

kret@ifpan.edu.pl

Abstract. We present structure and spatially resolved composition studies by TEM (Transmission Electron Microscopy) and XANES (X-ray Absorption Near Edge Structure) of InAlN MOVPE (Metal-Organic Vapor Phase Epitaxy) epilayers containing 16-27 at% of indium. Investigations of the In L_3 edge by synchrotron radiation absorption show a significant change of the post-edge structure depending on the indium content. We attribute this to the solubility limit and phase separation in this system. Our measurements suggest that the critical composition is 18% for our growth conditions. HRTEM cross-sectional and EDX investigations confirm such phase separation as well as the changing of the structure from 2D growth to columnar like growth for the sample with the highest indium content.

1. Introduction

InAlN is a promising material for optoelectronic and electronic devices especially for HEMT (High Electron Mobility Transistor) and optic elements of InAlN based lasers – bottom cladding for green laser diodes and Bragg reflectors. For indium content of about 18at % the lattice parameter *a* of InAlN matches the GaN substrate. From theoretical predictions spinodal decomposition in this system is expected [1]. There are not many reports showing the spatial scale and amplitude of this decomposition and its dependence of the growing conditions. For the MBE (Molecular Beam Epitaxy) grown samples [2,3] the honeycomb structure with indium rich domain boundaries are observed. For the MOVPE (Metal Organic Vapor Phase Epitaxy) grown samples there is no systematic TEM investigation of these phenomena. We present the TEM and XANES studies of the phase separation which occurs in MOVPE grown InAlN layers with different composition.

2. Experimental procedures

We investigate a series of 4 samples of c-plane wurtzite $In_xAl_{1-x}N$ films grown by MOVPE on sapphire substrates using 1μm-thick GaN buffer layers. InAlN was deposited at temperatures between 800 and 850 °C. Indium contents in the A, B, C and D samples were determined by X-ray diffraction and are equal to 15.9 +/- 0.3%, 18.0 +/- 0.5%, 20.5 +/- 0.5%, 27.5 +/- 0.3%, respectively. The cross-sectional TEM samples of the Al_2O_3/GaN/InAlN were prepared and thinned to 40μm through various polishing and dimpling stages. The last step of the preparation was performed by the ion milling of the specimen using a Precision Ion Polishing System (PIPS) in dual mode beam modulation for the cross section and in off mode for the plan view. The ion milling took place under nitrogen cooling to reduce the sample's temperature during the treatment and achieve herewith a more gentle surface removal and less radiation damage. The low temperature milling was finished at 1kV and then the sample was heated up to room temperature. The samples' surfaces final cleaning was performed using an acceleration voltage of 200 V for 2h. It was found that this procedure gives a contamination and radiation defect free thin foil with large electron transparent areas. TEM investigations were performed using a JEOL 2000EX microscope equipped with a LaB_6 electron gun operating at 200kV and FEI TITAN Cubed 80-300 microscope equipped with Cs-corrector, HAADF (High Angle Annular Dark Field) and EDX (Energy-dispersive X-ray Spectroscopy) detector at 300kV. TEM observations were performed along the $<11\bar{2}0>$ and $<1\bar{1}00>$ zone axes. The XANES spectra were measured at L_3 In edge (Hasylab, A1 station). The principal component analysis (PCA) of the XANES spectra was performed with ATHENA program, part of IFEFFIT package.

3. Results and discussion

Low magnification images (fig. 1a) obtained by conventional TEM (BF and DF) as well as HAADF–STEM confirm that the A sample has homogenous composition and structure in micro and nanometric scale. In the case of the B sample small fluctuations in diffraction contrast can be observed. These fluctuations have 1 nanometer scale but near the top a layer of few nanometer inhomogeneity can be observed (fig. 1b). For the C sample, containing more than 20% of indium, initially only short range disorder at few atomic distances can be observed (not shown here). However, on the top part of the epilayer lateral structural inhomogeneity is observed. The columnar like growth starts to be dominating. Nevertheless, growing columns preserve the same crystal orientation (fig. 1c).

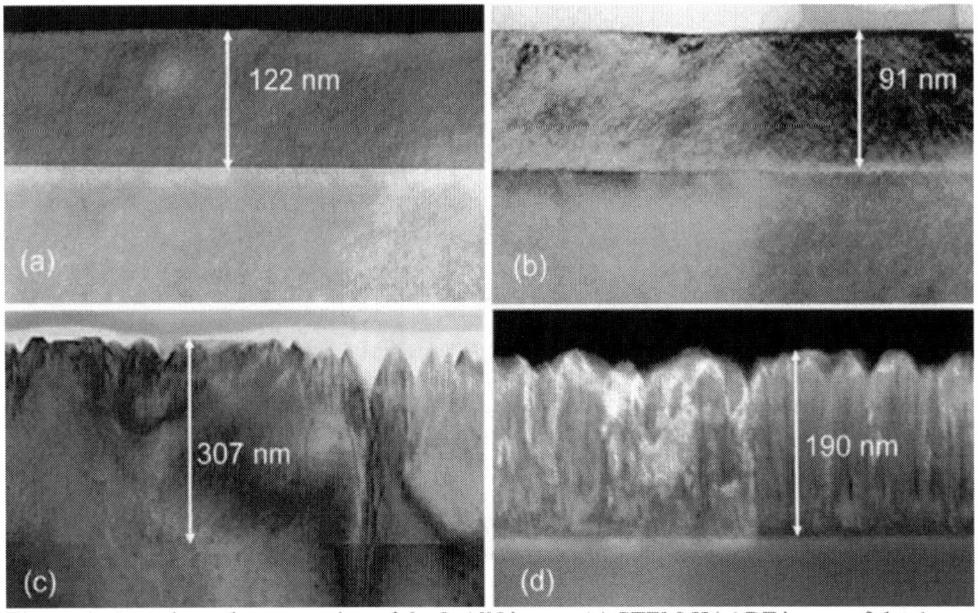

Figure 1. Overview of cross-section of the InAlN layers: (a) STEM-HAADF image of the A sample; (b) BF image of the B sample; (c) BF image of the C sample; (d) STEM-HAADF image of the D sample.

HAADF-STEM images of the D sample clearly show that columnar like growth starts few nanometers from interface and strong Z-contrast suggesting elemental segregation is observed. The HRTEM images (fig. 2) confirm that even for the D sample epilayers have almost perfect crystalline structure with only few defect like basal plane stacking faults and small inclusion of a cubic phase. High magnification Cs-corrected HRTEM images show the disorder of the structure at atomic scale. In the indium rich areas the bright and dark "columns" appear next to each other compared to neighbouring areas where contrast is more homogenous.

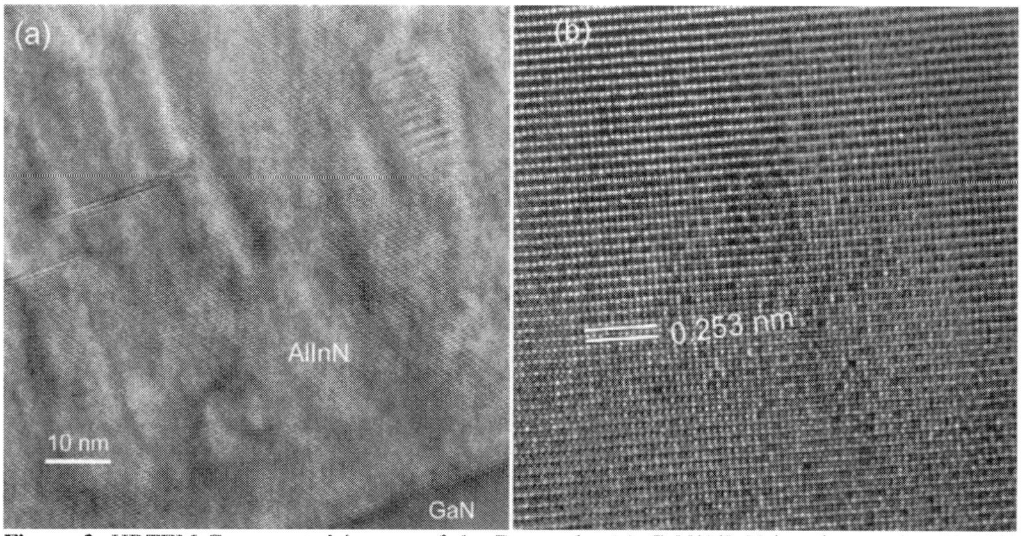

Figure 2. HRTEM-Cs corrected images of the D sample: (a) GaN/AlInN interface region [1-210] zone axis; (b) detail of local contrast variation in [1-110] zone axis.

EDX investigation confirms that we have strong fluctuation of the composition. Figure 3 shows examples of the EDX mapping of the near interface area of the D sample. The 2D map of counts from In-L and Al-K was used to calculate the In/Al ratio in this region. Figure 3b shows the spatial amplitude of such fluctuations in the range of 2-15 nm. The maximum indium signal correlates well with the brighter Z-contrast; dark Z-contrast corresponds to higher aluminium content. On the base of Z-contrast and additional EDX investigations we can say that the aluminium maximum is observed in the central part of the columns and indium concentration is higher on the top of the layer.

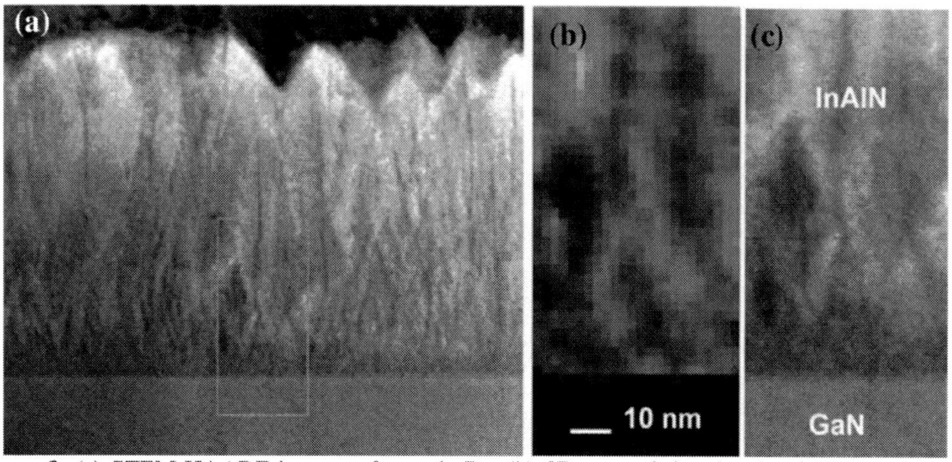

Figure 3. (a) STEM-HAADF images of sample D, (b) 2D map of signal ratio of EDX In-L/Al-K, (c) zoomed part of (a)

XANES is sensitive to the local structure around a selected element; therefore, it is often called a fingerprint of a compound. The shape of the spectrum can be treated as a linear combination of all the phases existing in the investigated material which makes it possible to perform the PCA. In case of the InAlN spectra as one of the reference compounds the InN thin layer representing the InN inclusions was taken. As the other reference, the A sample, which was found to be the perfect alloy without any structural inhomogeneities, was used. Both reference spectra were used in fitting the spectra of the B, C and D samples and the percentage of InN in each spectrum was found out to be equal to 1.7 +/- 1.5%, 11.2 +/- 1.2% and 34.0 +/- 2.1%, respectively. After recalculating these values to the indium content for each sample, we determined that 17.7 % (B), 18.2 % (C), 18.2 % (D) of In atoms are built in the AlN matrix. This leads to the conclusion that the solubility limits of In in the investigated system seems to be around 18%. The excess of the In atoms forms different phase - InN in this case and is accumulated in coherently segregating indium rich areas.

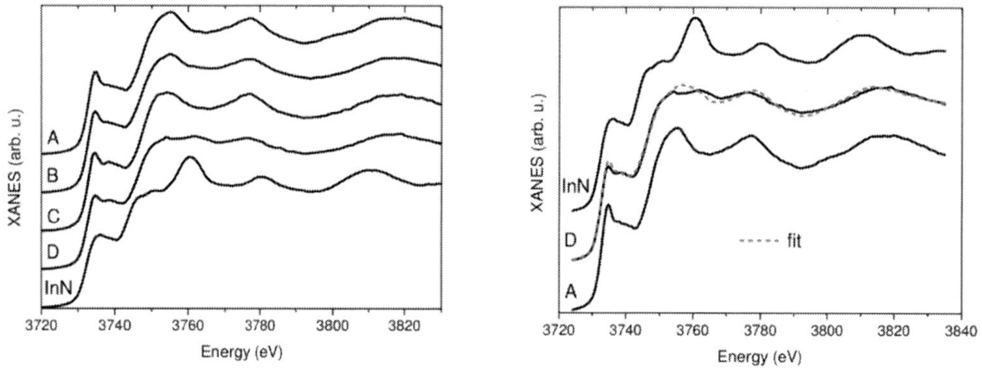

Figure 4. (a) XANES spectra at the L_3 In absorption edge for the A, B, C, D samples and InN layer; (b) reference spectra and result of fitting for the D sample. The spectra are shifted vertically for clarity.

4. Conclusions

We present evidence of the strong dependence of the indium content on the structure of the MOVPE grown InAlN layers. The solubility limit found out by XANES is around 18 % which is close to the MBE grown InAlN [3] where the EDX studies determined 15at% in domain centers. The phase separation can occur even for almost lattice matched compositions during the growth of more than 100 nm thick layers. Future investigations are planned to quantify the disorder in atomic scale and their role in the generation of the nanometric scale phase separation.

This work was partially supported by the European Community's Marie Curie Research Training Network RAINBOW 213238-2 as well as by the Ministry of Science and Higher Education (Poland) through grant N N202 204438. The synchrotron based research received funding from the European Community's Seventh Framework Programme (FP7/2007-2013) under grant agreement n°226716.

References

[1] Matsuoka T 1997 *Appl. Phys. Lett.* **71**, 105
[2] Zhou L, Smith DJ, McCartney MR, Katzer DS and Storm DF 2007 *Appl. Phys. Lett.* **90**, 081917
[3] Sahonta SL, Dimitrakopulos GP, Kehagias T, Kioseoglou J, Adikimenakis A, Iliopoulos E, Georgakilas A, Kirmse H, Neumann W and Komninou P 2009 *Appl. Phys. Lett.* **95**,021913

(V)EELS Characterization of InAlN/GaN Distributed Bragg Reflectors

A Eljarrat[1], Ž Gačević[2], S Fernández-Garrido[2,4], E Calleja[2], C Magén[3,5], S Estradé[1,6] and F Peiró[1]

[1] Laboratory of Electron NanoScopies, LENS-MIND-IN2UB, Dept. Electrònica, Universitat de Barcelona, Spain

[2] Instituto de Sistemas Optoelectrónicos y Microtecnología, ISOM, Universidad Politécnica de Madrid, Spain

[3] Laboratorio de Microscopías Avanzadas (LMA) - Instituto de Nanociencia de Aragón (INA), and Departamento de Física de la Materia Condensada, Universidad de Zaragoza, Zaragoza 50018, Spain

Email to: aeljarrat@el.ub.es

Abstract. Ten-period InAlN/GaN distributed Bragg reflectors are examined by aberration corrected scanning transmission electron microscopy and by valence electron energy-loss spectroscopy (VEELS) with sub-nanometric spatial resolution and sub-eV energy dispersion. Deconvolution and peak subtraction methods, implemented in Matlab routines, are applied to the low loss region of the obtained VEEL spectra to retrieve information about the band gap energy and chemical composition, whereas a Kramers-Kronig transformation is used to retrieve the complex dielectric function of the examined material. The VEEL measurements reveal significant compositional variations in InAlN layers and show a ~2nm thick InAlN layer with high indium content at each GaN/InAlN interface.

1. Introduction

The high potential of III-nitride semiconductors has been attested by their success in violet/blue lightening. However, high *in-plane* lattice mismatch between the three binaries (AlN, GaN and InN) is a serious constraint for further progress as it leads to defect formation at heterostructures' interfaces, affecting detrimentally their (opto)electronic properties. The InAlN ternary compound, with 17-18% indium content, grows *in-plane* lattice-matched to GaN and opens the possibility for fabrication of strain-free III-nitride heterostructures. Still, several problems remain concerning the growth, crystalline quality and homogeneity of InAlN epitaxial films [1-4]. A precise knowledge of the InAlN

[4] Also at: Paul-Drude-Institute for Solid State Electronics, Hausvogteiplatz 5-10117 Berlin, Germany

[5] Fundación ARAID, 50004 Zaragoza, Spain

[6] Also at: TEM-MAT, (SCT), Universitat de Barcelona, Solé I Sabarís 1, Barcelona, Spain

band gap value and its dependence on In content is a critical issue. Due to ternary composition fluctuations on the nanometer scale, traditional optical techniques used for band gap determination are prone to errors, since they are limited to the micron-range spatial resolution. Aberration corrected scanning transmission electron microscopy (STEM), with sub-nanometric spatial resolution, combined with monochromated electron energy loss spectroscopy (EELS), with < 0.2 eV energy resolution, is a promising tool to unravel optoelectronic properties of the novel material. In particular, Valence-Electron Energy Loss Spectroscopy (VEELS), is an excellent technique which focuses on the lower energy loss (low-loss) part of the EEL spectrum (typically < 50 eV), that exhibits the most important features regarding the valence and conduction band of the examined material.

2. Experimental

Ten-period nearly lattice-matched InAlN/GaN DBRs, with peak reflectivity around 400 nm, were grown in a RIBER Compact 21 MBE system equipped with a radio-frequency plasma nitrogen source and standard Knudsen cells for gallium, aluminium and indium. A procedure to obtain good quality InAlN/GaN periods with ~18% In content in the InAlN semi period has been described in previous studies [3]). Preparation of the samples for (S)TEM-EELS observation was done in cross-section geometry by polishing down to 50 μm, followed by a dimpling down to 25 μm and a final Ar+ bombardment at V=5kV with an incident angle of 7° using a PIPS Gatan ion mill. TEM experiments were conducted in a JEOL 2010F (S)TEM and in a probe corrected FEI Titan (S)TEM instrument operated at 300 kV with a Wien monochromator and Gatan Tridiem 866 ERS energy filter/spectrometer (collection angle was 17 mrad)

3. Data treatment

Figure 1 illustrates the complete VEELS analysis work-flow carried out on a single raw spectrum. Originated from the elastically scattered part of the electron beam, the zero loss peak (ZLP, see Figure 1a for an example) is used to calibrate the origin of energies of the spectra. The FWHM of the ZLP indicates the energy resolution limit (analogous to an optical system signal widening) of our data: our experimental set-up provided EEL spectrum lines with an energy resolution below 0.2 eV (quantitatively determined using Gaussian fitting of each individual spectrum) and a spatial dispersion of 0.5 nm. Because of low signal-to-noise ratio in low-loss monochromated beam experiments, sometimes, it has been necessary to average 10 spectra (corresponding to about 5 nm of linear displacement). However, when this is done, the spectra are always taken from a homogeneous region far from any interface.

One typical individual VEEL raw spectrum obtained in the GaN layer of the first semiperiod is shown in Figure 1b. Significant information contained in the low-loss regime spectrum can almost be directly examined by naked eye. Besides the ZLP, traces of signals associated with band gap transitions are also noticed. A prominent peak at ~19 eV, the plasmon peak, can be attributed to the collective oscillation of the electrons. The following intensity maxima have been attributed to the presence of the d-band in GaN [5].

Low-loss EELS data processing started with the deconvolution of the ZLP and plural scattering by a Fourier-log routine (according to Egerton [6]). In optimal conditions and with a careful preparation of the spectra for Fourier analysis to obtain best results, the F-log routine will not only separate the ZLP from the rest of the spectrum, but it will also get rid of the part of the signal coming from plural scattering events (through Poissonian modeling). Then, the resulting signal will be the so-called single scattering distribution (SSD, see Figure 1.c), a signal modeled in the theory by Kröger for electron beams traversing thin foils [6].

Before continuing performing transformations on the signal, it is instructive to take a look at the plasmon peak. We can precisely determine the energy and spatial position of the peak through the layers using a peak-fitting procedure to locate the plasmon energy position in an individual spectrum (see Figure 1.d for an example of a Lorentzian peak-fitting procedure in action).

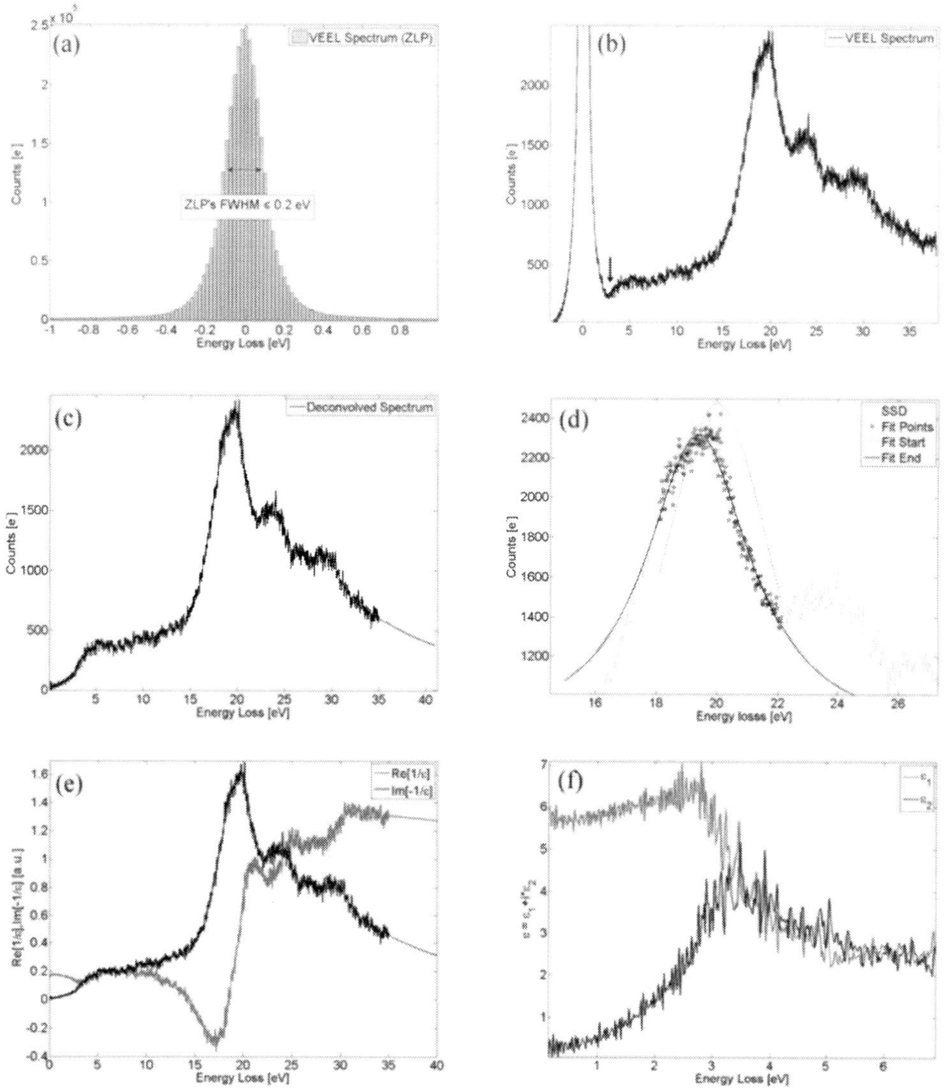

Figure 1. A typical VEELS analysis work-flow, performed on a raw individual spectrum from a GaN layer of one semi-period. These images show the evolution and transformation data from raw untreated spectra in images (a) and (b), to deconvolved spectra in (c) and (d), and finally, calculated complex dielectric functions in (e) and (f).

Once the SSD is identified a small preparation of the signal is enough to perform Kramers-Kronig analysis (KKA) on the spectra assuming the refractive index (n) of GaN as a known parameter. We have used n=2.4 according to [3]. The case of GaN is clear because its refractive index is known, and will allegedly remain constant in the layer. In the case of InAlN the choice of n is not so clear because of the suspected presence of In segregation in the layer [3], so this is why a GaN example spectrum is presented. In Figure 1.e the energy-loss function (ELF, Im(-1/ε)), along with its KK-complementary function Re(1/ε), are shown. This ε represents the complex dielectric function (CDF) of the material (part of which is portrayed in Figure 1.f); KKA makes it possible to use n to make a measurement of the foil thickness, and then use Kröger's equation to separate the ELF. Through Fourier analysis it then computes the KK transformation, retrieving the full CDF of the material. This iterative process will also deal with the surface-loss signal, leaving only the bulk features in the final result.

From Vegard's Law we know that for a ternary alloy like InAlN, the lattice constant is related to the ratio of the materials through a linear law (with or without a parabolic *bowing* term). Using the fact that the plasmon frequency for crystalline materials is also related to the lattice constant, if we precisely determine the energy and spatial position of the peak through the InAlN layer (using our peak-fitting routine) we shall be able to calculate the indium ratio through simple fitting of the plasmon energies for each spatial position to this equation (see Figure 3),

$$Ep_{InAlN} = \delta \cdot Ep_{InN} + (1-\delta) \cdot Ep_{AlN} - b \cdot \delta(1-\delta), \qquad \text{(eq. 1)}$$

in which δ is the In ratio and the *Ep*'s are the plasmon energies for binary components InN and AlN (15.7 and 21.1 eV, respectively) and measured for InAlN, *b* is the bowing parameter which is adjusted to obtain a mean In ratio equal to the 18% nominal one (*b*=-3.1eV).

The resulting ELF and CDF contain very valuable information about optoelectronic properties of the bulk material in the sample which can be compared with optically measured ones. Transitions on the electronic band structure of the bulk material related with its crystalline structure appear clearly as discernible features in these functions. The ELF is directly related to the joint density of states between the valence and conduction bands, a magnitude which can be modelled and calculated theoretically by ab-initio computer simulation. In this line, the analysis presented here has been extended to relate optoelectronic and structural properties presented in latter publications [1-3], through band-gap determination, plasmon peak related indium ratio and Cole-Cole plots of CDF.

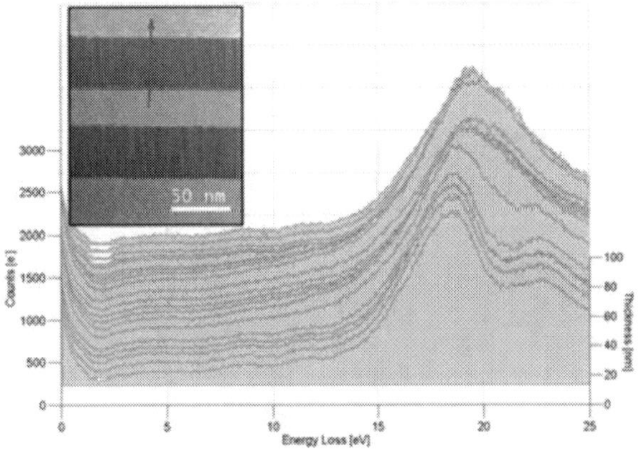

Figure 2. plasmon peak evolution along the line displayed in the HAADF micrograph (inset) which suggests the existence of an In rich layer.

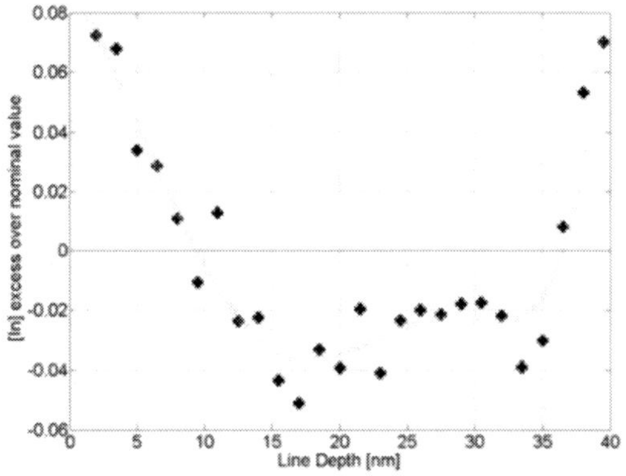

Figure 3. Indium excess concentration showing accumulation of InN in the borders of the InAlN layer.

References

[1] Fernández-Garrido S, Gačević Ž, and Calleja E 2008 *Appl. Phys. Lett.* **93** 191907

[2] Gačević Ž, Fernández-Garrido S, Calleja E, Luna E, and Trampert A 2009, *Phys. Stat. Sol. C* **6** S643

[3] Gačević Ž, Fernández-Garrido S, Hosseini D, Estradé S, Peiró F, and Calleja E 2010, *J. Appl. Phys.* **108** 113117

[4] Gačević Ž, Fernández-Garrido S, Rebled JM, Estradé S, Peiró F, and Calleja E 2011, accepted for publication in Appl. Phys. Lett.

[5] Lazar S, et al. 2003, *Ultramicroscopy* **96** 535

[6] Egerton R F 1996, *EELS in the Electron Microscope* (Plenum Press, New York)

Journal of Physics: Conference Series **326** (2011) 012015 doi:10.1088/1742-6596/326/1/012015

Structural characterization of GaN/AlN layers on 3*C*-SiC/Si(111) by TEM

L M Sorokin[1], A E Kalmykov[1], A V Myasoedov[1], N V Veselov[1], V N Bessolov[1], N A Feoktistov[1], A V Osipov[2] and S A Kukushkin[2]

[1] Ioffe Physico-Technical Institute, Russian Academy of Sciences, St. Petersburg, 194021 Russia

[2] Institute for Problems of Mechanical Engineering, Russian Academy of Sciences, St. Petersburg, 199178 Russia

Lev.Sorokin@mail.ioffe.ru

Abstract. In the present work 20-μm-thick GaN epilayers have been grown by hydride–chloride vapor-phase epitaxy on 1.5-inch Si(111) substrates with AlN and SiC interlayers. Silicon carbide has been obtained by an original method of solid phase epitaxy. The transmission electron microscopy analysis revealed the effect of AlN layer structure on the quality of the gallium nitride epilayer. It is established that pores with dimensions of several nanometres formed on an AlN/GaN interface favour the growth of an epitaxial weakly strained single crystalline GaN layer. A good quality of the obtained GaN epilayers (local dislocation density is about 4×10^{7} cm^{-2}, average density ~10^{9} cm^{-2}) is demonstrated. The silicon substrate structure features near to a SiC interlayer will be discussed.

1. Introduction

Gallium nitride (GaN)-based heterostructures are extensively used in light emitting devices, high frequency, high power devices, and high-electron-mobility transistor substrates. The absence of low-cost GaN substrates stimulates a search for materials appropriate for heteroepitaxy. The use of sapphire as a substrate results in a high density of dislocations because of a significant (~13%) lattice mismatch in the (0001) interface plane [1]. At the same time, silicon carbide (SiC) substrates that are more suitable from the standpoint of lattice matching, are too expensive and have small areas. In recent years, much attention has been given to the growth of GaN on silicon (Si) substrates [2]. This interest is related, first, to the prospects for integration of the gallium-nitride and silicon electronics technologies and, second, to the possibility of using inexpensive single crystal silicon substrates, available in large diameters (up to 150 mm) and with reasonable thermal and electrical conductivity. However, GaN layers grown on Si suffer from crack generation, possibly due to the strain induced by a large lattice mismatch (17%) and difference in thermal expansion coefficients (33%) between GaN and Si. On the other hand, the lattice mismatch between (0001) GaN and (111) 3*C*-SiC is only 3% [3]. Thus, high quality GaN is expected by employing 3*C*-SiC as intermediate layers (ILs) in the GaN/Si system. The aim of this work is to study the influence of ILs morphology on the structural quality of GaN layers.

2. Experimental details

A thin SiC layer (110 nm) was initially formed on a Si(111) substrate by means of solid phase epitaxy [4,5], and then AlN (260 nm) and GaN (15 μm) layers were sequentially grown over the silicon carbide interlayer by the method of hydride–chloride vapour-phase epitaxy (HVPE) (sample 1) using a procedure described in [6]. Sample 2 differed from sample 1 in thickness of SiC (350 nm), AlN (100 nm) and GaN (20 μm) layers. The microstructure of GaN/AlN/3C-SiC/Si(111) heterostructure was examined by cross-sectional transmission electron microscopy (XTEM). Samples for XTEM were prepared by mechanical thinning with the use of tripod and ion milling in Gatan Precision Ion Polishing system at 4 kV. Conventional TEM study was carried out with a Philips EM420 transmission electron microscope operating at 100 kV.

3. Results and discussion

Figure 1 shows the microstructure of sample 1. The analysis of the selected area electron diffraction pattern (SAED) recorded from area of interfaces has confirmed that layers are epitaxial (figure 1b). The TEM analysis revealed the epitaxial relationships:

$$(0001)\mathrm{GaN}\|(111)3C\text{-}\mathrm{Si}\|(111)\mathrm{Si} \text{ and } [11\bar{2}0]\mathrm{GaN}\|[1\bar{1}0]3C\text{-}\mathrm{SiC}\|[1\bar{1}0]\mathrm{Si}.$$

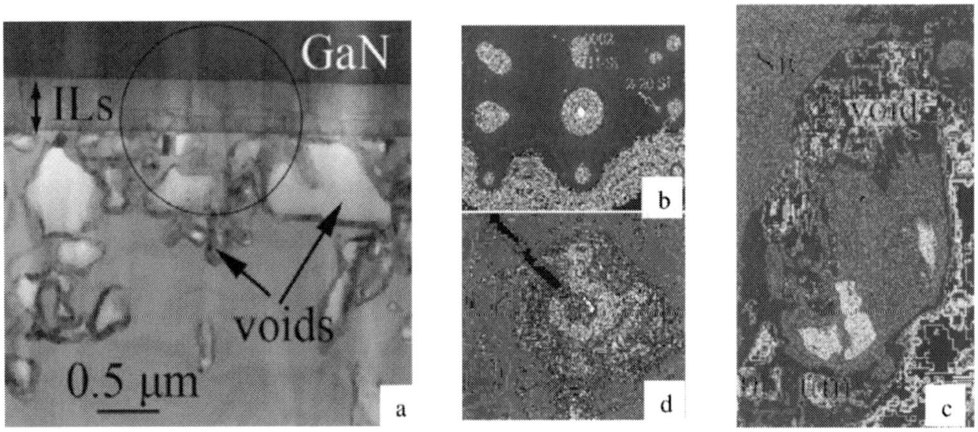

Figure 1. a: Cross-sectional TEM image of GaN/AlN/3*C*-SiC/Si(111) heterostructure (sample 1); b: SAED recorded from round area (figure 1). Diffraction spots of 0 0 0 2 (GaN and AlN) and 111 3*C*SiC are superimposed; c: a void in Si substrate; d: SAED recorded from the void.

The SiC interlayer was a product of chemical reaction between crystalline silicon and carbon monoxide [5].

$$2\mathrm{Si}(cr) + \mathrm{CO}(v) = \mathrm{SiC}(cr) + \mathrm{SiO}(v)\uparrow$$

Fort the formation of one molecule of SiC two atoms Si are necessary, and one of them volatilizes, which leads to the formation of voids in silicon near to interface Si/SiC [5,7]. The thickness of the defective layer in the substrate reaches 3 μm (figure 1). According to [5], the volume of pores must be approximately equal to the that of grown silicon carbide. In the electron micrograph, the total volume of voids seems to be somewhat greater, but it should be taken into account that the voids could be additionally increased in size as a result of etching with argon ions in the course of sample preparation. These voids improve the quality of a growing epilayer, since a decrease in the contact area leads to a reduction of elastic stresses at the interface that are caused by lattice mismatch and a difference in the coefficients of thermal expansion of the substrate and deposit layer.

TEM investigation has revealed that besides the SiC epilayer formed on the surface of a silicon substrate, inside of voids silicon carbide is also formed as a result of a chemical reaction between CO and the gaseous silicon (see streaky structure in figure 1c). Streaking of diffraction spots arises due to a large number of basal stacking faults (figure 1d).

In the SiC layer there are a lot of defects like stacking faults and dislocations, but there are no cracks. In [8], the presence or absence of cracks is connected to the thickness of the SiC layer: if its thickness is larger than x, the formation of cracks is suppressed. The value of x, according to [8], lies in the range of 0.05–1 μm. Though the thickness of the silicon carbide layer in our case (figure 2) also gets to this range, we consider that here the method of SiC growth has played a main part in the prevention of crack formation.

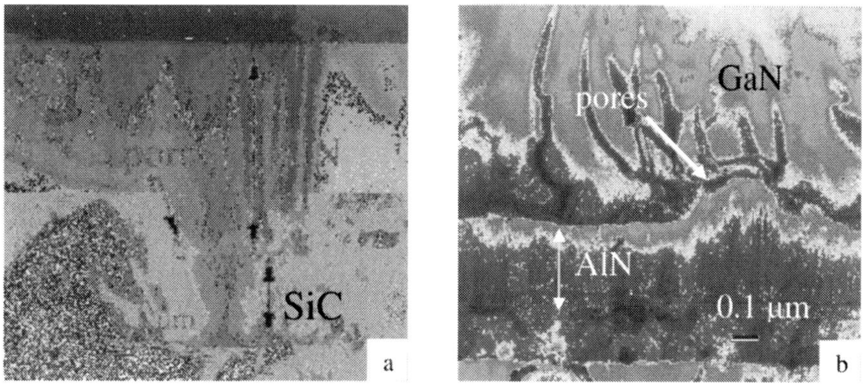

Figure 2. a — a pore in AlN layer at SiC/AlN interface; b — pores at AlN/GaN interface.

Generally, ILs have a smooth surface except for rare protrusions (figures 2,3). In the AlN layer, besides planar defects and dislocations, a small amount of pores in the size range of 10–20 nm is revealed. These pores are located on SiC/AlN and AlN/GaN interfaces.

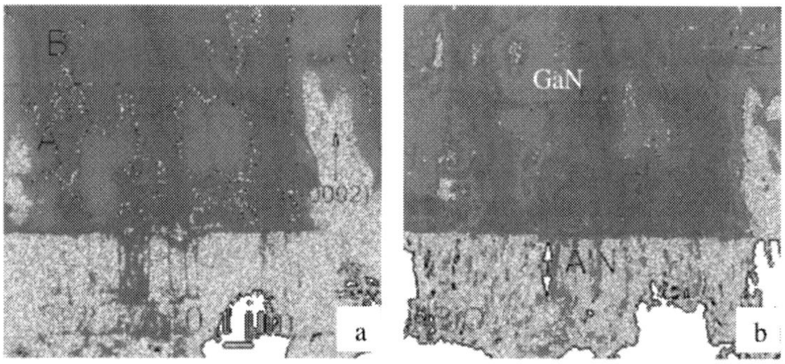

Figure 3. Two-beam images of the same area taken with different g vectors showing absence of c-screw dislocations. a: $\mathbf{g}$=(0002); b: $\mathbf{g}=(11\bar{2}0)$.

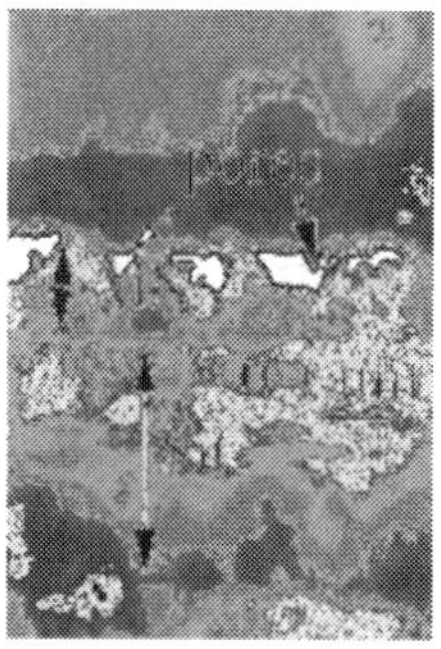

Figure 4. XTEM image of sample 2.

In the GaN layer presented in figure 3, it is possible to identify three areas. A very defective area (in figure not shown) of width 50 nm begins at the Si/SiC interface. The area A with a low density of dislocations follows further. In area B, the density is even lower. The reduction of the dislocation density is possible particularly because of a small number of threading dispositions with lines parallel to the c axis. Burgers vectors have been determined by two beam analysis. The dislocation density near the AlN layer is too high for the identification of these dislocations. The results of TEM analyses of distribution of different types of dislocations in sample 1 are summarized in Table 1.

Table 1. Distribution of different types of dislocations in different areas of sample 1 obtained by TEM analysis.

Type of dislocation	Burgers vector	Density (cm^{-2})	Area under investigation (figure 3a)
Screw	$a=1/3<11\bar{2}0>$	4×10^8	A+B
Mixed	$a=1/3<11\bar{2}0>$	1.4×10^8	A+B
Mixed	$a+c=1/3<11\bar{2}3>$	1×10^9	A+B
All types	$c=<0001>$	0	A+B
All types	*a*, *c* and *a+c*	3×10^9	A
All types	*a*, *c* and *a+c*	6×10^8	B

Figure 4 shows the structure of sample 2. The main difference of this structure from the structure of sample 1 is a presence of a system of pores on the AlN/GaN interface. The subsequent nucleation and growth of GaN grains starts on the tops of pyramids while pores are left between pyramids. The pores can be produced by the impurities collected on the growth fronts of the GaN layer [9]. These pores lower the stresses generated by mismatch between the lattice parameters, which leads to improvement of quality of the GaN layer. Results of measurements have shown that local dislocation density is about 4×10^7 cm^{-2}, average density ~10^9 cm^{-2}.

4. Conclusions

GaN epilayers grown by HVPE on 1.5-inch Si(111) substrates with AlN and SiC interlayers have been studied by TEM. A comparison of structures of samples with different morphology of interlayers has shown that a system of pores on the AlN/GaN interface lowered the dislocation density in the GaN layer. In the GaN layer the fraction of dislocation lines parallel to the *c* axis is small and $c=<0001>$ dislocations have not been revealed. Good quality GaN epilayers (with local dislocation density about 4×10^7 cm^{-2}, average density ~10^9 cm^{-2}, compare [2]) have been grown.

Acknowledgments

This study was supported by the Russian Foundation for Basic Research (project nos.10-03-00433, 09-03-00596, 11-02-00496 and 11-02-00662), the St. Petersburg Foundation for Support of Science and Education, and the Presidium of the Russian Academy of Sciences (Programs "Principles of Basic Research in Nanotechnologies and Nanomaterials" and "Support of Innovations and Developments").

References

[1] Rouviere J-L, Arlery M, Bourret A 1997 *Microscopy of Semiconducting Materials* 173–82
[2] Cordieretal Y *et al. Journal of Crysta lGrowth* **312** (2010) 2683–8
[3] http://www.ioffe.ru/SVA/NSM/
[4] Kukushkin S A, Osipov A V, Feoktistov N A Patent No 2008102398, filed on 22 January 2008
[5] Kukushkin S A, Osipov A V 2008 *Phys. Solid State.* **50** 1238–45
[6] Bessolov V N et al. 2006 *Tech. Phys. Let.* **32** 674–6
[7] Sorokin L M et al 2008 *Tech. Phys. Let.* **34** 992–4
[8] Komiyama J *et al* 2006 *Appl. Phys. Lett.* **88** 091901–3
[9] Liliental-Weber Z, Zakharov D, Wagner B, Davis R F 2006 *Proc. of SPIE* **6121** 612101–10

Strain-induced effects on the electronic structure and N K-edge ELNES of wurtzite AlN and $Al_xGa_{1-x}N$

M Petrov[1], D Holec[2,3], L Lymperakis[1], J Neugebauer[1] and C J Humphreys[3]

[1] Max-Planck-Institut für Eisenforschung GmbH, Max-Planck-Strasse 1, D-40237 Düsseldorf, Germany
[2] Department of Physical Metallurgy and Materials Testing, Montanuniversität Leoben, Franz-Josef-Strasse 18, A-8700 Leoben, Austria
[3] Department of Materials Science and Metallurgy, University of Cambridge, Pembroke Street, CB2 3QZ Cambridge, United Kingdom

david.holec@unileoben.ac.at

Abstract. Analysis of the electron energy loss near edge structure (ELNES) provides an experimental tool to probe the density of unoccupied states. Here we present a first principles study on the projected density of states (PDOS) of AlN, GaN, and $Al_xGa_{1-x}N$ alloys in order to investigate the impact of strain on the N K-edge ELNES. Uni-axial and bi-axial strain, volume conserving, and bi-axial stress deformation modes are calculated for the whole compositional range from AlN to GaN. Our results show that only the strain along the c-axis has a pronounced impact on the PDOS. Furthermore, we find that bi-axial stress in the basal plane, which is present in pseudomorphic polar heteroepitaxial layers, does not significantly influence the N K-edge spectra. However, strain-induced changes may appear for different deformation modes and/or specimen geometries.

1. Introduction

AlN, GaN, and their ternary alloys, $Al_xGa_{1-x}N$, are technologically important wide band gap semiconductors used in devices such as light emitting diodes (LEDs) and laser diodes (LDs) [1]. The large direct band gap of AlN (6.2 eV) makes high Al content $Al_xGa_{1-x}N$ alloys ideally suited for extending the spectral range of the optoelectronic devices into the deep ultraviolet range. However, due to the substantial lattice mismatch between AlN (a = 3.11 Å, c = 4.98 Å) and GaN (a = 3.19 Å, c = 5.11 Å) [2], residual strain is always present in heteroepitaxial $Al_xGa_{1-x}N$ structures.

The electron energy loss near edge structure (ELNES), a subset of the electron energy loss spectrum, reflects the density of unoccupied states and thus provides an experimental probe to the electronic structure [3]. The very high spatial resolution that can be achieved in modern transmission electron microscopes, allows e.g. to identify local compositional fluctuations. However, the interaction of the high energy probing electrons with lattice atoms in the specimen does not always have a straightforward interpretation. In order to understand the experimental data, measured EELS spectra need to be compared in detail with their calculated counterparts.

Recently, Keast *et al.* [4] performed an experimental and theoretical study of strain effects in the N K-edge of ELNES in $In_xGa_{1-x}N$ alloys. In that work, the edge onset was observed to shift towards lower energies with respect to the central peak due to the bi-axial stress. Surprisingly, none of the

Journal of Physics: Conference Series **326** (2011) 012016 doi:10.1088/1742-6596/326/1/012016

numerous papers on experimental and/or theoretical ELNES in AlN, GaN, and AlGaN [5–10] discusses strain effects. In the present work we therefore aim to fill this gap by performing first principle calculations on AlN, GaN and their ternary alloy AlGaN. Strain effects on the unoccupied density of states projected on N-sites for several deformation modes are investigated in order to predict the actual influence on the N K-edge.

2. Methodology

Calculations were carried out within density functional theory (DFT) in the plane-wave pseudopotential approach as implemented in the object-oriented DFT program library S/PHI/nX [11]. Specifically, Troullier-Martin pseudopotentials [12], a plane-wave cut-off of 70 Ry, the Perdew-Berke-Ernzerhof generalised gradient approximation [13] to describe exchange and correlation were used. 576 **k**-points in the irreducible part of the Brillouin zone were used. An energy convergence of 3×10^{-8} eV in the electronic minimisation was employed. Volume relaxation in accord with the deformation modes and relaxation of all atomic coordinates were fully taken into account. Various compositions of the ternary alloy were modelled with 2x2x1 supercells, where the atoms were distributed according to the special quasi-random structures approach as described in Ref. [14].

The N-K edge ELNES intensity is given by the joint density of the initial and final states. Since the electron is excited from a core N-*s* state with a very narrow (almost δ-function-like) energy range, the intensity profile is almost exclusively given by the projected density of states (PDOS) for the unoccupied N-*p* states. Therefore we use the unoccupied N-*p* PDOS to describe N K-edge ELNES. The calculated PDOS is broadened with a Gaussian having 0.8 eV full width at half maximum.

In order to gather a complete and consistent picture of the effect of strain on the N-*p* PDOS, several deformation modes were simulated. These were the uni-axial strain along the *c*-direction ($\varepsilon_{xx} = \varepsilon_{yy} = 0$, $-0.023 < \varepsilon_{zz} < 0.024$), bi-axial strain in the *c*-plane ($-0.012 < \varepsilon_{xx} = \varepsilon_{yy} < 0.012$, $\varepsilon_{zz} = 0$), deformation with the unit cell volume kept constant ($-0.012 < \varepsilon_{xx} = \varepsilon_{yy} < 0.012$, $0.024 > \varepsilon_{zz} > -0.023$), and the bi-axial stress case with $\sigma_{zz} = 0$ ($-0.012 < \varepsilon_{xx} = \varepsilon_{yy} < 0.012$, $0.007 > \varepsilon_{zz} > -0.006$) (see figure 1). Here, the uni- and bi-axial strain modes correspond to deformations with assumed Poisson ratio 0. The different deformations were achieved by varying the corresponding lattice parameters. The bi-axial stress mode corresponds to pseudomorphically grown heterostructures. Strain values of the order of 0.01 in the basal plane correspond approximately to pseudomorphically grown layers of $Al_{0.5}Ga_{0.5}N$ on either GaN or AlN.

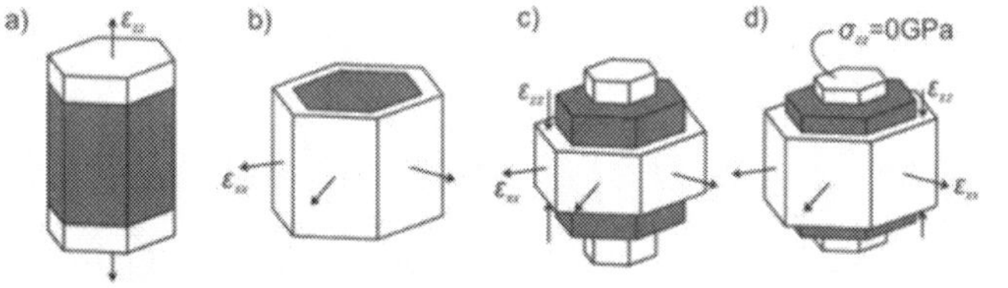

Figure 1. Deformation modes investigated in this paper: a) uni-axial strain along the *c*-direction, b) bi-axial strain in the *c*-plane, c) volume conserving deformations, and d) bi-axial stress with $\sigma_{zz} = 0$. The undeformed unit cells are shown in grey.

3. Results

The evolution of unoccupied N-*p* states with the amount of aluminium for the constant volume deformation mode is shown in figure 2. On the Al-rich side (figures 2a, b) it is clearly observed that when going from tension ($\varepsilon_{xx} > 0$) to compression ($\varepsilon_{xx} < 0$) in the *c*-plane, the first (at about 6 eV) and third (at about 11–12 eV) peaks sharpen. Furthermore, the second peak (at about 9–10 eV) diminishes. In contrast, inspecting the Ga-rich side (figures 2d, e) reveals (i) no strain-induced variations of the middle peak, and (ii) a small shift of the position of the third peak at about 11 eV.

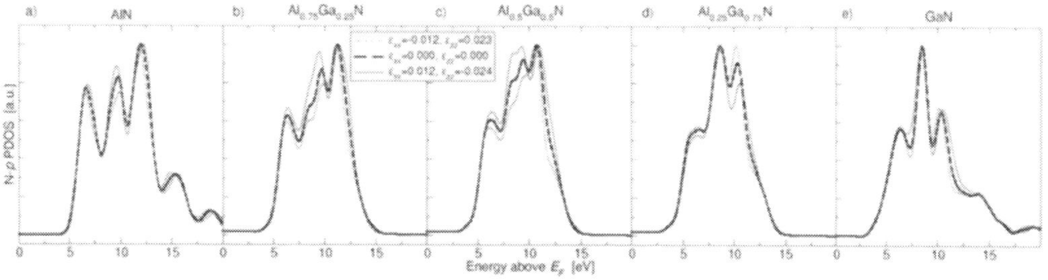

Figure 2. Unoccupied broadened N-p PDOS of $Al_xGa_{1-x}N$ for the volume conserving deformation mode. Blue solid, black dashed and red dash-dotted lines indicate compressive, zero, and tensile strains in the basal plane, respectively.

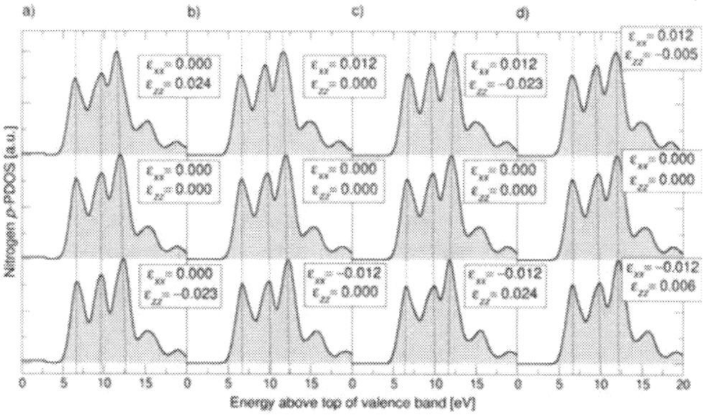

Figure 3. Effect of various strain/stress modes on the unoccupied broadened N-p PDOS of AlN: a) uni-axial strain along the c-direction, b) bi-axial strain in the c-plane, c) deformation with constant cell volume, and d) bi-axial stress in the c-plane with $\sigma_{zz} = 0$.

The unoccupied N-p states of AlN for the uni-axial strain along the c-axis and the bi-axial strain in the c-plane deformation modes are depicted in figures 3a and b. An interesting observation is that the middle peak intensity is much more sensitive to the uni-axial than the bi-axial strain case. Figure 3c shows the deformation mode with the constant unit cell volume and reveals the same behaviour as the uni-axial strain case. Based on these results it can be concluded, that it is the deformation along the c-axis which is responsible for the spectral changes in AlN. Lastly, figure 3d suggests that the PDOS variations in the bi-axial stress state are small. This is consistent with fact that for this case deformations along the c-axis are small.

Apart from the strain induced changes in the peak shapes, strain also shifts their positions. The distance between the second and third peaks in the uni-axial deformation mode varies as much as 0.7 eV (2.6 eV for $\varepsilon_{zz} = -0.02$, 1.9 eV for $\varepsilon_{zz} = 0.02$), while the distance between the first and the second peak, as well as the peak distances in the bi-axial strain mode remain almost unaffected. These trends combine in the constant unit cell volume case resulting in a similar behaviour to the uni-axial mode. Finally, in the case of the bi-axial stress deformation, the peak positions remain practically invariant.

4. Discussion

It has been shown that the unoccupied N-p_z states orientated along the c-axis contribute mainly to the first and third peak of the N K-edge ELNES, whereas the in-plane N-(p_x+p_y) states contribute to all three peaks [15,16]. This is consistent with our predicted strain response of the N-p PDOS. Deformations along the c-axis (e.g. the uni-axial mode) are expected to influence mainly the p_z states contributing to the first and the third peak. As a consequence, the relative intensities of the three peaks are expected to change. On the other hand, deformations within the c-plane (e.g. the bi-axial strain mode) affect preferably the p_x+p_y states. Thus the relative intensities are not expected to vary significantly.

The above described observations have several implications: (i) The second peak in the AlN triplet is highly sensitive to strain along the *c*-direction. (ii) Since the physically most relevant strain/stress state for polar (Al,Ga)N material, the bi-axial stress ($\sigma_{xx} = \sigma_{yy} \neq 0$ with $\sigma_{zz} = 0$ and thus the proper Poisson ratio), shows only a very weak dependence, the experimental N K-edge of $Al_xGa_{1-x}N$ is expected to be practically strain-independent. (iii) Additionally, the N-K edge onset will also shift depending on the band gap which decreases with the tensile in-plane strain (not discussed here). (iv) When switching from polar to non- and semi-polar materials, which are currently of a great interest, the actual strain-induced changes are predicted to emerge due to different (and much more pronounced) deformations along the *c*-axis. To actually observe this effect, much care in the sample design and preparation will be needed as strain will probably be reduced due to relaxation effects resulting from the thinness of the specimen.

5. Conclusions

First principle calculations on strained ternary $Al_xGa_{1-x}N$ alloys have been performed with a particular focus on the unoccupied N-*p* PDOS as an approximation to the N K-edge ELNES. Our results clearly indicate that applied strain has a substantial effect on peak shape and peak position. These effects are maximised for deformations along the *c*-axis. However, significantly less pronounced effects are observed when the in-plane strain is applied in the *c*-plane. We explain this behaviour by the fact, that the p_z states (influenced by the *c*-axis deformation) contribute mostly to the first and third peak, while the p_x+p_y states (affected by the in-plane strain) contribute to all peaks. Finally, we observe that the experimentally obtained bi-axial strain is not likely to influence samples with polar geometry (i.e. when stress is applied along the *c*-plane), while it may influence specimens with non- and semi-polar geometries.

References

[1] Jain S C, Willander M, Narayan J and Overstraeten R V 2000 *J. Appl. Phys.* **87** 965–1006
[2] Wright A and Nelson J 1995 *Phys. Rev. B* **51** 7866–7869
[3] Rez P and Muller D A 2008 *Annu. Rev. Mater. Res.* **38** 535–558
[4] Keast V J, Kappers M J and Humphreys C J 2003 *J. Microsc. (Oxford)* **210** 89–93
[5] Radtke G, Epicier T, Bayle-Guillemaud P and Le Bossé J C 2003 J. Microsc. 210 60-65 131
[6] Mizoguchi T 2003 *Micron* **34** 249–254
[7] Mizoguchi T, Tanaka I, Yoshioka S, Kunisu M, Yamamoto T and Ching W Y 2004 *Phys. Rev. B* **70** 045103
[8] Gao S P, Zhang A, Zhu J and Yuan J 2004 *Appl. Phys. Lett.* **84** 2784–2786
[9] Holec D, Costa P M F J, Cherns P D and Humphreys C J 2008 *Micron* **39** 690–697
[10] Holec D, Rachbauer R, Kiener D, Cherns P D, Costa P M F J, McAleese C, Mayrhofer P H and Humphreys C J 2011 *Phys. Rev. B* **83** 165122
[11] Boeck S, Freysoldt C, Dick A, Ismer L and Neugebauer J 2011 *Comput. Phys. Commun.* **182** 543–554
[12] Troullier N and Martins J L 1991 *Phys. Rev. B* **43** 1993–2006
[13] Perdew J P, Burke K and Ernzerhof M 1996 *Phys. Rev. Lett.* **77** 3865–3868
[14] Wei S H, Ferreira L, Bernard J and Zunger A 1990 *Phys. Rev. B* **42** 9622–9649
[15] Radtke G, Bayle-Guillemaud P and Thibault J 2004 *Electron Microscopy And Analysis 2003 conference (Institute of Physics Conference Series no 179)* pp 303–306
[16] Lawniczak-Jablonska K, Suski T, Gorczyca I, Christensen N E, Attenkofer K, Perera R C C, Gullikson E M, Underwood J H, Ederer D L and Liliental-Weber Z 2000 *Phys. Rev. B* **61** 16623–16632

Cathodoluminescence measurement of grain boundary recombination velocity in vapour grown *p*-CdTe

B G Mendis and L Bowen

Department of Physics, Durham University, South Road, Durham, DH1 3LE, UK

E-mail: b.g.mendis@durham.ac.uk

Abstract. A cathodoluminescence (CL) based method for measuring the minority carrier diffusion length (L) and reduced recombination velocity (S_{red}) of an individual grain boundary is presented. The technique is based on the van Roosbroeck model for the steady-state carrier distribution near a free surface in a semi-infinite solid. Values of L = 0.55±0.03 μm and S_{red} = 0.23±0.02 are obtained for vapour grown *p*-type CdTe. The effect of the electron beam generation volume on the accuracy of measurement is also discussed. Intermediate beam voltages (e.g. 15 kV for CdTe) are likely to produce the most reliable results. The method can be used to characterise the electrical activity of grain boundaries in thin-film solar cells.

1. Introduction

Grain boundaries have an important effect on the electrical properties of thin-film solar cells, such as CdTe and Cu(In,Ga)Se_2. Charge is accumulated at the grain boundary due to the energy difference between the quasi-neutrality level of the boundary and Fermi level in the bulk semiconductor. The built-in potential leads to band bending (upward bending for *n*-type semiconductors and downward bending for *p*-type), such that majority carriers are repelled and minority carriers attracted towards the boundary, at a rate given by the recombination velocity, where they undergo recombination. Grain boundaries with a high recombination velocity are sinks for minority carriers, as well as strong barriers for majority carrier flow, and therefore reduce solar cell efficiency. The electrical activity depends on the nature of the grain boundary (e.g. coincident site lattice or random grain boundary).

Electron beam induced current (EBIC) has previously been used to measure the *reduced* recombination velocity (i.e. the recombination velocity divided by the bulk diffusion velocity for minority carriers) of individual grain boundaries in silicon [1-3]. The reduced recombination velocity is extracted from curve fitting of the raw data to non-analytic functions involving several material and experimental parameters. The electron beam generation volume is modeled using a simplified analytic function (e.g. a spherically symmetric Gaussian [3]). The present authors however have developed a cathodoluminescence (CL) method for measuring the reduced recombination velocity [4], based on the van Roosbroeck [7] model for steady-state carrier distribution near a free surface. The advantages of this method are that it does not need any electrical contacts (cf. EBIC) and the parameters of interest are extracted directly from the experimental data, without the need for curve fitting. However, the electron beam generation volume is assumed to be a uniform plane parallel to the boundary, which is unrealistic. In this paper the effect of the electron beam energy (and hence generation volume) on accuracy of the CL measurement is discussed. For completeness the theory behind the CL technique and experimental data on *p*-type CdTe are also presented, although a more detailed report on these aspects can be found in [5].

2. Background theory

The van Roosbroeck model [5] describes the steady-state carrier distribution in a semi-infinite solid containing a free surface. Time independent, uniform carrier generation in a plane parallel to and at a distance x_b from the free surface is assumed. If a grain boundary is substituted for the free surface the CL intensity, $I(x_b)$, for an electron beam incident at position x_b is:

$$I(x_b) = k\int_0^{\infty}\eta(x)\left[\exp\left(-\frac{|x-x_b|}{L}\right)-\frac{S_{red}-1}{S_{red}+1}\exp\left(-\frac{(x+x_b)}{L}\right)\right]dx \qquad (1)$$

where L is the minority carrier diffusion length, S_{red} the reduced recombination velocity, η the CL emission efficiency and k a constant relating to CL detection efficiency. The term within the square brackets is the van Roosbroeck steady-state carrier distribution [5], i.e. it is assumed that the CL intensity is directly proportional to the carrier density. This assumption may be invalid for strongly excitonic materials or at high injection levels, but was experimentally verified for CdTe by demonstrating a linear relationship between the CL intensity and electron beam current.

If η is assumed to be independent of position x, equation (1) simplifies as:

$$\ln[\Delta I(x_b)] = \ln\left(\frac{S_{red}}{S_{red}+1}\right)-\frac{x_b}{L} \ ; \ \Delta I(x_b) = 1-\frac{I(x_b)}{I(x_b=\infty)} \qquad (2)$$

A plot of $\ln[\Delta I(x_b)]$ vs x_b should therefore be a straight line. The gradient of the straight line gives the diffusion length L and the intercept gives the reduced recombination velocity S_{red}.

3. Experimental procedure

Panchromatic CL images were obtained from a ~70 μm thick, p-type CdTe thin-film vapour grown on a Ge substrate. A Hitachi SU-70 scanning electron microscope (SEM) operating at 15 keV and 1.6 nA beam current was used together with a Gatan MonoCL system for CL detection. The pixel dwell time for the panchromatic CL image was 300 μs, i.e. significantly longer than the electron minority carrier lifetime (τ) of 20 ns [6], thus ensuring steady-state conditions.

2D Monte-Carlo simulations of the electron beam generation 'volume' were carried out using screened Rutherford cross-sections, and from this the steady-state minority carrier distribution in bulk p-type CdTe was determined by solving the continuity equation using finite difference methods. In this paper we refer to this carrier distribution as the 'steady-state source distribution'. The simulation parameters are L = 500 nm (section 4.1), τ = 20 ns [6] and 1.6 nA electron beam current. A free surface recombination velocity of 5×10^5 cm/s [7] was applied as the boundary condition.

4. Results and Discussion

4.1. Experimental data

Figure 1(a) shows a panchromatic CL image of a near end-on grain boundary in p-CdTe. The intensity normal to the grain boundary is extracted from the box region in figure 1(a), and is shown in figure 1(b). The intensity profile is used to plot a $\ln[\Delta I(x_b)]$ vs x_b graph (figure 1(c)), which shows overall linear behavior except for data points closer to the grain boundary. The deviation from linearity could be due to variations in the CL emission parameter η at the grain boundary compared to the bulk (see [6] for a more detailed discussion on the effects of η). A straight line was least squares fitted to figure 1(c), ignoring the first four data points closest to the grain boundary, and values of L = 0.55 ± 0.03 μm, S_{red} = 0.23 ± 0.02 extracted from the gradient and intercept respectively. The diffusion length L is similar to values reported in the literature [9]. Using τ = 20 ns [6], the measured values of L and S_{red} give a recombination velocity of ~630 cm/s. This is small compared to a 5×10^5 cm/s recombination velocity for a free surface in p-CdTe [7] or a value of ~10^4 cm/s for grain boundaries in Si [1-2]. However, our observations as well as previous reports [8] indicate that vapour grown CdTe contains low misorientation, sub-grain boundaries, which could explain the small recombination velocity.

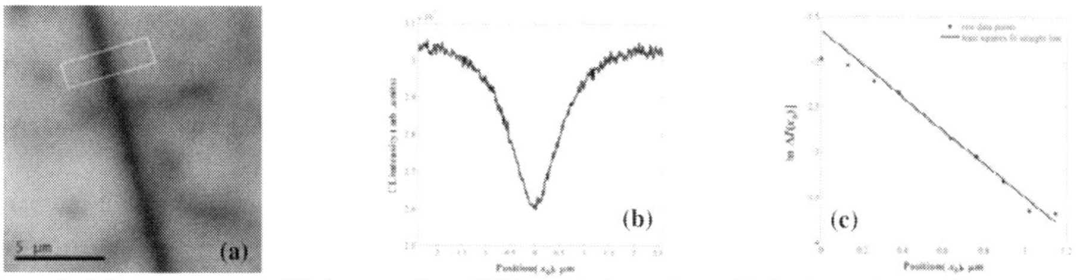

Figure 1. (a) panchromatic CL image of a p-CdTe grain boundary, (b) the intensity profile normal to the grain boundary extracted from the box region in figure 1(a) and (c) the $\ln[\Delta I(x_b)]$ vs x_b plot.

4.2. Numerical simulations

Although a near-linear relationship was observed between $\ln[\Delta I(x_b)]$ and x_b (figure 1(c)), carrier generation by the electron beam needs to be analysed in more detail. In the time independent van Roosbroeck model uniform carrier generation in an infinite plane parallel to the grain boundary is assumed. Since the CL data are acquired under steady-state conditions, the steady-state source distribution (section 3) effectively provides the excess carriers. Simulated steady-state source distributions for 5, 15 and 25 keV beam energies in p-CdTe are shown in figure 2. The spatial extent of the steady-state source distribution increases with beam energy.

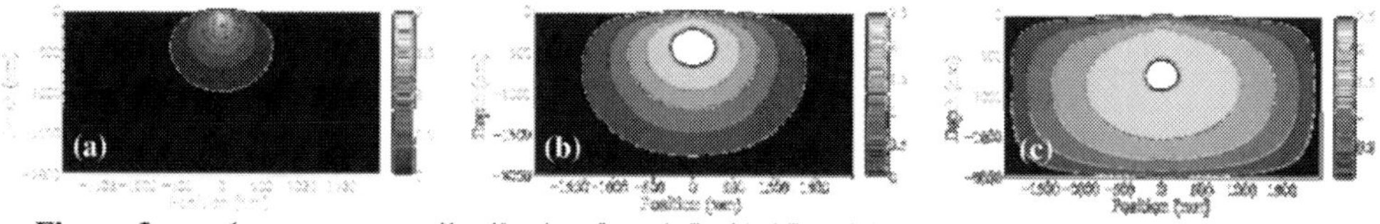

Figure 2. steady-state source distribution for (a) 5, (b) 15 and (c) 25 keV electron beams in CdTe. Excess carriers are plotted on a logarithmic scale (note intensity scale for different figures).

The van Roosbroeck criterion of an infinite carrier generation plane clearly cannot be satisfied. A more reasonable criterion is that the dimensions of the steady-state source distribution must be at least $2L$. The full width at half maximum (FWHM) of the steady-state source distribution along a given direction is arbitrarily defined as being the characteristic dimension along that direction. Hence the plots in figure 2 must have a FWHM of at least $2L$ along the vertical and horizontal axes. For the 15 keV beam the FWHM is 820 and 1070 nm along the vertical and horizontal axes respectively (the FWHM was calculated through the most intense region of the steady-state source distribution using a linear scale for the excess carriers). Since L~500 nm (section 4.1), the steady-state source distribution size is satisfactory for the 15 keV beam. Beam energies below 15 keV (e.g. 5 keV) have steady-state source distributions that are more point-like and are therefore unsuitable, and vice-versa for beam energies above 15 keV (e.g. 25 keV).

The symmetry of the plots in figure 2 about the vertical beam direction means that if the steady-state source distribution is uniformly broadened by 2δ ($\delta \geq L$) along the horizontal axis, then the plane of excess carriers in the van Roosbroeck model must be replaced by a uniform slab of thickness 2δ. If the slab is centred at the incident beam position x_b the steady-state carrier distribution, $c'(x)$, at distance x from the boundary is obtained from superposition of the van Roosbroeck result over the entire slab thickness, i.e.:

$$c'(x) = \frac{1}{2\delta}\int_{x_b-\delta}^{x_b+\delta}\left[\exp\left(-\frac{|x-x_b|}{L}\right) - \frac{S_{\mathrm{red}}-1}{S_{\mathrm{red}}+1}\exp\left(-\frac{(x+x_b)}{L}\right)\right]dx_b \quad \text{or} \quad c'(x) = c(x)\left[\frac{\sinh(\delta/L)}{(\delta/L)}\right] \qquad (3)$$

where $c(x)$ is the van Roosbroeck distribution for an excess carrier plane at position x_b (i.e. $c(x)$ is equal to the integrand in equation (3)). The $1/(2\delta)$ pre-integral factor is a normalisation constant. Equation (3) indicates that a uniform slab of excess carriers does not change the shape of the steady-state carrier distribution from the van Roosbroeck model, but simply increases its magnitude by a constant factor. This has no effect on the extracted values of L and S_{red} or the linear relationship between $\ln[\Delta I(x_b)]$ and x_b. It must be noted however that this is only an approximation, since the true steady-state source distribution is more complicated than a uniform slab of constant thickness.

Figure 3 shows the steady-state carrier distribution for a 15 keV electron beam incident at a distance of $0.5L$, or 250 nm, away from a grain boundary, which is at the arbitrary position of 0 nm. It is assumed that carrier generation by the electron beam is not significantly modified at the grain boundary. In figure 3(a) the grain boundary recombination velocity is 630 cm/s (value obtained in this study), while in figure 3(b) it is 10^4 cm/s (value for a high angle grain boundary in Si [1-2]). It is clear that excess carriers can be present on either side of the grain boundary, even if the recombination velocity is relatively large. The van Roosbroeck model however is for a free surface and hence excess carriers can only be present on one side of the boundary, meaning that the $\ln[\Delta I(x_b)]$ vs x_b plot will deviate from linearity for electron beam positions close to the grain boundary. Since the steady-state source distribution size increases monotonically with beam energy, higher beam energies give rise to larger deviations from linear behaviour. Note that the CL emission parameter η could also vary close to the grain boundary and is a further reason for non-linear behaviour.

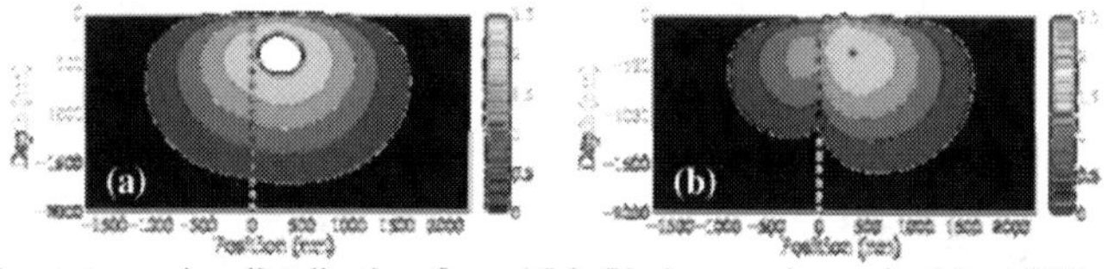

Figure 3. steady-state carrier distribution for a 15 keV electron beam incident 250 nm away from a grain boundary (broken line in the figure), which is at the arbitrary position of 0 nm. In (a) the grain boundary recombination velocity is 630 cm/s, while in (b) it is 10^4 cm/s.

5. Conclusions

A CL method, based on the van Roosbroeck model, for measuring the reduced recombination velocity of a single grain boundary is described and applied to *p*-type CdTe. The role of electron beam energy on the accuracy of the measurement has also been investigated. Low beam energies are unsuitable as the steady-state source distribution is more similar to a point-source, rather than a plane of excess carriers. High beam energies however generate carriers on either side of the grain boundary, while in the model it is assumed that the carriers are only present on one side. Hence an intermediate beam energy (e.g. 15 keV for CdTe) is likely to give the most reliable results. Further experiments are required to confirm this. Furthermore a comparison with Donolato's EBIC method [1], applied to the same grain boundary, is necessary in order to determine the accuracy of the reduced recombination velocity measured using the CL technique.

References

[1] Donolato C 1983 *J. Appl. Phys.* **54** 1314.
[2] Romanowski A, Buczkowski A 1985 *Solid State Electr.* **28** 645.
[3] Donolato C 1994 *Mat. Sci. Eng. B* **24** 61.
[4] Mendis BG, Bowen L and Jiang QZ 2010 *Appl. Phys. Lett.* **97** 092112.
[5] van Roosbroeck W 1955 *J. Appl. Phys.* **26** 380.
[6] Wight DR, Bradley D, Williams G, Astles M, Irvine SJC and Jones CA 1982 *J. Cryst. Growth* **59** 323.
[7] Cohen R, Lyahovitskaya V, Poles E, Liu A and Rosenwaks Y 1998 *Appl. Phys. Lett.* **73** 1400.
[8] Watson CCR and Durose K 1993 *J. Cryst. Growth* **126** 325.

Journal of Physics: Conference Series **326** (2011) 012018 doi:10.1088/1742-6596/326/1/012018

Aberration-corrected electron microscopy of MnAs and As nanocrystals and voids in annealed (Ga,Mn)As

A Kovács[1], T Kasama[1], J Sadowski[2,3], T Dietl[3,4] and R E Dunin-Borkowski[5,1]

[1] Center for Electron Nanoscopy, Technical University of Denmark, DK-2800 Kgs. Lyngby, Denmark
[2] MAX-Lab, Lund University, S-221 00 Lund, Sweden
[3] Institute of Physics, Polish Academy of Science, PL-02-668 Warsaw, Poland
[4] Institute of Theoretical Physics, University of Warsaw, PL-00-681 Warsaw, Poland
[5] Ernst Ruska Centre for Microscopy and Spectroscopy with Electrons, Forschungszentrum Jülich, D-52425 Jülich, Germany

E-mail: ak@cen.dtu.dk

Abstract. Aberration-corrected transmission electron microscopy is used to study voids and nano-crystalline MnAs and As phases formed during the annealing of Mn-doped GaAs. The effects of defocus and inner annular dark-field detector semi-angle on contrast of the nanocrystals are discussed.

1. Introduction

A composite system comprising ferromagnetic MnAs in GaAs is of interest for combining the properties of semiconductors and magnetic materials [1]. The formation of MnAs nanocrystals in GaAs can be achieved by the low-temperature growth of a Mn-doped GaAs layer followed by annealing at a temperature above 773 K [1]. The densities and sizes of the MnAs nanocrystals are determined by the initial Mn concentration and the kinetics of the heat treatment. As a result of the small sizes of the nanocrystals and the sensitivity of GaAs to specimen preparation for transmission electron microscopy (TEM), the detailed chemical and physical properties of this composite system are not yet fully understood. Here, we use both image-aberration-corrected TEM and probe-aberration-corrected scanning TEM (STEM) to study voids and both MnAs and As nanocrystals in (Ga,Mn)As annealed at 560 and 630 °C.

2. Experimental

The samples described below were deposited in a KRYOVAK molecular beam epitaxy (MBE) system onto a GaAs (001) substrate. (Ga,Mn)As layers were grown at 270 °C using an As_2 flux generated by a DCA valve cracker effusion cell with an As/Ga flux ratio of ~2. The Mn concentration was varied between 0.1 and 2 %. After MBE growth, the samples were annealed at 400, 560 or 630 °C in ultra-high vacuum conditions. Annealing at the highest temperature of 630 °C was carried out in the presence of an As_2 flux to prevent degradation the surfaces of the samples due to As desorption.

Structural and chemical characterisation was performed on cross-sectional TEM specimens that had been prepared for TEM examination using conventional mechanical polishing and low-energy (< 1 keV) Ar ion milling. Great care was taken during TEM specimen preparation to minimise Ar-ion-beam induced artefacts. Both image-aberration-corrected TEM and probe-aberration-corrected STEM studies were carried out using FEI Titan microscopes operated at 300 kV. Objective lens aberrations were corrected up to fourth order and a small negative value of the spherical aberration coefficient (*Cs*) was used. The beam convergence semi-angle used for STEM was 15.7 mrad. The inner detector semi-angle for annular dark-field (ADF) imaging was varied between 20.2 and 99.2 mrad when collecting low-angle ADF (LAADF) and high-angle ADF (HAADF) signals. Chemical compositions were determined using energy-dispersive X-ray spectroscopy (EDXS) with an INCA X-sight Oxford Instruments detector.

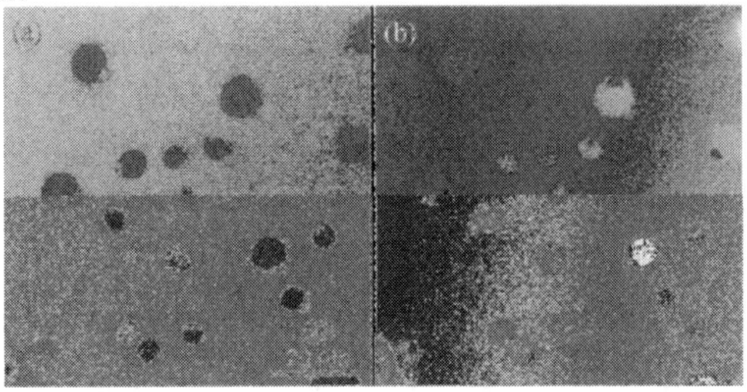

Figure 1. (a) Bright-field and (b) LAADF STEM images of a (Ga,Mn)As layer that had been annealed at 630 °C. The inner detector semi-angle used was 30.9 mrad. The dark contrast adjacent the nanocrystals in (b) corresponds to voids.

3. Results and Discussion

Figure 1 shows representative low magnification bright-field (BF) and LAADF STEM images of a (Ga,Mn)As sample that had been annealed at 630 °C. The nanocrystals were identified to have either hexagonal (NiAs-type, space group 194) or cubic (ZnS-type, space group 216) crystal structures using electron diffraction [2]. In the samples that had been annealed at 560 °C, As nanocrystals were identified, as described below. The MnAs nanocrystals in figure 1 appear dark in the BF image and bright in the LAADF image. Voids adjacent to the nanocrystals are barely visible in the BF image, whereas they are readily apparent in the LAADF image. The bright band contrast around each void in the LAADF image may be associated with strain and is expected to depend sensitively on the collection angle and sample thickness [3].

3.1. TEM imaging

Figure 2 shows aberration-corrected high-resolution TEM images of a hexagonal MnAs nanocrystal and an associated void in the GaAs host, recorded using different objective lens defocus values. The images illustrate the complexity of TEM image interpretation, despite the fact that the image delocalization that would be present at the interfaces between the void or the nanocrystal and GaAs is almost absent due to the small values of *Cs* (~ -5 μm) and defocus used. Although the presence of the void results in an effective difference in specimen thickness, it is still barely visible in figure 2 (a). A change in defocus by 20 nm results in different image contrast at the position of the void, as shown in figure 2 (b). Interestingly, the apparent size of the MnAs nanocrystal measured from each image was different (5.1x6.1 nm vs. 6.9x6.6 nm). In contrast, the associated Moiré fringe pattern was present in each image making the image interpretation difficult.

Figure 2. Aberration-corrected TEM images of a hexagonal MnAs nanocrystal and associated void recorded using defocus values that differ from each other by 20 nm. The scale bar is 5 nm.

3.2. LAADF and HAADF STEM imaging

Figure 3 shows high-resolution aberration-corrected STEM images of a cubic nanocrystal and an adjacent void recorded using different inner ADF detector semi-angles. The nanocrystal is fully coherent with the GaAs matrix but has a slightly larger lattice parameter than that of GaAs, resulting in a strain field, associated with bright contrast in the LAADF image shown in figure 3 (a). The presence of lighter Mn atoms (Z=25) in the nanocrystal results in a slightly darker contrast than that of the GaAs host (Z=31, Z=33) in the HAADF image, shown figure 3 (b). A steep decrease in intensity of the nanocrystal relative to that of GaAs was observed when the inner detector semi-angle was increased to ~30 mrad, followed by a reversal of its contrast and a gradual further decrease in intensity, as shown in figure 3 (c). A quantitative interpretation of the effect of strain on such images requires a detailed comparison the experimental results with simulations [3, 4].

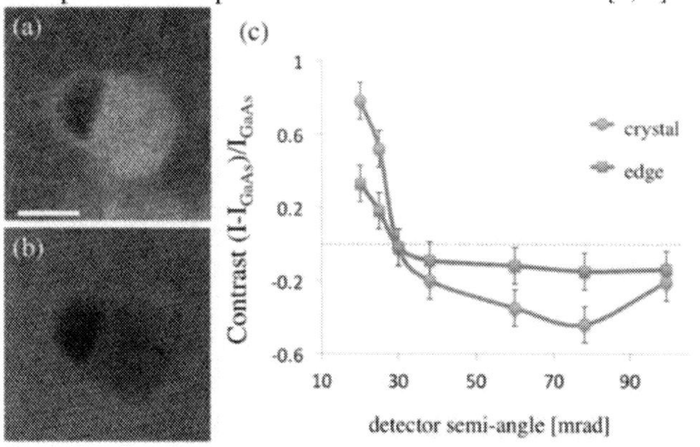

Figure 3. High-resolution (a) LAADF and (b) HAADF STEM images of a cubic (Ga,Mn)As nanocrystal. The inner detector semi-angles used were (a) 24.5 and (b) 78.4 mrad. The scale bar is 5 nm. (c) Contrast of the nanocrystal and the edge of the void relative to GaAs measured using different ADF inner detector semi-angles.

Unexpectedly, rhombohedral and orthorhombic As nanocrystals were identified adjacent to hexagonal MnAs crystals in the sample that had been annealed at 560 °C. The identification of As relied on excluding the possible presence of Mn in the nanocrystals. Figure 4 shows a high-resolution aberration-corrected ADF STEM image and EDXS spectra recorded from MnAs and As nanocrystals adjacent to a void. A Mn signal was measured at the position of the MnAs nanocrystal, while an increasing As was measured at the position of the As nanocrystal, as shown in figure 4 (b) and (c).

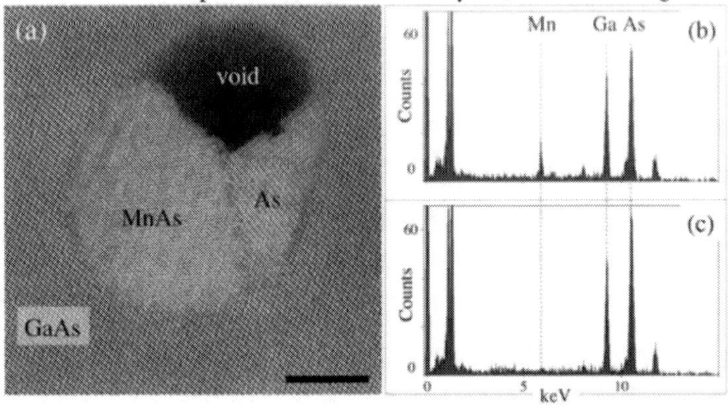

Figure 4. (a) ADF STEM image acquired using inner detector semi-angle 47.4 mrad from a hexagonal MnAs, a rhombohedral As crystals and a void in a GaAs host. The scale bar is 5 nm. (b) and (c) EDXS measurements obtained from the MnAs and As crystals, respectively. Mn, Ga and As K lines are marked.

4. Conclusions

A composite system comprising embedded voids and both MnAs and As nanocrystals in a GaAs matrix was studied using both probe- and image-aberration-corrected electron microscopy. High-resolution TEM images of the MnAs nanocrystals were found to be highly sensitive to imaging parameters such as defocus values, however the limiting factor of the image interpretation is still the Moiré fringes. ADF STEM images recorded as a function of camera length suggest the presence of strain fields around the embedded MnAs nanocrystals. Future work will involve the use of image simulations to understand the observed contrast quantitatively.

References

[1] De Boeck J, Oesterholt R, Van Esch A, Bender H, Bruynseraede C, Van Hoof C and Borghs G 1996 *Appl. Phys. Lett.* **68**, 2744
[2] Kovács A, Sadowski J, Kasama T, Domagala J, Mathieu R, Dietl T and Dunin-Borkowski R E 2011 *J. Appl Phys* in press
[3] Yu Z, Muller D A and Silcox J 2004 *J. Appl. Phys.* **95** 3362
[4] Grillo V, Carlino E and Glas F 2008 *Phys. Rev. B* **77** 54103

Acknowledgments

This work was supported by the "FunDMS" Advanced Grant of the ERC within the "Ideas" 7th Framework Programme of the EU.

Structure and electronic properties of μc-SiC:H for photovoltaic applications

A Heidt, T Chen, R Carius, F Finger, U Rau
Institute of Energy and Climate Research, Research Centre Jülich, 52425, Germany

J Mayer
General Facility for Electron Microscopy, RWTH Aachen, Ahornstrasse 55, 52074 Aachen, Germany

M Luysberg[1]
Peter Grünberg Institute 5 and Ernst Ruska-Centre for Microscopy and Spectroscopy with Electrons, Research Centre Jülich, 52425, Germany

E-mail: a.heidt@fz-juelich.de, m.luysberg@fz-juelich.de

Abstract. Thin microcrystalline silicon carbide films (μc-SiC:H) were grown by the hot-wire deposition technique for implementation as a window layer into thin-film solar cells. The μc-SiC:H layers are characterised by a complex microstructure, which is dominated by crystalline columns with a high density of planar defects. In order to understand the relation between structural and optical properties of the μc-SiC:H layers, low-loss electron energy loss spectra were recorded. At heterogeneous grain boundaries and within the nucleation zone, which is characterised by a high density of smaller grains, we observe that the volume plasmon peak position shifts to lower energy compared to measurements within larger grains of the microcrystalline SiC:H film. This shift of the plasmon energy is attributed to a higher degree of disorder, which alters the electronic properties.

1. Introduction

Silicon carbide is an attractive material for optoelectronic applications, because of its wide bandgap, high mobility, high thermal conductivity and resistance to high temperature oxidation [1]. These properties make SiC particularly interesting for photovoltaic applications, i.e. as a window layer within microcrystalline silicon (μc-Si) solar cells. The thin film deposition of hydrogenated, microcrystalline silicon carbide (μc-SiC:H) by various established chemical vapour deposition (CVD) techniques has been proven to be difficult, since unwanted phase mixtures of Si and SiC occur [2]. Using hot-wire (HW) CVD, μc-SiC:H films containing crystalline SiC (c-SiC) grains have been successfully grown [3-7]. The optimum conditions for crystalline growth are governed by the growth conditions, e.g. pressure during growth [8] or dilution of monomethylsilane (MMS) in hydrogen [4]. Typically these films are of microcrystalline nature, showing a columnar growth of SiC crystals. In the paper presented here we report on local variations of the structural and electronic properties in Al-

[1] To whom any correspondence should be addressed.

doped μc-SiC:H films, which are measured by transmission electron microscopy (TEM) and electron energy loss spectroscopy (EELS), respectively.

2. Experimental

μc-SiC:H thin films were co-deposited on glass (Corning 1737), glass coated with chromium (Cr) and polished Si wafer in an ultra high-vacuum HWCVD chamber with Rhenium filaments heated to 2000 °C. The substrate temperature was kept at about 360 °C during deposition. A gas mixture of (MMS) diluted in hydrogen with a concentration of 0.6 % was used. Trimethylaluminum (TMAl), $[Al_2(CH_3)_6)]$ is the Al-doping source. The TMAl/MMS ratio was adjusted by controlling the flow rate of carrier gas hydrogen. Details of the growth procedure are described elsewhere [7]. TEM specimens have been prepared by focused ion beam thinning method (FIB) using 15 kV and subsequently 5 kV Ga ions, followed by a low angle Ar ion milling (2 kV) for some minutes.

For analysis of structural and electronic properties a Tecnai F20 TEM operated at 200 kV and equipped with a Gatan imaging filter (Tridiem 866 ERS) has been used. Energy filtered TEM (EFTEM) imaging series have been recorded with an energy selective slit of 0.2 eV, a step size of 0.1 eV and a spectral range of 14 to 26 eV. Electron energy loss spectra were acquired with a dispersion of 0.05 eV/channel, a 2 mm GIF entrance aperture and parallel illumination.

3. Results and Discussion

The structural properties of microcrystalline SiC:H are quite complex. Apart from amorphous regions, crystalline grains may vary in composition and in crystalline structure, since SiC is known to crystallize in many polytypes. Bright field and dark field images of a μc-SiC:H film with a relative doping level of 0.001 (figures 1a, b) reveal a columnar structure varying between 8 nm and 30 nm in width. The film is grown on Si substrate covered by platinum due to the FIB lamella preparation. The contrast within the bright field images varies according to Bragg conditions, which differ as the crystallographic orientation changes with respect to the electron beam. If a particular orientation is selected by choosing a set of reflections with an objective aperture, grains of similar orientation are highlighted (figure 1b). The columnar crystals grow mostly perpendicular to the substrate. A high density of planar faults, mostly twin lamella, is indentified by the fringe-like contrasts. Crystalline grains are observed to form directly at the substrate, which evolve into larger columns at later stages of the growth. This behaviour is typical for microcrystalline growth of Si [9] and SiC [8]: First a nucleation layer consisting of small crystallites and amorphous regions is formed. By competitive growth smaller crystallites are overgrown by larger ones. In general, this competitive growth is not related to orientation dependent growth rates, i.e. a random distribution of crystallographic orientations for μc-Si [9] and μc-SiC:H [8] is commonly observed. However, for the samples investigated here, textured growth along the [111] direction is revealed by the diffraction pattern

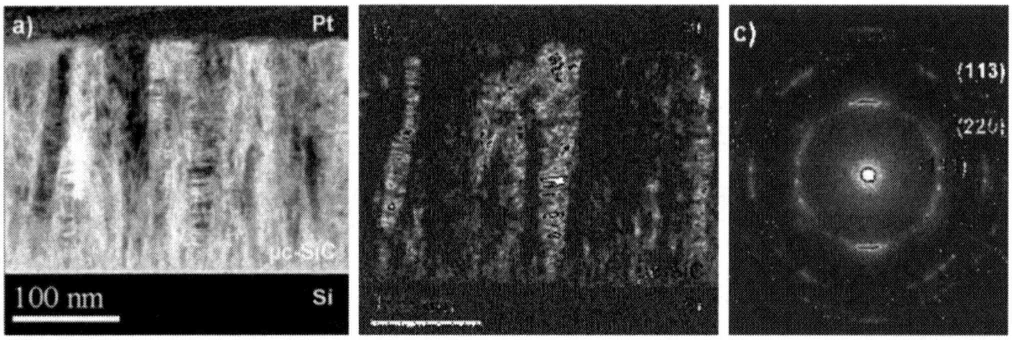

Figure 1: Bright-field (a), dark-field TEM image (b) and electron diffraction pattern (c) of a μc-SiC:H film with a relative doping level of 0.001. A columnar structure of crystalline grains is observed. According to the diffraction pattern, the cubic phase prevails. The arc like intensity distribution along the (111) ring indicates a preferential orientation of the (111) planes perpendicular to the growth direction.

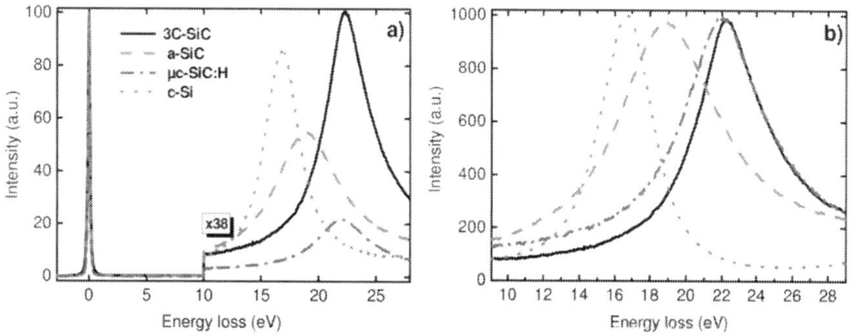

Figure 2: (a) Overview of low loss EEL spectra containing two dominant features: the zero energy loss peak (zlp) and the volume plasmon peak. The spectra are calibrated to the height of zlp, the intervall between 10 and 28 eV is enlarged by a factor of 38 for better visualisation. (b) Same data calibrated to the maximum of the plasmon peak. Following peak positions are measured: 16.8 eV for c-Si (green, dotted), 18.8 eV for a-SiC (red, dashed), 22.0 eV for µc-SiC (blue, broken), and 22.25 eV for 3C-SiC (black, solid). Note that the FWHM is broadest for a-SiC phase.

showing an arc-like intensity distribution along the (111) ring. The diffraction pattern is dominated by reflections, which can be assigned to the cubic phase of SiC (3C-SiC). Occasionally hexagonal inclusions are observed, too (not shown here). For all samples investigated, only slight variations in microstructure are detected, e.g. with respect to the size of the columns. This implies that neither the doping level nor the type of substrate show significant influence on the structural properties.

In order to investigate the electronic properties low-loss EELS spectra are recorded. In figure 2 spectra obtained from µc-SiC:H and different reference materials are compared: amorphous SiC (a-SiC), crystalline Si (c-Si) and cubic SiC. According to differences in specimen thickness the height of the volume plasmon peaks vary. To visualize the peak positions, figure 2b shows the same data normalized to the maximum of plasmon peak. Two major differences between the spectra are evident: (i) the peak position and (ii) the full width of half maximum, the latter one being largest for the amorphous sample. The peak of the µc-SiC:H sample almost coincides with that of the cubic SiC. However, the width of the peak is markedly larger, which may be an indication of higher structural disorder allowing a broader spectrum of plasmonic modes to be excited.

The local variation of the electronic properties is investigated by EFTEM. Figure 3a shows a bright field TEM image of a thin sample region which is surrounded by vacuum, where the Si substrate was removed by the Ar ion thinning procedure. The corresponding map of the plasmon peak position, which has been determined from fitting Gaussians to the spectra of the EFTEM series, is displayed in figure 3c. A drift correction procedure developed by B. Schaffer et al. is employed prior to the fit [10]. Special care was taken in order to minimize artefacts due to non-isochromaticity. The peak position in figure 3c is seen to vary between 21 eV and 22.3 eV, being largest within the centre of individual crystalline grains. However, some of the grains with strong Bragg contrasts, which arise from twin lamellas, show fluctuations of the plasmon peak position. These features are considered as artefacts, since they are related to a residual Bragg contrast. At the grain boundaries between adjacent grains, the plasmon peak position clearly decreases. The same holds for the edges of the specimen, which consist mostly of amorphous material due to ion milling. In order to investigate possible shifts of the plasmon peak position within growth direction, a thicker sample area was chosen (figures 3b, d). Since the grains overlap in viewing direction neither distinct Bragg contrasts nor well-defined grain boundaries can be observed. The peak position of the plasmons clearly increases by about 0.8 eV with increasing distance from the interface between µc-SiC layer and the Si substrate. Hence, it is concluded, that the red-shift of the volume plasmon energy is related to structural disorder, which appears at the heterogeneous grain boundaries and within the nucleation zone.

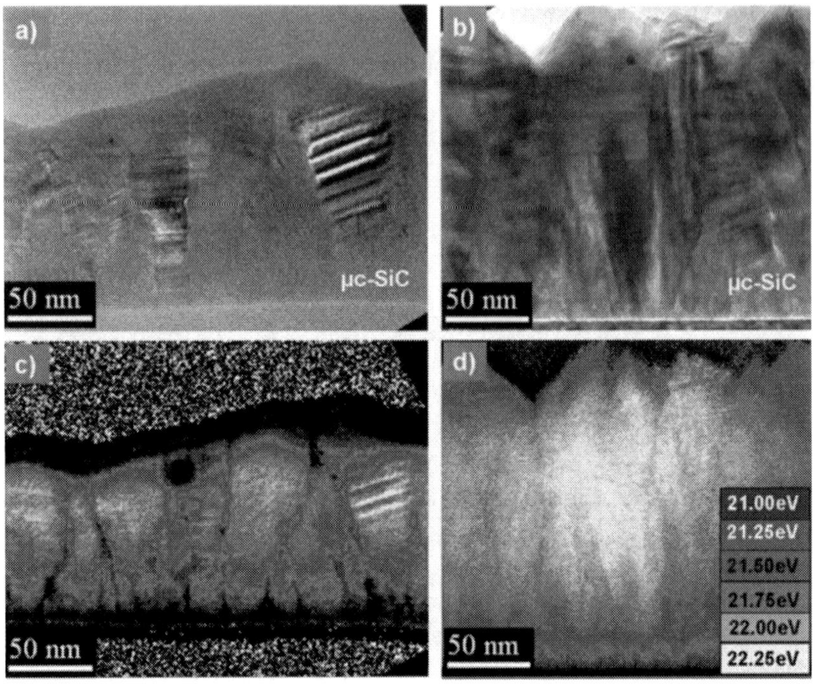

Figure 3: Bright field images and corresponding maps of the plasmon peak position measured from EFTEM series. (a), (c): thin specimen area surrounded by vacuum, (b), (d), thick area of the film on Si substrate. Clearly, a red shift of the plasmon peak position is observed at heterogeneous grain boundaries and close to the substrate.

In literature a red shift of the plasmon energy has been reported for Al-implanted SiC [11]. The authors attribute this effect to a volume swelling, which effectively lowers the density of free electrons. However, in our case the differences in electronic properties of the various components have to be considered as well, namely of amorphous and crystalline phases and of the heterogeneous grain boundaries, which are expected to contain high densities of hydrogen incorporated during growth.

In conclusion, μc-SiC:H is characterized by a complex microstructure containing a high degree of disorder within the nucleation zone and at heterogeneous grain boundaries. The local structural variations cause changes of the electronic properties. A red shift of the plasmon energy is observed in disordered regions.

References

[1] Bhatnagar M and Baliga B J 1993 *IEEE Trans. Electron Devices* **40** 645
[2] Dasgupta A, Saha S C and Ray S 1999 *J. Mat. Res.* **14** 2554
[3] Dasgupta A, Klein S, Houben L, Carius R, Finger F and Luysberg M 2008 *Thin Solid Films* **516** 618
[4] Klein S, Carius R, Finger F and Houben L 2006 *Thin Solid Films* **501** 169
[5] Miyajima S, Yamada A and Konagai M 2006 *Thin Solid Films* **501** 189
[6] Tabata A and Komura Y 2007 *Surf. Coat.Technol.* **201** 8986
[7] Chen T, Huang Y, Yang D, Carius R and Finger F 2011 *Thin Solid Films* In Press
[8] Dasgupta A, Huang Y, Houben L, Klein S, Finger F, Carius R and Luysberg M 2008 *Thin Solid Films* **516** 622
[9] Houben L, Luysberg M, Hapke P, Carius R, Finger F and Wagner H 1998 *Phil. Mag. A* **77** 1447
[10] Schaffer B, Grogger W and G Klothleitner 2004 *Ultramicr.* **102** 27
[11] Wang C M, Zhang Y, Weber W J, Jiang W and Thomas L E 2003 *J. Mat. Res.* **18** 772

Influence of annealing temperature on the structural, optical and electrical properties of amorphous Zinc Sulfide thin films

F Göde [1], E Güneri [2], A Kariper [2], C Ulutaş [3], F Kirmizigül [3] and C Gümüş [3]

[1] Department of Physics, Mehmet Akif Ersoy University, 15030 Burdur, Turkey
[2] Department of Primary Education, Erciyes University, 38039 Kayseri, Turkey
[3] Department of Physics, Cukurova University, 01330 Adana, Turkey

E-mail: ftmgode@gmail.com

Abstract. Zinc sulfide films have been deposited on glass substrates at room temperature by the chemical bath deposition technique. The growth mechanism is studied using X-ray diffraction, scanning electron microscopy, optical absorption spectra and electrical measurements. The as-deposited film was given thermal annealing treatment in air atmosphere at various temperatures (100, 200, 300 400 and 500 °C) for 1 h. The annealed film was also characterized by structural, optical and electrical studies. The structural analyses revealed that the as-deposited film was amorphous, but after being annealed at 500 °C, it changed to polycrystalline. The optical band gap is direct with a value of 4.01 eV, but this value decreased to 3.74 eV with annealing temperature, except for the 500 °C anneal where it only decreased to 3.82 eV. The refractive index (n), extinction coefficient (k), and real (ε_1) and imaginary (ε_2) parts of the dielectric constant are evaluated. Raman peaks appearing at ~478 cm^{-1}, ~546 cm^{-1}, ~778 cm^{-1} and ~1082 cm^{-1} for the annealed film (500 °C) were attributed to $[TO_l+LA]_\Sigma$, $2TO_\Gamma$, $2LO$, $3LO$ phonons of ZnS. The electrical conductivities of both as-deposited and annealed films have been calculated to be of the order of ~10^{-10} (Ω cm)$^{-1}$.

1. Introduction

Zinc sulphide (ZnS) is an important optoelectronic material for use in the violet and blue spectral regions because of its wide band gap. This II–VI compound semiconductor material has been studied extensively for a variety of applications, for example, in optical coatings, solid state solar window layers, electro-optic modulators, photo conductors, field effect transistors, optical sensors, photo catalysts, electroluminescent materials [1], phosphors and other light emitting materials [2, 3]. ZnS is characterized by an n-type direct wide band gap in the range of 3.79–3.93 eV [4] for the wurtzite phase and 3.6 eV for the cubic phase. Besides, it has become very attractive due to its low toxicity compared to other semiconductors. A wide variety of synthesis methods have been used to prepare ZnS nanocrystals, such as solvothermal [5], hydrothermal and thermal evaporation methods [6], RF sputtering [7], molecular beam epitaxy (MBE) [8] and chemical bath deposition (CBD) [9]. Among these, chemical bath deposition is the least costly low-temperature technique to deposit large-area thin films of semiconductors.

In this work, the structural, optical and electrical properties of an as-deposited ZnS film grown by the CBD technique and the crystallographic, optical and electrical properties of the annealed ZnS film have been studied. The refractive index, extinction coefficient and dielectric constants as a function of wavelength for both as-deposited and annealed ZnS films were investigated from the absorption, transmittance and reflection data, and the results are discussed. Moreover, the I-V measurements were taken to calculate electrical conductivities.

2. Experimental procedure

Thin films of ZnS were prepared by the CBD method, using the deposition solution reported previously [10]. Such films were deposited from the deposition solution that was prepared as follows: 1.25 ml of 1 M $ZnSO_4 \cdot 7H_2O$, 1.5 ml of 3.75 M triethanolamine [$(HOCH_2CH_2)_3N$], 1 ml of ammonia/ammonium chloride (NH_3/NH_4Cl, pH = 10.55) buffer solution, 0.05 ml of 0.66 M tri-sodium citrate ($C_6H_5Na_3O_7$), 0.5 ml of 1 M thioacetamide (CH_3CSNH_2) were mixed; then enough deionized water was added to get a total solution volume of 25.05 ml. The deposition was made on a glass substrate at room temperature. ZnS nucleated onto the submerged surfaces including the beaker walls over about 24 h for a single deposition run. For two deposition runs, the thin film obtained from the first run was submerged into freshly prepared solution for another 24 hours. In order to see how annealing will affect the structural, optical and electrical properties of the film, the obtained film was heated to 100, 200, 300, 400, and 500 °C for 1 h in air atmosphere.

The X-ray diffraction (XRD) spectra for as-deposited and annealed film were obtained using an X-ray diffractometer (Rigaku Rint 2000) with Cu-Kα_1 monochromatic radiation at 40 kV and 30 mA. Surface morphology of both as-deposited and annealed films was performed by scanning electron microscopy (SEM) in a JEOL JSM-5500LV. The Raman spectroscopy was performed at room temperature using a Raman spectrometer (Bruker Senterra) with a 532 nm (20 mW) Ne laser line as an excitation source. Optical transmission data were obtained by a UV-visible spectrophotometer (Shimadzu UV-2101PC).

3. Results and discussion

No peaks corresponding to the hexagonal phase were observed for the as-deposited film, which confirmed the amorphous nature of the as-deposited ZnS thin film. Figure 1 (a) and (b) show the SEM micrographs of the as-deposited and the annealed ZnS film deposited at 30 ºC for 96 h. It is clearly seen that the film after annealing looks somewhat different from the as-deposited one. The as-deposited ZnS thin film shows uniform, continuous and dense microstructures. However, annealed film has some cracks originated from tensile strain. The tensile strain developed in the annealed film may be due to the differences in thermal expansion coefficients of the substrate and deposited material and/or the crystallization process during annealing which may cause vacancies to diffuse to the surface of the film [11].

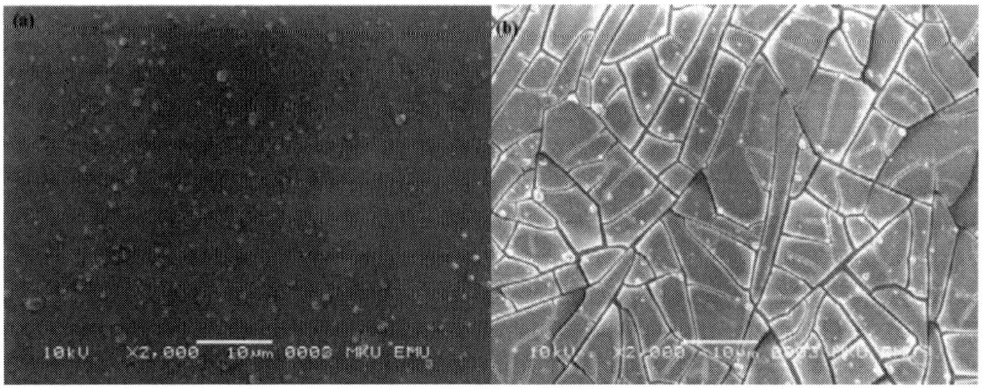

Figure 1. (a) SEM micrographs for ZnS films (a) as-deposited at 30 °C; (b) annealed in air atmosphere at 500 °C for 1 h.

Figure 2 shows the XRD pattern of the ZnS thin film annealed at 500 °C for 1 h in air atmosphere. As seen from this figure, the deposited film has become polycrystalline after being annealed at 500 ºC. The annealed film exhibits only one weak diffraction peak corresponding to the (008) plane of the hexagonal ZnS based on PDF card no: 39-1363. The structural data are also in agreement with our previous work [12] and that reported by Cheng *et al* [13]. It can be concluded that post-deposition annealing slightly improves the crystallinity.

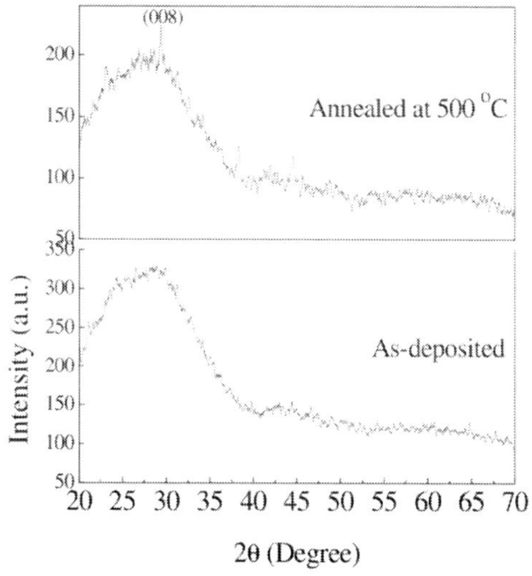

Figure 2. X-ray diffraction pattern of ZnS thin films a) as-deposited b) annealed at 500 °C.

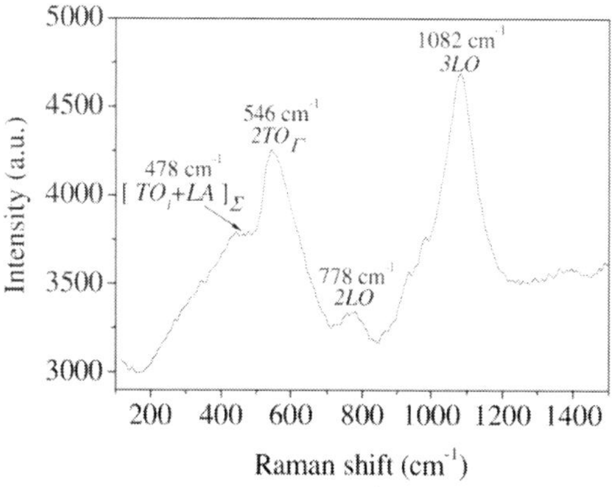

Figure 3. Raman spectra from 100 to 1500 cm^{-1} of the ZnS film annealed at 500 °C

The Raman spectrum of the annealed ZnS film is shown in figure 3. The first order Raman phonon observed at ~478 cm^{-1} corresponds to the $[TO_l + LA]_\Sigma$. The second order Raman phonons observed at ~546 cm^{-1} and ~778 cm^{-1} might originate from the $2TO_\Gamma$ and $2LO$ respectively. The third order Raman phonons observed at ~1082 cm^{-1} can be assigned as the $3LO$ mode. The Raman phonon modes reported here are in good agreement with our previous work [10, 14] and data reported by Serrano *et al* [15] and Zhang *et al* [16]. From figure 4, the transmittance in the visible region of the as-deposited ZnS film was ~58 % and when annealed in air for 1 h at different annealing temperatures from 100 to 500 °C, it decreased to ~40 %.

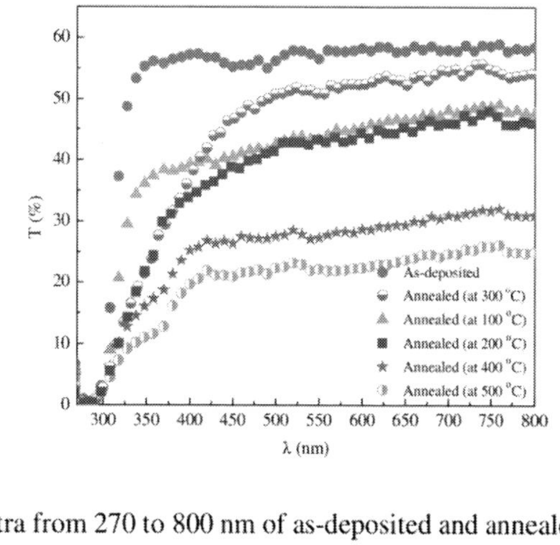

Figure 4. Transmission spectra from 270 to 800 nm of as-deposited and annealed ZnS thin films.

The absorption coefficients (α) for as-deposited and annealed ZnS films were calculated by the following relation [10]: $\alpha = 1/t \ \ln[(1-R)/T]$. The absorption coefficient for the amorphous film was ~10^5 cm^{-1}, it increased to ~10^6 cm^{-1} for the crystalline film. Figure 5 shows the variation of $(\alpha h\nu)^2$ vs. $h\nu$ of the as-deposited and annealed ZnS thin films at annealing temperatures ranging from 100 to 500 °C. The band gap energy was derived by extrapolating the straight-line portion of the $(\alpha h\nu)^2$ vs.

$h\nu$ plot to zero absorption. The optical band gap of the as-deposited ZnS thin film was 4.01 eV. From figure 6, it can be seen there is a decrease (to ~3.8 eV) of the band gap for all temperatures.

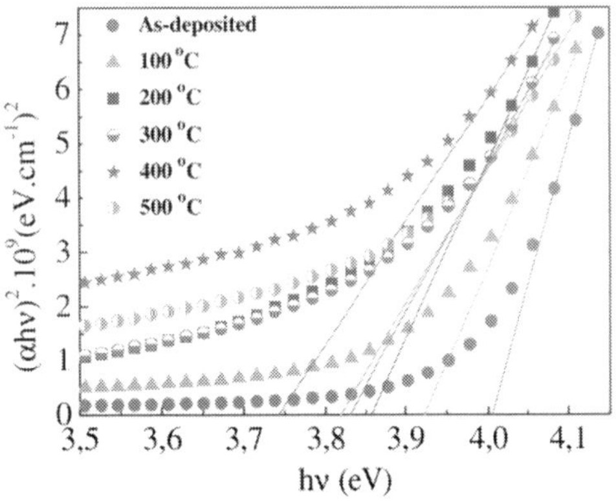

Figure 5. The plots of $(\alpha h\nu)^2$ versus $h\nu$ for as-deposited and annealed ZnS thin films.

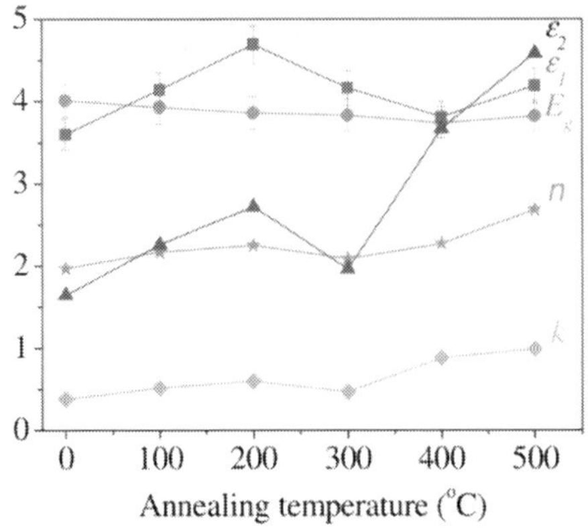

Figure 6. E_g, n, k, ε_1 and ε_2 for as-deposited and annealed ZnS films versus anneal temp.

Figure 6 also shows the variation of the band gap (E_g), refractive index (n), extinction coefficient (k), and real (ε_1) and imaginary (ε_2) parts of the dielectric constant as a function of the annealing temperature. The detailed calculations of these parameters also are given in previous studies [10, 17]. Moreover, the electrical conductivities of as-deposited and annealed films with Au contacts have been measured to be of the order of $\sim 10^{-10}$ $(\Omega\ \mathrm{cm})^{-1}$. The detailed information about calculating electrical conductivity of both as-deposited and annealed ZnS films are given in [4, 10]. The calculated electrical conductivity values are in agreement with other work [18].

4. Conclusion

In summary, X-ray diffraction analysis indicated that as-deposited ZnS films were amorphous, and heat treatments improved the structure. The Raman spectra of annealed ZnS samples showed first, second and third order Raman phonons. The optical band gap of optical transition has been determined to be 4.01 eV for direct transition, and it decreased from this value to ~3.8 eV with an increase in annealing temperature. The electrical conductivities for as-deposited and annealed ZnS films with Au contacts were found to be of the order of 10^{-10} $(\Omega\ \mathrm{cm})^{-1}$.

References

[1] Manzoor K, Vadera S R, Kumar N and Kutty T R N 2004 *Appl. Phys. Lett.* **84** (2) 284.
[2] Klausch A et al., 2010 *J. Lum.* **130** 692.
[3] Peng W Q, Cong G W, Qu S C and Wang Z G 2006 *Opt. Mater.* **29** 313.
[4] Göde F, Gümüş C and Zor M 2007 *J. Cryst. Growth* **299** 136.
[5] Li Y, Liu Y, Shen W, Yang Y, Wen Y and Wang M 2011 *Mater. Lett.* **65** 2518.
[6] Xu Y, Shi L, Zhang X, Wong K and Li Q 2011 *Micron* **42** 290.
[7] Arenas O L, Nair M T S and Nair P K, 1997 *Semicond. Sci. Technol.* **12** 1323.
[8] Ichino K, Onishi T, Kawakami Y and Fujita S, 1994 *J Crystal Growth* **138** 28.
[9] Allouche N K, Nasr T B, Kamoun N T and Guasch C 2010 *Mater. Chem. Phys.* **123** 620.
[10] Göde F 2011 *Physica B*, **406** 1653.
[11] Tu KN and Rosenberg R 1988 *Analytical Techniques for Thin Films*, Academic Press, San Diego
[12] Göde F and Gümüş C 2009 *Optoelectron. Adv. Mat.* **11** (4) 429.
[13] Cheng J, Fan DB, Wang H, Liu BW, Zhang YC and Yan H 2003 *Semicond. Sci. Technol.* **18** 676
[14] Göde F 2010 *Balkan Phys. Lett.* **18** (181057) 422.
[15] Serrano J et al., 2004 *Phys. Rev. B* **69** 014301.
[16] Zhang Y J, Xu C S, Liu Y C, Liu Y X, Wang G R and Fan X W, 2006 *J. Lum.* **242** 119.
[17] Göde F, Gümüş C and Zor M 2007 *Optoelectron. Adv. Mat.* **9** (7) 2186.
[18] Marquardt P, Nimtz G and Muhlschlegel B, 1988 *Solid State Commun.* **65** (6) 539.

Shape transformation of $Si_{1-x}Ge_x$ structures on ultra clean Si(5 5 7) and Si(5 5 12) surfaces

J K Dash[1], A Rath[1], R R Juluri[1], P Santhana Raman[1], K Müller[2], M Schowalter[2], R Imlau[2], A Rosenauer[2] and P V Satyam[1,2*]

[1]Institute of Physics, Sachivalaya Marg, Bhubaneswar, India – 751005

[2]Institute of Solid State Physics, University of Bremen, Otto-Hahn-Allee 1, 28359 Bremen, Germany

*Email: satyam@iopb.res.in, pvsatyam22@gmail.com

Abstract. We report growth of Ge nano/micro structures on ultra clean, high vicinal silicon surfaces of Si(5 5 7) and Si(5 5 12) under two substrate heating conditions: direct current (DC) and radiative heating (RH). These were grown under ultra high vacuum conditions while keeping the substrate at a temperature of 600°C. The results for 10 monolayer (ML) and 12 ML thick Ge deposited on the above surfaces show spherical island structures for RH conditions while aligned trapezoidal structures were observed under DC conditions of heating. We find that the longer side of trapezoid structures are along <$\bar{1}$10> irrespective of DC current direction. In the case of 10 ML Ge deposited on Si (5 5 7), elongated Si_xGe_{1-x} nanostructures with an average length of ≈300 nm and a length/width ratio of ≈3.3 have been formed along the step edges. Under similar conditions for 10 ML Ge growth on Si(5 5 12), we found aligned Si_xGe_{1-x} trapezoidal microstructures of length ≈6.25 μm and an aspect ratio of ≈3.0. Scanning transmission electron microscopy (STEM) measurements showed the mixing of Ge and Si at the interface and throughout the over-layer. Detailed electron microscopy studies (scanning electron microscopy (SEM) and STEM) reveal the structural aspects of these microstructures.

1. Introduction

The use of high index silicon surfaces as substrates can enhance the formation of aligned nanostructures with self-assembled growth [1–5], which is not so easily achievable with lithography tools. Due to its natural process- and damage-free features, structural self-organization on high index surfaces has received much attention in the formation of coherent 1D and 2D semiconductor / metal nanostructures. High index and vicinal surfaces are found to have reconstructions into regular hill (step) and valley (terrace) structures with periods ranging from several nanometers to about one hundred nanometers. The periodicity of such structures depends mainly on the substrate orientation and miscut or vicinal angles. These self assembled nano-structures can be used as templates for the growth of low dimensional semiconductor and metallic nanostructures with controlled size and periodicity. In the past, the growth of Ge nanostructures on low index silicon substrates (such as Si(100), Si(111)) has been studied extensively. However, less attention has been paid to Ge growth on high-index Si substrates. Hence, the study of self-assembly of Ge on high-index silicon surfaces would be an important field of research in obtaining various types of nanostructures that may be useful in optoelectronic devices. The lattice mismatch between Ge and Si (≈ 4.2%) is an important property of such quantum devices, as it determines the maximum thickness of smooth films (quantum wells) and controls band structures, mobility of carriers, and quantum confinement potential.

Study of Ge growth on high index surfaces, such as Si(5 5 7), Si(5 5 12) surfaces, is an area where very limited research work has been carried out [1–5]. Among the high index silicon surfaces oriented between (001) and (111), Si(5 5 12) exhibits a relatively stable reconstruction having 1D symmetry with a $(1\bar{1}0)$ mirror plane [5, 6]. Si(5 5 12) is oriented 30.5° away from (001) towards (111) with one-dimensional periodicity over a large unit cell [5]. Si(5 5 7) has a vicinal angle of 9.45° from (111) towards $(11\bar{2})$ [7].The highly corrugated triple step structure of the Si(5 5 7) surface is easier to detect than other long-range surface reconstructions. In both the Si(5 5 12) and (5 5 7), the step edges are parallel to the $\langle 1\bar{1}0 \rangle$ direction.

In this paper, we report the comparative study of aligned Si-Ge structures formed at the interface of ultra clean reconstructed Si(5 5 12) and Si(5 5 7) surfaces and Ge, where the substrates were heated by direct current (DC) and radiative heating (RH) during deposition of Ge. The size of the aligned Si-Ge Structures depends on the substrate orientation and over-layer thickness.

2. Experimental Details

The Ge depositions on Si(5 5 7) and Si(5 5 12) were carried out under ultrahigh vacuum (UHV) condition in a custom-designed molecular beam epitaxy (MBE) system with a base pressure of $\approx 2.5 \times 10^{-10}$ mbar in the growth chamber. The substrates used were of size 8 × 3 mm^2 cut from p-type (boron doped) single crystal Si wafers with a resistivity of 10 to 15 Ω cm. Atomically clean, reconstructed surfaces were prepared by following the usual thermal cleaning procedure: the samples were first degassed at about 600 °C for 12–14 h followed by a repeated flashing for 30 s at≈1250 °C. The reconstructed surfaces were confirmed by *in-situ* reflection high energy electron diffraction (RHEED). Ge was deposited to various thicknesses (3.0 – 12.0 ML) at a typical deposition rate of 0.6 ML min^{-1} (one ML corresponds to 7.8×10^{14} atoms cm^{-2} for a Si(111) surface). The substrates were kept under three substrate heating conditions: (i) at room temperature, (ii) at 600°C where RH is achieved through a filament underneath, and (iii) at 600 °C but with DC heating achieved by passing a direct current. In the second case, the sample was annealed at a temperature of 600 °C for 15 min after deposition using RH, and for case (iii) the annealing was done by passing the direct current through the sample with a power of 7.95 W (1.5 A and 5.3 V) corresponding to a temperature of 600 °C . The post-growth characterization of the samples was carried out ex-situ by field emission gun scanning electron microscopy (FEGSEM) and scanning transmission electron microscopy (STEM) based x-ray energy dispersive spectrometry (EDS). The FEGSEM measurements were carried out with 20 kV electrons using a Neon 40 cross-beam system (M/S Carl Zeiss GmbH) at Institute of Physics, Bhubaneswar, India. STEM and TEM measurements were carried out with 300 keV electrons using a Cs-corrected FEI Titan 80/300 system at University of Bremen, Germany. The cross-sectional TEM (XTEM) specimen was prepared for 12 ML Ge grown on Si(5 5 12) under DC heating conditions. The preparation of the specimen to electron transparency was achieved by focused ion beam milling followed by low-energy-ion milling.

3. Results and Discussion

Here we present results for the 10 ML and 12 ML Ge film thickness. For the case of RT deposition of 10 ML thick Ge on Si(5 5 12), the average island diameter was found to be 504 ± 38 nm (data not shown). Similar random structures are observed for 10 ML Ge on Si(5 5 7) under RT conditions. Figure 1(a) and 1(b) show the morphology for 10 ML Ge on Si(5 5 12) and Si(5 5 7), respectively, at a substrate temperature of 600°C followed by a post-anneal of 15 min at the same temperature with RH heating. In this case, a distribution of bigger circular Ge islands surrounded by a large number of smaller islands was observed. This type of distribution is consistent with Ostwald ripening. Figure 1(c) shows the typical particle size histogram for 10 ML Ge/Si(5 5 12)-RH depicting a bimodal size distribution with an average value for the size of the smaller islands of 15.8 ± 1.5 nm and for the bigger islands of 54.0±7.0 nm. In figure 1(d), a bimodal distribution of particle size is shown for the spherical Ge islands grown on Si(5 5 7) in RH condition as well. In this case, the number of bigger islands is also significant, unlike in the case of Si(5 5 12). The average values for the size of the smaller islands and bigger islands are 12.8 ± 0.5 nm and 37.2 ± 2.1 nm, respectively.

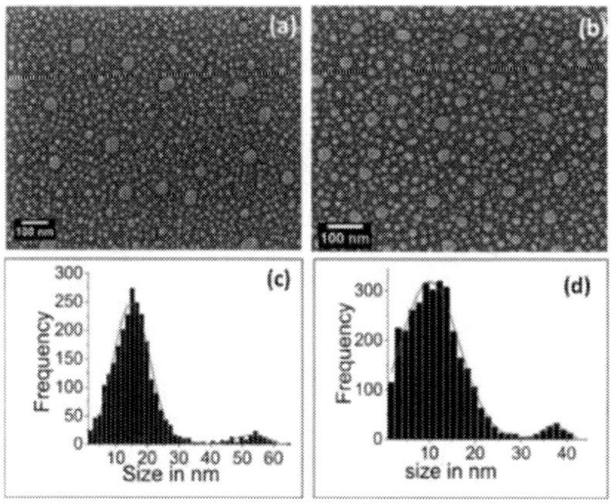

Figure 1.(a) and (c) SEM image and islands size histogram of 10 ML Ge/Si(5512) at 600 °C – RH. (b) and (d) SEM image and island size histogram of Ge/Si(557) at 600°C-RH

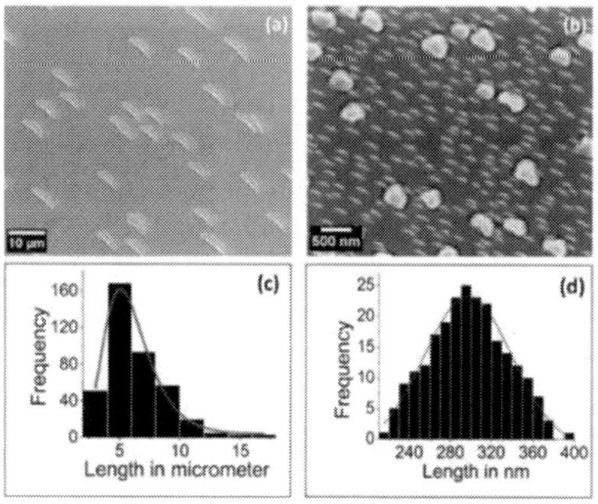

Figure 2.(a) and (c) SEM image and histogram/length of DC heated 10 ML Ge/Si(5512) at 600 °C. (b) and (d) SEM micrograph and histogram (length) of Ge/Si(557) at 600 °C-DC

We have shown circular shapes of Ge structures for the RH conditions in figures 1(a) and (b), but for the case of DC heating, we found a different shape evolution of the islands [8]. Figure 2(a) and (b) depict the formation of ordered trapezoidal structures for both Ge/ Si(5 5 12) and Ge/Si(5 5 7) cases in which direct current (DC) was passed through the sample. The histogram for length of the trapezoidal structures is shown in figure 2(c) and (d). The length of the aligned structures for Ge/Si(5 5 12) and Ge/Si(5 5 7) systems are (6.2 ± 0.3) μm and (297 ± 11.2) nm, respectively . Though the reason for the larger island structures is not yet understood, our results show that substrate orientation indeed makes a difference in size of self-assembled structures. Interestingly, similar length/width ratio [(≈3.1 for Ge/Si(5 5 12) and ≈3.3 for Ge/Si(5 5 7)] were found. The elongated trapezoidal structures are formed along the step edges irrespective of the DC current direction. In all the cases in which the direction of DC current was changed, we have found that the trapezoidal structures aligned in the step direction $<\bar{1}10>$ only [8].

In the following, we present the results on the composition determination of Si_xGe_{1-x} layer using STEM-EDS measurements with 300 keV electron beams. STEM measurements were carried out on 12 ML Ge deposited on Si(5 5 12) substrate under DC heating conditions. Figure 3(a) shows a STEM image taken from this specimen. As shown in Figure 2(a), the SiGe trapezoidal structures are of an average length of 6.2 μm. Figure 3(a) shows a cross-section image of this island, which is a small part of the bigger island. For making a cross-sectional FIB specimen, two kinds of protective Pt depositions were made (one using e-beam deposition and a larger thick one with FIB based gas injection system). Pt shown in Fig. 3(a) is deposited by gas injection with e-beam. The energy

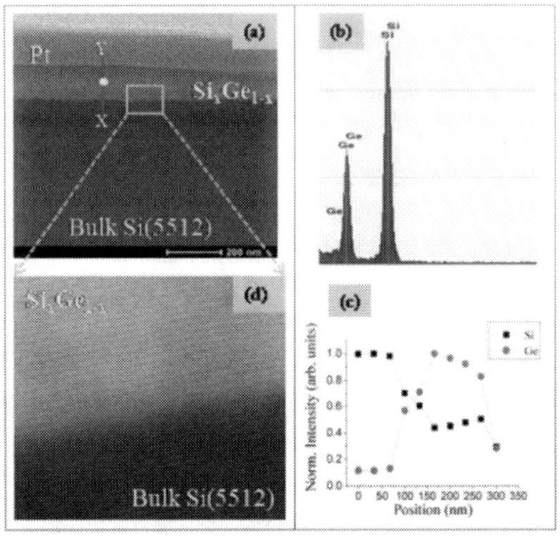

Figure 3.(a) and (d) STEM image (low and high magnification) of XTEM specimen corresponding to 12 ML Ge on Si(5 5 12). Figure (b) shows EDS at the middle of line XY. Figure (c) is the concentration variation of Si and Ge along XY.

dispersive spectrum in (b) shows Ge and Si signals from the region that is shown with a dot on the line XY in Figure 3(a). Figure 3(c) depicts the integrated and normalized (to maximum value of Si and Ge signals, respectively) intensity of Si and Ge signals obtained at various points along the line XY (300 nm length, shown in figure 3 (a)). The intensity line profile confirms the variable composition of Si and Ge in the island layer, in agreement with our ion scattering measurements [8]. Figure 3(d) shows a high resolution STEM image. From this figure it is clear that the film of Si_xGe_{1-x} is crystalline in nature and grown epitaxially on Si (5 5 12). These results show that it is possible to make aligned structures of variable sizes and graded composition. In general SiGe alloy layers are formed by either co-deposition of Si and Ge or by depositing buffer-layers of Si before the Ge deposition and followed by vacuum annealing.

4. Summary

In this work, we presented a one-step procedure to form a graded composition of Si_xGe_{1-x} epitaxially grown on clean high index surfaces. Trapezoidal structures of similar length/width ratio ($\approx$ 3.1 – 3.3) were observed. The longer side of the trapezoidal structures is $\approx$ 21 times larger in case of Si(5 5 12) compared to Si(5 5 7).

Acknowledgements

P V Satyam would like to thank the Department of Atomic Energy, Government of India for granting FEGSEM (Project no::11-R&D-IOP-5.09-0100) and University of Bremen, Germany for his sabbatical visiting program.

References

[1] Zhachuk R A, Teys S A, Dolbak A E and Olshanetsky B Z 2004 *Surf. Sci.* **565**, 37
[2] Omi H and Ogino T 1999 *Phys. Rev. B* **59** 7521
[3] Kim H, Li H and Seo J M 2008 *Surf. Sci.* **602** 2563
[4] Ware M E and Nemanich R J 2010 *Thin Solid Films* **518** 1990
[5] Baski A A, Erwin S C and Whiteman E J 1995 *Science* **269** 1556
[6] Kim H, Li H, Zhu Y-Z, Hann J R and Seo J M 2007 *Surf. Sci* **601** 1831
[7] Kirakosian A,Bennewitz R, Crain J N and Fauster Th 2001 *Appl. Phys. Lett.* **79**, 1608
[8] Dash J K et al. 2011 *J. Phys.: Condens. Matter* **23** 135002.

Measuring strain on HR-STEM images: application to threading dislocations in $Al_{0.8}In_{0.2}N$

J L Rouvière[1], A Mouti[2] and P Stadelmann[2]

[1] CEA-INAC/UJF-Grenoble1 UMR-E, SP2M, LEMMA, Minatec Grenoble, 17 rue des Martyrs 38054 France
[2] CIME, EPFL, Lausanne, Switzerland

E-mail: jean-luc.rouviere@cea.fr

Abstract. The Geometrical Phase Analysis (GPA) is an efficient method to measure strain from High Resolution Transmission Electron Microscopy (HRTEM) images. Here we show that it can also be applied efficiently to High Resolution Scanning Transmission Electron Microscopy (HR-STEM) images, which have several advantages: (i) patterns in HR-STEM do not change with thickness and chemical composition (ii) thicker samples can be analysed and (iii) strain and composition can be simultaneously determined. In many situations, the distortions due to the scanning of the beam can be corrected. The strain fields around different threading dislocations in an AlInN layer have been determined from plan view samples prepared by focus ion beam (FIB). Experimental strain maps were compared to analytical calculations that take into account the strain field of dislocations and of the In segregation. Mixed type dislocations are always terminated by an inverse hexagonal pyramidal pit, at the sample surface. The edges of the inverse pyramid are indium rich. The dislocation core is not situated at the centre of the inverse pyramid, which is indium-rich, but slightly shifted.

1. Introduction

Determining the strain locally is in many cases important as strain can change the physical and chemical characteristics of the material. For instance, microelectronics devices use stress to enhance their transport properties. Consequently, it is important to develop tools to measure locally at a nanometer scale strain to fully understand the property of materials. GPA is an effective techniques to analyse HR-TEM images [1,2,3,4]. Recently it has been tested on HR-STEM images [5, 6, 7] as this technique is becoming more popular. In this paper, we report some advantages of using HR-STEM images to measure strain and apply the technique to a 100 nm thick $Al_{0.8}In_{0.2}N$ layer lattice matched on top of a GaN buffer layer.

2. Experimental details

HR-STEM images were acquired at 300kV on a TITAN microscope equipped with a probe Cs-corrector. Image analysis was performed in Digital Micrograph, by using home made scripts. GPA was used and provided an easy way to correct the distortions due to scanning. The studied sample is a 100 nm thick $Al_{0.8}In_{0.2}N$ layer nearly latticed matched to a 2 μm thick GaN buffer layer grown on a sapphire substrate. Layers were grown along the c-axis using the conditions reported in [8]. Plan view samples were prepared by focused ion beam (FIB). HR-STEM images, imaging only the remaining top AlInN layer, were realised around different threading dislocations coming from the GaN buffer layer. Edge dislocations, mixed type dislocations and screw dislocation having respectively Burgers vectors equal to ***a***=1/3[2,-1,-1,0], ***a+c***=1/3[2,-1,-1,0] and [0001] were analysed.

3. Correction of scanning distortions

As stated in [6] and as can be seen in figure 1, the main source of error in strain measurement of HR-STEM images is the shift that can happen between different scanned lines. For instance, the phase image of figure 1b, obtained with the $\boldsymbol{g}$ = (1,0,-1,0) beam, contains horizontal broad lines that correspond to shift in the STEM images, as phase is proportional. In figure 1c, a phase shift of 2π corresponds to a displacement of one (1,0,-1,0) interplanar distance, ie. 0.2741 nm. At the bottom of figure 1b, i.e. right part of figure 1c, the peak corresponds to a phase shift of $\pi/2$, i.e. to a shift in the image of 0.7 nm. As proposed in [6], one solution to minimize these scanning errors is to align the scanning direction parallel to the principal axis of the strain and select frequencies that are close to the scanning direction. For instance, in the case of epitaxial layers, the growth direction and the scanning direction should be parallel.

Here we propose a simple way to correct for these scanning errors. It can be applied when the image has a large region that can be used as a reference, which is generally the case. Firstly the scanning direction should be chosen such that the reference region covers at least a narrow vertical strip in the image. In the case of epitaxial layers, this happens when, as before, the growth direction and the scanning direction are parallel. In figure 2b, the reference region we used is outlined by a dotted rectangle. Secondly the phase jumps in the vertical strip should be suppressed either by changing the origin of the phase or by unwrapping the phase. Thirdly, the average profile over this vertical strip should be calculated (figure 1c) and extended uniformly along the scanning direction (figure 1d). Finally, this last image should be subtracted from the original phase. As can be seen in figure 1e, most of the horizontal lines have been suppressed. Figures 1d and 1e show that by using the corrected phase, the strain ε_{gg} parallel to the selected g-vector contains few scanning errors. In [6], these scanning distortions are attributed to "flyback error" of the electron beam when the electron beam moves from the end of a line to the beginning of the next one. They could also be correlated to mechanical/thermal instabilities.

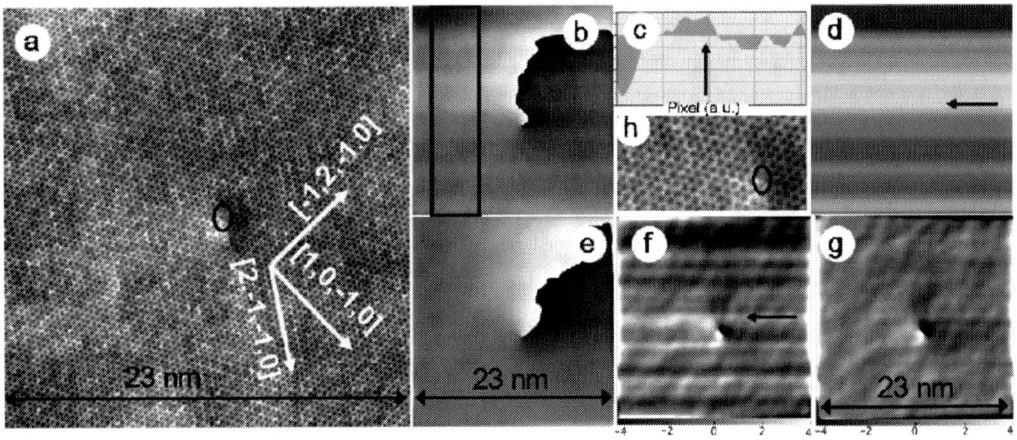

Figure 1. a) Initial HAADF-STEM image. A pure edge dislocation is present at the center of the black ellipse. b) Initial phase image obtained by selecting the $\boldsymbol{g}$ = (1,0,-1,0) beam. c) Phase profile horizontally averaged in the vertical strip of figure 1b. So the horizontal pixels in figure 1c are vertical in figure 1b. d) Image made from the average profile of fig. 1c. The horizontal arrow points to a scan error that is pointed by a vertical arrow in figure 1c. e) Corrected phase image f) ε_{gg} strain image (grey scale in %) calculated from the initial phase. The scan errors are clearly seen. g) ε_{gg} strain image (in %) calculated from the corrected phase. h) Zoom into the dislocation core of figure 1a.

4. Application to $Al_{0.8}In_{0.2}N$

In the studied sample, threading dislocations with screw components terminate at the surface of the sample by inverse pyramidal pits that have hexagonal basis [9, 10]. On plan view images, these pyramidal pits look like hexagonal stars (figure 2 and 4) whose arms are elongated along the six <1,-1,0,0> directions. In HAADF imaging these arms, which are in fact the edges of the inverse pyramid, are bright indicating that they are indium rich. Pure edge dislocation, which have a Burgers vector equal to 1/3<2,-1,-1,0> do not form pits (figure 1a). The analysis of the strain fields around these 3 types of dislocations are summarized in figure 2, 3 and 4. In figure 2 and 3 the positions of the dislocation cores were found by locating either the extremity of the additional plane well defined in "{1,0,-1,0} filtered images" (see figures 2b and 2c). It is found that in mixed type dislocation, the centre of the pit, which is In-rich, does not coincide with the dislocation core. As in pure edge dislocation, In tends to segregate on the tensile region of the dislocation. We applied a model presented in [10, 11] to calculate the In concentration around the dislocation core. For pure-edge dislocation taking into account the segregation of In, the calculated stress field reproduces quite well the experimental one, especially in the In rich region (figure 3).

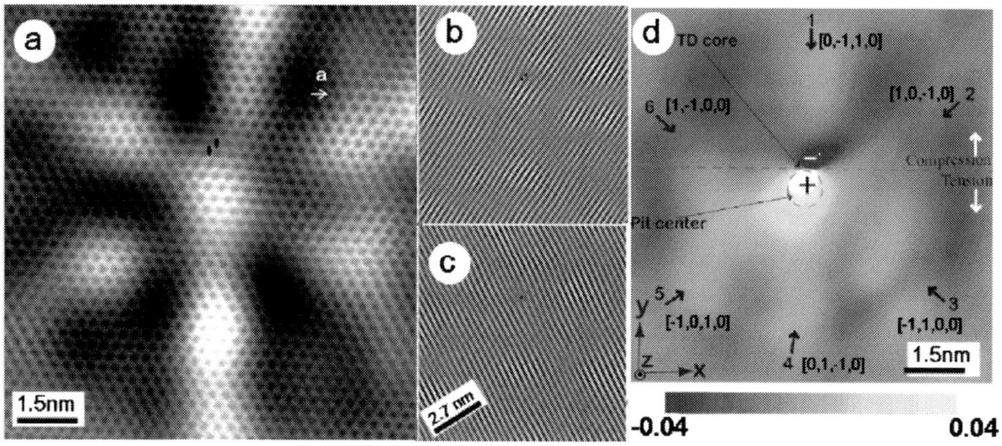

Figure 2. a) Smoothed HAADF-STEM image of a mixed type dislocation (***a+c***) in $Al_{0.8}In_{0.2}N$. The edge component of the (***a+c***) Burgers vector is represented by the white **a** vector. The two black spots, as determined in figure2b and 2c, locate the dislocation core. b) Filtered image of figure 2a obtained by selecting the (1,-1,0,0) and (-1,1,0,0) spots. The place of the dislocation is determined by the upper black spot that is reported in figure 2a. c) Filtered image of figure 2a obtained with the (1,0,-1,0) and (-1,0,1,0) spots. d) ε_{xx} strain map.

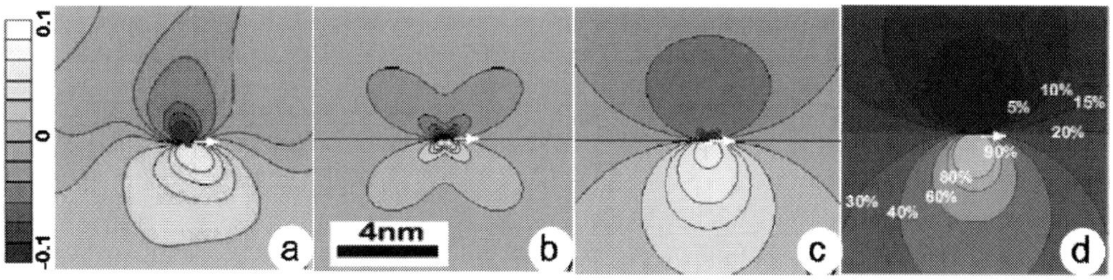

Figure 3. a) ε_{xx} strain of the pure edge dislocation shown in figure 1a, which has been rotated in order that the Burgers vector ***a*** = 1/3[2,-1,-1,0] becomes horizontal i.e. parallel to the x-axis. b-c) Simulated ε_{xx} strain fields of a pure edge dislocation in AlInN without (image b) and with (image c) In segregation. d) In concentration as determined by the analytical model.

Even for screw dislocations, a strain field was measured around a dislocation core. It is attributed to the variation of In concentration around the screw dislocation core, even though there are no really tensile or compressed regions, but mainly shear regions around the core. Indeed, our model indicates that shear components lead to In segregation (figure 4, [10]). Wider lattice parameters are found at the dislocation core and along the edges of the inverse pyramid, which is qualitatively in agreement with the contrast of STEM images that are bright, i.e In-rich, along these edges.

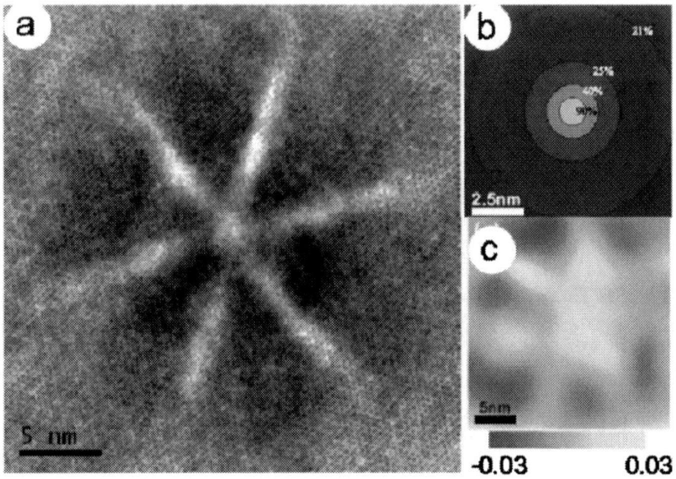

Figure 4.
a) HAADF-STEM image of a screw dislocation. The star shape is due to presence of an inverse pyramidal pit at the surface of the sample.
b)Calculated In-concentration around the dislocation core. The model described in [10] does not include the presence of the hexagonal pit and so the calculated concentration is symmetric around the dislocation core.
c) $\varepsilon_{xx}+\varepsilon_{yy}$ strain field as determined from figure 4a.

5. Conclusion

By correcting the scanning distortions of HR-STEM images, we were able to determine the strain fields around 3 types of threading dislocations. Such analysis would have been more difficult with HR-TEM images due to the thickness variations around the dislocation cores that have a pyramidal pit and because the studied areas were rather thick. The Z-contrast of STEM images, that gives the In-content, was qualitatively correlated to the measured strain fields. Further work would be needed to determine more quantitatively from the Z-contrast the In-content around threading dislocations. In principle, HR-STEM images allow to measure simultaneously, quantitatively and independently, strain and composition, which is not the case of HR-TEM images.

References

[1] Hÿtch M J and Gandais M 1995 *Phil. Mag. A* **72** 619
[2] Takeda M and Suzuki J 1996 *J. Opt. Soc. Amer. A* **13** 1495
[3] Hÿtch M J, Snoeck E and Kilaas R 1998 *Ultramicroscopy* **74** 131
[4] Rouviere J-L and Sarigiannidou E 2005 *Ultramicroscopy* **106** 1
[5] Rouviere J-L, Sarigiannidou E and Daudin B 2006 *Proceeding of the International Microscopy Congress IMC16 Sapporo, Japan* **vol. 2**, *p. 605*
[6] Chung J, Lian G and Rabenberg L 2010 *Electron Device Letters, IEEE* **31** n°. 8, 854
[7] Guerrero E, Galindo P L P,Yáñez L, Pizarro J, Guerrero-Lebrero M P and Molina S I 2009 *J. Appl. Phys.* **95** 143126
[8] Carlin J-F and Ilegems M 2003 *Appl. Phys. Lett.* **83** 668
[9] Mouti A, Rouviere J-L, Cantoni M, Francois Carlin J-F, Feltin E, Grandjean N and Stadelmann P accepted in *Phys. Rev. B*
[10] Kehagias T, Dimitrakopulos G P, Kioseoglou J, Kirmse H, Giesen C, Heuken M, Georgakilas M, Neumann W, Karakostas T and Komninou P 2009 *Appl. Phys. Lett.* **95** 071905
[11] Hirth J and Lothe J 1982 *Theory of dislocations* (Krieger)

A TEM study of Ge-on-(111)Si structures for potential use in high performance PMOS device technology

DJ Norris[1], IM Ross[1], A Dobbie[2], M Myronov[2], TE Whall[2], EHC Parker[2], DR Leadley[2] and T Walther[1]

[1] Kroto Centre for High Resolution Imaging and Spectroscopy, University of Sheffield, Sir Frederick Mappin Building, Mappin Street, Sheffield, S1 3JD, UK

[2] Department of Physics, University of Warwick, Coventry, CV4 7AL, UK

Corresponding author: d.j.norris@sheffield.ac.uk

Abstract. We have performed a transmission electron microscopical study of samples where pure Ge has been grown on differently oriented Si wafers by reduced pressure chemical vapour deposition. In the present study we examine pure Ge layers grown on (111)Si wafers by annular dark-field (ADF) scanning transmission electron microscopy (STEM) and by high-resolution electron microscopy (HREM). There is evidence of immediate strain relaxation taking place by dislocations and twinning. A novel configuration of micro-twins has been observed at the interface between the pure Ge and the underlying (111)Si wafer. Growth mechanisms and models to explain the interfacial configurations of this type of wafer are suggested.

1. Introduction

In recent years, there has been an increased interest in using Ge as a higher charge carrier mobility alternative to Si, for advanced complementary metal-oxide-semiconductor (CMOS) device applications, based on hole transport [1].

One consideration is the choice, type and orientation of the substrate used. In recent years, Ge, or an alloy of SiGe, has been grown commonly on (001)Si wafer orientation; however, enhancements in hole mobility have been theoretically found if alternative orientations of the surface normal of the Si wafers are utilised; namely growth on (101)Si or (111)Si wafers [2-4]. However, growth upon these latter two orientations can impose difficulties in terms of the final microstructure observed. Growth on these latter types of wafers is very challenging because it is the way in which the layer relaxes by the nucleation of dislocations into the final microstructure that can have an impact on the electrical properties [5]. Ideally we would like to grow layers onto virtual substrates; however, the microstructure of the virtual substrate is non-ideal with significant defects and non-planar uniformity of the surface of the substrate. Micro-twins and extended stacking faults are prevalent in structures grown on non standard orientated wafers [5]. The main consideration is the orientations of the various (111) slip planes in Si or Ge grown on top of the wafers. In (001)Si there is a symmetry in the orientation of the various (111) planes which can balance the effects of misfit relief and also the tilt components in the final microstructure. One compromise would be to grow pure Ge, or alloy SiGe, layers directly onto the differently oriented wafers which circumvents the need to fabricate potentially defective virtual substrates.

In the present work, we focus on growth of pure Ge on (111)Si wafers through transmission electron microscopy studies. We will show that relaxation of the pure Ge layer is evident and also an unusual twinning effect appears at the boundary between the pure Ge layer and the underlying Si.

2. Experimental Details

The Ge layers investigated in this study were grown by reduced pressure chemical vapour deposition (RP-CVD) using an ASM Epsilon 2000 reactor on (111) orientated p-Si substrates. The Si substrates were cleaned using a high temperature *in-situ* H bake immediately prior to epitaxial growth to remove any native oxide. The Ge layers were all grown at 400°C using a GeH_4 precursor gas. The growth time was 15 mins. The control of precursors, carrier gas flows and chamber pressure resulted in a GeH_4 partial pressure of around 10mTorr.

[110] cross-sectional specimens, for TEM, were prepared in the usual manner with 3mm disks thinned mechanically to ~30-40μm followed by Ar^+ ion thinning to electron transparency. Finished specimens were then analysed using a JEOL 2010F field-emission gun instrument, operating at 197kV) and a JEOL Z3100 R005 (300kV) aberration corrected cold-FEG scanning (S)TEM, both equipped with Gatan Imaging Filters as well as bright-field (BF) and annular dark-field (ADF) detectors.

3. Results and Discussion

As an overview of the morphology of Ge layers grown onto (111)Si wafers, a low magnification annular dark-field image is provided in figure 1. The average thickness of the Ge layer on (111)Si of 8 nm corresponds to an average growth rate of 0.5nm/min. It is clear that the surface of this 'pseudo'-continuous layer is rough with an amplitude of roughness of ~5nm.

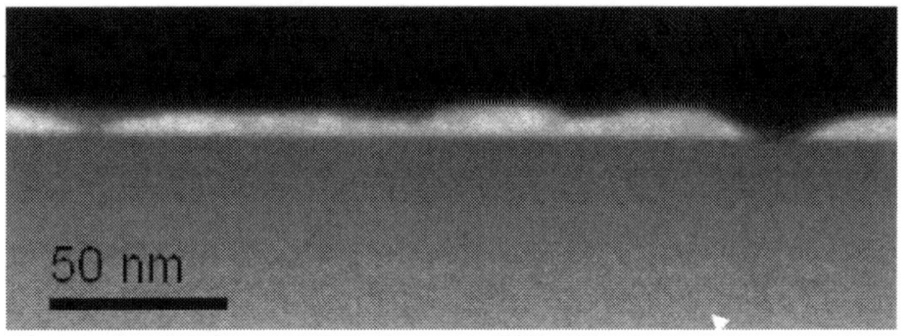

Figure 1. ADF image of Ge layer on Si(111)

If the thinnest region of the TEM specimen is now focused upon, where the material is suitable for high resolution phase contrast imaging, we observe discrete islands such as the one shown in figure 2. This image is obtained using the JEOL 2010F analytical TEM and shows clearly the atomic columns close to the island/wafer interface. There appear to be small regions on this boundary that are amorphous-like in appearance. If a Burgers circuit is drawn around these amorphous regions, as in the two regions of figure 2, we find that the centres of these amorphous regions contain dislocation cores.

One interesting feature of the present (111)Si grown wafer is the occurrence of twinning. Since the basal plane is {111} type, there is a possibility that the (001) direction of deposited material can be oriented in one of two ways. Either the (001) direction can follow the (001) direction of the substrate wafer. Or, alternatively, a twinned configuration can be established whereby the (001) of the deposited material is mirrored about the interface plane. An example of this can be seen in figure 3. Here, we see a grain (indexed B) which is a mirror twin of the underlying substrate, and also the surrounding material (grains A and C).

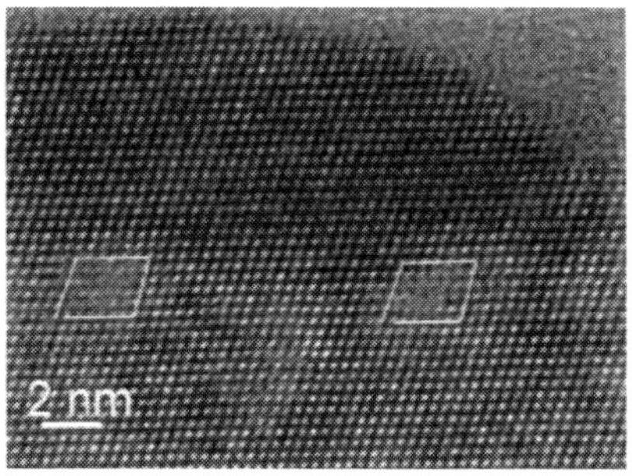

Figure 2. Lattice images of boundary between Ge island (top) and underlying (111)Si. Burgers circuits show non-closure.

Figure 3. HREM image showing twinned (B) and untwinned (A+C) grains.

In addition, edge dislocations are observed in those grains grown epitaxially without the occurrence of twinning; whereas no dislocations are observed at the grain-substrate interface for the twinned configuration. This would indicate that twinning or, more precisely, the introduction of twin boundaries, may be regarded as an alternative to the introduction of misfit dislocations; however, the situation is more complicated. If the microscope point resolution is sufficient to resolve individual atomic columns of the diamond structure along the <110> zone axis (so-called dumb-bells), i.e. better than 0.14 nm, we can determine at the atomic level the orientation of these dumb-bells, which are aligned along the (001) orientation of the local crystal lattice.

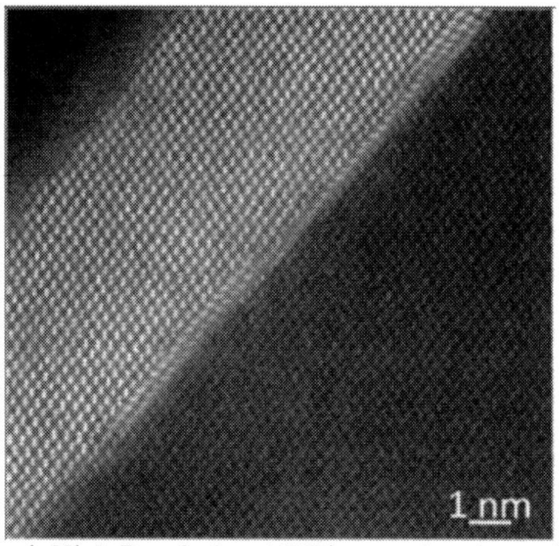

Figure 4. High Resolution ADF image showing twinned configuration at island substrate interface.

In the ADF image shown in figure 4, taken using the JEOL Z3100 R005, the Ge atomic columns appear much brighter than the Si columns, in agreement with the contrast expected for pure Ge/Si [6]. It can be clearly seen that there is a switching of orientation of the Ge dumb-bells at various regions of the interface (these areas are marked transparent yellow in figure 4). Upon further growth of the island the orientation of the (004)Ge lattice planes reverts back to the correct alignment where they are parallel to those dumb-bells of the underlying Si wafer. These micro-twins observed at the boundary between the deposited island and the underlying wafer are very narrow and found to be always exactly 2 monolayers (= ½ unit cell), 4 monolayers (=1 unit cell) or 6 monolayers (=1½ unit cell) in thickness. Their origin is still unclear, although it is likely due to the energy of formation of the twinned orientation being very similar to the untwinned because growth on a (111)Si wafer has the option of adopting a non-mirror or a mirrored orientation.

In the case of growth upon the (111)Si surface, we observed what seems to be an array of dislocations at the island/substrate interface. A Burgers circuit around the amorphous like regions showed non-closure; however, the closing vector does not seem to represent the full Burgers vector. It appears that this closure is simply the edge component of the Burgers vector observed in projection.

Upon analysis, we find that this dislocation is not a pure edge dislocation such that the total Burger's vector is perpendicular to the beam direction. If this were the case then the Burgers vector would not be a normal translation vector. Instead, the Burgers vector is inclined to the beam direction to give a conventional ***a***/2{110} vector. So, what is observed in the image (figure 2) is a projection of the Burgers' vector along the beam direction. Moreover, these dislocations are very efficient at relieving interfacial misfit because the edge component of the Burgers vector is actually in the interface plane and not inclined to it as would be seen in growth in the other two systems ((001) and (101)).

If there is a (111) glide plane parallel to the sample surface then it is possible to introduce dislocations quite efficiently at the edges of this plane. This is reminiscent of dislocations introduced in materials which grow via the Volmer-Weber growth mode where growth proceeds immediately in a 3D islanding mode and relieves misfit strain energy quite efficiently by introducing dislocations at the edges of islands as the island size increases. What is clear is that misfit relief in the (111)Si system proceeds by introducing dislocations in a glide plane parallel to the island/wafer interface. In any case, this may have important implications in that it may be difficult to produce pseudomorphically strained layers on such (111)Si wafers.

The interfacial twinning observed in growth of Ge on (111)Si is interesting in that this behaviour would not have been observed without the use of an aberration corrected instrument with the necessary resolution to resolve the (004)Ge 'dumb-bells'. What is clear is that the (111) surface acts as a mirror plane and allows growth in one of two configurations where (004)Ge is aligned with that of the substrate or grows in a mirror related twin configuration. Both alternatives seem energetically equally favourable and are perhaps evidence of a misfit relieving mechanism in this system. Moreover, it appears that with further growth the required (001)Si//(001)Ge epitaxial relationship is obtained as the micro-twins achieve a lateral length of only 6-7 nm and so cover only a small fraction of the substrate. These twin structures appear to be confined to the film-substrate interface and do not have components which extend up to the final free surface.

4. Conclusions

On (111) surfaces it has been shown through high resolution images that an array of misfit relieving dislocations are present at the island/substrate interface indicating that the islands are not fully strained. This has been explained in terms of the presence of a slip/glide plane parallel to the film substrate interface along which misfit dislocations can be introduced. These dislocations are mixed type and have the usual ***a***/2<110> Burgers vector. A novel configuration of micro-twins has also been observed at a portion of the Ge/Si(111) interface where the twins appear to be confined to the interface and do not extend up to the surface.

Acknowledgements

The authors would like to thank the Engineering and Physical Sciences Research Council for their financial support of this work under grant number EP/F033893/1 "Renaissance Germanium".

References

[1] On Chui C et al. 2002 *IEEE Electron Dev. Lett.* **23**, 473.
[2] Hartmann JM et al. 2008 *J. Cryst. Growth* **310**, 5287.
[3] Arimoto K et al., 2009 *Solid State Electronics* **53**, 1135.
[4] Lee ML, Antoniadis DA and Fitzgerald EA, 2006 *Thin Solid Films* **508**, 136.
[5] Arimoto K et al., 2009 *J. Cryst. Growth* **311**, 809.
[6] Walther T 2006 *J. Microscopy* **221**, 137.

Evaluation of the substitutional carbon content in annealed Si/SiGeC superlattices by dark-field electron holography

T Denneulin[1], J L Rouvière[2], A Béché[3], M Py[1], J P Barnes[1], J M Hartmann[1] and D Cooper[1]

[1] CEA-LETI, Minatec Campus, 17 rue des Martyrs, 38054 Grenoble Cedex 9, France
[2] CEA-INAC/UJF-Grenoble1 UMR-E, SP2M, Minatec Grenoble, 38054 France
[3] FEI Company, Achtseweg Noord 5, 5651 GG Eindhoven, The Netherlands

E-mail: thibaud.denneulin@cea.fr

Abstract. Si/SiGeC superlattices can be used as core structure for the fabrication of new microelectronic architectures with enhanced electrical properties such as multichannel transistors. The introduction of carbon in SiGe allows for compensation of the compressive strain and to avoid plastic relaxation. However, the formation of incoherent β-SiC clusters during annealing at high temperature limits the processability of SiGeC. This precipitation leads to a strong modification of the stress in the alloy due to the reduction of the substitutional carbon content. Here, we investigated the variation of the substitutional C content in annealed Si/SiGeC superlattices using a comparison of dark-field electron holography and finite element simulations.

1. Introduction

Multichannel devices have recently been proposed as an alternative architecture of metal oxide semiconductor field effect transistors (MOSFETs) for future microelectronic technologies [1]. They offer a high current drivability and have a relative immunity to short-channel effects. Their fabrication relies first on the epitaxy of Si/SiGe superlattices (SLs) [2]. After deposition on silicon-on-insulator substrate, the SLs are patterned to define the source and drain regions. The SiGe layers are then selectively etched and the voids left between the suspended Si beams are filled with a gate stack.

Alloying the SiGe layers with carbon is interesting because it allows the germanium concentration and the strain in the layers to be independently controlled. A high germanium content can be used while maintaining a low level of strain with a small amount of carbon in substitutional position. However, during annealing at high temperature (> 900°C), the strain behaviour of SiGeC can be modified by the formation of β-SiC clusters [3]. These clusters are incoherent with the surrounding matrix [4] and the C atoms incorporated are therefore lost from the point of view of the global strain distribution. Several characterization techniques can be used to monitor the reduction of the substitutional C content in SiGeC during thermal processing. Most common are Fourier transformed infrared spectrometry (FTIR), X-ray diffraction (XRD), Rutherford and resonant backscattering spectroscopy (RBS/rBS). However, all of them have a relatively poor lateral resolution. Here, we propose a method that could be applied to nanometric structures. It consists of a combination of a recent transmission electron microscopy (TEM) technique: dark-field electron holography (DFEH) [5] with finite element simulation (FEM). DFEH has been shown to open new possibilities for the measurement of strain in semiconductor devices. It provides a large field-of-view (up to 1×0.25 μm^2) and a spatial resolution of 10 to 4 nm [5]. Moreover, the comparison with simulations is direct which allows structural and compositional information to be rapidly linked.

2. Experimental procedure

The $Si/Si_{0.744}Ge_{0.244}C_{0.012}$ superlattices (SLs) consist of four periods epitaxially grown by reduced pressure chemical vapour deposition (RP-CVD) on <100> Si wafers [2]. Intrinsic $Si/Si_{0.786}Ge_{0.214}$ SLs were also created and were used as a reference for simulations. The growth was conducted in an Epi Centura cluster tool from Applied Materials at 650°C and 20 Torr. The gaseous precursors were SiH_2Cl_2 and GeH_4 for the SiGe layers and SiH_4 for the Si layers. $SiCH_6$ was added for the SiGeC layers. Samples were then annealed in H_2 for two minutes at different temperatures. The following conditions are described in this paper: as deposited and annealed at 1050°C for the Si/SiGe SLs; as deposited, annealed at 950°C and 1050°C for the Si/SiGeC SLs. The distribution of Ge and C was determined by time of flight secondary ion mass spectrometry (ToF-SIMS) using an ION-TOF TOF SIMS V instrument. Ge depth profiles are shown in figure 1.

Dark-field electron holography was performed on a FEI Titan transmission electron microscope with an operating voltage of 200 kV. The objective lens was switched off and the holograms were acquired using a Lorentz lens. The biprism was biased with a voltage of 180 V corresponding to a fringe spacing of 2 nm. Information in the growth direction was obtained by selecting the (004) diffraction spot with a 10 μm objective aperture. Acquisitions were performed during 64 s for signal-to-noise improvement. More details about the experimental conditions and post-treatment method are described in Béché et al. [6]. Samples were prepared into [110] parallel-sided cross sections using a dual beam FEI Strata 400 with a final milling step at 8 kV.

Finite element simulations of the strain in the lamellas were carried out using the structural mechanics module of COMSOL Multiphysics. The plane strain approximation was made to simulate the samples in 2D and save computing time. The lamella geometry was constructed, taking into account the crystalline thickness, determined by convergent beam electron diffraction (CBED) and the dimensions of the layers measured by X-ray reflectivity. The SiGe and SiGeC alloys stiffness coefficients were calculated according to Vegard's law. The SiGe lattice parameter a_{SiGe} was estimated using an empirical law based on Dismukes's results [7]:

$$a_{Si_{(1-x)}Ge_x}(\text{Å}) = a_{Si} + 0.2x + 0.026x^2 \tag{1}$$

The SiGeC lattice parameter a_{SiGeC} was calculated according to a law proposed by De Salvador et al. [8]:

$$a_{Si_{(1-x-y)}Ge_xC_y}(\text{Å}) = (1-x-y)a_{Si} + xa_{Ge} + ya_C - 0.026x(1-x) - 0.59y(1-y) + 0.06xy \tag{2}$$

where a_E is the lattice parameter of the element E. x and y are respectively the Ge and C contents. The concentrations were taken directly from the ToF-SIMS depth profiles using a MATLAB program. Only substitutional C atoms were taken into account in the model. It was supposed that non-substitutional C atoms have a negligible influence on the strain state [8]. Several simulations were performed for different substitutional content until a good fit was obtained with the experiment.

3. Results

3.1. *Dark-field electron holography and ToF-SIMS profiling*

Strain maps reconstructed from dark-field electron holograms are shown in figure 2. Strain profiles extracted from the maps in the growth direction are represented as continuous lines in figure 3. In the case of the intrinsic Si/SiGe SL (figure 3(a-b)), the distribution of the strain becomes smoother and broader after annealing at 1050°C which is due to the germanium interdiffusion (figure 1(a)). It induces a decrease of the maximum strain level in the SiGe layers from 1.2 to 0.9%.

Figure 3(c) shows the strain profiles corresponding to the Si/SiGeC superlattices. For the as-grown sample, the strain level in the SiGeC layers is equal to 0.9% which is less than the value measured in the intrinsic superlattice (1.2%). Whereas the germanium content is higher in the C doped layers (24% against 21% in the intrinsic layers), the strain level is lower, meaning that it is effectively reduced by the presence of carbon. The diffusion of the germanium is more important in these samples (figure 1(b)) than in the intrinsic ones. This comes from the presence of the carbon which strongly reduces the activation energy of the Si-Ge diffusion process. Although the strain distribution is broader at 1050°C than at 950°C, the level of the strain is also slightly higher. It indicates that the strain compensation induced by the substitutional C atoms is reduced, which is linked to the formation of β-SiC clusters.

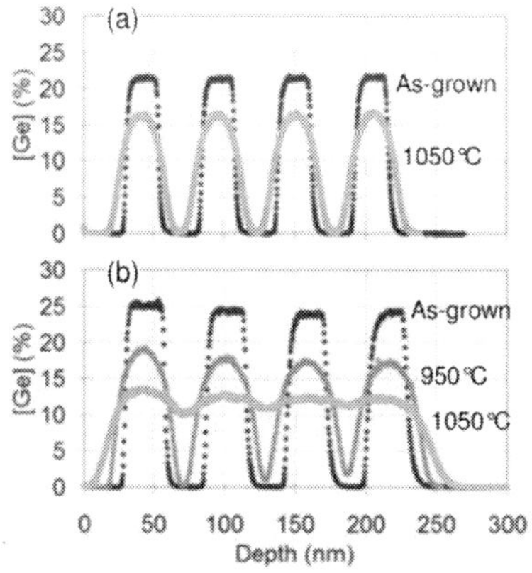

Figure 1. Ge ToF-SIMS concentration profiles for the (a) Si/SiGe and (b) Si/SiGeC superlattices.

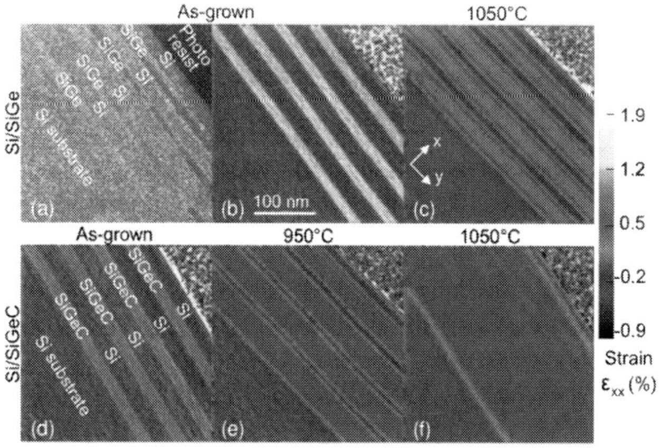

Figure 2. (a) Example of dark-field electron hologram. (b-f) Reconstructed strain maps of samples: (b) Si/SiGe as deposited, (c) Si/SiGe annealed at 1050°C, (d) Si/SiGeC as deposited, (e) Si/SiGeC annealed at 950°C and (f) Si/SiGeC annealed at 1050°C.

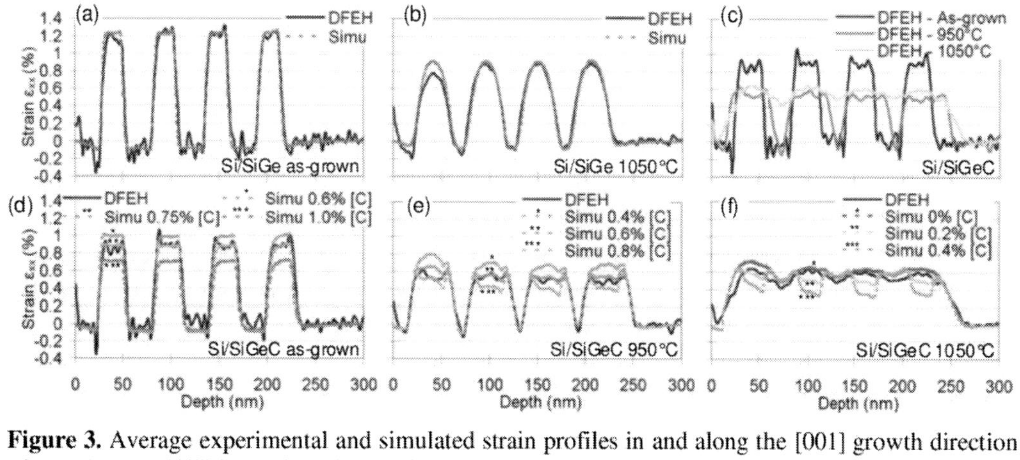

Figure 3. Average experimental and simulated strain profiles in and along the [001] growth direction of samples: (a) Si/SiGe as deposited and (b) Si/SiGe annealed at 1050°C. (c) Si/SiGeC experimental strain profiles. (d-e-f) Simulations for different substitutional C contents, compared to the experiment.

3.2. *Finite element simulations*

The simulations corresponding to the Si/SiGe superlattices are shown as dashed lines in figure 3(a-b). Globally, the experiment was found to be in excellent agreement with the simulated strain profiles, for both the as-grown and annealed SiGe samples. The strain in the uppermost layer is however slightly lower than in the other layers especially in the case of the annealed sample which is not confirmed by the simulation. It can be linked to a variation of the foil thickness close to the edge of the specimen.

Simulations corresponding to the C doped samples and for three different substitutional C contents are represented in figure 3(d-e-f). In the case of the as-grown sample (d), the experimental profile is in accordance with the simulation for a substitutional C content of 0.75% ± 0.05%. Compared to the total C concentration of 1.24%, it indicates that 61% ± 4% of the C content is in substitutional position. Similarly, at 950°C, the agreement between experiment and simulation was obtained for a substitutional C content of 0.60% ± 0.05% (or 48% ± 4% of the total C concentration). For the sample annealed at 1050°C, the match was obtained for 0% of substitutional C. We note that when inputting $y = 0$ in the equation (2), it leads to the Dismukes expression (1) which is known to be valid for intrinsic SiGe. It shows that from the point of view of strain, the sample behaves like pure SiGe.

4. Discussion

Relying on the conclusions deduced from the comparison of DFEH and simulations, the behaviour of the $Si/Si_{0.744}Ge_{0.244}C_{0.012}$ superlattice during annealing can be explained in greater detail. It was noticed that although there is an important diffusion of the germanium, the level of the strain is slightly higher at 1050°C than at 950°C. This can be explained by the massive decrease of the substitutional C content for the highest annealing temperature. At 950°C, there is still 48% of the total C content in substitutional position (compared to 61% for the as-grown sample) and the strain in the superlattice is reduced due to the strong interdiffusion of the germanium. On the contrary, at 1050°C, there is no more substitutional C and the related strain compensation is therefore lost, allowing the strain level to reincrease. However, because this phenomenon is in competition with the diffusion of the germanium, the reincrease is not so important and the effective strain level is almost the same at 950°C and 1050°C.

5. Conclusion

The influence of thermal processing at high temperature on the strain in Si/SiGeC superlattices has been studied by dark-field electron holography. The strain behaviour of the SL was found to be coherent with the formation of SiC clusters. Finite element modelling was then used to simulate the distribution of the strain in the TEM lamellas. As a good correlation was obtained between experiment and simulation for intrinsic Si/SiGe superlattices, the same method was applied to the investigation of the Si/SiGeC SLs. It was found that the most important loss of substitutional C occurs at 1050°C and it eliminates the effect of C on the strain. At this temperature, the strain is governed essentially by the distribution of the germanium which is highly diffused in comparison with the intrinsic superlattice.

This study demonstrates the possibilities offered by the combination of DFEH strain mapping and finite element simulations for linking structural and compositional information at the nm-scale. The method could be applied to the investigation of the substitutional C content in SiGeC recessed source and drain transistors for example. This would help for the understanding of the strain level in the channel and therefore the electric properties of the devices.

References

[1] Bernard E, Ernst T, Guillaumot B, Vulliet N, Garros X, Maffini-Alvaro M, Coronel P, Skotnicki T and Deleonibus S 2008 *Solid State Electron.* **52** 1297–302
[2] Hartmann J M, Papon A M, Barnes J P and Billon T 2009 *J. Cryst. Growth* **311** 3152–7
[3] Warren P, Mi J, Overney F and Dutoit M 1995 *J. Cryst. Growth* **157** 414–9
[4] Powell A R, LeGoues F K and Iyer S S 1994 *Appl. Phys. Lett.* **64** 324–6
[5] Hÿtch M, Houdellier F, Hüe F and Snoeck E 2008 *Nature* **453** 1086–9
[6] Béché A, Rouvière J L, Cooper D and Barnes J P 2010 *Ultramicroscopy* **111** 227-38
[7] Dismukes J P, Ekstrom L and Paff R J 1964 *J. Phys. Chem.* **68** 3021-7
[8] De Salvador D et al. 2000 *Phys. Rev.* B **61** 13005–13

Strain measurement for the semiconductor industry with nm-scale resolution by dark field electron holography and nanobeam electron diffraction

D Cooper[1], *A Béché[2], JM Hartmann[1], L Hutin[1], C Le Royer[1] and JL Rouviere[2]

[1] CEA-LETI, Minatec, 17 rue des Martyrs, F38054 Grenoble, France
[2] CEA-INAC, Minatec, 17 rue des Martyrs, F38054 Grenoble, France

Email: david.cooper@cea.fr

Abstract. Strain is routinely used in state-of-the-art semiconductor devices in order to improve their electrical performance. Here we present experimental strain measurements obtained by two different Transmission Electron Microscopy (TEM) based techniques. Dark field electron holography and nanobeam electron diffraction (NBED) have been applied to a range of semiconductor devices including recessed SiGe source and drain devices and SiGe pFETs on SOI. Here we demonstrate how the spatial resolution and sensitivity of these two techniques has been improved by using a state-of-the-art, probe corrected Titan TEM, and have benchmarked the capabilities of the two different techniques and highlight which method is most appropriate for different types of semiconductor specimens.

1. Strain measurement in the TEM

Although the effects of introducing strain in semiconductor devices is well known from an electrical point of view, little is known about the distribution of strain. Indeed, there was until recently no way of measuring the strain with the required level of spatial resolution and field of view. Dark field electron holography [1] and Nanobeam Electron Diffraction (NBED) [2] are two TEM based techniques that have been recently developed in order to solve this problem. Figure 1 shows the process for dark field electron holography. An electron biprism is used to interfere an electron wave that has passed through a strained region of interest with an electron wave that has passed through an unstrained reference, such as the substrate. By using an objective aperture to select a diffracted beam, a phase image can be reconstructed which corresponds to a displacement field for the chosen set of lattice parameters. A strain map can then be calculated using a Geometrical Phase Analysis algorithm. Dark field holography is well adapted for the semiconductor industry as 2D strain maps can rapidly be produced with a field of view of up to a micron and a spatial resolution of 4 nm. We have benefited here from the stability of the Titan TEM to acquire electron holograms for up to one minute to provide strain maps with a sensitivity of +/- 0.02 % [4]. Figure 2 shows the process for nanobeam electron diffraction. A near-parallel electron beam is used to provide diffraction patterns from which the local strain can be determined from the shifts in the diffraction spots relative to a reference pattern. By using the three condenser lens system in the FEI Titan TEM we can now provide an electron beam with a convergence angle of 0.2 mrads and a FWHM of 6 nm for 200 kV electrons. The electron beam is scanned across the region of interest in STEM mode and the patterns are automatically processed using in-house software to provide strain profiles. This technique is extremely easy to perform and 1D strain profiles with a strain sensitivity of 0.06 % and a spatial resolution of 3 nm have now been demonstrated [4].

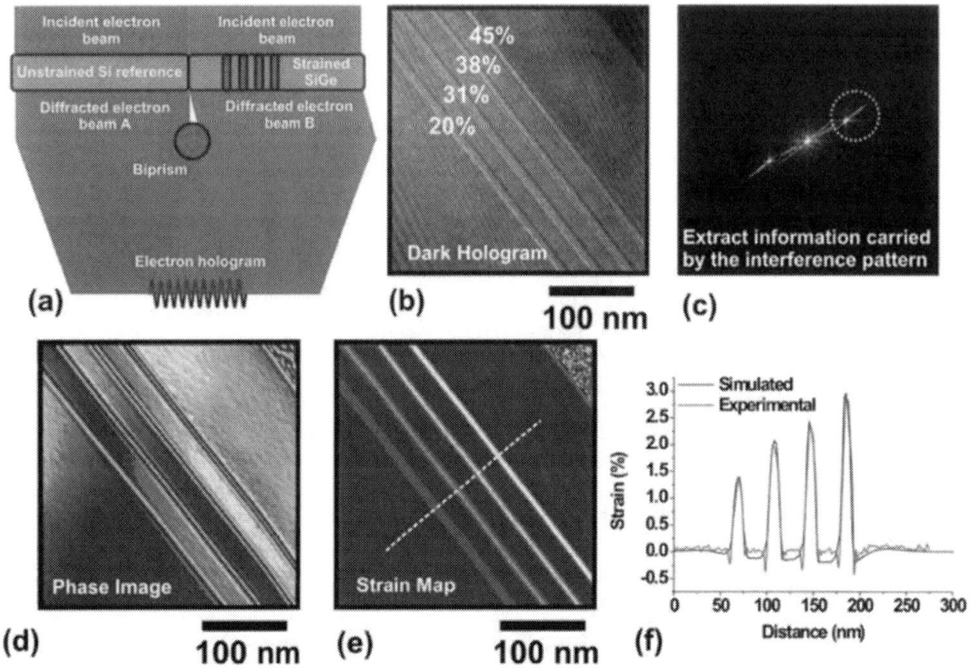

Figure 1. (a) schematic of the TEM used in dark field holography. (b) a {004} dark field electron hologram of a Si calibration specimen containing 10-nm-thick SiGe layers with different Ge concentrations. (c) A Fourier transform of the hologram is performed and the sideband is selected from which (d) a phase image can be reconstructed. GPA is used to calculate (e) a strain map showing the SiGe strain in the {004} growth direction. (f) A profile of the experimental results can be compared to finite element strain simulations demonstrating that this technique is fully quantitative.

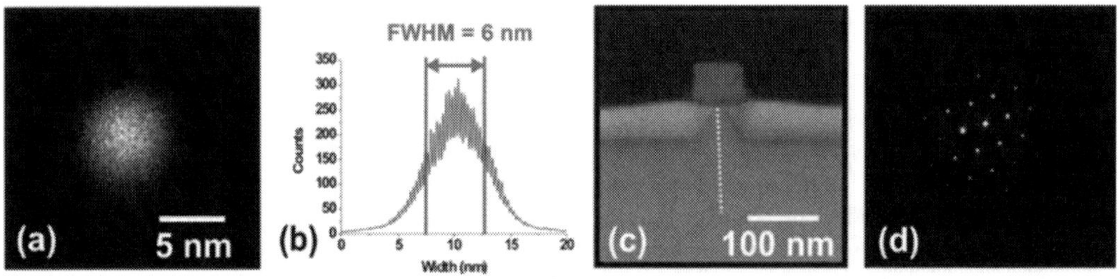

Figure 2. Schematic of the NBED procedure. (a) A 200 kV near-parallel electron probe on a {110} Si TEM specimen (b) The FWHM here is 6 nm. (c) The electron beam is scanned across the specimen in STEM mode and a diffraction pattern is recorded at each point from which the strain can be determined from the shift in the spots automatically using a peak fitting software developed at CEA/INAC.

2. Strain mapping of semiconductor devices

Figure 3 shows how both techniques have been applied to a 27-nm-gate length $Si_{0.77}Ge_{0.23}$ recessed source and drain device-like structure to measure the compressive strain in the Si conduction channel. Six {220} strain profiles are shown for each technique acquired at different times on different specimens to show the reproducibility of the technique. A compressive strain of 1.2 % has been measured and the experimental results shown here are clearly consistent with each other. Dark field holography has been used to determine the effect of changing the gate length in SiGe recessed source and drain devices. Dark field electron holography can now be performed routinely for strain characterization and has recently been used to track the strain evolution in SiGe pMOS devices during the silicidation process [5].

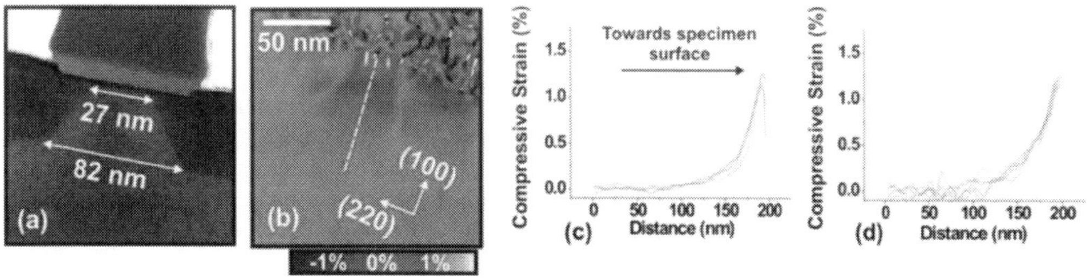

Figure 3. (a) HRTEM image of a 27-nm-gate length $Si_{0.77}Ge_{0.23}$ recessed sources and drains device-like structure. (b) A {220} strain map showing the compression in the conduction channel. (c) Six different {220} strain profiles acquired by dark field holography showing the reproducibility of the technique. The profiles have been averaged over just 3 nm. (d) Six different {220} strain profiles acquired by NBED. The results are consistent between the two techniques.

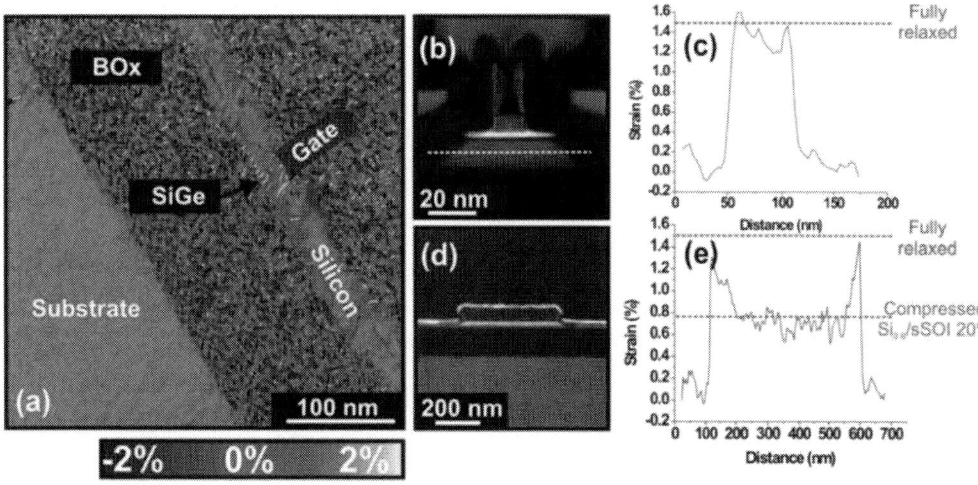

Figure 4. (a) {220} Strain map of a 17-nm-gate $Si_{0.6}Ge_{0.4}$ pFET grown on sSOI. The strain map suggests that the SiGe region is relaxed. (b) HAADF STEM image of the same pFET device. (c) A NBED profile taken from across the region indicated by the dashed line in (b). The SiGe layer has an in-plane lattice parameter that is 1.5 % larger than the Si reference, demonstrating that it is fully relaxed. (d) A HAADF STEM image of a 500-nm-gate pFET device. (e) The NBED {220} strain profile shows that the device is relaxed to around 70 nm each side of the device. The central regions of the SiGe are by contrast compressively strained on the 1.3 GPa tensily-strained Silicon-On-Insulator substrate.

Although we have shown that it is relatively straightforward to apply dark field holography to bulk CMOS devices, it is more complicated for thin film devices grown on strained silicon on insulator substrates (sSOI). Here the problem is the rotational mismatch between the region of interest and the substrate which leads to difficulties in acquiring a dark field hologram. Figure 4 shows a {220} strain map of a 17-nm-gate SiGe pFET device. Although the strain map suggests that the SiGe region under the gate is relaxed, the map is noisy and it is difficult to extract quantitative information. NBED is well adapted for these thin films and the strain profile in Figure 4(c) shows directly that the 17-nm-gate device is indeed fully relaxed. NBED has also been applied to a 500-nm-gate SiGe pFET to assess the relaxation distance on either side of these devices and Figure 4(e) shows that this device is fully compressed 70 nm from each end. Therefore pMOS devices that are grown in this way need to have a gate length of at least 140 nm if the effects of strain on the electrical mobility are to successfully introduced.

3. Conclusion

Dark field electron holography and NBED are two TEM-based techniques that can be used to measure the strain in semiconductor specimens with nm-scale resolution. Dark field electron holography can be used to provide 2D strain maps with a spatial resolution of approximately 5 nm and a sensitivity of 0.02 %. This technique is very well suited for semiconductor devices grown on bulk silicon. Disadvantages of dark field holography are that it is a difficult technique to perform and that the specimen needs to be tilted to a two-beam condition that limits the spatial resolution that can be achieved in a thick TEM sample. For NBED a spatial resolution as low as 3 nm has been demonstrated with a sensitivity of 0.06 %. Although NBED provides strain profiles and not maps, it is more versatile than dark field electron holography as there is no need to have a reference with the same crystallographic orientation as the region of interest. For example, NBED seems very well adapted for the characterization of specimens grown on SOI substrates. In addition, NBED is a very simple technique to perform. In summary, it is useful to be able to perform both techniques on different specimens in order to check that the values of strain that are measured in semiconductor specimens are correct.

Acknowledgements

This work was partially supported through the "Recherche Technologique de Base" Program and performed within the IBM – ST Microelectronics – LETI alliance.

References

[1] Hytch M, Houdellier F, Hue F and Snoeck E, 2008 *Nature.* **453** 1086
[2] Usada K, Numata T and Takagi S, 2005 *Mat. Sci. Eng. B.* **124** 143
[3] Cooper D, Hartmann JM, Béché A and Rouviere JL, 2009 *Appl. Phys. Lett.* **95** 053501
[4] Béché A, Rouviere JL, Clement L and Hartmann JM, 2009 *Appl. Phys. Lett.* **95** 123114
[5] Cooper D, Béché A, Hartmann JM and Rouviere JL, 2010 *Appl. Phys. Lett.* **96** 113508

TEM 3-beam study of annealing effects in InGaNAs using ab-initio structure factors for strain-relaxed supercells

K Müller[1], M Schowalter[1], O Rubel[2], D Z Hu[3], D M Schaadt[3], M Hetterich[3], P Gilet[4], R Fritz[5], K Volz[5] and A Rosenauer[1]

[1] Universität Bremen, Otto-Hahn-Allee 1, 28359 Bremen, Germany
[2] Thunder Bay Regional Research Institute, 290 Munro Street, Thunder Bay, Canada
[3] Karlsruhe Inst. of Technology, Wolfgang-Gaede-Str. 1, 76131 Karlsruhe, Germany
[4] CEA LETI Minatec Campus, 17 Av. des Martyrs, 38054 Grenoble Cedex 9, France
[5] Philipps Universität Marburg, Hans-Meerwein-Str., 35032 Marburg, Germany

Abstract. We report on a Transmission Electron Microscopy 3-beam technique based on the interference of 000, 200 and 220. Nonlinear imaging artefacts are eliminated by Fourier filtering, yielding 200 and 220 lattice fringe images, from which chemically sensitive contrast and strain are measured, respectively. In this way, In and N composition can be mapped at atomic scale in quaternary InGaNAs by comparison with simulated reference data. Our Bloch wave simulations are based on structure factors derived from supercells with 10^6 atoms, which have been strain-relaxed by valence force field methods. Additionally, the influence of electron redistributions due to chemical bonding is accounted for by modified atomic scattering amplitudes derived from density functional theory. By comparing local compositions in an annealed $In_{0.28}Ga_{0.72}N_{0.025}As_{0.975}$ sample with its as-grown counterpart, we find homogenisation of InGaNAs layer thickness and –stoichiometry upon annealing.

1. Introduction

Due to their potential for near-infrared signal transmission in optical fibres, dilute nitrides such as quaternary $In_xGa_{1-x}N_yAs_{1-y}$ are intensely studied for more than a decade. As to the optical properties of InGaNAs, room temperature photoluminescence (PL) in the desired spectral range at wavelengths of 1.3 and 1.55μm can indeed be realised, but PL peaks are very weak and broad. Common to most reports is the observation of a drastic increase of PL intensity accompanied by a blue-shift of several tens of nanometres after thermal annealing [1]. To explain this behaviour, enormous theoretical effort has been spent on studying responsible atomic-scale scenarios, of which the most prominent either focus on long-range diffusion processes [2,3] or modifications of preferential coordinations of In and N atoms [1,4]. However, it is difficult to clarify this issue from the experimental point of view with available transmission electron microscopy (TEM) techniques, because unique atomic-scale mapping of both local In and local N concentration is necessary.

In the following, we firstly introduce a TEM technique which exploits local chemically sensitive contrast and local strain contained in an image formed by the beams 000, 200 and 220. In this way, subsequent superposition of two images taken under different specimen tilts [5] is avoided to assure availability of strain and contrast from the same position. Because accurate simulations for the 200 fringe amplitude are required as reference data, we secondly report on modelling InGaNAs structure

[1] Any correspondence should be addressed to Knut Müller (mueller@ifp.uni-bremen.de).

factors in order to include chemical bonding via modified atomic scattering amplitudes (MASA) obtained from density functional theory (DFT), as well as static atomic displacements (SAD) via valence force field (VFF). Thirdly, thermal annealing effects on $In_{0.28}Ga_{0.72}N_{0.025}As_{0.975}$ are studied using the 3-beam imaging method and reference data accounting for bonding and SAD.

2. The three-beam imaging method

An L-shaped objective aperture has been fabricated in a focused ion beam facility from a 7 µm Pt-foil and was inserted in an FEI Titan 80/300 equipped with an aberration corrector for imaging. The method is summarised in Fig. 1. The aperture shown in Fig. 1a leads to the 3-beam image shown in Fig. 1b. As can be seen in the diffractogram in Fig. 1c, 220, 200 and 020 fringes occur, the latter being caused by nonlinear imaging.

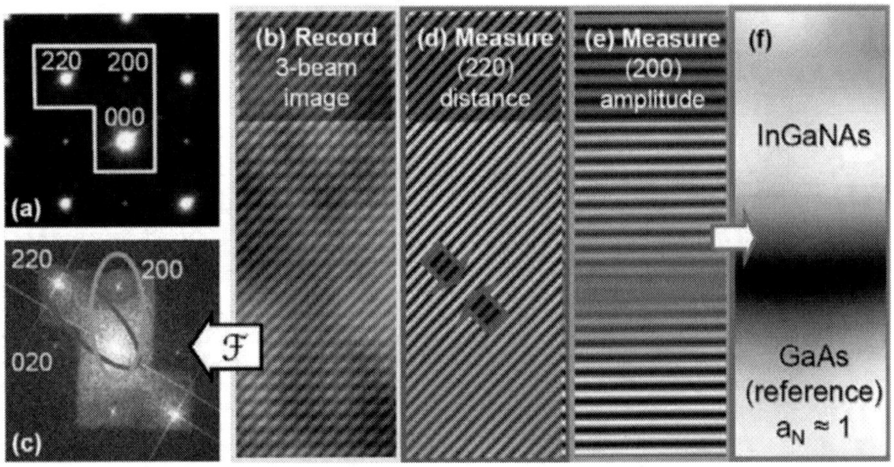

Figure 1. Formation and evaluation of TEM 3-beam images. (a) FIB-aperture in diffraction plane. (b) 3-beam image taken with Laue circle centre at (0 4.2 0). (c) Diffractogram. (d) 220 and (e) 200 lattice planes constructed from (b,c). (f) 200 fringe amplitude a_N showing an InGaNAs/GaAs interface.

Using beams $\mathbf{g}_1 = (000), \mathbf{g}_2 = (200)$ and $\mathbf{g}_3 = (220)$, the image intensity reads

$$I = A_1^2 + A_2^2 + A_3^2 + 2A_1A_2T_{12}\cos(2\pi\mathbf{g}_2\mathbf{r} - \phi_{12} - \varphi_{12}) + 2A_1A_3T_{13}\cos(2\pi\mathbf{g}_3\mathbf{r} - \phi_{13} - \varphi_{13}) +$$
$$2A_2A_3T_{23}\cos(2\pi(\mathbf{g}_2 - \mathbf{g}_3)\mathbf{r} - \phi_{23} - \varphi_{23}), \quad (1)$$

where $A_n\exp(i\varphi_n)$ is the amplitude of diffracted beam $\mathbf{g}_n$, $\varphi_{nm} = \varphi_n - \varphi_m$ is the phase difference between beams $\mathbf{g}_n$, $\mathbf{g}_m$ and $T_{nm}\exp(i\phi_{nm})$ is the corresponding transmission cross coefficient. Hence the occurrence of the 020 reflection in Fig. 1c is the only nonlinear contribution, meaning that amplitudes and phases of the other diffractogram reflections are equal to the respective two-beam results. Indeed, including any further diffracted beam in the imaging process would impose its thickness dependence on amplitudes and phases of the 200 and 220 fringes, making their interpretation more complicated or impossible in case specimen thickness is inaccurately known. However, according to Figs. 1d and 1e, two-beam 220 and 200 lattice fringe images can be obtained by Fourier filtering, the former providing strain and the latter chemically sensitive contrast. Note that strain data is available also at the minimum of the 200 fringe amplitude. As indicated in Fig. 1f, the 200 fringe amplitude is normalised to a background fitted to regions with known composition.

Figures 1d and 1f define both strain and amplitude for each lattice point, as depicted on the left of Fig. 2. Together with the simulated counterparts, this defines an intersection of an isostrain and an isoamplitude contour in the reference data on the right of Fig. 2, yielding both local In content x and local N content y. Reference 200 fringe amplitudes correspond to Bloch wave simulations [6] with a Laue circle centre at (0 4.2 0), for which the thickness dependence was found to be minimal.

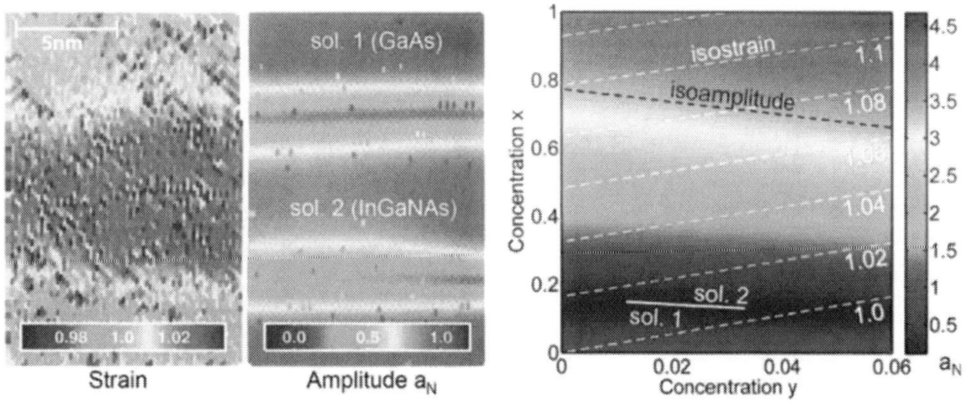

Figure 2. *Left:* Normalised fringe distance and amplitude determined from Figs. 1d, f. *Right:* Simulation of strain (dashed white isolines) and normalised 200 fringe amplitude a_N (colour).

3. Modelling bonding and SAD effects on electron structure factors

Even in binary GaAs and InAs electron redistribution due to chemical bonding severely affects low-order structure factors such as 200, so that it must definitely be taken into account in InGaNAs alloys. However, differences in atomic radii of N and In with respect to atoms of the host sublattices cause SAD. These are usually taken into account by considering supercells with about 10^6 atoms and performing VFF calculations. Since bonding is typically treated by DFT which applies to cells with about 10^2 atoms only, bonding effects need to be modelled.

For the Bloch wave simulations in Fig. 2 we used atomistic modelling of bonding based on MASA. These are indeed calculated by DFT, but only for all *binary* zinc-blende compounds that can be constructed from $In_xGa_{1-x}N_yAs_{1-y}$. Composition dependence and thin foil relaxation are introduced by calculating MASA from tetragonally distorted Bravais cells. For details, see Refs. [7,8]. In the latter report, we found that SAD from VFF are accurate to a few picometres, being less than mean thermal vibrational amplitudes. It was furthermore pointed out that, for supercells with 216 atoms, modelling bonding by MASA and SAD by VFF is in good agreement with full DFT calculations. However, large errors in structure factor amplitudes and phases may result if neglecting bonding and SAD.

The actual procedure for calculating structure factors is explained in Fig. 3.

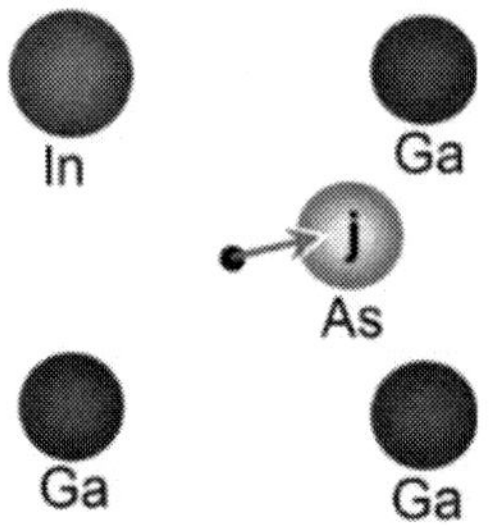

Figure 3. Schematic to demonstrate the treatment of bonding and SAD for structure factor calculation. First, large supercells (10^6 atoms) are strain-relaxed using VFF calculations. Second, structure factors are calculated for this cell. The MASA at lattice site j is modelled by determining nearest neighbours, defining the type of (binary) bonds involved. In the example on the left, As at site j has the MASA

$$f_{As,j}^{InGaNAs}(\mathbf{g}) = \tfrac{1}{4}[f_{As}^{InAs}(\mathbf{g}) + 3 f_{As}^{GaAs}(\mathbf{g})] \ ,$$

as 3 bonds of type As in GaAs and one of type As in InAs are present.

4. Annealing effects on $In_{0.28}Ga_{0.72}N_{0.025}As_{0.975}$

The 3-beam method in combination with reference data from the previous section has been applied to an $In_xGa_{1-x}N_yAs_{1-y}$ sample with nominal contents $x = 0.28$ and $y = 0.025$ before and after thermal annealing for 15 min under N_2 atmosphere at 600K. The PL of the annealed sample is shifted by 65nm towards smaller wavelengths and 20 times more intense. Composition maps and profiles are depicted in Fig. 4, exhibiting island formation in the as-grown sample. Inside the island, In enrichment of 0.03-0.05 is found, being anticorrelated to an N enrichment of 0.02-0.03 outside. In contrast, nearly homogeneous layer thickness and –stoichiometry is found after annealing. Error bars in Fig. 4c correspond to statistical lateral fluctuations, whereas the grey corridor indicates the error introduced by inaccurately known specimen thickness. As-grown profiles in Fig. 4f show both anticorrelation of In and N content as well as changes in InGaNAs layer thickness with respect to positions A, B in Fig. 4d.

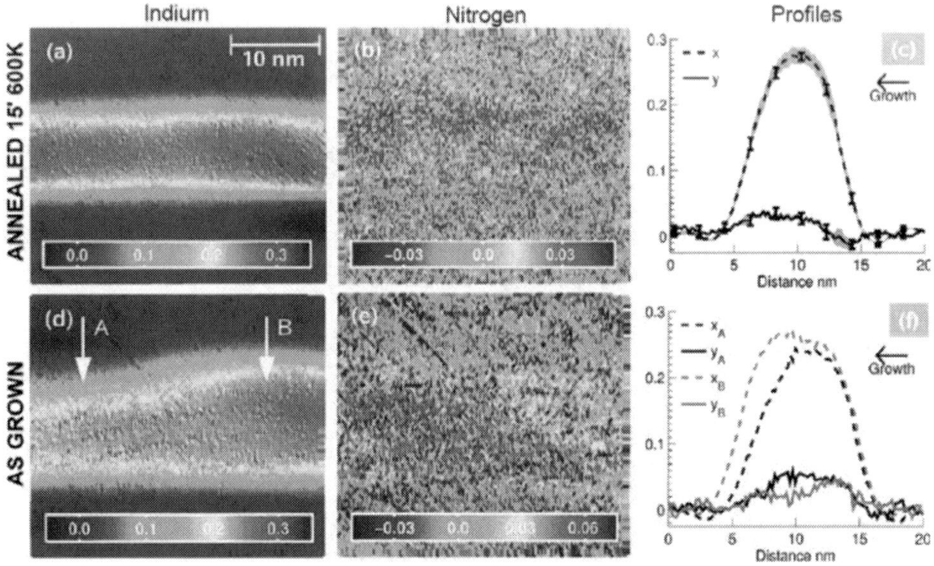

Figure 4. (a-c) In and N contents after annealing. The grey corridor in (c) includes all evaluations for thicknesses between 10 and 80 nm. (d-f) In and N contents for the as-grown counterpart. Profiles in (f) are taken at paths indicated in (d).

5. Conclusions

TEM 3-beam imaging can be used to subsequently recover 200 and 220 2-beam lattice fringe images for which linear imaging theory applies. Both chemically sensitive amplitude and strain can thus be measured independently, enabling atomic-scale concentration mapping in quaternary (zinc-blende) alloys by comparison with strain and amplitude reference data. The latter includes bonding via DFT-MASA and SAD from VFF. Disappearance of In-N anticorrelations in an $In_{0.28}Ga_{0.72}N_{0.025}As_{0.975}$ layer after annealing suggests that both long-range diffusion and increase of the frequency of In-N bonds ought to be responsible for a blue-shift of 65 nm observed in PL spectra in this study.

References

[1] Klar P J et al. 2001 *Phys. Rev. B* **64** 121203.
[2] Djie H S, Wang D-N, Ooi B S, Hwang J C M, Fang X M, Wu Y, Fastenau J M, Liu W K 2007 *Thin Solid Films* **515**(10) 4344-4347.
[3] Peng C S, Pavelescu E M, Jouhti T, Konttinen J, Pessa M 2003 *Solid-State Electronics* **47**(3) 431-435.
[4] Jenichen A, Engler C, Leibiger G, Gottschalch V 2005 *Surface Science* **574**(2-3) 144-152.
[5] Grillo V, Albrecht M, Remmele T, Strunk H P, Egorov A Y, Riechert H 2001 *Journal of Applied Physics* **90**(8) 3792-3798.
[6] Müller K, Schowalter M, Jansen J, Tsuda K, Titantah J, Lamoen D, Rosenauer A 2009 *Ultramicroscopy* **109** 820-814.
[7] Rosenauer A, Schowalter M, Glas F, Lamoen D 2005 *Phys. Rev. B* **72** 085326.
[8] Müller K, Schowalter M, Rosenauer A, Rubel O, Volz K 2010 *Phys. Rev. B* **81** 075315.
[9] This work was supported by the DFG under contracts RO 2057/4, SCHO 1196/3 and VO 805/4,5.

Low voltage TEM for semiconductor analysis

M Stöger-Pollach[1], A Steiger-Thirsfeld and S Schwarz

University Service Center for TEM, Vienna University of Technology, Wiedner Hauptstr. 8-10/052 A-1040 Vienna, AUSTRIA

Email: stoeger@ustem.tuwien.ac.at

Abstract. Valence electron energy loss spectrometry (VEELS) is more and more employed for the quantification of dielectric properties in semiconductor research. The improvement of the energy resolution of the thermo-ionic emitter (lanthanum-hexaboride) is obtained by decreasing the acceleration-to-Wehnelt voltage ratio and by inserting a smaller Wehnelt diaphragm. These modifications lead to an improved energy resolution of 0.25 eV full width at half maximum and 3.03 eV full width at thousandth maximum at 20 keV beam energy, which is pretty close to the one of a monochromated field emission gun. Beside the improved energy resolution we demonstrate the accuracy of low voltage VEELS in terms of band gap determination of $Al_xGa_{1-x}As$, with x varying from zero to 0.58. The experimental results are compared with a well established semi-empirical model commonly used for the calculation of the properties of III-V semiconductors.

1. Introduction

In recent years a renaissance of low beam energies for operating transmission electron microscopes (TEMs) can be observed. The main driving force behind this development is the reduced beam damage based on the knock-on mechanism. The second benefit of employing low beam energies is the enhanced contrast in TEM and STEM imaging, due to the beam energy dependent scattering behaviour. Therefore all main TEM and scanning TEM (STEM) manufacturers apply nowadays low voltage instruments with different ranges of available beam energies. Two of these TEMs are outstanding, because they combine best imaging capabilities (by employing C_s-correctors) and high-resolution EELS at very low voltages. These are the JEOL microscope at AIST [1], Japan, equipped with a probe and an image C_s-corrector and EELS capabilities from 30 kV to 60 kV using a post column GATAN ENFINA spectrometer. The second one is the recently inaugurated ZEISS SALVE microscope [2] equipped with an image C_s-corrector and EELS capabilities from 20 kV to 200 kV using an in-column Ω-type energy filter. In the actual work we use a FEI TECNAI 20-200 operated at 20 kV equipped with a GATAN GIF2001 energy filter. The TECNAI 20-200 has a thermo ionic electron emitter [3,4]. Its energy resolution was improved by introducing a smaller Wehnelt aperture with a diameter of 0.3 mm thus achieving 0.6 eV full width at half maximum (FWHM) in the zero loss peak (ZLP) at 200 kV and only 0.25 eV FWHM at 20 kV operation voltage.

In the actual work, we employ low beam energies for having another benefit: we want to avoid relativistic energy losses, like Čerenkov losses and light guiding modes [5-8] which alter the low loss spectrum. Thus we are able to perform Kramers-Kronig Analyses (KKA) without the need for taking into account relativistic effects [8,9]. From KKA we gain the energy loss function and further the optical constants, like refractive index, absorption coefficient, band gap, complex dielectric function –

[1] To whom any correspondence should be addressed.

and all of them for a much broader energy range than would be measurable by employing optical methods. So the advantage of electron beam techniques is clear: the spectral range and the spatial resolution are much better compared to optical methods. The latter benefit is limited by the Coulomb delocalization [4], which is in the range of several nanometres for very low losses. However, the delocalization is also reduced by using lower beam energies.

2. Experimental

The specimen was a layered structure of two $Al_xGa_{1-x}As$ layers on a GaAs substrate. The chemical composition was characterized by means of energy dispersive X-ray analysis (EDX). Table 1 gives the chemical compositions of the layers. The layer thickness was 200 nm for layer 1 and 400 nm for layer 2. The sample was prepared by using a focused ion beam. A final Ar^+-ion polishing step was performed. The final lamella thickness in the investigated area varied from 77 nm – 82 nm.

substrate	layer 1	layer 2
GaAs	Al[illegible]Ga[illegible]As	Al[illegible]Ga[illegible]As

Table 1: Chemical composition of the investigated layers.

The valence EELS (VEELS) experiments were performed on a conventional FEI TECNAI 20-200 operated at 20 kV. The measured energy resolution was 0.25 eV in the FWHM of the ZLP using a fully saturated filament (see Figure 1). The full width at thousandth maximum is 3.03 eV, which is the more important value for band gap studies compared to the FWHM.

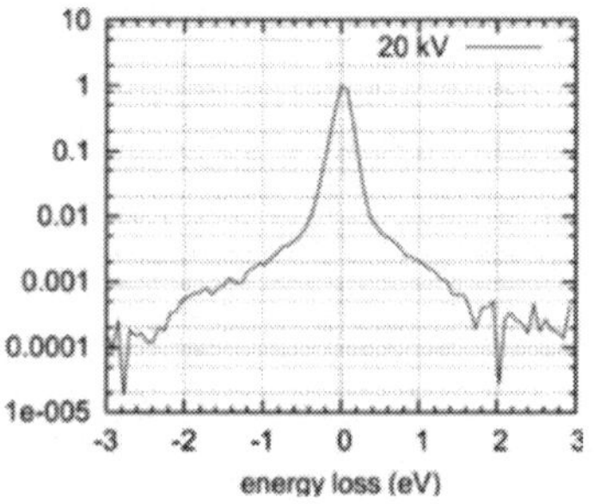

Figure 1: ZLP recorded at 20 kV.

Due to the fact that Čerenkov losses are strong in GaAs [6,7] and also in $Al_xGa_{1-x}As$, the beam energy was reduced such, that any relativistic energy losses were avoided experimentally. This was achieved by using a 20 keV electron beam. The microscope itself was operated in image mode, because the smallest available selected area aperture is 260nm back-projected onto the sample. For the thinnest layer (layer 1) this would be too large by a factor of two. For the VEELS experiments the collection semi-angle was kept small, in order to be able to compare the results with optical data, which are intrinsically limited to zero momentum transfer. The selected collection semi-angle was 2.8 mrad, which is due to the smallest available objective aperture (OA). Table 2 gives the relative size of the smallest OA in comparison with the dimensions of the Brillouin zone of GaAs for 20 kV. However, the strong band bending of GaAs must be considered when comparing the VEELS results with optical data.

location	distance[a] [k]	β/k[b]
$\Gamma = 000$	-	-
$X = 100$	$\lvert\Gamma X\rvert = 1$	0.373
$L = \frac{1}{2}\frac{1}{2}\frac{1}{2}$	$\lvert\Gamma L\rvert = \sqrt{3/4}$	0.431
$K = \frac{3}{4}\frac{3}{4}0$	$\lvert\Gamma K\rvert = \sqrt{9/8}$	0.352

[a] Distances in dimensionless "q-space" units
[b] $\beta = 2.83$ mrad, $E_0 = 20$ keV

Table 2: Relative size of the collection aperture β (2.8 mrad) to the Brillouin zone of GaAs for 20 kV.

3. Results on $Al_xGa_{1-x}As$

In the following we compare the experimental results by the theoretical model established by Adachi [10,11], where the refractive index $n(\omega)$ in zincblende materials below the direct band gap edge can be expressed as

$$n(\omega) = \sqrt{A_0 \cdot \left(f(\chi) + 0.5 \cdot \left(\frac{E_0}{E_0 + \Delta_0}\right)^{3/2} \cdot f(\chi_{so})\right) + B_0}$$

with

$$f(\chi) = \frac{2 - \sqrt{1+\chi} - \sqrt{1-\chi}}{\chi^2}$$

$$\chi(\omega) = \frac{\omega}{E_0}$$

$$\chi so(\omega) = \frac{\omega}{E_0 + \Delta_0}$$

where A_0, B_0, E_0, and Δ_0 are given in [11]. Δ_0 represents the spin-orbit coupling energy, E_0 is the fundamental band gap at the Γ-point – which is plotted as a function of the Al concentration x in Figure 2 (left). This figure also shows the energy separation in the X and L point of the Brillouin zone. As long as the Γ-line (full line) is the lowest one, $Al_xGa_{1-x}As$ has a direct band gap, but as soon as the X-line (dashed line) has crossed, $Al_xGa_{1-x}As$ has an indirect gap. The right hand side of Figure 2 shows the experimentally obtained band gaps of the three layers.

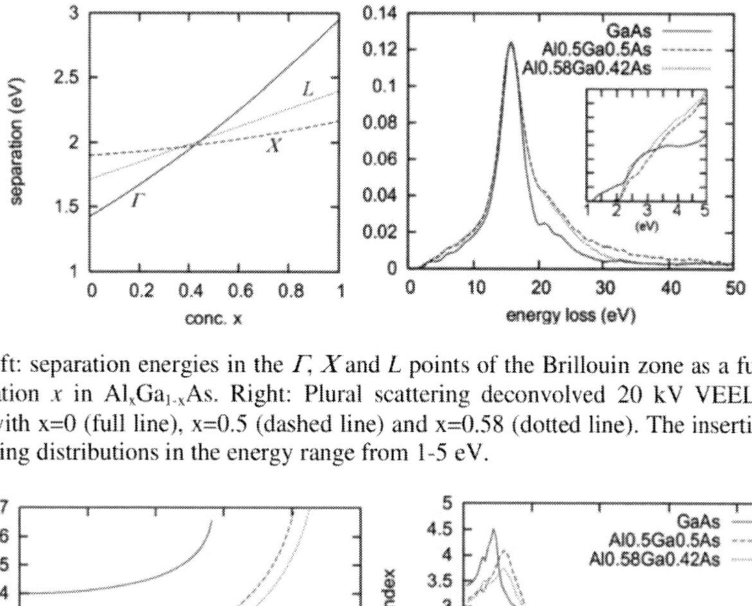

Figure 2: Left: separation energies in the Γ, X and L points of the Brillouin zone as a function of the Al concentration x in $Al_xGa_{1-x}As$. Right: Plural scattering deconvolved 20 kV VEELS spectra of $Al_xGa_{1-x}As$ with x=0 (full line), x=0.5 (dashed line) and x=0.58 (dotted line). The insertion shows the single scattering distributions in the energy range from 1-5 eV.

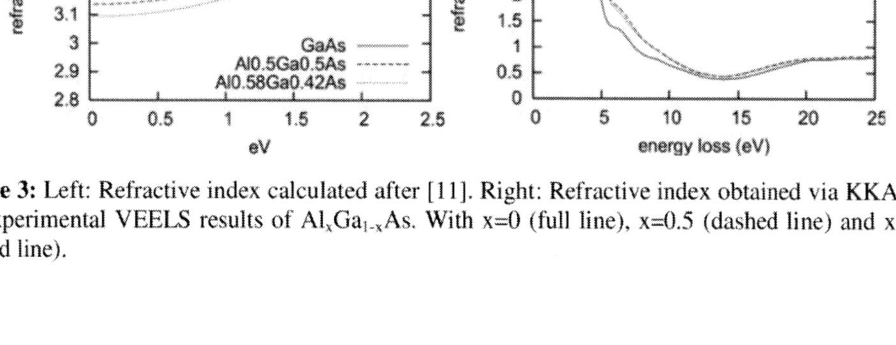

Figure 3: Left: Refractive index calculated after [11]. Right: Refractive index obtained via KKA from the experimental VEELS results of $Al_xGa_{1-x}As$. With x=0 (full line), x=0.5 (dashed line) and x=0.58 (dotted line).

3.1. Kramers-Kronig Analysis (KKA)

Before KKA the experimental spectra were deconvolved from plural scattering using the matrix approach [12]. Further the surface loss contributions were removed from the single scattering distribution following Kröger's fully relativistic theory [5] and the spectra were normalized using the sample thickness approach [8]. The results fit very well the calculated ones. On the left hand side of Figure 3 the calculated values of the refractive index for the band gap [11] are displayed and on the right hand side the experimental obtained values for a much larger energy range are shown. As displayed in Figure 3 (right) the refractive index fits also very well the calculated data.

3.2. Band gaps

Due to the small collection semi angle the indirect band gaps cannot be observed. The measured band gaps therefore follow the Γ-line in Figure 2. In Figure 4 the experimental (bullets with error bars) and calculated (full line) band gaps are displayed. The uncertainty stems first from the spectral broadening due to the initial energy width of the electron beam and second from the signal to noise ratio (SNR). Due to the fact that in the band gap region the signal is very low, the SNR is worse than in the rest of the VEELS spectrum. The indirect band gap was outside the collection aperture and can therefore not be measured.

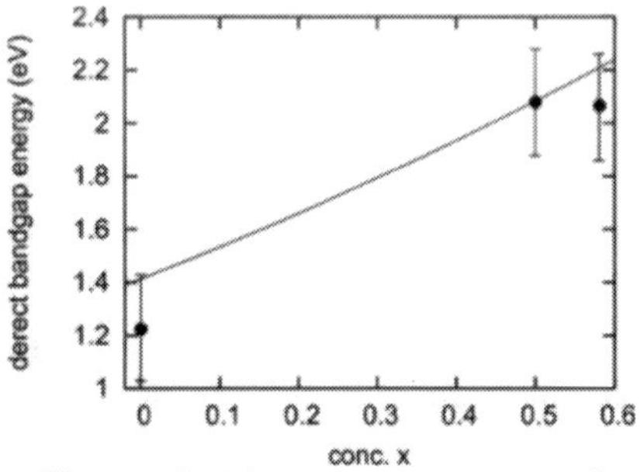

Figure 4: Direct band gap energies: Experimental – bullets with error bars, calculated Γ-line – full line.

4. Discussion

In this work we exemplarily showed the accuracy of the determination of dielectric properties by employing VEELS. Under the condition that relativistic effects can be avoided VEELS is a reliable tool. Due to the fact that a small collection semi angle is chosen for having a better comparability with optical methods, indirect band gaps cannot be determined. However, the direct transitions are measured very accurately. For measuring indirect band gaps in $Al_xGa_{1-x}As$ the collection aperture must be located at the X-point of the first Brillouin zone. This kind of dark field VEELS was done for Silicon in [13]. Because Čerenkov losses and light guided modes induce strong forward scattering ($\theta_{Čerenkov} < 200\ \mu rad$), their signature cannot be recorded in dark field VEELS thus allowing higher beam energies. However, such results are not comparable with the ones from optical methods.

References

[1] Suenaga K et al. 2009 *Nature Chemistry* **1**, pp. 415-418
[2] Kaiser U et al. 2011, *Ultramicroscopy*, accepted
[3] Stöger-Pollach M and Pongratz P 2010 *J. Phys.: Conf. Ser.* **209**, pp. 012031
[4] Stöger-Pollach M 2010 *Micron* **41**, pp. 577-584
[5] Kröger E 1968 *Z. Phys.* **216**, pp. 215-235
[6] von Festenberg C 1969 *Z. Phys.* **227**, pp. 453-481
[7] Stöger-Pollach M et al. 2006 *Micron* **37**, pp. 396-402
[8] Stöger-Pollach M 2008 *Micron* **39**, pp. 1092-1110
[9] Stöger-Pollach M et al. 2008 *Ultramicroscopy* **108**, pp. 439-444
[10] Adachi S 1982 *J. Appl. Phys.* **53**, 5863
[11] Adachi S 1985 *J. Appl. Phys.* **58**, R1-29
[12] Schattschneider P 1983 *Ultramicroscopy* **11**, pp. 321-322
[13] Gu L et al. 2007 *Phys. Rev. B*, pp. 195214 1-8
[14] The Hochschuljubiläumsstiftung der Stad Wien, contract Nbr. H-01585/2007, is kindly acknowldged for funding.

Accurate calibration for the quantification of the Al content in AlGaN epitaxial layers by energy-dispersive X-ray spectroscopy in a Transmission Electron Microscope

H Amari[1], L Lari[1], H Y Zhang[1], L Geelhaar[2], C Chèze[2], M J Kappers[3], C McAleese[3], C J Humphreys[3] and T Walther[1]

[1] Department of Electronic & Electrical Engineering, University of Sheffield, Sir Frederick Mappin Building, Mappin Street, Sheffield S1 3JD, UK
[2] Paul-Drude-Institut für Festkörperelektronik, 10117 Berlin, Germany
[3] Department of Materials Science & Metallurgy, University of Cambridge, Cambridge CB2 3QZ, UK

E-mail: h.amari@sheffield.ac.uk

Abstract. Since the band structure of group III- nitrides presents a direct electronic transition with a band-gap energy covering the range from 3.4 eV for (GaN) to 6.2 eV (for AlN) at room temperature as well as a high thermal conductivity, aluminium gallium nitride (AlGaN) is a strong candidate for high-power and high-temperature electronic devices and short-wavelength (visible and ultraviolet) optoelectronic devices. We report here a study by energy-filtered transmission electron microscopy (EFTEM) and energy-dispersive X-ray spectroscopy (EDXS) of the microstructure and elemental distribution in different aluminium gallium nitride epitaxial layers grown by different research groups. A calibration procedure is out-lined that yields the Al content from EDXS to within ~1 at % precision.

1. Introduction

A range of electron microscopy techniques can be used to monitor changes in elemental distributions, thicknesses of layers and the effects of different types of dislocations. Reliable elemental distribution mapping by EDXS in a (scanning) transmission electron microscope (TEM/STEM) depends upon knowledge of different parameters for each line, such as the absorption *(A)* [1] and the fluorescence *(F)* [2] coefficients. This approach also includes atomic number *(Z)* [3] effects and has become known as *ZAF* correction. This correction to energy-dispersive X-ray microanalysis is important for accurate quantification, in particular for low atomic number materials. In order to calibrate EDXS for determining the chemical composition of AlGaN accurately within ~ 1 at%, this project compared experimental measurements of X-ray line intensity ratios obtained by EDXS from four different AlGaN layer samples that covered a wide compositional range from Ga-rich ($Al_{0.23}Ga_{0.77}N$) to Al-rich ($Al_{0.87}Ga_{0.13}N$). Particular attention is drawn to the effects on the intensities of the Al-K, Ga-K and Ga-L lines by stray X-rays from the Al_2O_3 (corundum) substrate and by potential Ga implantation during focused Ga^+-ion beam thinning. Usually, the thickness of a specimen for TEM is not precisely known but can influence the *ZAF* correction significantly. Hence, unknown sample thicknesses are a major source of error in quantitative EDXS. Here, we follow an approach where the X-ray signals are measured as function of known thicknesses to eliminate this source of errors [4].

2. Sample growth

AlGaN/GaN and AlGaN/AlN heterostructures were grown by different methods on α-alumina (corundum) substrates. To reduce the dislocation density in the Ga-rich sample ($Al_{0.23}Ga_{0.77}N$), a 140 nm GaN buffer layer was grown on top of (0001) Al_2O_3 at a temperature of 700°C by molecular beam epitaxy (MBE), followed by the $Al_{0.23}Ga_{0.77}N$ layer at 800 °C [5]. For the other three samples with higher Al content a thick AlN buffer layer (~ 4 μm) was deposited on (0001) Al_2O_3 at a temperature of 1130 °C by metal-organic chemical vapour deposition (MOCVD), followed by 2 μm $Al_xGa_{1-x}N$ layers (x = 0.44, 0.72, and 0.87) at 1090 °C [6]. The Ga and Al concentrations were investigated by Rutherford Back Scattering (RBS) or X-ray diffraction (XRD).

3. TEM experiments

The focused ion beam (FIB) technique has been employed to prepare electron transparent specimens. This technique allowed us to mill site-specific windows into the lamella samples of known sample thicknesses in the range 50 – 700 nm. The TEM analysis was carried out in a JEOL 2010F field-emission gun TEM/STEM microscope operated at 197 keV and equipped with an energy dispersive X-ray spectrometer (Oxford Instruments Si:Li detector with ultra thin window and Link ISIS 300 acquisition software), annular dark field (ADF) (55 mrad inner angle) and bright field (BF) STEM detectors coupled with a Gatan Digiscan system, and a Gatan GIF 2000 Imaging Filter.

4. Results and Discussion

4.1. STEM and EFTEM analyses

The FIB milling was employed for sample preparation in order to obtain regions with different calibrated thicknesses (t) as shown in Fig. 1. The BF images in Fig. 1 show the lamella structures studied in this work. In Fig.1a) the 134 nm GaN and the 139 nm AlGaN layers appear darker because of their higher mean atomic number with respect to corundum (substrate) on the bottom and the carbon region on the top. Fig. 2 shows columnar grain growth in both GaN buffer layer and AlGaN layer, which are separated by a rough interface.

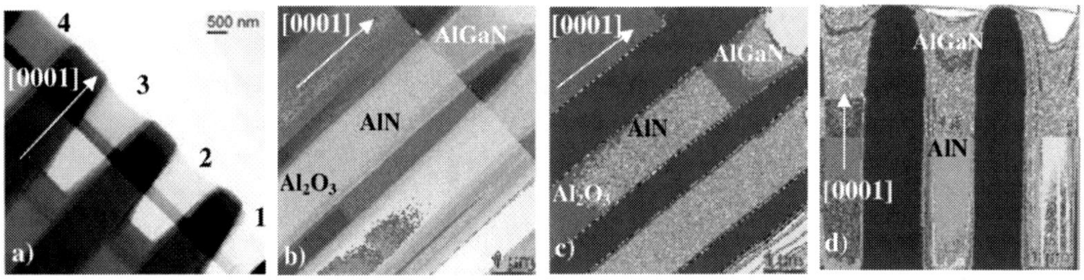

Figure 1. BF images showing the lamella structures including AlGaN/GaN or AlN/Al_2O_3 for (a) $Al_{0.23}Ga_{0.77}N$, (b) $Al_{0.44}Ga_{0.56}N$, (c) $Al_{0.72}Ga_{0.28}N$, and (d) $Al_{0.87}Ga_{0.13}N$. Numbers indicate regions of different thicknesses.

The sample thickness (t) was obtained using Energy-Filtered Transmission Electron Microscopy (EFTEM). It was measured locally, as shown in Fig. 2(b), in units of the total mean free path, λ, for inelastic scattering, obtained via the EELS routines of Gatan Digital Micrograph.

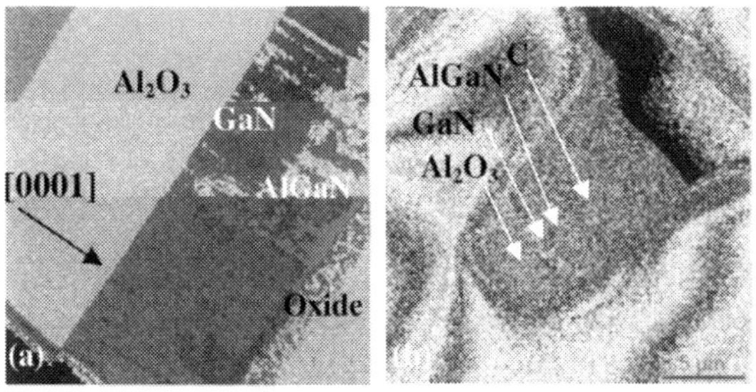

Figure 2. (a) BF image showing $Al_{0.23}Ga_{0.77}N/GaN/Al_2O_3$ interfaces. (b) EFTEM thickness map of region 1 of the first structure (arrows indicate positions of thicknesses measurements and EDXS point analyses reported below). Grey levels: Black: $t/\lambda = 0$, White: $t/\lambda = 4.76$.

Table 1. Thicknesses measured from relative thickness maps using the inelastic mean free paths of the materials for a collection angle of $\beta = 40$ mrad employing Egerton's approximation [7].

		window number				
sample		*1*	*2*	*3*	*4*	*5*
$Al_{0.23}Ga_{0.77}N$	t (nm)	86	160	280	427	700
$Al_{0.44}Ga_{0.56}N$		64	82	127	214	390
$Al_{0.72}Ga_{0.28}N$		105	142	250	416	600
$Al_{0.87}Ga_{0.13}N$		103	184	221	316	443

4.2. Determination of local composition by EDXS

Operating at 197 keV, the minimal electron probe of the JEOL 2010F is around 0.25 nm diameter, but in order to investigate the Al distribution in individual thin layers by a series of EDX point analyses in STEM mode, a probe size of 1 nm was used which, along the beam direction through the sample, will broaden to several nm. All the EDX spectra were acquired for acquisition times of 300 s, an energy resolution of 120 eV, and an energy dispersion of 20eV/channel. The concentration of element *j* was obtained using the ISIS 300 X-ray quantification routine [8], based on the following formula:

$$x_j = \frac{I_j \times k_j \times a_j / A_j}{\sum_i [I_i \times k_i \times a_i / A_i]} \quad (1)$$

where *x* is the elemental concentration in atomic percent (at %), *I* the integral intensity of the peak of interest (background subtracted), and *k*, *a*, *A* are *k*-factor (for weight %), absorption and the atomic weight respectively. Since EDX analysis for low atomic number elements such as N is not accurately enough, only Al-K, Ga-K and Ga-L peaks were considered. In Table 2, we list *k*-factors and absorption factors relative to Si-K as a standard which are used in our calculations. Fig. 3 presents typical EDX spectra of two different structures: $Al_{0.23}Ga_{0.77}N/GaN/Al_2O_3$ (left side), and $Al_{0.87}Ga_{0.13}N/AlN/Al_2O_3$. Even though the accuracy of EDX analysis for light elements is poor, a quantitative analysis of Al, Ga, N, and O has been attempted. It can be seen that the layers contain all these elements; Cu and C signals are due to the specimen support grid and surface contamination. In the EDX spectrum obtained from Al_2O_3, we attributed the weak Ga peak (~ 0.3 at %) to Ga implantation related to the FIB sample preparation method. Also, because the $Al_{0.23}Ga_{0.77}N$ layers are thinner than the Al-richer AlGaN samples, the presence of the apparent Al peak in the GaN spectrum is due to beam broadening and the vicinity of the corundum substrate.

Journal of Physics: Conference Series **326** (2011) 012028 doi:10.1088/1742-6596/326/1/012028

Table 2: list of *k*-factors and absorption factors relative to Si-K that ISIS uses for Al and Ga ratio calculations for different thicknesses with an effective take-off angle of 45° (25° + 20° tilt of specimen towards detectors). AlGaN densities are; 5.45 for $Al_{0.23}Ga_{0.77}N$, 4.8 for $Al_{0.44}Ga_{0.56}N$, 4.1 for $Al_{0.72}Ga_{0.28}N$, and 3.7 for $Al_{0.87}Ga_{0.13}N$.

		t (nm)				
		100	200	400	600	700
line	*k*-factor	abs. fact	abs. fact	abs. fact	abs. fact	abs. fact
Al-K	1.03	1.007	0.995	0.952	0.956	1.009
Ga-K	1.609	0.966	0.969	0.898	0.589	0.407
Ga-L	1.654	0.955	0.989	0.846	0.743	0.680

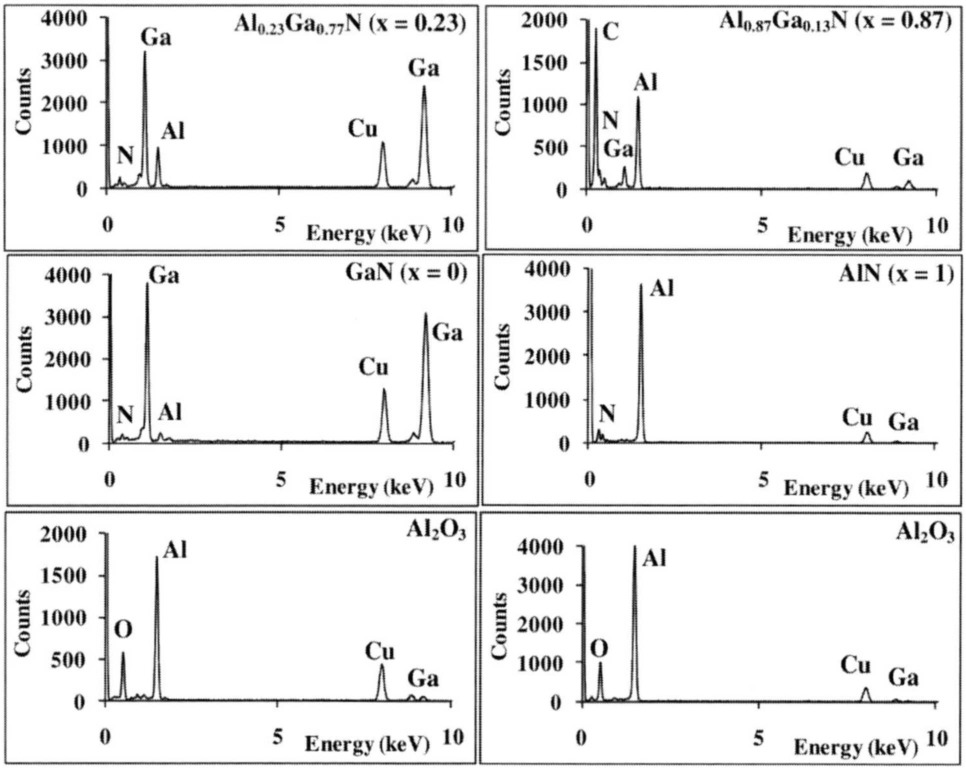

Figure 3. Typical EDX spectra acquired from Al_2O_3 substrate (below), GaN or AlN buffer layers (middle) and AlGaN epitaxial layers (above) respectively; (left side: $Al_{0.23}Ga_{0.77}N$/GaN at $t \approx 90$ nm, right side: $Al_{0.87}Ga_{0.13}N$/AlN at $t \approx 100$ nm).

The composition ratio of Al to Ga is close to the nominal stoichiometric values for all AlGaN layers except the narrowest $Al_{0.23}Ga_{0.77}N$ and GaN layers, where the analyses points were much closer to the substrate. To estimate the combination of beam broadening and stray X-rays, EDX spectra were acquired in the GaN buffer layer, as shown in Fig. 4. The Al contribution from Al_2O_3 is fitted by a polynomial extrapolation of the curve in Fig. 4 to $t = 0$ nm. It was made sure by electron energy-loss spectroscopy (EELS) that the GaN layer did not actually have any noticeable Al contamination. The quadratic behaviour shown in Fig. 4 agrees with the electron beam broadening successively with larger specimen thicknesses, which yields more signal from the substrate region.

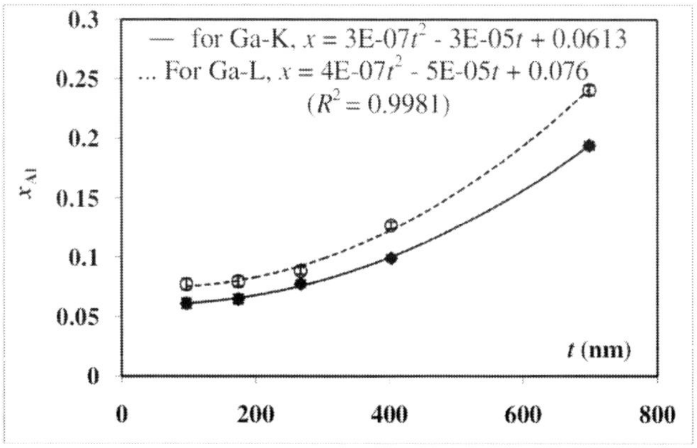

Figure 4. EDX point analyses of apparent Al content in GaN.

Assuming to a first approximation that the amount of Al-K stray X-rays from the substrate depends on the sample thickness as shown and linearly on the Ga-content of the layer investigated (as Ga is the heaviest element in our samples and mostly responsible for beam broadening and stray X-rays), we can calculate that a certain fraction δI_{Al} of the intensity of the Al-K X-rays is due to stray scattering:

$$\delta I_{Al} = (bt^2 + ct + d) \times x_{Ga} \quad (2)$$

where b, c and d are fitting constants, t is thickness and $x_{Ga} = 1 - x_{Al}$ the nominal Ga content. Also, as mentioned above, Ga implantation by the FIB sample preparation method needs to be considered. The estimated contribution (δI_{Ga}) was obtained by the investigation of Ga in AlN buffer and Al_2O_3 layers. The results showed that a value of 0.3 at% should be subtracted from the Ga signal for improved calculations. To obtain corrected Al ratios ($\tilde{x}_{Al}^{K,L}$), the following equation was applied:

$$\tilde{x}_{Al}^{K,L} = \frac{(I_{Al} - \delta I_{Al}) \times k_{Al} \times a_{Al} / A_{Al}}{[(I_{Al} - \delta I_{Al}) \times k_{Al} \times a_{Al} / A_{Al}] + [(I_{Ga} - \delta I_{Ga}) \times k_{Ga} \times a_{Ga} / A_{Ga}]} \quad (3)$$

Fig. 5 plots the results of the EDXS quantification for the different AlGaN layers obtained after the application of equation (3). As we can see we get a linear dependence of $\tilde{x}_{Al}$ on thickness, which is weak for Al-rich AlGaN but strong for Ga-rich AlGaN.

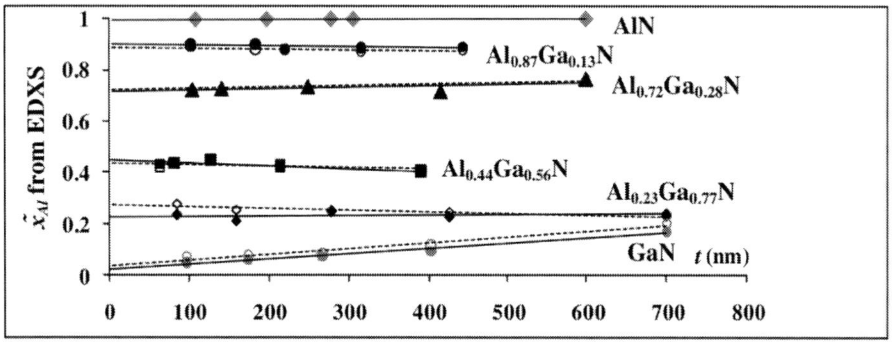

Figure 5. EDX point analyses of Al fraction $\tilde{x}_{Al}$ in AlGaN layers, from $Al_{0.23}Ga_{0.77}N$, $Al_{0.44}Ga_{0.56}N$, $Al_{0.72}Ga_{0.28}N$, and $Al_{0.87}Ga_{0.13}N$ after using equation (3).

Using a linear fit down to $t = 0$ as a final correction, we plot the Al-concentration $\tilde{\tilde{x}}_{Al}$ from EDXS vs. that concentration from RBS/XRD in Fig. 6 c).

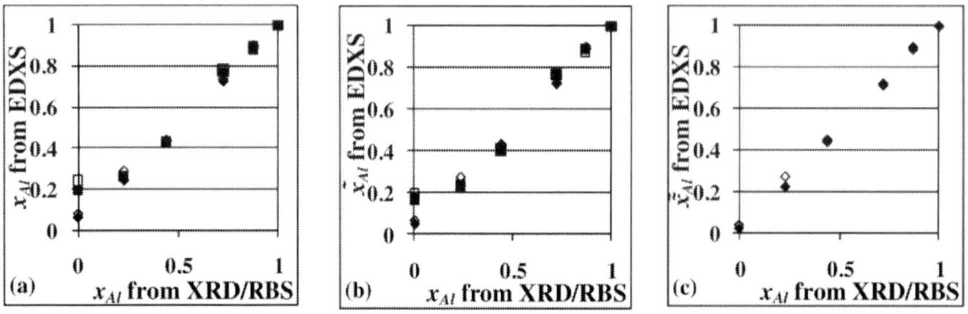

Figure 6. Al fractions obtained by EDXS for thinnest (diamond symbols) and thickest (square symbols) regions for Al-K/Ga-K (solid symbols) and Al-K/Ga-L (open symbols) vs. concentrations from RBS/XRD (a) without any correction, (b) using equation (3), and (c) using equation (3) and extrapolation to t= 0.

Table 3: Quantitative comparison of Al concentration from X-ray spectra of samples of different thicknesses after linear fitting of data.

		Ga-line	slopes	y-offsets
x_{Al} without correction	thinnest regions	Ga-K	0.962 ± 0.025	0.038 ± 0.017
		Ga-L	0.930 ± 0.027	0.064 ± 0.018
	thickest regions	Ga-K	0.868 ± 0.071	0.120 ± 0.046
		Ga-L	0.831 ± 0.089	0.147 ± 0.057
$\tilde{x}_{Al}$ with correction	thinnest regions	Ga-K	0.973 ± 0.025	0.027 ± 0.016
		Ga-L	0.939 ± 0.029	0.053 ± 0.019
	thickest regions	Ga-K	0.898 ± 0.077	0.089 ± 0.050
		Ga-L	0.870 ± 0.088	0.107 ± 0.057
$\tilde{\tilde{x}}_{Al}$ with correction	$t = 0$	Ga-K	0.989 ± 0.019	0.013 ± 0.012
		Ga-L	0.959 ± 0.017	0.037 ± 0.011

5. Conclusion

It has been demonstrated that the EDX analysis applying commercial software is not accurate enough to obtain quantification within ± 1 at%. This can be improved using our method of calibration, by

1. analyzing samples under identical experimental conditions,
2. taking systematic experimental measurements of X-ray line intensity ratios of Al-K, Ga-K and Ga-L, as a function of known sample thickness,
3. subtracting the elemental contributions from substrate (for Al), and ion implantation (for Ga),
4. extrapolating all corrected Al concentration values to t = 0 nm from a thickness series.

We determine the Al concentration with a precision of ~ ± 1 at %.

HA and LL acknowledge funding by EPSRC under grant EP/F02374X/1.

References

[1] J Philibert 1963 *Proc. Int. Symp. X-ray Optics and X-ray Microanalysis (Academic Press, New York, USA)* ed HH Pattee, VE Cosslett and E Engstrom, p. 379
[2] SJB Reed 1965 *Brit. J. Appl. Phys.* **16**, 13
[3] HA Bethe 1930 *Ann. Phys.* **397**, 325
[4] T Walther, 2010, *J. Phys. Conf. Ser.* **209**, 012029; *ibid.*, **241** 012016
[5] H Riechert *et al.* 1997, *Proc. Mater. Res. Soc. Symp.* **449** pp. 149
[6] TY Chang *et al. 2010, J. Appl. Phys.* **108**, 123522 (2010)
[7] RF Egerton: *EELS in the Electron Microscope*, Plenum (1996, New York) p. 304
[8] EDX LINK ISIS 300 software Rev. 3.2, manufactured by Oxford Instruments plc

A (S)TEM and atom probe tomography study of InGaN

Thorsten Mehrtens[1,3], Stephanie Bley[1], Marco Schowalter[1], Kathrin Sebald[1], Moritz Seyfried[1], Jürgen Gutowski[1], Stephan S A Gerstl[2], Pyuck-Pa Choi[2], Dierk Raabe[2] and Andreas Rosenauer[1]

[1] Institute of Solid State Physics, University of Bremen, 28359 Bremen, Germany
[2] Max-Planck-Institut für Eisenforschung GmbH, 40237 Düsseldorf, Germany

E-mail: mehrtens@ifp.uni-bremen.de

Abstract. In this work we show how the indium concentration in high indium content $In_xGa_{1-x}N$ quantum wells, as they are commonly used in blue and green light emitting diodes, can be deduced from high-angle annular dark-field scanning transmission electron microscopy (HAADF-STEM) images. This method bases on introducing normalized intensities which can be compared with multislice simulations to determine the specimen thickness or the indium concentration. The evaluated concentrations are compared with atom probe tomography measurements. It is also demonstrated how the quality of focused ion beam prepared TEM-lamellas can be improved by an additional etching with low energy ions.

1. Introduction

$In_xGa_{1-x}N$ is a well suited material for opto-electronic devices such as light emitting diodes and laser diodes, due to its variable bandgap which allows for light emission over the whole visible and part of the ultraviolet spectral range. Despite a high dislocation density leading to non-radiative recombinations, $In_xGa_{1-x}N$ based LEDs or laser diodes show a very good lighting efficiency. The reason for this is still under discussion, but small fluctuations of the indium concentration [1] or the layer thickness [2] are assumed to cause bandgap minima which work as traps for carriers and thus prevent them from moving to the dislocations. The precise determination of the indium content in such structures is still challenging. In this contribution we will introduce a method how the indium concentration can be directly derived from HAADF-STEM images.

2. Investigated specimen and optical properties

The investigated specimen was an InGaN/GaN test structure consisting of high indium content $In_xGa_{1-x}N$ quantum wells separated by GaN barriers. It was grown by metalorganic vapour phase epitaxy (MOVPE) on a low defect-density GaN substrate. Optical properties of the specimen were investigated via μ-photoluminescence (μ-PL) at a temperature of 4 K (figure 1). Beside typical contribution of the GaN barriers to the PL spectrum as the D^0X (3,469eV) and the LO-phonon replica the PL maximum of the spectrum is centred at 525 nm (2.36 eV). This bandgap energy corresponds to an In concentration of 29% for bulk material InGaN.

---
[3] Corresponding author.

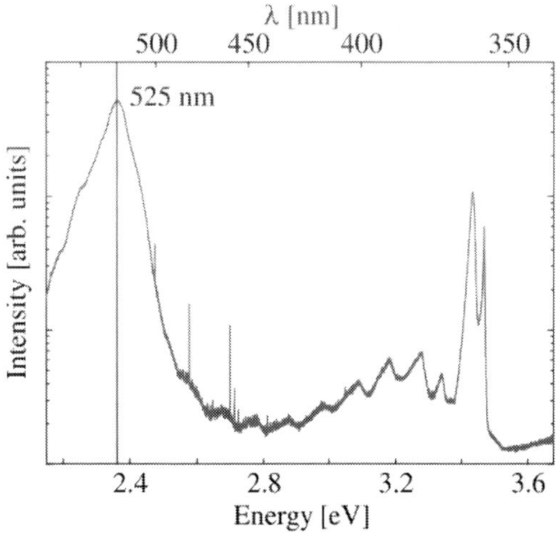

Figure 1. μ-PL spectrum with maximum at 525nm measured at 4K.

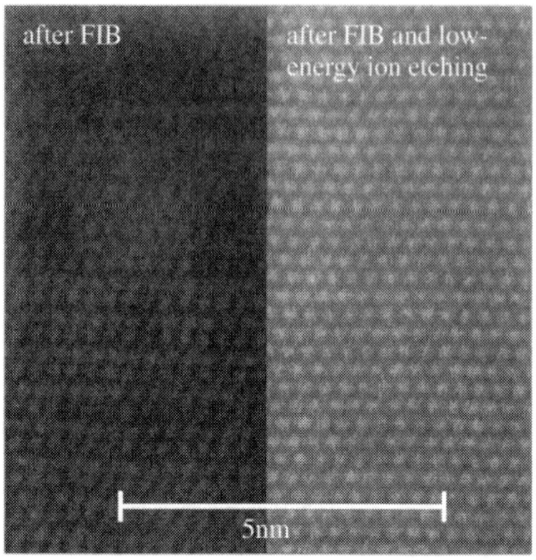

Figure 2. HAADF-STEM images before and after low-energy ion etching.

3. HAADF-STEM

3.1 Optimisation of preparation process

In recent years the FIB-liftout technique [3,4] has become one of the most common preparation methods for TEM-samples. Unfortunately, this procedure induces an amorphous surface layer, which leads to a stained image contrast and reduces resolution (figure 2 left side). Especially if specimens are investigated for concentration fluctuations or interface sharpness this is not desired.

We have minimised the amorphous surface layer by applying another etching step after FIB-preparation with low-energy ions (E < 1keV) using a Technoorg Linda GM IV5 system. The right side of figure 2 shows the best result that was found for an ion energy of 400eV leading to a reduction of the amorphous surface layer and thereby to a higher resolution. Furthermore, it is easier to identify concentration fluctuations.

3.2 Specimen thickness and concentration determination

To determine the indium concentration in the quantum wells HAADF-STEM images were taken with an FEI Titan 80-300 operated at 300kV. The integration angle of the detector was set to 33-200mrad. Before all measurements the detector was scanned in STEM imaging mode without any specimen inserted into the beam. This is done to adjust the detector in such a way that it shows a linear amplification characteristic and to measure the intensity of the scanning electron probe.

The probe intensity is used to normalize the measured image intensities following LeBeau and Stemmer [5], which allows quantitative analysis. These normalized intensities are afterwards compared with reference values simulated with multislice calculations in the frozen lattice approach using the STEMsim software [6]. Static atomic displacements were taken into account and the simulations were carried out in dependence on specimen thickness and indium concentration (figure 3). The specimen thickness is deduced by comparing the normalized intensities in regions of known concentration (GaN regions in this case) with the reference values for the according concentration (0% indium in this case) [7]. The specimen thickness is then interpolated over the regions of unknown concentration (figure 4).

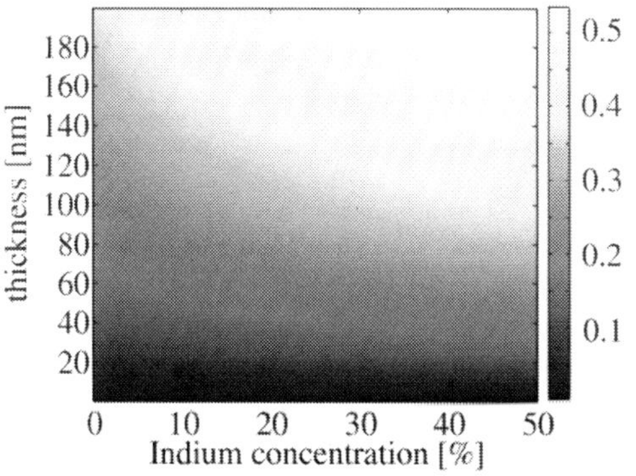

Figure 3. Reference values obtained via multislice calculations.

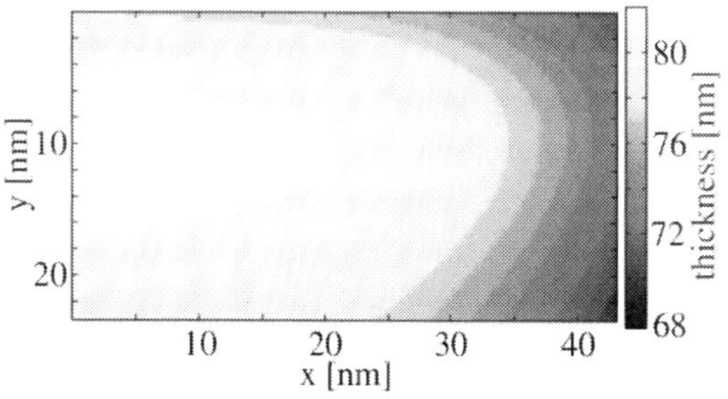

Figure 4. Specimen thickness derived from GaN regions and interpolated over the image.

To evaluate the indium concentration, the normalized intensities in quantum wells are compared with the reference values, this time taking the interpolated thickness of figure 4 into account. This leads to the calculated In concentration map shown in figure 5. The indium content of the two shown layers increases up to values around 23%. Averaged along the monolayers in growth direction ((0002)-planes) an average concentration of (17.7±1.7)% for the first (QW1) and (18.9±1.8)% for the second quantum well (QW2) can be found for this image, but these values fluctuate over the specimen. We observe regions where the quantum wells are interrupted and regions where they possess a lower concentration.

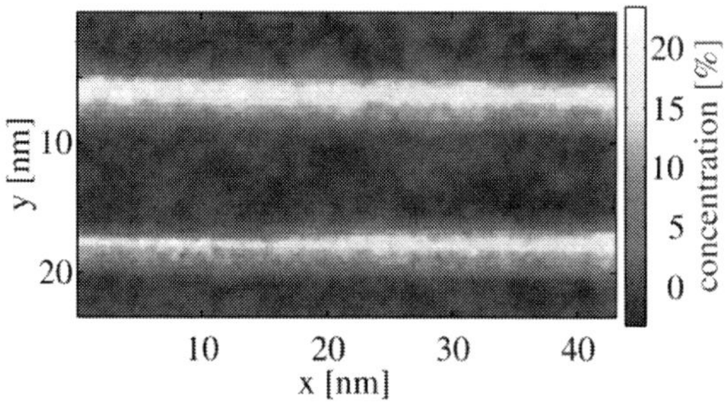

Figure 5. Indium concentration map (HAADF-STEM).

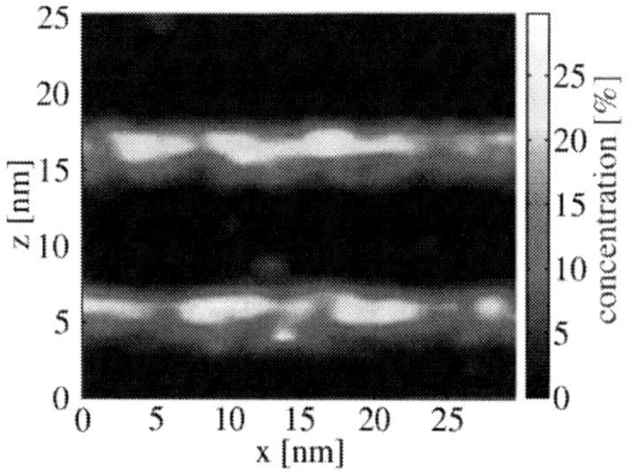

Figure 6. Indium concentration map (Atom probe tomography).

4. Atom probe tomography (APT)

APT measurements were carried out on a laser assisted Imago LEAP 3000 system and the specimen was reconstructed with the IVAS software. The reconstruction was analysed for the average indium concentration and for small-range indium fluctuations. For the average indium concentration all atoms within bins of 0.1nm lying in planes perpendicular to the growth direction were taken into account. A maximum indium concentration of around 25% was measured, which is significantly higher than the values obtained from HAADF-STEM. Most probable reasons for this deviation are long range fluctuations of the In concentration that lead to a decrease of the average composition in projection, and the broadening of the electron beam within the specimen.

Figure 6 shows a concentration map of one reconstruction averaged over a thickness of 1nm. There seem to be indium clusters with a diameter of around 5nm present, especially in the bottom layer. In these clusters the indium concentration increases to nearly 30%. This is in good agreement with the concentration calculated from the PL spectrum. However, it is not quite sure if these concentration fluctuations are not only due to statistical fluctuations in the ternary compound.

5. Conclusion

In this work an InGaN based multi quantum well test structure was investigated via HAADF-STEM and APT. It was shown how the indium concentration and specimen thickness can be directly deduced from HAADF-STEM images by introducing normalized intensities. The concentrations measured via HAADF-STEM are significantly lower than the results obtained via APT and calculated from the PL spectrum. This may be due to long-ranging fluctuations in the indium concentrations of the quantum wells and layer interruptions which were observed but not shown here. Especially if these fluctuations are present in electron beam direction, they could be responsible for the discrepancy.

The atom probe reconstructions were also investigated for fluctuations on smaller length scales. Indium clusters with a diameter of around 5nm seem to be present, but it is still an open question if they are only of statistical origin.

Acknowledgements

This work was supported by the Deutsche Forschungsgemeinschaft under contract RO 2057/8-1.

References

[1] Chichibu S, Sota T, Wada K and Nakamura S 1998 *J. Vac. Sci. Technol.* B **16** 2204-14

[2] Graham D M, Soltani-Vala A, Dawson P, Godfrey M J, Smeeton T M, Barnard J S, Kappers M J, Humphreys C J and Thrush E J 2005 *J. Appl. Phys.* **97** 103508

[3] Overwijk F M H, van den Heuvel F C and Bulle-Lieuwma C W T 1993 *J. Vac. Sci. Technol.* B **11** 2021-24

[4] Giannuzzi L A and Stevie F A 1999 *Micron* **30** 197-204

[5] LeBeau J M and Stemmer S 2008 *Ultramicroscopy* **108** 1653-58

[6] Rosenauer A and Schowalter M 2007 STEMsim - a New Software Tool for Simulation of STEM HAADF Z-Contrast Imaging *Microscopy of Semiconducting Materials 2007* ed A G Cullis and P A Midgley (Springer Netherlands) pp 170-172

[7] Rosenauer A, Gries K, Müller K, Pretorius A, Schowalter M, Avramescu A, Engl K and Lutgen S 2009 *Ultramicroscopy* **109** 1171-82

Toward atom probe tomography of microelectronic devices

D J Larson[1,*], D Lawrence[1], W Lefebvre[2], D Olson[1], T J Prosa[1], D A Reinhard[1], R M Ulfig[1], P H Clifton[1], J H Bunton[1], D Lenz[1], J D Olson[1], L Renaud[3], I Martin[3] and T F Kelly[1]

[1] Cameca Instruments Inc., 5500 Nobel Drive, Madison, WI 53711 USA
[2] Université de Rouen, Saint Etienne du Rouvray 76801 FRANCE
[3] Cameca SAS, 29 Quai des Grésillons, Gennevilliers 92622 FRANCE

Summary: Atom probe tomography and scanning transmission electron microscopy has been used to analyze a commercial microelectronics device prepared by depackaging and focused ion beam milling. Chemical and morphological data are presented from the source, drain and channel regions, and part of the gate oxide region of an Intel ® i5-650 p-FET device demonstrating feasibility in using these techniques to investigate commercial chips.

1. Introduction

Shrinking feature sizes continue to provide one of the main challenges for physical metrology methods. From the 2009 ITRS [1]: "To achieve desired device scaling, metrology tools must be capable of measurement of properties on atomic distances." Atom probe tomography nominally has this capability and has been used in materials characterization and materials science for more than 40 years [2]. Although using a laser to assist field evaporation of materials in the atom probe was developed ~30 years ago [3], only recently has APT begun to see widespread usage in the areas of semiconductors [4-7] and ceramics [8-12]. Although APT has also recently been applied to transistor and FINFET-type structures [13-15], these structures usually have been stopped at some point in the semiconductor build process in order to accommodate the APT analysis. As it is a complex, multi-step task to analyze a fully processed microelectronic device using APT, we will take a step back and evaluate the three steps that are necessary to result in a successful APT analysis of *any* structure.

Number one is specimen preparation. For APT, the use of focused ion beam (FIB) instruments has revolutionized specimen preparation (for a recent review see [16]). For the case of microelectronic device analysis, FIB preparation to isolate <50nm heterogeneous devices in arbitrary XYZ orientations is critical. This occurs only after suitable deprocessing (may consist of depackaging, dry and wet electrochemistry for selective material etching, etc.) has been successfully achieved. Since a protective layer is needed if near-surface features are to be analyzed, proper "cap matching" becomes important. The important requirements are relatively low temperature deposition (to avoid inducing undesired diffusion), matching of evaporation field with target material (to minimize reconstruction artifacts) and good adhesion.

Number two is specimen yield during data collection. Even though a specimen may be prepared which contains the region of interest, specimen yield must be adequate (for the individual user's purpose or specific laboratory environment). Clearly the details of specimen preparation are intricately related to yield, but data collection parameters may also have a significant effect on yield, and there is often a trade-off between yield and data quality (primarily mass spectral and background noise quality).

* David.larson@ametek.com

Number three is APT data reconstruction, which is not discussed significantly in this document. Data may be collected with an adequate yield, but if the resulting data reconstruction is either not accurate or not precise enough, then the analysis may fail to meet requirements. The main problem with APT data reconstruction, from the perspective of devices, is that field evaporation of heterogeneous structures often leads to evaporated surfaces that are far from hemispherical [17] and a hemispherical end form is a fundamental assumption employed in the majority of current reconstruction algorithms.

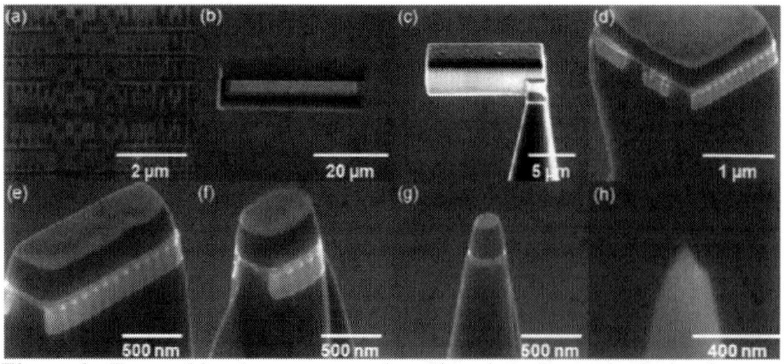

Figure 1. Focused ion beam preparation of a microelectronic device for APT analysis: a) initial surface, b) FIB-milled trenches, c) the extracted coupon, d) pre-tip cut from the coupon, and e-h) the stages of focused-ion-beam annular sharpening.

This work presents the status of efforts to prepare, analyze, and reconstruct data from a commercial microelectronics device (32 nm node Intel® i5-650 nFET). This is a work in progress, and we acknowledge that satisfactory yield has not yet been obtained. Also, issues relating to APT data reconstruction of transistors are non-trivial and will not be discussed further herein.

2. Experimental

Specimen preparation was carried out in an FEI Novalab dual-beam focused ion beam instrument with an Omniprobe AP200 *in-situ* micromanipulator. STEM observations were performed along the [110] orientation of the Si substrate with a JEM ARM 200F operating at 200kV equipped with a Schottky field emission gun. APT data collection was performed on a LEAP ® 4000XHR from Cameca Instruments Inc. The atom probe was operated in a 200 kHz pulsed laser mode with an energy of 100 pJ into an estimated spot size (four sigma) at the specimen of ~3 µm. The specimen temperature was 50K and the ion detection rate was 0.20% (1 ion detected in every 500 laser pulses).

The specimen examined in this work is a device from a commercial 32-nm technology chip (Intel ® i5-650) which was purchased at a retail outlet. Following depackaging to (approximately) metal-1 [18], focused-ion-beam preparation [19] was used to create specimens. Figure 1 shows a sequence of images throughout the process from the initial surface of the wafer after depackaging (figure 1a) through to the final state of focused-ion-beam annular sharpening [20] of the tip (figure 1h).

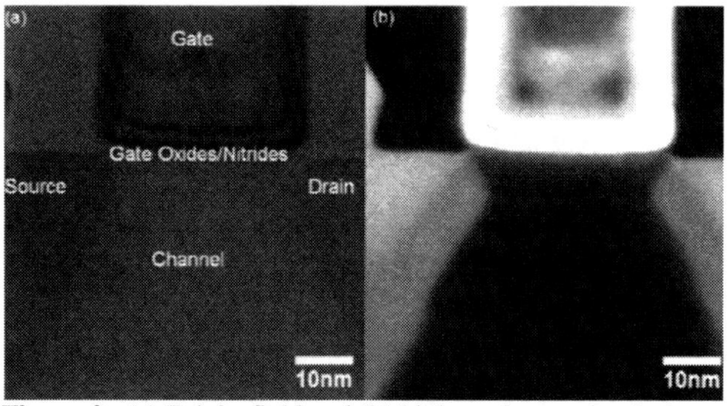

Figure 2. (a) Bright-field and (b) high-angle annular-dark-field images of the Intel® i5-650 device.

3. Results and Discussion

The device structure analyzed with APT is shown in figure 2, which contains a bright-field, high-angle annular-dark-field pair of images. In figure 2a a multilayered gate oxide structure is clearly visible while in figure 2b the undercut and angled regions between the SiGeB source/drain regions and the channel become much more obvious.

The APT analysis obtained from a device in a region near the one shown in figure 2 is presented

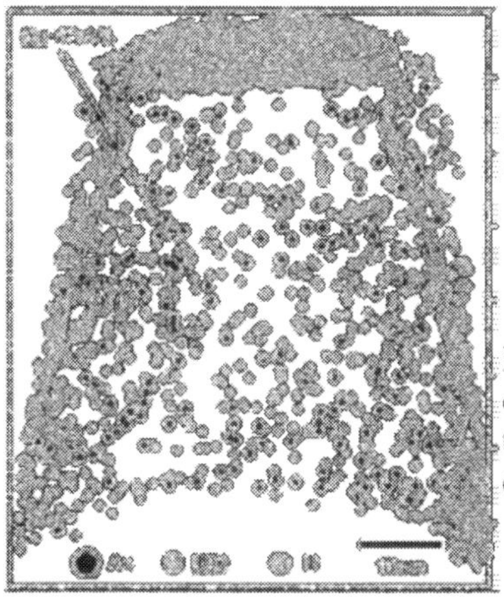

Figure 3. APT atom map containing As (large black spheres), B (small dark grey spheres), and HfO (small light grey spheres).

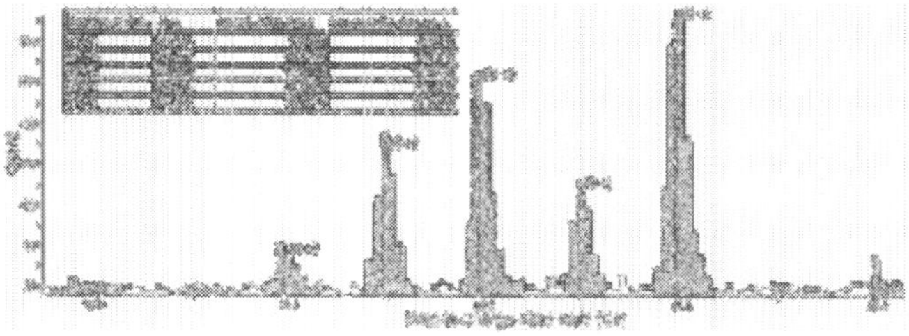

Figure 4. HfO complex molecule peaks detected in the 2+ charge state over the range of 95 to 99 Da. The inset table shows the comparison of expected isotopic abundances (using >0.2at.% only) to the detected abundances.

in figure 3, which is an atom map containing As (large black spheres), B (small dark grey spheres), and HfO (small light grey spheres) together with a 12at.% Ge isoconcentration surface (arrowed). The Hf is detected entirely in HfO complex molecule peaks, shown in figure 4, which are detected in the 2+ charge state over the range of 95 to 99 Da. Measured at the peak at 98 Da, the mass resolving power is >1000 (FWHM) and >500 (FWTM), which demonstrates the capability to achieve good spectral resolution, even on very complicated heterogeneous structures. The Hf atoms are detected at the top of the image in figure 3, in a region estimated to have a composition of approximately 80at.% (Hf+O) and 20at.% Si. (Note that the apparent HfO molecules detected in the channel region are spectral noise due to the large number of mass ranges required for the HfO isotopes.) The maximum level of Ge (positioned along the left and right edges of the image shown in figure 3 is ~25at.%. The Ge isoconcentration surface shown in figure 3 (arrowed) delineates the source/drain regions from the channel region and has a shape which resembles the undercut shape of the channel region near the gate oxide shown in figure 2b.

This shape correlation may be used to create a composite image of the STEM and APT data, figure 5, in which the APT image was scaled to match the STEM image. Note the qualitative evidence of carbon atoms clustering together in the lower center portion of the image. The estimated concentration of carbon in the APT data is ~0.10at.%

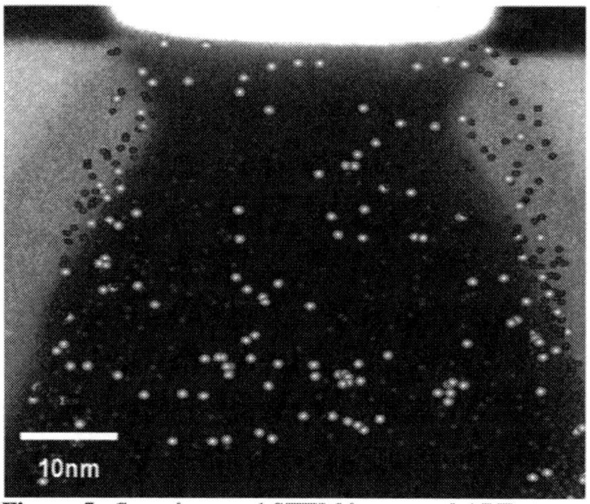

Figure 5. Superimposed STEM image and APT atom map containing B (small dark grey spheres), and carbon (large light grey spheres) atoms only. Note the indications of carbon atoms clustering together. (The ATP data are 20nm in thickness into the page.)

Although these data are not from exactly the same volume of material (something which has been done very few times to date in the literature [21]), the exercise is still useful. It provides us with information on the accuracy of the APT data reconstruction as well as perhaps a glimpse of the future of correlative microscopy. Indeed, TEM and APT are complementary techniques that will likely develop a closer relationship in the future where the spatial fidelity of TEM is combined with the analytical sensitivity and three dimensionality of APT. Instruments that combine these techniques into a single instrument are currently being pursued by adding a STEM to a LEAP [22] and a LEAP to a STEM [23].

This work demonstrates that a wealth of high quality information may be obtained from site-specific atom probe analysis of post-production microelectronic devices. Adequate yields (>50%) need to be realized and APT reconstruction methods improved going forward, but certainly at this time, feasibility has been shown.

Acknowledgements

The authors would like to thank our colleagues at Cameca Instruments Inc., who assisted in assembling the materials presented in this manuscript, including B. Geiser, J. Olson, J. Shepard, T. Payne, E. Strennen, E. Oltman, T. Gribb, D. Rauls, J. Watson and S. Gerstl (currently at Eidgenössische Technische Hochschule Zürich). We would especially like to thank P. Ronsheim (IBM) for helpful discussions.

References

[1] International Technology Roadmap for Semiconductors (http://www.itrs.net/)
[2] Müller E W, Panitz J A and McLane S B 1968 *Review of Scientific Instruments* **39** 83-6
[3] Kellogg G L and Tsong T T 1980 *Journal of Applied Physics* **51** 1184-94
[4] Kelly T F, Larson D J, Thompson K, Alvis R L, Bunton J H, Olson J D and Gorman B P 2007 *Annual Review of Materials Research* **37** 681-727
[5] Lauhon L J, Adusumilli P, Ronsheim P, Flaitz P L and Lawrence D 2009 *MRS Bulletin* **34** 738-43
[6] Larson D J, Prosa T J, Lawrence D, Geiser B P, Jones C M and Kelly T F 2011 *Handbook of Instrumentation and Techniques for Semiconductor Nanostructure Characterization*, ed R Haight, *et al.* (London: World Scientific Publishing/Imperial College Press)
[7] Mutas S, Klein C and Gerstl S S A 2011 *Ultramicroscopy* **in press**
[8] Larson D J, Alvis R A, Lawrence D F, Prosa T J, Ulfig R M, Reinhard D A, Clifton P H, Gerstl S S A, Bunton J H, Lenz D R, Kelly T F and Stiller K 2008 *Microscopy and Microanalysis* **14** 1254-5
[9] Chen Y M, Ohkubo T, Kodzuka M, Morita K and Hono K 2009 *Scripta Materialia* **61** 693–6
[10] Marquis E A, Yahya N A, Larson D J, Miller M K and Todd R I 2010 *Materials Today* **13(10)** 42-4
[11] Li F, Ohkubo T, Chen Y M, Kodzuka M, Ye F, Ou D R, Mori T and Hono K 2010 *Scripta Materialia* **63** 332-5
[12] Payne D J and Marquis E A 2011 *Chemistry of Materials* **23** 1085-7
[13] Moore J S, Jones K S, Kennel H and Corcoran S 2008 *Ultramicroscopy* **108** 536–9
[14] Inoue K, Yano F, Nishida A, Takamizawa H, Tsunomura T, Nagai Y and Hasegawa M 2009 *Ultramicroscopy* **109** 1479-84
[15] Kambham A K, Mody J, Gilbert M, Koelling S and Vandervorst W 2010 *Ultramicroscopy* **in press**
[16] Miller M K, Russell K F, Thompson K, Alvis R and Larson D J 2007 *Microscopy and Microanalysis* **13** 428-36
[17] Marquis E A, Geiser B P, Prosa T J and Larson D J 2011 *Journal of Microscopy* **241** 255
[18] SVTC Technologies (http://www.svtc.com/)
[19] Thompson K, Lawrence D J, Larson D J, Olson J D, Kelly T F and Gorman B 2007 *Ultramicroscopy* **107** 131-9
[20] Larson D J, Foord D T, Petford-Long A K, Liew H, Blamire M G, Cerezo A and Smith G D W 1999 *Ultramicroscopy* **79** 287-93
[21] Arslan I, Marquis E A, Homer M, Hekmaty M A and Bartelt N C 2008 *Ultramicroscopy* **108** 1579-1585
[22] Gorman, B P, et al., *Microscopy and Microanalysis* (2011) Supplement 2, **in press.**
[23] Kelly, T F, et al., *Microscopy and Microanalysis* (2011) Supplement 2, **in press.**

Accuracy of pulsed laser atom probe tomography for compound semiconductor analysis

M Müller[1], B Gault[1,2], G D W Smith[1] and C R M Grovenor[1]

[1]Department of Materials, University of Oxford, Parks Road, Oxford OX1 3PH, UK

chris.grovenor@materials.ox.ac.uk

Abstract. Atom probe tomography has recently experienced a renaissance, strongly promoted by the revival of pulsed laser atom probe. The technique is now widely employed to study semiconductor materials at the nanometre level. This paper summarises some aspects of the accuracy of pulsed laser atom probe relevant to semiconductor applications. It is shown that laser pulsing can reduce the lateral resolution due to thermally stimulated surface migration. Moreover, the commonly observed cluster ions can undergo field dissociation which results in an increased probability of ion loss due to pile-up effects at the detector. Field dissociation can also induce a new type of local magnification that increases spatial inaccuracy in the data reconstruction. These effects can be reduced by an appropriate choice of experimental parameters. Despite these difficulties, the atom probe technique can provide unparalleled insight into the nanoscale structure and chemistry of a wide range of semiconductor devices.

1. Introduction

Although first conceived and realized in the 1980's [1], pulsed laser atom probe (PLAP) resurfaced just a few years ago, driven by the implementation of reliable, high repetition rate, ultra-short pulse width laser systems. It was established that the pulsing mechanism in PLAP is thermal, where a locally and temporally confined heating of the specimen apex triggers the field evaporation of surface atoms. Under normal operating conditions, the temperature rise should only moderately heat the apex from the cryogenic range to near room temperature [2,3] with decay times of only few tens of nanoseconds. The use of laser pulsing has unlocked atom probe tomography (APT) investigations of low electrical conductivity materials, and PLAP has since been applied successfully to a variety of materials that had previously been inaccessible. Many of those new applications are in the field of semiconductor analysis ranging from silicon-based materials and device structures, to semiconductor nanowires and compound semiconductor nanostructures [4,5]. The great success of PLAP applications in this field has somewhat outpaced fundamental studies addressing how the choice of experimental parameters affects the accuracy of the data obtained and how the performance of contemporary PLAP compares to conventional voltage-pulsed APT of metallic materials. This paper aims at providing a brief review of studies addressing those issues and assesses the relevance of PLAP for semiconductor analysis.

[2] Present address: Australian Centre for Microscopy & Microanalysis, Madsen Building F09, The University of Sydney, NSW 2006 Australia.

2. Spatial accuracy

As for all microscopy techniques, the spatial resolution is probably the most important benchmark, ultimately defining the instrument performance. However, it is not always trivial to quantify this measure. It is thus not surprising that there had been no procedure to define and measure the spatial resolution in APT data until very recently. Gault *et al.* [6] deployed a complex multi-step procedure to derive a measure for the spatial resolution and showed that contemporary voltage-pulsed APT can in fact yield data with sub-nanometer resolution in all three dimensions for pure metals like W and Al at low temperatures. In a subsequent study, the evaluation of the spatial resolution was extended to PLAP data [7]. The results show that in particular the lateral resolution is sensitive to the application of laser pulsing and monotonically deteriorates with increasing pulse energy. On the other hand, the in-depth resolution seems more tolerant towards laser pulsing. This observation can be explained by an increased likelihood of surface migration prior to field evaporation promoted by sample heating. The migration of surface atoms impedes the accurate reconstruction of the original crystal lattice. It is also important to mention that the evolution of the spatial resolution is material-dependent. In studies carried out on specimens held at equivalent fractions of their melting temperature, it degrades faster in FCC Al than in BCC W, likely due to more rapid surface atom migration on FCC surfaces [7].

All these studies were dealing with pure metals only. Similar studies for semiconductors are difficult, since the assessment procedure employed by Gault *et al.* [6] seems not easily applicable. However, Cadel *et al.* [8] and Geiser *et al.* [9], amongst others, were able to reveal lattice planes in APT datasets of silicon. With the help of further data analysis procedures, the *depth* resolution was estimated to be better than 200 pm under laser pulsing [8]. Very recently, Shimizu *et al.* [10] employed silicon isotope superlattices [11] to further evaluate the spatial resolution of PLAP. This method also allows for the assessment of the lateral resolution, if the growth direction of the stack is oriented perpendicular to the atom probe specimen axis. The authors estimated the depth resolution for a fixed laser pulse energy to be $\approx$ 400 pm and the lateral resolution to be $\approx$ 1000 pm [10]. More work still needs to be done, in particular to investigate how the spatial resolution depends on the pulse energy. Furthermore, no investigations seem yet to exist on the spatial resolution of PLAP for compound semiconductors, which have become an important area of application.

The studies published so far suggest that the *lateral* resolution for semiconductors may no longer be on an atomic level, which would explain why no real space crystallographic information can be generated to determine the lateral resolution with the procedure described in [6]. One possible explanation has already been presented, namely the application of laser pulses, which is expected to reduce the spatial resolution a priori. It is thus also conceivable that the spatial resolution in semiconductors deteriorates with increasing laser pulse energy as observed for pure metals [7]. Another aspect might be the different nature of the bonds in semiconductors. For pure metals with their delocalized bonds and electrons, detailed models for the field evaporation process exist [2]. However, less is known about the field evaporation of semiconductors with their highly directional, strong covalent bonds. This difference in the bonding character might impede single atom-by-atom field evaporation and thus reduce the spatial resolution.

Another interesting phenomenon potentially affecting the spatial accuracy of the data reconstruction is the change in the specimen apex shape during PLAP experiments [12,13]. Koelling *et al.* reported a unidirectional absorption regime for PLAP on silicon for laser wavelengths at which the optical skin depth is smaller than the specimen thickness [13]. Together with equally small thermalisation depths [14], this can result in inhomogeneous laser heating and the formation of pronounced apex asymmetries. Such asymmetries in the local radius of curvature cannot be easily accounted for in standard data reconstruction procedures [15] and are likely to cause artifacts and inaccuracies in the 3-dimensional datasets.

3. Cluster ion detection

A seemingly inevitable feature of PLAP studies of compound semiconductors is the detection of cluster ions X_n, for compound semiconductors most abundantly detected as, for example, homonuclear group-V cluster ions. Recently observed cluster sizes n are up to 6 or beyond for arsenic [16], up to 4-5 for antimony [17,18], up to 5 for phosphorous [16,19], and 3 for nitrogen [20]. Furthermore, these clusters can exhibit different charge states commonly ranging from 1+ to 3+. Besides homonuclear group-V cluster ions, heteronuclear clusters of III-V [17] or V-V elements are also possible [21].

The formation mechanism of cluster ions during the field evaporation process of (compound) semiconductors is not very well understood. Cerezo *et al.* demonstrated that the concentration of larger arsenic clusters increased with increasing base temperature in voltage-pulsed atom probe of GaAs [3]. This indicates a temperature mediated process. Drachsel *et al.* also assumed high, temperature-mediated surface mobilities for the materials that showed cluster ions in their laser-assisted field evaporation experiments [22]. It is conceivable that enhanced surface migration at higher temperatures [7] promotes the agglomeration of, e.g., single, mobile group-V atoms and their field evaporation in form of a cluster ion. For PLAP, this would mean that with increasing effective pulse fraction, i.e., increased laser heating more and larger cluster ions should be detected. This has been observed for antimony clusters in PLAP studies of GaSb [17]. Tsong [23] argued that electric fields could penetrate much deeper into semiconductor materials than is the case for metals. Consequently, not only the outermost atoms, but also deeper atom layers might experience sufficient field strengths for field evaporation, potentially detaching larger agglomerates from the surface. The detachment of those larger building blocks might also be the origin for cluster ions in the mass spectrum. Finally, mass spectra from liquid metal ion sources also show an abundance of cluster ions [24,25] and their formation processes may also be relevant to the cluster ion formation under atom probe conditions. The presence of cluster ions in PLAP data has several consequences affecting the data accuracy.

3.1. Mass peak overlaps

A very obvious issue is the overlap of peaks in the mass spectrum. The position of each peak is given by the isotope mass m divided by the ion charge z. Consequently, there can be more than one permutation of m and z giving the same mass-to-charge state-ratio m / z. For example, the mass peaks for As_2^{1+} and As_4^{2+} have an identical m/n ratio of 150 Da and thus fully overlap. This issue is most pronounced for mono-isotopic group-V elements for which the overlap cannot be deconvoluted easily and compositional inaccuracies appear inevitable. However, for multi-isotopic elements like Sb the formation of cluster ions results in characteristic, additional mass peaks, for example $^{122}Sb_2^{2+}$, which can be employed to deconvolute mass peak overlaps.

3.2. Field Dissociation

We recently investigated the cluster detection in PLAP studies of GaSb [17]. One of the main observations was that the size and concentration of cluster ions detected increased with increasing effective pulse fraction (PF_{eff}) and with decreasing field strength as shown in Figure 1 (a). At low PF_{eff} and thus high field strengths the majority of Sb ions detected were atomic Sb. In contrast, for higher PF_{eff} the Sb_3 cluster ions are predominant and overall more Sb_n clusters were detected. This pronounced change could be explained in two ways, firstly by thermally promoted cluster formation as described earlier or secondly, by reduced field dissociation. Careful investigations of the data suggest that field dissociation is the predominant process under the experimental conditions considered. An increase in laser pulse energy heats the sample more, but equally reduces the field strengths required for field evaporation. This drop in field strength is assumed to diminish field dissociation of antimony clusters, so they appear in higher concentrations and larger sizes n in the atom probe data. However, the assessment of cluster ion detection in atom probe is determined by two distinct processes, firstly the formation of the cluster ions, and secondly their field dissociation. Consequently, the high concentrations and larger sizes of cluster ions at high effective pulse fractions [17] would be a combined result of enhanced cluster ion formation and suppressed cluster ion dissociation. Unfortunately those two processes cannot easily be distinguished and studied separately.

3.2.1. *Effects on chemical accuracy.* Field dissociation of cluster ions may result in a multiplication of detector events [17]. Assuming, for example, a large Sb_5^{1+} cluster undergoing complete field dissociation into five Sb^{1+} fragments, this would mean that five ions instead of only one are moving towards the atom probe detector. Figure 1 (b) demonstrates that PLAP studies of GaSb can exhibit a substantial fraction of multiple detector events. When operating at low effective pulse fraction, up to 45 % of all ions detected arrive within multiple events. The size of those events can be as big as 14 ions per pulse [17]. Given a detection efficiency of the instrument of 40 %, this implies that more than 30 ions were field evaporated during a single pulse. The proportion of multiple events reduces drastically with increasing pulse fractions indicating suppressed field dissociation [17]. Materials not normally forming cluster ions and thus not experiencing field dissociation like, for example, aluminium and tungsten have multiple fractions well below 5 %. If too many fragmentation products strike the detector in very close succession or proximity to each other then the detector signals they produce might no longer be resolved and some of those ions are simply lost. This is referred to as detector ion pile-up [2]. Since mainly group-V elements form cluster ions, their compositional accuracy will thus be affected more strongly i.e. their concentrations are likely to be underestimated. In addition, field dissociation may produce neutral fragmentation products. This has been indirectly proved for PLAP of GaN, where the field dissociation of GaN_3 led to Ga^{1+}, N^{1+} and neutral N_2 [20]. Any neutrals which are formed normally do not possess sufficient energy to trigger a detector signal or are lost in the instrument, thus they will also be lost from the chemical analysis.

(a) (b)

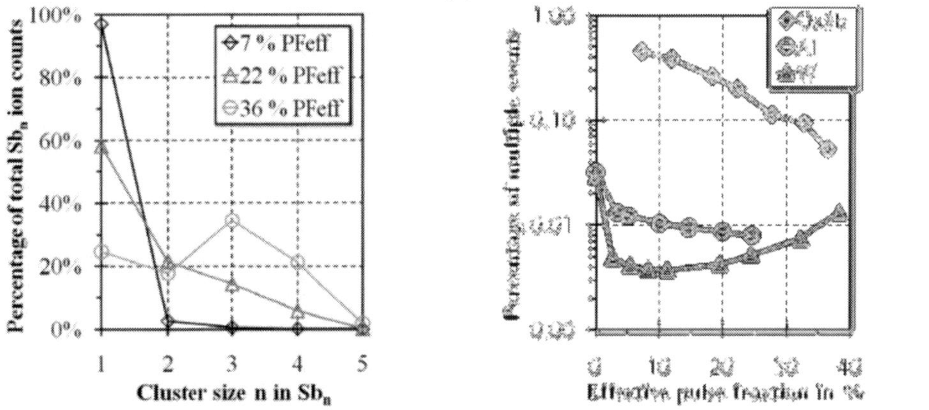

Figure 1. (a) Abundances of the various Sb cluster ions of size *n*, expressed as percentages of the overall number of antimony ion counts, for three selected effective pulse fractions (PF_{eff}). (b) Percentage of multiple events within the overall number of detector events for three different materials studied at various effective pulse fractions. Higher effective pulse fractions correspond to higher laser pulse energies and lower evaporation field strengths. Lines as guide for the eye.

3.2.2. *Effects on spatial accuracy.* Although hardly any knowledge exists on the trajectories of fragmentation products, one could assume that field dissociation leads to divergent ion trajectories. In particular so, since most fragments will possess positive charges either due to a split of multiple charges during dissociation [26] or post-ionisation, which result in electrostatic repulsion. An in-depth study of the distance distribution of detector events in GaSb atom probe data [17] revealed that antimony multiple detector hits have a particularly wide distance distribution. This is also shown in Figure 2 in direct comparison with a pure metal. The few multiple hits in aluminium are assumed to originate exclusively from correlated field evaporation [27] and their distance distribution on the detector is indeed less wide compared to that of the Ga or Sb multiple hits [17]. The distance distribution of Sb-Sb multiple hits shows a shoulder at ≈ 0.5 mm, the peak position of the distance distribution of aluminium. This can be attributed to co-evaporation of adjacent Sb ions. However, the distance distribution also possesses a major peak at ≈ 1.1 mm which is believed to describe the more widespread impact of fragmentation products on the detector caused by trajectory aberrations. This means a large antimony cluster starting out as a point source on the sample surface is "blurred" into a finite disc on the detector. Applying basic data reconstruction algorithms to these widespread detector events results in a misallocation of antimony in the reconstruction. The reconstruction procedure translates the distance on the detector into a distance on the specimen surface, whereas in fact the ions originated at the very same point, i.e., where the cluster field evaporated. This results in a decrease in spatial accuracy. Interestingly, the distance distribution of Ga-Ga multiples does not exhibit such a pronounced peak, presumably due to the fact that Ga does not form stable cluster ions [22] which could experience strong field dissociation. Thus, the inaccuracies mentioned are expected to be most pronounced for the spatial allocation of antimony ions during the data reconstruction.

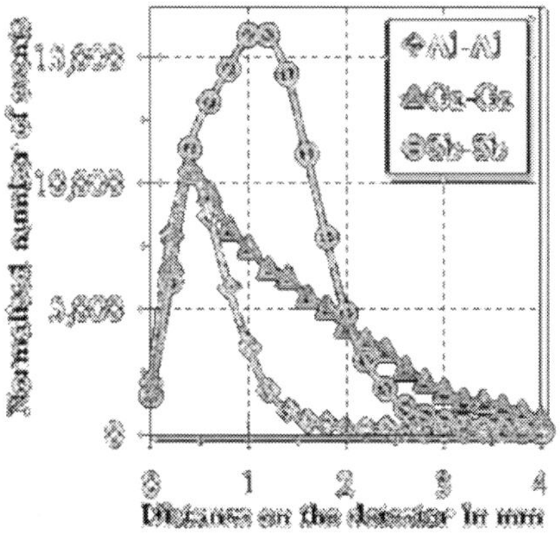

Figure 2. Distance distribution of Al-Al in an Al6XXX reference material (-◇-), Ga-Ga in GaSb (-△-) and Sb-Sb in GaSb (-○-) multiple hit pairs on the position-sensitive detector. The numbers of hits were normalised to the position of the shoulder in the Sb-Sb distribution. Data acquired at low laser pulse energies. Lines as guide for the eye.

4. Optimisation strategies

A suitable choice of experimental settings can reduce the effects described here. As outlined in reference [7], the lower the laser pulse energy and the induced sample heating, the better the lateral resolution. Hence, it appears plausible that operating at lowest pulse energies and sample base temperatures can, to some extent, counteract the deterioration in spatial resolution. However, the colder the sample, the higher the field strength required for field evaporation, which in turn promotes field dissociation [17]. In order to reduce field dissociation and its deleterious effects, it is thus preferable to operate at higher laser pulse energies. Obviously, the two objectives are somewhat in conflict and a compromise needs to be established. This could mean applying intermediate laser pulse energies, reducing field dissociation significantly, but accepting that some degree of surface migration reduces the spatial resolution. We also showed that the compositional accuracy is highest when operating with intermediate laser pulse energies [17].

More work needs to be dedicated to optimising experimental settings, quantifying the spatial resolution and comprehending the formation and detection of cluster ions. However, it needs to be stressed that PLAP is indeed a very promising tool for quantitative tomography of compound semiconductors, with unrivalled chemical sensitivity and spatial resolution.

Acknowledgements

The authors would like to thank A. Cerezo and D. Saxey for fruitful discussions. M.M. is highly indebted to the Engineering and Physical Sciences Research Council (Doctoral Training Account, grant no.: EP/P502667/1 and PhD Plus Award, grant no.: EP/P503876/1), the German Academic Exchange Service (DAAD, grant no.: D/07/48437), the German National Academic Foundation ("Studienstiftung des Deutschen Volkes"), the former Imago Scientific Instruments Ltd. and the Oxford University Vice Chancellor's Fund for the generous scholarships provided.

References

[1] Kellogg G L, Tsong T T. 1980 *J. Appl. Phys.* **51** 1184-1193
[2] Miller M K, Cerezo A, Hetherington M G, Smith G D W. 1996 *Atom Probe Field Ion Microscopy* (Oxford Science Publications)
[3] Cerezo A, Grovenor C R M, Smith G D W. 1986 *J. Microsc.* **141** 155-170
[4] Müller M, Cerezo A, Smith G D W, Chang L, Gerstl S S A. 2008 *Appl. Phys. Lett.* **92** 233115-1
[5] Galtrey M J, Oliver R A, Kappers M J, Humphreys C J, Stokes D J, Clifton P H, Cerezo A. 2007 *Appl. Phys. Lett.* **90** 061903-1
[6] Gault B, Moody M P, De Geuser F, Haley D, Stephenson L T, Ringer S P. 2009 *Appl. Phys. Lett.* **95** 034103-1
[7] Gault B, Müller M, La Fontaine A, Moody M P, Shariq A, Cerezo A, Ringer S P, Smith G D W. 2010 *J. Appl. Phys.* **108** 044904-1
[8] Cadel E, Vurpillot F, Lardé R, Duguay S, Deconihout B. 2009 *J. Appl. Phys.* **106** 044908-1
[9] Geiser B P, Kelly T F, Larson D J, Schneir J, Roberts J P. 2007 *Microsc. Microanal.* **13** 437-447
[10] Shimizu Y, et al. 2011 *J. Appl. Phys.* **109**
[11] Shimizu Y, Kawamura Y, Uematsu M, Itoh K M, Tomita M, Sasaki M, Uchida H, Takahashi M. 2009 *J. Appl. Phys.* **106** 076102-1
[12] Sha G, Cerezo A, Smith G D W. 2008 *Appl. Phys. Lett.* **92** 043503-1
[13] Koelling S, Innocenti N, Schulze A, Gilbert M, Kambham A K, Vandervorst W. 2011 *J. Appl. Phys.* **109** 104909
[14] Meyer J R, Kruer M R, Bartoli F J. 1980 *J. Appl. Phys.* **51** 5513-5522
[15] Gault B, Haley D, de Geuser F, Moody M P, Marquis E A, Larson D J, Geiser B P. 2011 *Ultramicroscopy* **111** 448-457
[16] Gorman B P, Norman A G, Yan T. 2007 *Microsc. Microanal.* **13** 493-502
[17] Müller M, Saxey D W, Smith G D W, Gault B. 2011 *Ultramicroscopy* **111** 487-492
[18] Gault B, Marquis E A, Saxey D W, Hughes G M, Mangelinck D, Toberer E S, Snyder G J. 2010 *Scripta Mater.* **63** 784-787
[19] Liddle J A, Cerezo A, Grovenor C R M. 1989 *J. Phys. Colloques* **50** C8-437
[20] Bennett S E, Saxey D W, Oliver R A, Cerezo A, Humphreys C J. 2010 *52nd IFES, Sydney, Australia* 05/07-08/07/2010
[21] Grovenor C R M, Cerezo A, Liddle J A, Smith G D W. 1987 *Inst. Phys. Conf. Ser.* **87** 665
[22] Drachsel W, Jentsch T, Gingerich K A, Block J H. 1985 *Surf. Sci.* **156** 173-182
[23] Tsong T T. 1979 *Surf.Sci.* **81** 28-42
[24] Gamo K, Ukegawa T, Inomoto Y, Ochiai Y, Namba S. 1981 *J. Vac. Sci. Technol.* **19** 1182-1185
[25] Wilkens B, Venkatesan T. 1983 *J. Vac. Sci. Technol. B* **1** 1132-1136
[26] Bréchignac C, Cahuzac P, Carlier F, De Frutos M, Leygnier J, Roux J P. 1995 *J.Chem.Phys.* **102** 763-769
[27] De Geuser F, Gault B, Bostel A, Vurpillot F. 2007 *Surf. Sci.* **601** 536-543

Counting Tm dopant atoms around GaN dots using high-angle annular dark field images

J-L Rouvière, H Okuno, P H Jouneau, P Bayle-Guillemaud and B Daudin

CEA-INAC/UJF-Grenoble UMR-E, SP2M, Minatec, 17 rue des Martyrs 38054 Grenoble France

E-mail: jean-luc.rouviere@cea.fr

Abstract. High resolution Z-contrast STEM imaging is used to study the Tm doping of GaN quantum dots grown in AlN by molecular beam epitaxy (MBE). High-angle annular dark field (HAADF) imaging allows us to visualize directly individual Tm atoms in the AlN matrix and even to count the number of Tm atoms in a given AlN atomic column. A new visibility coefficient to determine quantitatively the number of Tm atoms in a given atomic column is introduced. It is based on locally integrated intensities rather than on peak intensities of HAADF images. STEM image simulations shows that this new visibility is less sensitive to the defocus-induced blurring or to the position of the Tm atom within the thin lamella. Most of the Tm atoms diffuse out of GaN dots. Tm atoms are found at different positions in the AlN matrix. (i) Above the wetting layer, Tm atoms are spread within a thickness of 14 AlN monolayers (MLs). (ii) Above the quantum dots all the Tm are located in the same plane situated at 2-3 MLs above the apex of the GaN dot, i.e. at a distance of 14 MLs from the wetting layer. (iii) In addition, Tm can diffuse very far from the GaN dot by following threading dislocations lines.

1. Introduction

In order to improve the efficiency of GaN based devices, it has been proposed to use rare-earth doped GaN quantum dots (QDs) in order to confine carriers in the QDs, far from defects [1]. In this case, the accurate incorporation and localization of dopants in the dots is crucial. In this work, high resolution Z-contrast STEM imaging is used to study the Tm doping of GaN quantum dots grown in AlN by molecular beam epitaxy (MBE). HAADF STEM allows us to visualize directly individual Tm atoms in the AlN matrix.

2. Experimental details

The studied sample is composed of several layers of GaN dots embedded in AlN. The nitride layers were grown on sapphire substrate by molecular beam epitaxy (MBE) at 720 °C in a chamber equipped with a Tm effusion cell [2]. According to Stranski-Krastanow growth mode, QDs appear after deposition of GaN beyond a critical thickness of about 2.4 monolayers. The Tm cell shutter was opened only during the growth of GaN in order to in-situ dope the QDs, and closed during the growth of the capping AlN layers. Samples have been prepared in cross sections for STEM observations along [2, 1, -1, 0] and [0,1,-1,0] zone axis, but quantitative information was obtained only for the first direction. Cross-section samples were prepared by tripod polishing. Sample preparation is a key parameter to visualize and count individual dopant atoms [3-5]. The use of ion milling, which significantly damaged the surfaces by introducing amorphous layers, gave blurred STEM contrast. The optimum sample thickness for quantitative analysis was about 10 nm. STEM images were acquired at 300kV on a TITAN microscope equipped with a probe Cs-corrector. STEM image simulations were performed using Kirkland's code [6]. Experimental parameters (convergence angle α, inner detection angle β) were optimised as explained in part 3.

3. Experimental results

Figure 1 shows HAADF STEM images of a Tm doped GaN QDs. Clearly the AlN regions on these images contain bright spots that are attributed to Tm atoms, as these bright spots are not present when QDs are not Tm doped. Indeed the HAADF contrast of a material of averaged atomic number Z is empirically well described by a Z^{δ} law where δ is a constant for a given experimental setup. Thus atomic columns containing Tm atoms are expected to be brighter.

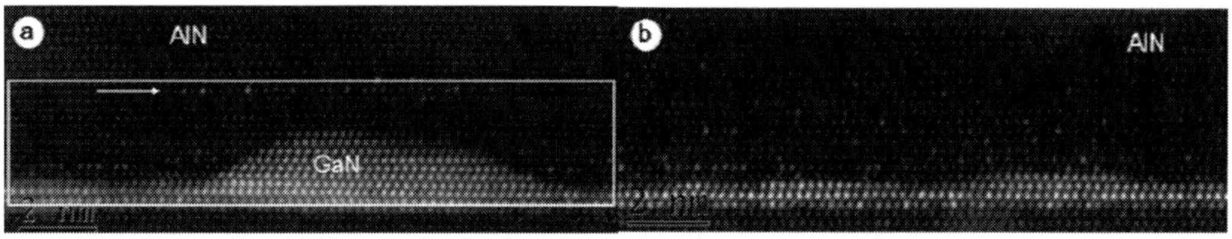

Figure 1. a) HAADF-STEM image of a Tm doped GaN dot embedded in AlN. The white box has a height of 14 monolayers of AlN. b) HAADF-STEM image showing the region above the GaN wetting layer.

3.1. Optimisation of the experimental parameters

The convergence angle α was optimized by analyzing series of simulated HAADF images obtained with the Kirkland code. Our criteria was to obtain a Tm visibility on an atomic column i that does not change with other experimental parameters such as defocus or Tm position within the thin lamella. So we selected a convergence angle that does not favour three dimensional sectioning and choose $\alpha = 15$ mrad. As introduced in [3], we found that a visibility coefficient based on locally integrated intensities, that we named integrated visibility and denote by $V_{int}(i)$, is better than using only the visibility based on the maximum intensity (denoted by V_{max}) :

$$V_{max}(i) = (I_{max}(i) - I_{vac})/(I_{max}(AlN) - I_{vac}) \quad \text{and} \quad V_{int}(i) = (I_{int}(i) - I_{vac})/(I_{int}(AlN) - I_{vac}) \qquad (1)$$

$I_{max}(i)$ is the maximum intensity value on the studied column i. I_{vac} is the intensity in the vacuum. $I_{int}(i)$ is the intensity averaged on a small disk of 0.3 nm in diameter around the maximum intensity. As explained in [3], when the beam is focused on the Tm atom, the intensity of the column is maximum. When the beam is focused either below or above the Tm atom the intensity of the atomic column is more diffused. In fact the intensity of the atomic column is simply "spread" differently. The integrated intensities on the disks are almost constant and are better suitable for defining a visibility coefficient. In fact a similar approach was used in [7]. The inner angle β was chosen in order that the contrast is dominated by incoherent effects, and not by elastic scattering. For that purpose we used a Si reference sample which contains a highly As doped thick layer (figure 2). The As concentration was measured by SIMS and found equal to 1.52%. Using this concentration and assuming a Z^{δ} law, the δ parameters for different experimental settings were fitted (figure 2). These curves have several surprising characteristics whose discussions go beyond the scope of this paper (see [9] for a good starting point). For instance, we can note that the δ parameter is very sensitive to the orientation of the sample, and that higher contrast is obtained in low magnification images when the sample is tilted 2° degrees off the on-axis geometry (Figure 2b). Most importantly for this paper, in the on-axis geometry, which is the case for high-resolution images, β reaches the stable value of δ=2.0±0.05 slightly before β=60 mrad. Thus, the contrast should be due to incoherent effects for β=60 mrad.

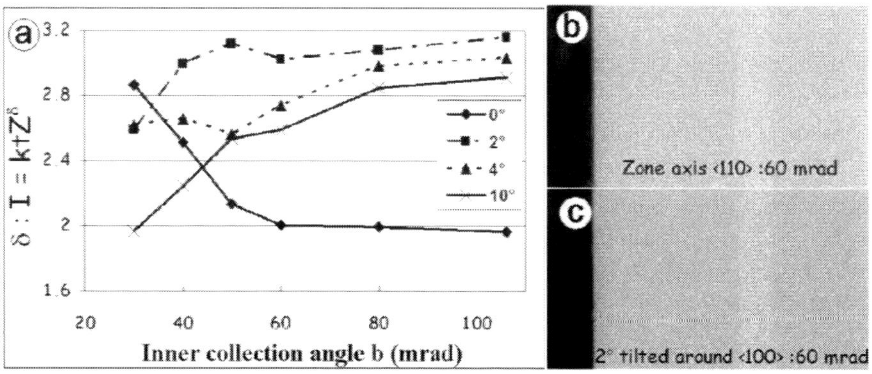

Figure 2. a) The δ parameter of the Z^{δ} law as a function of the inner collection angle β, at different tilt angles as calculated on the As doped sample. Tilt of 0° means on <110> zone. b) On axis HAADF-STEM image of the As-doped sample. c) 2° tilted HAAD-STEM image of the As-doped sample.

3.2. Quantitative analysis of Tm content

Figure 3 shows the integrated visibility for different atomic columns of Figure 1a taken aside the horizontal white arrow. Four groups of values are clearly distinguished: 1.02, 1.14, 1.34 and 1.58. We tried to reproduce these 4 values by using the Z^{δ} law with δ=2.0 as determined above, and by assuming that there were respectively n=0, 1, 2, 3 Tm atoms in these different locations. As discussed in [8-9], we tested two models. In the first one, the coherence along the beam is supposed to be perfect and the image intensity should follow equation 2 where N is the total number of atoms in the column and n is the number of Tm atoms. In the second model the coherence along the beam is supposed to be absent and the intensity from each atom has to be added (equation 3). The thickness of the sample was first estimated by energy loss spectroscopy using a "t/λ map". Then it was adjusted by fitting the 4 averaged values found in figure 3. The exact model and thickness are the ones that reproduce the experimental values (table 1), i.e. the model is "coherent" and the thickness is about 10 nm.

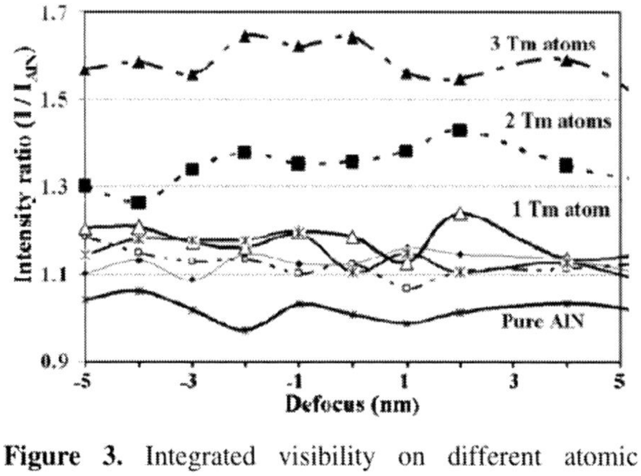

Figure 3. Integrated visibility on different atomic columns at different defoci.

Table 1. Calculated visibility coefficients for different thicknesses and number n of Tm atoms (n=0,1,2,3)

Thickness (nm)	$V_{coherent}$ (n) $(Z_1 + Z_2)^{\delta}$	$V_{incoherent}$ (n) $Z_1^{\delta} + Z_2^{\delta}$
8	1 - 1.23 1.49 - 1.77	1 - 1.8 2.65 - 3.48
10	**1 - 1.18** **1.38 - 1.59**	1 - 1.66 2.3 - 2.97
12	1 - 1.15 1.31 - 1.48	1 - 1.55 2.1 - 2.65
14	. 1 - 1.2 1.42 - 166	1 - 1.47 1.94 - 2.4
experiment	**1.02** (n=0) - **1.14** (n=1) **1.34** (n=2) - **1.58** (n=3)	

$$V_{coherent}(n) = \left(\mathrm{N}(\mathrm{Z}_{\mathrm{Al}} + \mathrm{Z}_{\mathrm{N}}) + \mathrm{n}(\mathrm{Z}_{\mathrm{Tm}} - \mathrm{Z}_{\mathrm{Al}})\right)^{\delta} / \left(\mathrm{N}(\mathrm{Z}_{\mathrm{Al}} + \mathrm{Z}_{\mathrm{N}})\right)^{\delta} \tag{2}$$

$$V_{incoherent}(n) = \left(\mathrm{N}\,(\mathrm{Z}_{\mathrm{Al}}{}^{\delta} + \mathrm{Z}_{\mathrm{N}}{}^{\delta}) + \mathrm{n}(\mathrm{Z}_{\mathrm{Tm}}{}^{\delta} - \mathrm{Z}_{\mathrm{Al}}{}^{\delta})\right) / \left(\mathrm{N}\,(\mathrm{Z}_{\mathrm{Al}}{}^{\delta} + \mathrm{Z}_{\mathrm{N}}{}^{\delta})\right) \tag{3}$$

3.3. Qualitative distribution of Tm atoms in the sample

In figure 1, it is interesting to note that the distribution of Tm atoms in AlN is not uniform. Above the wetting layer (Figure 1b), the dopant atoms are uniformly distributed within 13-14 MLs. By contrast, above QDs, Tm atoms are located in a single plane situated 3 or 4 MLs above the QDs (figure 1a). This phenomenon can be explained by the capping mechanism of GaN dots by AlN, as reported in [2, 3]. In addition, as shown in figure 4, Tm can diffuse very far from the GaN WL by following threading dislocations lines.

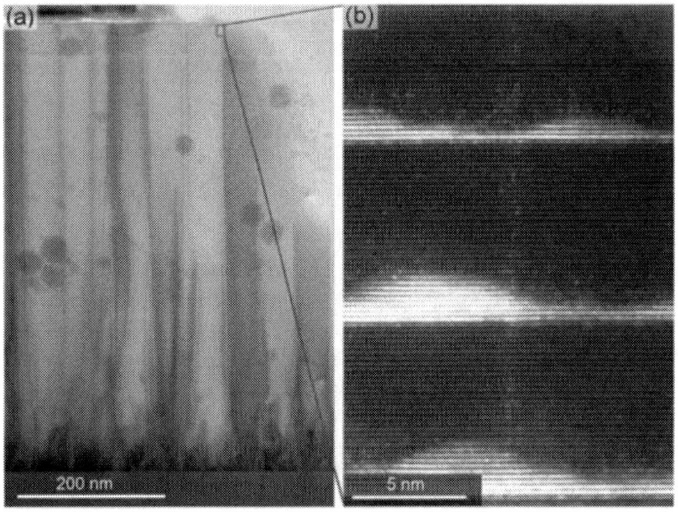

Figure 4. a) General view of the sample showing the sapphire substrate, the thick AlN buffer layer and the layers of GaN dots. Several threading dislocations can be seen.
b) High resolution STEM image taken on the area of figure 4a marked by a red rectangle. Tm atoms (white spots) are located along a dislocation line.

4. Conclusion

Using HAADF-STEM images, we located individual Tm atoms in the AlN matrix. Most of the Tm atoms located initially in the GaN dots and wetting layers diffuse out of GaN. Above the wetting layer and above the dots, the diffusion of Tm is limited to a distance of 14 AlN monolayers from the wetting layer. On the other hand, Tm can diffuse very far from the GaN WL by following threading dislocations lines. A fairly simple coherent model reproduces the experimental visibility coefficients. Better understanding will be needed and will require better comparisons with simulations, particularly using Lebeau et al's calibration [10, 11].

References

[1] Hori Y et al. 2004 *Appl. Phys. Lett.* **84** 206.
[2] Coraux J et al. 2006 *Phys. Rev.* **B 74** 195302.
[3] Okuno H et al. 2010 *Appl. Phys. Lett.* **96** 251908
[4] Okuno H, Takeguchi M, Mitsuishi K, Guo XJ and K Furuya 2008 *J. Elec. Microsc.* **57** 1
[5] Voyles P M, Grazul J L and Muller D A, 2003 *Ultramicroscopy* **96** 251
[6] Kirkland EJ, Roane RF and Silcox J 1987 *Ultramicroscopy* **23** 77
[7] Grillo V, Carline E and Glas F 2008 *Phys. Rev.* B **77** 054103
[8] Nellist P, Scanning Transmission Electron Microscopy *Science of Microscopy* chap 2 pp65-132 (ed Hawkes)
[9] Walther T 2006 *J. Microscopy* **221** 137
[10] LeBeau J M and Stemmer S 2008 *Ultramicroscopy* **108** 1653
[11] Rosenauer A, Gries K, Müller K, Pretorius A, Schowalter M, Avramescu A, Engl K, Lutgen S, 2009 *Ultramicroscopy* **109** 117

Determination of Nitrogen Concentration in Dilute GaNAs by STEM HAADF Z-Contrast Imaging

Tim Grieb[1], Knut Müller[1], Oleg Rubel[2,3,4], Rafael Fritz[4], Claas Gloistein[1], Nils Neugebohrn[1], Marco Schowalter[1], Kerstin Volz[4] and Andreas Rosenauer[1]

[1] University of Bremen, Otto Hahn-Allee 1, 28359 Bremen, Germany
[2] Thunder Bay Regional Research Institute, 290 Munro Street, Thunder Bay, Canada ON P7A 7T1
[3] Lakehead University, 955 Oliver Road, Thunder Bay, Canada ON P7B 5E1
[4] Philipps University Marburg, Hans-Meerwein-Straße, 35032 Marburg, Germany

E-mail: grieb@ifp.uni-bremen.de

Abstract. We determine the nitrogen concentration of $GaN_{0.01\leq x\leq 0.05}As_{1-x}$ quantum wells by evaluation of high resolution scanning transmission electron microscopy (STEM) images using a high-angle annular dark field detector. Although nitrogen has a smaller atomic number than Ga the image intensity increases with the nitrogen content. This is explained by the influence of static atomic displacements by comparison with frozen lattice simulations. The resulting nitrogen concentrations agree with high resolution X-ray diffraction measurements and strain state analysis applied to STEM images.

1. Introduction

As small amounts of nitrogen reduce band gaps of ternary GaNAs and quaternary InGaNAs, these semiconductors find many applications in the fields of electronics and optoelectronics. The large bowing parameters allow tuning over a wide spectral range by setting the nitrogen content. For all those applications an accurate understanding of the structural and chemical morphology is important. For investigation of dilute semiconductors high-precision techniques are essential. In this proceeding we employ high resolution (HR) scanning transmission electron microscopy (STEM) imaging of GaNAs/GaAs using a high-angle annular dark field (HAADF) detector. The HAADF intensity mainly stems from thermal diffuse scattered electrons, thus depending on the atomic number of the scattering atoms [1]. In STEM images intensity maxima of the HR pattern can directly be used to measure positions of atom columns, which is an advantage over conventional transmission electron microscopy (TEM) where positions of extremal image intensity depend on imaging parameters, lens aberrations and gradients of specimen thickness and composition. We show that the correct interpretation of STEM intensity requires to take into account the influence of static atomic displacements (SADs) that occur due to the small covalent radius of nitrogen atoms compared with gallium or arsenic [2, 3].

2. Experimental Procedures

The sphalerite type GaNAs/GaAs sample was grown by metal-organic vapor phase epitaxy. Five GaN_xAs_{1-x} quantum wells (QWs) with varying nitrogen concentrations x between 1 and 5 percent were embedded in GaAs. The nominal values of nitrogen concentration were derived from high-resolution X-ray diffraction (HRXRD) measurements. The QWs are separated by 80 nm spacers and have a width of approximately 10 nm. We prepared an about 160 nm thick TEM lamella by focussed ion beam thinning using an FEI dual beam system Nova Nanolab 200. Final thinning step was performed by ion milling at 400 V using a Technoorg Linda Gentle Mill IV5. All STEM measurements were performed with an FEI Titan 80/300 microscope operated at 300 kV. The specimens were viewed along the [001] direction. We used a camera length of 196 mm where the Fischione 3000 HAADF detector covers an angular range between 33 mrad and 200 mrad. Imaging conditions were chosen as suggested in [4]. These conditions avoid saturation of the detector and provide a linear signal transfer of the amplifier. We scanned 2048×2048 image points with a dwell time of 16 µs, the scan direction was chosen parallel to the growth direction to minimize the influence of specimen drift on the measurement of atom column positions.

3. Simulation of Reference Data

STEM reference data were computed with the STEMsim program [5] using the frozen-lattice multislice approach with supercells of 9×9 unit cells size perpendicular to the electron beam direction. Each unit cell was scanned with a resolution of 20×20 picture elements (pixels). The convergence angle of the electron beam was set to 9 mrad, the spherical aberration constant of the condensor lens was defined to be 1.2 mm. Images were simulated for supercells with nitrogen concentrations between 0 and 8 % in steps of 2 %. Intermediate concentrations and thicknesses were interpolated by polynomial fitting. We implemented SADs by relaxation of the supercells using a valence force field model [2, 6]. Debye Waller temperature factors were taken from Ref. [7]. The non-homogeneous sensitivity of our HAADF detector was taken into account. To suppress biasing of measured concentrations towards higher values in regions with a concentration of zero due to noise, the reference data set was extrapolated towards negative nitrogen concentrations.

4. Evaluation of nitrogen concentrations

For determining the nitrogen concentration within a GaNAs QW the QW itself as well as part of the GaAs region was imaged. In a first step the positions of the atom columns are found. For this the original STEM image was Wiener filtered and short-range fluctuations were reduced by applying a low-pass filter in Fourier space. All further steps are performed with the original unfiltered image. Next, the image area is segmented into Wigner-Seitz cells. Then the image intensity is averaged within each cell which gives the local mean intensity. From the HAADF intensity of the pure GaAs we deduce the specimen thickness which is extrapolated to the nitrogen containing region by a polynomial. For doing this the ratio of HAADF intensity to incident electron beam intensity is to be measured. Latter one was determined by a detector scan. With known specimen thickness the HAADF intensity in the QW region is then directly associated to a specific nitrogen concentration given by the simulated reference data.

5. Results and Discussion

The approach is described exemplarily on the basis of an HAADF STEM image containing a QW with a nominal nitrogen concentration of 3.3 % at a magnification of 3.6 Mx. The applied magnification was chosen to achieve adequate high resolution (the pixel size is about 50 pm) on the one hand and a large GaAs reference region on the other hand. Latter one is important because the regions next to the QW show decreased intensity and can therefore not be used as reference signal. This effect is due to surface strain relaxation [8] and depends highly on the amount of nitrogen in the well. The mean thickness obtained from the STEM image was 165 nm.

Figure 1 a) shows an HAADF image of all five QWs of the investigated specimen. Considering only scattering depending on atomic number, one would expect an inverse contrast ratio: the

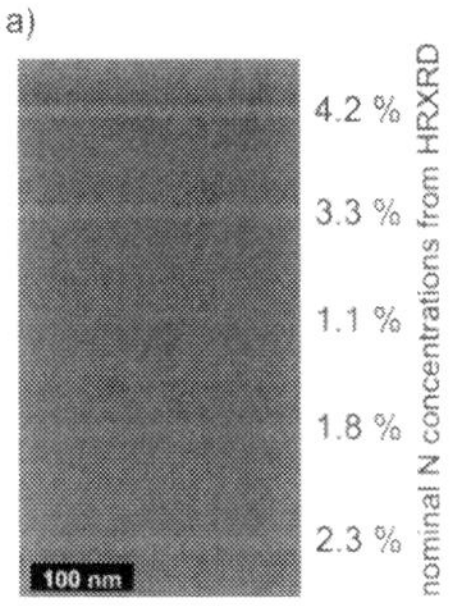

Figure 1. a) HAADF image of all GaNAs QWs encased in GaAs. The N containing QWs show higher scattering abilities compared to GaAs. b) Simulated data of thickness dependent HAADF signal normalized to GaAs for selected nitrogen concentrations. Including SADs HAADF intensity above 50 nm increases with N content. Intensity decreases slightly if SADs are not taken into account.

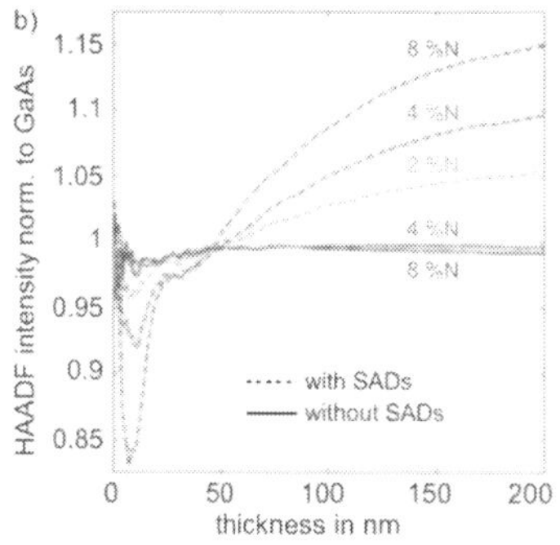

imaged GaAs should be brighter compared to the nitrogen containing QW (e.g. observed by Herrera *et al.* [9]). However, taking SADs into account, we observe a different behavior: for typical STEM specimen thicknesses above 50 nm, the HAADF signal increases with nitrogen concentration. Figure 1 b) shows the simulated HAADF intensity normalized to GaAs for selected concentrations.

Evaluation of the nitrogen concentration as described in section 4 yielded the concentration map depicted in Figure 2 a). Averaging the concentration values along atomic rows parallel to the QW gives the concentration profile shown in Figure 2 b). The errorbars involve the standard deviation caused by intensity fluctuations as well as actual concentration deviations divided by the root of the number of positions that account for the calculated mean value (standard error).

We find that the nitrogen distribution is well confined to the QW area with rather sharp edges. The regions next to the QW cannot easily be evaluated because surface strain fields influence the HAADF contrast and the concentrations in these areas do not belong to real nitrogen contents. For defining the mean concentration value from Figure 2 we trust the values close to the center of the QW because there surface strain can be neglected due to symmetry. Here we find a nitrogen concentration of $2.8 \pm 0.6\,\%$ which agrees with the nominal value of $3.3\,\%$. This effect was well reproduced in simulated images where elastic relaxation and lattice plane bending were taken into account. These results will be published elsewhere.

Figure 3 a) shows the N concentrations derived from HAADF intensity evaluation for all five QWs. We measured at four different positions on the specimen for each QW. Shown are the mean values of these positions, the errorbars involve the standard deviation only. Taking the errorbars of each single measurement into account the accuracy of the concentration determination is about $1\,\%$ N. To obtain further values for comparison we evaluated the same images in terms of a strain

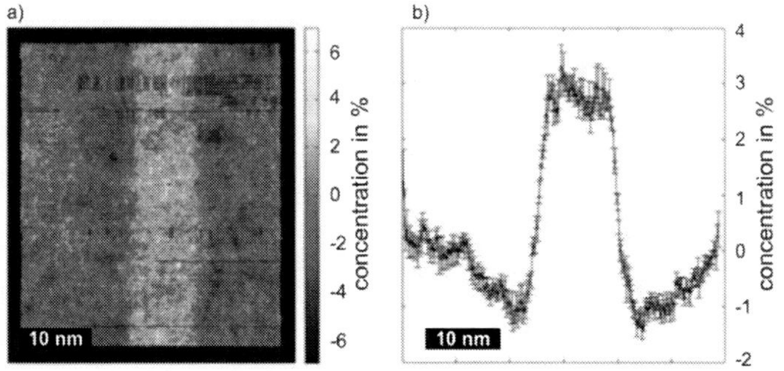

Figure 2. a) Concentration map achieved from evaluating HAADF intensity. b) Arithmetic mean of linescans in growth direction. The errorbars involve standard error.

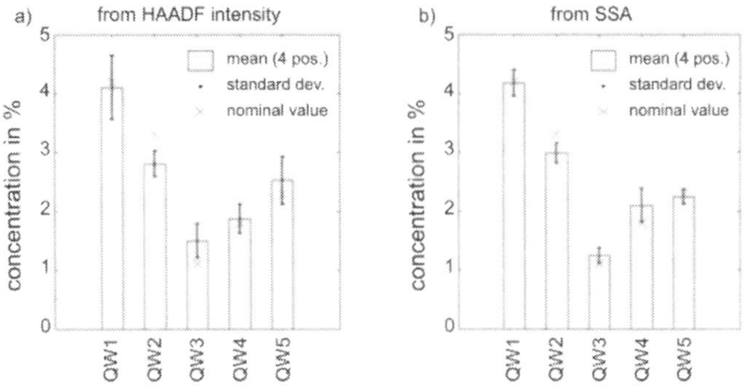

Figure 3. Concentrations measured for all five QWs a) by evaluation of HAADF intensity and b) from strain state analysis. Shown are the mean values of four evaluated areas in each case. The errorbars represent the standard deviations and do not contain the error of the single measurement.

state analysis (SSA). The nitrogen in the QW leads to a smaller lattice distance compared with GaAs reference regions. From the image we determine the tetragonal distortion for each position of the atomic columns from which the nitrogen concentrations are calculated applying Vegard's law and assuming complete tetragonal distortion (thick specimen limit). The concentrations derived from SSA are shown in Figure 3 b) where the values are presented in the same way as in Figure 3 a). Concentration from SSA agree with those from intensity evaluation and with nominal values when taking single measurement accuracy into account.
So a combination of SSA and HAADF intensity evaluation provides a concentration determination with enhanced reliability.

6. Conclusion
By taking static atomic displacements into account, the comparison of STEM HAADF images of dilute GaNAs quantum wells with simulated reference data allows a nitrogen concentration determination with an accuracy of 1 % N. We verified our results by evaluating the same images with strain state analysis.

Acknowledgements
This work was supported by the Deutsche Forschungsgemeinschaft (DFG) under contract number SCHO 1196/3-1 and RO 2057/8-1-3008074.

References
[1] Crewe A V, Wall J, and Langmore J P 1970 Visibility of single atoms *Science* **168**, 1338
[2] Glas F 2004 The effect of the static atomic displacements on the structure factors of weak reflections in cubic semiconductor alloys *Philos. Mag.* **84**, 2055
[3] Grillo V, Carlino E, and Glas F 2008 Influence of the static atomic displacement on atomic resolution Z-contrast imaging *Phys. Rev. B* **77**, 054103
[4] Rosenauer A, Gries K, Mueller K, Pretorius A, Schowalter M, Avramescu A, Engl K and Lutgen S 2009 Ultramicroscopy **109**, 1171
[5] Rosenauer A, and Schowalter M 2007 STEMSIM - a new software tool for simulation of STEM HAADF Z-contrast imaging *Springer Proceedings in Physics* **120**, 169
[6] Keating P N 1966 Effect of Invariance Requirements on the Elastic Strain Energy of Crystals with Application to the Diamond Structure *Phys. Rev.* **145**, 637
[7] Schowalter M, Rosenauer A, Titantah J T, and Lamoen D 2009 Computation and parametrization of the temperature dependence of Debye–Waller factors for group IV, III–V and II–VI semiconductors *Acta Crystallogr., Sect. A* **65**, 5
[8] Grillo V 2009 The effect of surface strain relaxation on HAADF imaging *Ultramicroscopy* **109**, 1453
[9] Herrera M, Ramasse Q M, Browning N D, Pizarro J, Galindo P, Gonzalez D, Garcia R, Du M W, and Zhang S B, and Hopkinson M 2008 High Resolution HAADF-STEM Imaging Analysis of N related defects in GaNAs Quantum Wells *Microsc. Microanal.* **14**, 318

Journal of Physics: Conference Series 326 (2011) 012034 doi:10.1088/1742-6596/326/1/012034

2D chemical characterization of complex semiconductor structures

P L Potapov[1] **and H J Engelmann**
Globalfoundries Dresden Module 1,
Wilschdorfer Landstr.101, 01109 Dresden, Germany
E-mail: Pavel.Potapov@globalfoundries.com

Abstract. We introduce a method to characterize the chemical distribution in nanostructures by using STEM and affiliated spectroscopy techniques. The method is applicable to any nanostructure where continuous layers of arbitrary geometry and dimensions can be identified. The key feature of the suggested approach is digital warping of the original STEM image into a quasi 1D image. The chemical profiles of high resolution and high signal-to-noise ratio can be extracted from the minimal set of the STEM spectroscopy data while minimizing material damage during acquisitions. Finally, the 2D chemical maps of the area of interest are reconstructed.

1. Introduction

Scanning transmission electron microscopy combined with spectroscopy methods is proven to be an excellent tool to characterize distribution of chemicals in nanostructures, e.g. semiconductor devices, biological specimens, environmental specimens etc. Spectroscopy implies irradiation of a given probed area with a sufficiently high electron dose in order to deliver spectra having the appropriate signal-to-nose ratio. As a consequence, 2D spectroscopy imaging is quite time consuming and sensitive to the sample or beam drift during mapping. Recently introduced STEM instruments with probe Cs correctors or high-brightness FEG can partially solve this problem by a drastic increase of the electron current within a small probe [1,2]. That however raises another severe problem: sample damage during the acquisition. A high electron dose concentrated within a small area tends to alter the sample and distort the results of chemical analysis [3]. That puts a limit on the minimal probe size for a given electron current and a dwell time. Thus, the choice of acquisition parameters in spectrum imaging is often dictated by a compromise between the signal-to-nose ratio and spatial resolution.

In the present paper, we suggest a method [4] that overcomes the mentioned problems and delivers chemical maps with the high signal-to-noise ratio while minimizing the overall dose applied to the sample. This method is applicable to one (although widely common) class of nanostructures, namely nanostructures composed of layers with a variable chemical content. To introduce the basic principle consider a structure with perfectly flat layers (figure 1a). The conventional spectrum imaging implies placing a number of probes over the whole investigated area. The sampling theory states that the sampling frequency must be at least twice the maximal transmission band [5], i.e. the probes should be placed at the distance not greater than a half probe diameter from each other. That would invariably result in strong *cumulative* damage effect from the neighbouring probes as illustrated in figure 1b. It is however easy to see that this structure can be characterized without compromising the electron dose

[1] To whom any correspondence should be addressed.

and spatial resolution. Several line scans each with sufficiently large sampling step might be performed across the layers and then projected onto one summed profile (figure 1c-d). For instance, in case when each line scan is four times undersampled, the summation of the four line scans shifted relative to each other by a quarter step would provide the four times denser sampling. Taking more line scans might not improve the spatial resolution but increase the signal-to-noise ratio. In this way, one can obtain any desired sampling frequency and/or any desired signal-to-noise ratio while keeping the probe positions sufficiently far from each other.

The projection of the dispersed probes onto the desired direction is not an entirely new approach. The variations of this principle were employed in stripe STEM [6], some examples of Smart acquisition [7] and in the earlier work by Hunt et al.[8]. The basic difficulty in wide application of this idea is the fact that few if any nanostructures show the perfectly flat layers. Even the intrinsically straight lattice fringes appeared to be curved in the real STEM images due to scanning instabilities, sample and beam drift et cetera. One solution is to warp digitally the original STEM image such a way that all visible layers appear flat. Digital warping was already devised to correct the scanning distortions and improve the quality of high resolution STEM images [9]. Within this method, we suggest more extensive application of digital warping for the accurate alignment of the spectroscopy information and extraction of the chemical profiles combining the high spatial resolution and the high signal-to-noise ratio as shown in the example in figure 2. Furthermore, 2D chemical maps can be reconstructed by reverse warping.

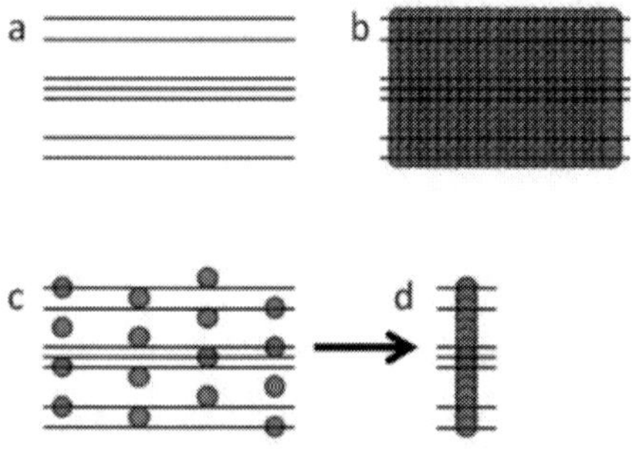

Figure 1. (a) Nanostructure with perfectly flat layers schematically shown by horizontal lines and (b) its probing pattern following the conventional concept of spectroscopy imaging. (c) Probing with few line scans afterwards projected (d) onto one line.

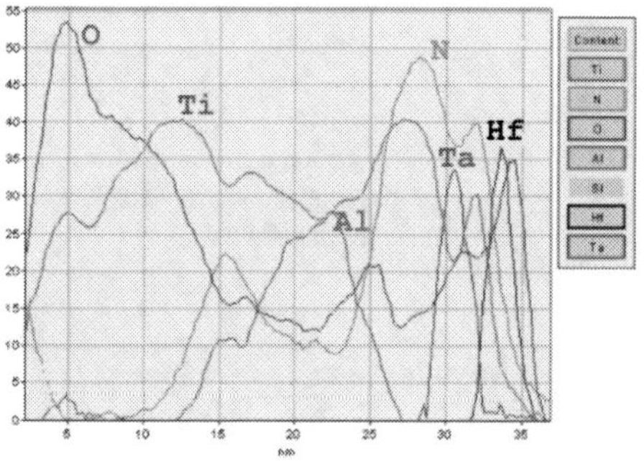

Figure 2. Examples of chemical profiles obtained by the projection and summation of spectroscopy data from several undersampled line scans. Content is depicted in atomic %.

2. Digital warping of STEM images

Digital warping as a specific part of image processing has received a lot of interest for making special effects in photography and cinema. A good review of warping techniques can be found in [10,11]. In the present work we would not focus on any particular warping algorithm as various algorithms may be implemented for realization of our method. However, as warping is used here for the reconstruction of information, not for the decorative purpose, some necessary conditions for application of the method should be formulated:

- a correlation between the acquired STEM signal and the changes in chemical composition is to be reasonably assumed, in particular, it means that HAADF STEM detector not BF one should be applied;
- the destination of warping should be a structure where all visible layers in a STEM image appear as flat as possible;

- the warping deformation should be continuous and topology preserving;
- the reverse warping formula, i.e. the transformation returning the warped image back to the original morphology should be available.

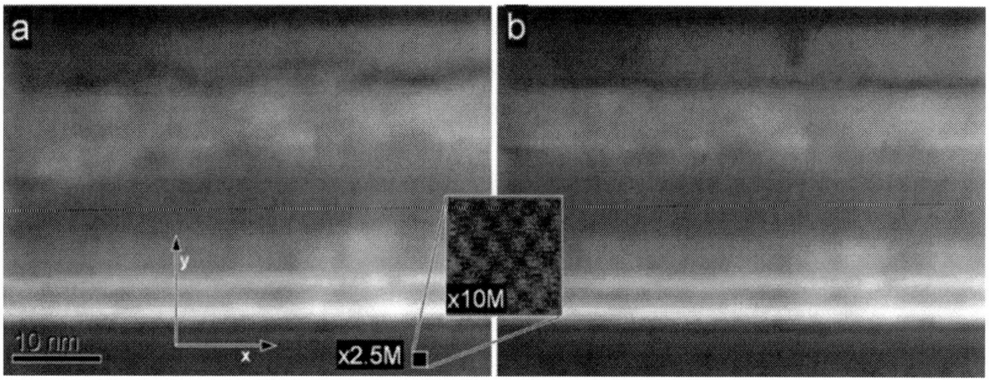

Figure 3. (a) Original STEM image of the metal gate stack and (b) its warping into quasi 1D image.

3. Treatment flow with an example of a metal gate stack

(I) A HAADF STEM image covering the whole area of interest is acquired and evaluated. The method can be applied only in case when the visible continuous layers are clearly detected, as in figure 3a where the high-k dielectric, work function barrier, metal electrode and other layers are seen in the nanostructure. The applied probe size approached atomic resolution, as seen in the inset.

(II) Few line scans across the visible layers are performed with acquiring EELS and/or EDXS spectra. The low sampling frequency along each line scan is applied to avoid the cumulative damage effect from two neighbouring probes. The line scans go approximately perpendicular to the visible layers in the examined structure although their exact positions are rather arbitrary. The STEM signal is collected during line scanning, which allows us to align accurately the position of each line scan relative to the original STEM image. Therefore, the results are not affected by sample/beam drift.

(III) The original STEM image is warped in order to obtain as flat layers as possible, i.e. a nearly 1D layer morphology (figure 3b). With use of the found warping transformation, the coordinates of the probes in each line scan are transformed into the coordinates of the warped image. As the warped image is supposed to be quasi one-dimensional (no or insignificant variations in the x-direction), the straightforward summation of the data from all line scans is possible. The sampling step in such a summed line profile is chosen to achieve the desired combination of spatial resolution and signal-to-noise ratio. The calculated probe coordinates from each line scan would apparently slightly deviate from the exact sampling positions in the summed line profile. That can be easily corrected by linear interpolation of the spectra between two nearest points. Taking more line scans than minimally necessary for the optimal sampling would improve the signal-to-noise ratio of the summed profile.

(IV) The summed spectroscopic line scan is then treated in the common way to extract profiles of the chemical elements. Our method however offers one remarkable advantage: spectra acquired covering different energy ranges are automatically set into the exact positional correspondence. This solves, in particular, the problem of EELS quantification that requires both core-loss and low-loss spectra collected from exactly the same place. In case of sub-nm STEM probes, it is almost impossible to satisfy accurately this requirement for two subsequent EELS acquisitions unless dualEELS [12] is applied. The present method aligns automatically the low-loss spectra to spectra taken in any other energy range no matter what was the time gap between their acquisitions. For the given example, figure 2 shows the obtained chemical profiles of Ti, N, Al, O, Hf and Ta along the *y*-axis of figure 3b.

(V) The obtained chemical profiles are expanded in the *x*-direction to form a kind of 2D maps with no chemical changes along the *x*-axis. These maps do not contain extra information as compared with the summed chemical profiles in figure 2, however their 2D format make them analogous to the warped STEM image in figure 3b.

(VI) Finally, the reverse warping of these quasi-1D maps generates the real-space chemical maps shown in figure 4. Comparison with the original STEM image in figure 3a suggests that the maps reflect quite realistically the expected distribution of chemicals in the nanostructure.

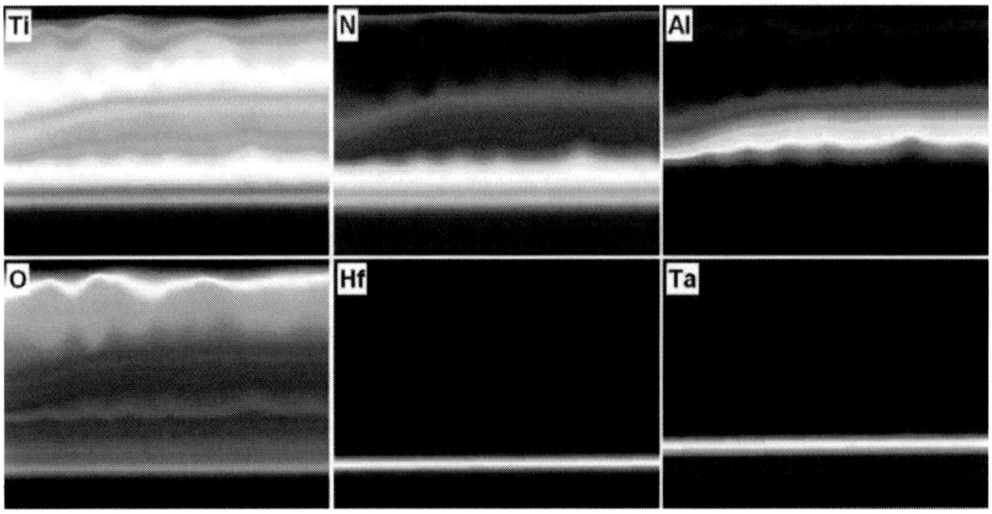

Figure 4. Chemical maps obtained by the extension of the chemical profiles from figure 2 in the *x*-direction followed by the reverse warping transformation.

References

[1] Muller D A, Kourkoutis L F, Murfitt M, Song JH, Hwang H Y, Silcox J, Dellby N and Krivanek O L 2008 *Science* **319** 1073
[2] Schlossmacher P, Klenov D O, Freitag B, von Harrach S and Steinbach A 2010 *Microscopy and Microanalysis* **24** S5
[3] Egerton R F, Li P and Malac M 2004 *Micron* **35** 399
[4] Potapov P L, Engelmann H J Application US patent filed 18.03.2011
[5] Markus R J 1991 *Introduction to Shannon sampling and interpolation theory* (New York: Springler Verlag)
[6] Heidelmann M, Barthel J and Houben L 2009 *Ultramicroscopy* **109** 1447
[7] Sader K, Schaffer B, Vaughan G and Bridson R 2010 *Ultramicroscopy* **110** 998
[8] Hunt J A, Disco M M, Behal S K and Leapman R D 1995 *Ultramicroscopy* **58** 55
[9] Recnik A, Möbus G and Sturm S 2005 *Ultramicroscopy* **103** 285
[10] Wolberg G 1992 *Digital image warping* (Los Alamitos: IEEE Computer Society Press)
[11] Glasbey C A and Mardia K V 1998 *J. Appl. Statistics* **25** 155
[12] Scott J et al 2009 *Ultramicroscopy* **108** 1586

Investigation of diffusion in AlAs/GaAs distributed Bragg reflectors using HAADF STEM imaging

M Schowalter[1], M Tewes[1], K Frank[1], R Imlau[1], A Rosenauer[1], H S Lee[2], A Rastelli[3], O G Schmidt[3], M Tavast[4], T Leinonen[4] and M Guina[4]

[1] Institut für Festkörperphysik, Universität Bremen, Otto-Hahn-Allee 1, D-28359 Bremen, Germany

[2] Advanced Photonics Research Institute, Gwangju Institute of Science and Technology, Gwangju 500-712, Republic of Korea

[3] Institute for Integrative Nanosciences, IFW Dresden e.V., Helmholtzstr. 20, D-01069 Dresden, Germany

[4] Optoelectronics Research Centre, Tampere University of Technology, Korkeakoulunkatu 3, FI-33720 Tampere, Finland

E-mail: `schowalter@ifp.uni-bremen.de`

Abstract. In this contribution we have studied the diffusion of Al in AlAs/GaAs distributed Bragg-reflectors using the high angle annular dark field (HAADF) scanning transmission electron microscopy (STEM) intensity. The measured intensity is normalized to the intensity of the incoming electron beam using a detector scan. The normalized intensity can be directly compared with a set of frozen lattice simulations yielding specimen thickness in regions with known composition or concentration in regions with known thickness. The thickness was evaluated both from GaAs and AlAs regions yielding that the specimen was about 15 nm thinner in AlAs regions due to oxidation. For the concentration evaluation the thickness was derived from GaAs regions and concentrations up to 1.2 were found due to the overestimated thickness. Concentration profiles were scaled down to 1.0 and fitted to the solution of Fick's laws.

1. Introduction

Optoelectronic devices containing quantum dot structures have gained a lot of interest in recent years due to their improved and novel properties. A rather simple way to produce quantum dots is by exploiting the Stranski-Krastanow growth mode. However, such ensembles of quantum dots exhibit rather broad size and composition distributions. Using a rapid thermal annealing (RTA) treatment the optical properties, especially the energetic position of the ensemble of dots or of an isolated dot can be influenced [1, 2]. For fabrication of vertical-cavity surface-emitting lasers (VCSELs) the quantum dots have to be grown embedded between two distributed Bragg reflectors (DBRs). A subsequent RTA treatment influences the properties of the quantum dots, but may also influence the reflectivity of the DBRs due to diffusion. Most of the previous investigations have focused on the effect of RTA on the active layers (quantum dots or also quantum wells, see e.g. [3, 4, 5]).

In this contribution we investigated the diffusion of Al in AlAs/GaAs DBRs due to a thermal treatment at temperatures as high as 1100 °C for 180 s by a quantitative analysis of the high angle annular dark field (HAADF) scanning transmission electron microscopy (STEM) intensity. Derived Al composition profiles were fitted by solving Fick's laws.

2. Growth, annealing treatment and preparation of specimen

The 24-pair and 25-pair AlAs/GaAs DBRs were grown by solid source molecular beam epitaxy (SS-MBE) using valved phosphorous and arsenic cracking sources. The sample manipulator temperature was kept close to the typical GaAs growth setting yielding sample temperatures in the range between 580 and 600 °C. The growth rates for both AlAs and GaAs layers were about 1.0 μm/h. The 25-pair AlAs/GaAs DBR was treated by RTA at 1100 °C for 180 s. TEM

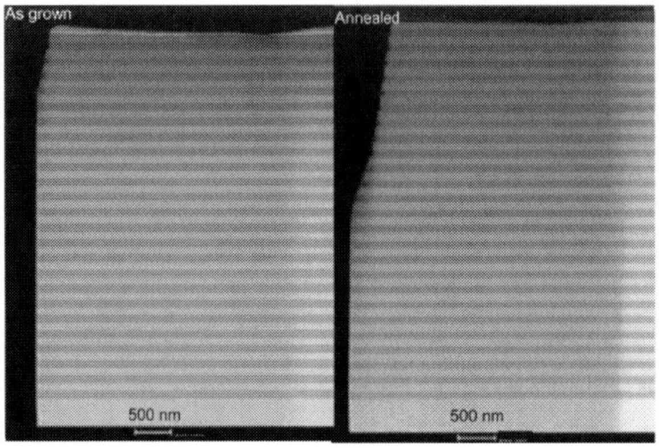

Figure 1. Overview STEM micrograph of the as-grown and the annealed structure.

specimens were prepared for a <110> zone axis orientation by cutting lamellas out of the wafer using a FEI dual beam focused ion beam(FIB) NOVA200. The lamellas were lifted out using a Kleindieck manipulator needle and fixed to a Cu ring. STEM HAADF images of the structures are depicted in Fig. 1. TEM specimens were directly transported from the FIB to the TEM or were stored in vacuum in order to prevent oxidation of the AlAs layers as much as possible. TEM investigations were carried out in a FEI Titan 80/300ST equipped with a corrector for spherical aberration of the objective lens, an EDAX EDX detector, a Fishione HAADF detector and a Gatan imaging filter (GIF).

3. Method for quantitative evaluation of HAADF intensity

Here we only briefly describe the basics of the quantitative evaluation technique, which is explained in more detail in Ref. [6]. Prior to the acquisition of the HAADF images an image of the detector is recorded by scanning the electron beam in image mode over the detector. In this way the intensity in the HAADF STEM images can be normalized to the incoming electron beam intensity and therefore can directly be compared with simulations.

A series of HAADF-STEM images were computed using the STEMsim software package [7] for concentrations from 0 to 100% and thicknesses of smaller than 350 nm. To take into account thermal diffuse scattering we used the "frozen lattice approximation", in which it is assumed that all atoms vibrate independently. For each scan point in the HAADF images the intensity was averaged over 10 different "frozen lattice" configurations. Static atomic displacements [8] were not taken into account, since the ionic radii of Al and Ga are nearly the same (lattice parameter a_{GaAs}=0.5653 nm and a_{AlAs}=0.5661 nm). The HAADF STEM intensity as a function of thickness and composition is depicted in Fig. 2. It exhibits strong thickness dependency and a linear concentration dependency.

The specimen thickness can be determined from reference regions with known composition (e.g. a region in the GaAs substrate) by directly comparing experimental and computed HAADF-STEM intensities, both normalized to the incoming beam intensity. In regions with

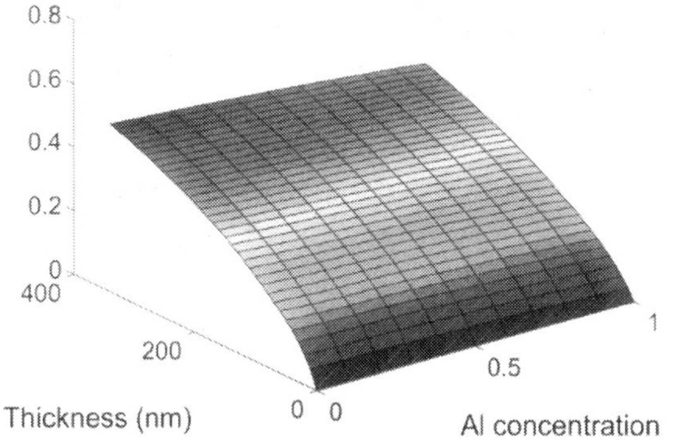

Figure 2. Simulated intensity on the HAADF detector as function of Al concentration and specimen thickness.

unknown composition, the specimen thickness is interpolated. Then the intensity ratio is determined by normalizing to the reference regions. Finally, the composition is deduced by a comparison with the theoretical data.

4. Results and discussion

Fig. 1 shows overview images of the as-grown and the annealed samples. From a first visual inspection it becomes clear that the interfaces in the as-grown specimen are sharper than interfaces in the annealed one indicating that Al is diffusing out of the AlAs layers. In order to ensure that in the middle of the GaAs the Al concentration (and in the middle of AlAs the Ga concentration) is negligible we carried out energy dispersive X-ray spectroscopy (EDX) on several points lying on lines parallel to the GaAs (AlAs) layers. We found that the Ga (Al) concentration in the middle of the AlAs (GaAs) layer is below 1%. For further investigations we assumed that the Al concentration in the centre of the GaAs layer (Ga concentration in the AlAs layer) is zero.

In a further step we acquired HAADF STEM images and performed linescans in growth direction. We selected the GaAs substrate and regions in the centre of the GaAs layers for the determination of the specimen thickness (see sec. 3). For the lammella of the annealed specimen shown in Fig. 1 we found a specimen thickness ranging from 70 nm to about 260 nm. The specimen thickness was interpolated within the AlAs layers using a 3rd order polynomial. We also determined the specimen thickness from the centre of the AlAs layers yielding a range from 60 nm to 230 nm. The found thickness dependencies are depicted in Fig. 3. We found that the specimen is about 15±4 nm thinner in AlAs than in GaAs regions. This result was confirmed by thickness maps derived from energy filtered images. This also becomes obvious in Fig. 1, where at the edge of the lamella the GaAs layers reach further into the vacuum than the AlAs layers. Under the assumption of the specimen thickness derived from the GaAs regions we found Al concentrations up to 120%. For this purpose the reference data set shown in Fig. 2 was extrapolated linearly up to a concentration of 150%. The determination of these unphysical Al concentrations can be explained by the overestimation of specimen thickness in the AlAs regions. In the EDX experiments we found that the Ga-concentration with the AlAs layers was negligible and therefore, we rescaled the measured concentration profiles down to an average Al concentration of 100% in the centre of the AlAs layers.

Rescaled concentration profiles were fitted by solving Fick's laws for square concentration profiles (see Fig. 4). For the as grown specimen only small deviations from a rectangular concentration profile are observed, which could be attributed to electron beam broadening or segregation of Al during growth of the sample. Rosenauer et al. [9] found a low segregation (R=0.5) efficiency in this material, so that the main contribution should come from electron beam broadening. To estimate the broadening the profile was fitted with a diffusion profile yielding a diffusion length of about 1 nm. In the annealed specimen we found a diffusion length of about 10 nm, showing that the beam broadening should only influence the determination of the diffusion length in a minor way. The diffusion coefficients were found to be $1.93 \cdot 10^{-15} cm^2/s$, respectively with a standard deviation of $0.25 \cdot 10^{-15} cm^2/s$. Cai et al. [10] measured the diffusion coefficient at 900°C and found $2 \cdot 10^{-17} cm^2/s$, which is about two orders of magnitude smaller than our value. However, assuming the activation energy of 3.5 ± 0.1 eV from Bracht et al. [11] one would expect a diffusion coefficient of $2.46 \cdot 10^{-15} cm^2/s$ at a temperature of 1100 °C, which is in agreement with our value.

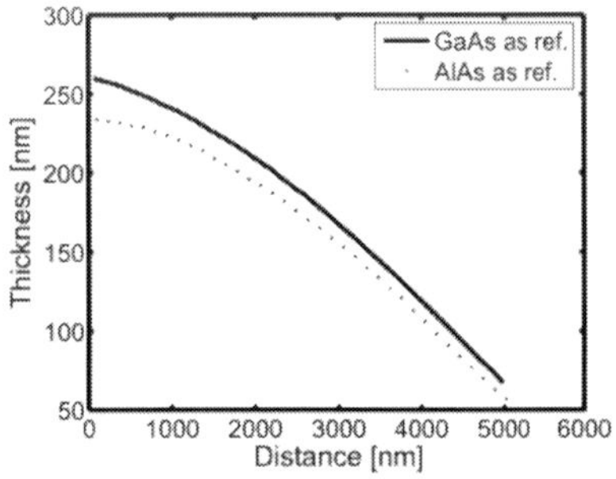

Figure 3. Comparison of specimen thickness as a function of position on the TEM lamella. The specimen thickness was once evaluated from GaAs and from AlAs regions.

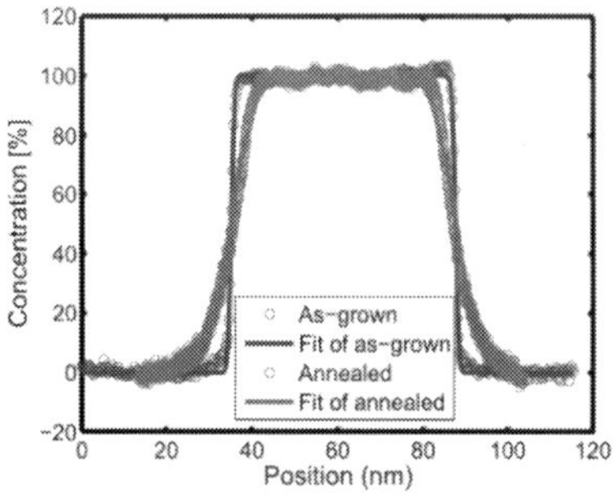

Figure 4. Comparison of concentration profiles measured in the as-grown and the annealed sample. Both profiles were fitted with a diffusion model.

5. Summary

In summary, we quantitatively evaluated Al concentrations from HAADF-STEM images of AlAs/GaAs DBR structures. The evaluation of as-grown and annealed specimens yielded a diffusion coefficient of $1.93 \cdot 10^{-15} cm^2/s$ for Al at 1100 °C.

References

[1] Leon R, Kim Y, Jagadish C, Gal M, Zou J and Cockayne D J H 1996 *Appl. Phys. Lett.* **69** 1888–1890
[2] Rastelli A, Ulrich S M, Pavelescu E, Leinonen T, Pessa M, Michler P and Schmidt O 2004 *Superlattices and Microstructures* **36** 181–191
[3] Hulko O, Thompson D A and Simmons J G 2009 *Semicond. Sci.* **24** 045015
[4] Deppe D G, Guido L J, Holonyak N, Hsieh K C, Burnham R D, Thornton R L and Paoli T L 1986 *Appl. Phys. Lett.* **49** 510–512
[5] Sugawara Y and Miyamoto T *Electronics Letters* **45** 167–168
[6] Rosenauer A, Gries K, Müller K, Pretorius A, Schowalter M, Avramescu A, Engl K and Lutgen S 2009 *Ultramicroscopy* **109** 1171–1182
[7] Rosenauer A and Schowalter M 2007 *Springer Proceedings in Physics* vol 120 ed Cullis A G and Midgley P A (Springer) pp 169–172
[8] Grillo V, Carlino E and Glas F 2008 *Phys. Rev. B* **77** 054103
[9] Rosenauer A, Van-Dyck D, Arzberger M and Abstreiter G 2001 *Ultramicroscopy* **88** 51–61
[10] Cai D, Zou J, Chen G, Lu W, Chen X and Ringer S P 2002 *Scripta Materialia* **47** 279–283
[11] Bracht H, Haller E E, Eberl K and Cardona M 1999 *Appl. Phys. Lett.* **74** 49–51

An advanced study of the response of ADF detector

V Grillo[1,2]

[1]Centro S3-NANO, CNR Via Campi 213A, I-41125 Modena, Italy
[2]IMEM—CNR, Parco Area delle Scienze37/A, I-43010 Parma, Italy

Abstract: The detailed characteristics of the Annular Dark Field detector and acquisition system may play a strong role in quantitative analysis. This article discusses the role in high angle experiments of a non-uniform detection efficiency and of the saturation. The former can be neglected in most comparative experiment while the latter can produce hidden nonlinearities that reduce the reliability of quantitative analysis.

vincenzo.grillo@unimore.it

1. Introduction

The development of quantitative High Angle Annular Dark Field (HAADF) promises to be the most interesting tool in semiconductor compositional analysis on the nanoscale. At present two families of approaches exist for quantitative analysis based on relative (intensity comparison with a reference material) and absolute intensity measurements [1],[2]. The absolute methods are based on the characterization of the detector response and do not require a reference material. Conversely, the relative methods are more robust and more suitable for the comparison between neighbouring materials and in particular semiconductor heterostructures.

The aim of this contribution is to highlight a few technical and physical considerations that can justify using relative techniques for quantitative analysis. The role of the detector saturation and of non-uniform efficiency will be also critically analyzed. The calibration of angular detection efficiency may prove to be quite complicated in some JEOL machines and it is relevant to evaluate to which extent the inhomogeneous detector response can be ignored.

2. Theory

2.1 Saturation effects of the acquisition system

For this study images were acquired by use of a DIGISCAN2 [3] system connected to the analog output of the STEM, but in principle many acquisition systems present similar dynamic limitations. In the detection system attached to the microscope the light from the scintillator is converted to an electrical signal and subsequently amplified by an internal amplifier and directly transformed into a TV signal. With old systems operating with a direct analog visualization of the image the synchronous movement of the scanning coils in the STEM optical system and in the monitor creates the image on the monitor. In digital systems the signal is directly matched to the Analog Digital Converter (ADC) range that converts it to an integer to be stored in the memory of the controlling PC system. The ADC sampling is completed within a short time t_s (every 0.8 μs in a DIGISCAN2) and is repeated several times for each scanning position until the dwell time is reached. The details of the acquisition are actually system dependent. Hereafter, the whole analog part will be modeled as a linear system:

$$I_{analog}=g\, n_s+ b \qquad (1)$$

where the analog current I_{analog} is proportional to the number of detected electrons in the sampling time by means of an overall gain g and a DC offset b. Whereas many factors concur to the overall gain the proportionality factor can be always regulated by adjusting the contrast knob, unfortunately this has also a minor effect on b. Assuming that the ADC is 8 bit (a different number of bits does not change the results hereafter but only the discretization), I_{analog} is an integer between 0 and 255; signals below 0 or larger than 255 are clipped by the system. However, the typical response of a detector presents a smooth saturation curve (see ref. [2],[4],[5]). Whereas this dependence has been largely observed, a detailed modeling of the saturation process has not been proposed.

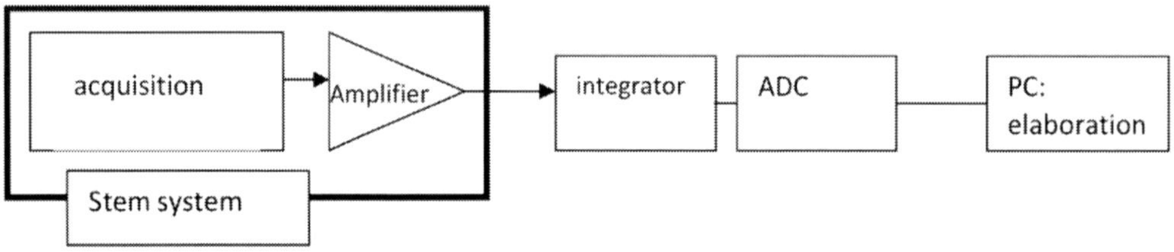

Figure 1. Schematic representation of acquisition system. The analog signal from the internal amplifier is driven to the ADC slot having its own amplifying system whose gain and DC offset can be also regulated by trim.

The reason for the smooth saturation is that many ADC acquisitions are summed together. Due to statistical fluctuations in the number of counts per acquisition in proximity of the maximum allowed signal, only part of the sum is saturated. In a given point the expected number of counts is $<n_s>$ and the counts are distributed according to Poisson statistics $P_{<n>}(i)$ with deviation $\sqrt{<n>}$. A convenient formula for the average intensity on the ADC (normalized to 255 levels) as function of $<n_s>$ is

$$I/255 = \sum_{n_0<i<n\max}(gi+b)\cdot P_{<n>}(i) + \sum_{i>n\max} p_{<n>}(i) + \sum_{i<n_0} b\cdot p_{<n>}(i) \tag{2}$$

where $n_{max} = \frac{1-b}{g}$ and $n_0 = \frac{-b}{g}$. Experimentally with a positive value of b its value can be gauged from regions without sample, provided that the relation between the read intensity and the ADC value is known. The value of g however may require information on the total beam current.

2.2 The relevance of angular detector efficiency in comparative experiments

Whereas the evaluation of the detector efficiency has been a strong progress in quantitative analysis, this is not strictly necessary for a quantitative analysis when relative methods of quantification are used based on the comparison between nearby regions. The justification is given by the assumption that $I=M(t,x)\sigma$ [6]. Here M contains the thickness dependence and σ is the cross section. The ratio of two intensities is therefore

$$R = \frac{I_A}{I_B} = \frac{M_A\sigma_A}{M_B\sigma_B} \tag{3}$$

As the dependence on the detection angle is only contained in the cross-section, this can be written as

$$R \propto \frac{\int g(s) f_A^2(s)(1-e^{-M_A s^2})ds}{\int g(s) f_B^2(s)(1-e^{-M_B s^2})ds} \approx \frac{Z_A^\alpha \int g(s) f^2(s)ds}{Z_B^\alpha \int g(s) f^2(s)ds} \tag{4}$$

where g is the angular efficiency of the detector, s is the in plane momentum and $f_{a,b}$ are the atomic scattering factors of atoms of type A,B. The simplification written here is valid for sufficiently high angles so that the exponential can be neglected. Moreover it has been assumed that the angular dependence of $f(s)$ on angle is unscreened Mott-Rutherford and therefore independent of Z. For sake of comparison the chemical contrast between InGaP and GaAs has been calculated for an annular detector starting in the range 75-210 mrad (corresponding to a camera length of 8 cm in our JEOL

2200 FS) with a flat and a slowly varying efficiency (fig. 2a). Simulations have been performed with an updated version of STEM_CELL [7] parallel computer software for Frozen Lattice multislice. It permits to store data for different angular sectors and add detection efficiency to the simulation post processing. The plot of the two almost superimposed curves is shown in fig. 2b. The additional curve is their ratio. The error in R is at most 3% at low thickness but otherwise about 0.3%. This explains the success of quantitative ADF methods not accounting for varying detection efficiency [1, 8].

2.3 Estimation of the thickness from comparative experiments

While it is not necessary for relative compositional HAADF analysis, the calibration of the detector efficiency may be useful to evaluate specimen thickness [4]. An alternative can be adopted in samples with strong thickness gradient and using a small outer detector angle (in this case 46-128 mrad). In such conditions it's easy to find a thickness for which the ADF intensity of relatively heavy materials (i.e. GaAs) reaches a maximum and the starts to decrease. According to simulations in fig. 2c this occurs at a thickness of about 450nm but in the case of gold this would be much less.

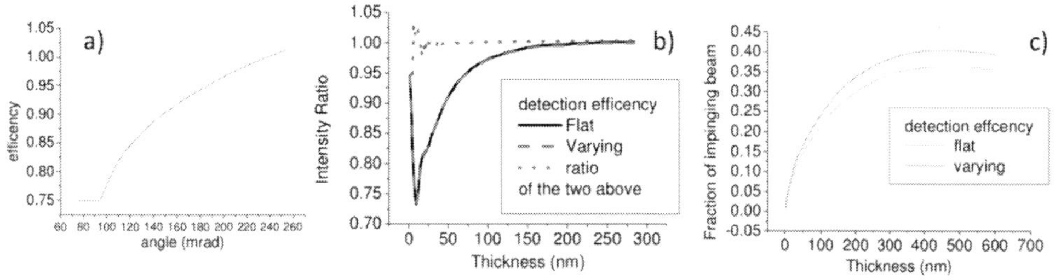

Figure 2. a) A hypothetic efficiency profile for a detector. b) intensity ratio between InGaP and GaAs for a detection efficiency as in fig. 2a compared to the flat case. The ratio between the two curves is also shown. c) intensity (normalized to impinging beam) for GaAs vs thickness with efficiency detection as in (a) and flat.

This critical thickness can be determined by simulations and used to gauge the ratio in regions with lower intensity. This is not an absolute intensity measurement of the beam, yet it can be useful to measure the thickness at any position in GaAs. In fact, the two curves in fig. 2b obtained for different angular efficiencies of the detector produce similar shapes and only differ by a constant factor. Therefore the ratio between the maximum intensity and the actual value at low thickness is largely independent of the detector response. This ratio can be used to gauge the specimen thickness.

3. Experimental

STEM experiments have been performed on a JEOL 2200 FS equipped with a Schottky FEG and working at 200keV. The spherical aberration of the objective lens was 0.5 mm. A nominal 0.7 nm spot was used to achieve high beam currents. Image acquisition has been performed by DIGISCAN2 attached to the lower ADF detector. The sample was a GaAs substrate with epitaxial InGaP grown by MOCVD which will not be discussed in this work. For STEM observation the sample has been thinned by mechanical grinding and final polishing in a GATAN DUO MILL.

4. Results and discussion

Fig. 3a shows an ADF image of the GaAs sample where two gain settings are used. The solid line corresponds to the unsaturated image while the dashed line corresponds to the saturated image. The saturated profile has been normalized in order to have 1 on the y-axis in the saturated region; the solid profile has been scaled in order to be coincident with the dashed profile in the low thickness region. A point to point comparison of saturated and unsaturated profiles gives a saturation curve as in fig. 3b. The most striking result is that saturation occurs before the actual flat region. The reason is the relatively low count in the time sampled of electrons scattered to the high angle regime (in this case 115-322 mrad). In the present case a good fit is obtained with g=10/255 while b has been set close to 0 and therefore neglected.

This means that the value of $<n_s>$ corresponding to the saturation signal is 25 electrons for each 0.8 μs sampled. The thickness at the saturation point is estimated to be roughly 200 nm and the detection corresponds to 15% of the impinging beam. This would correspond to a beam current of about 0.04 nA, compatible with a Schottky source. It is clear that a signal of 25 electrons per sample has a large statistical variability (20%) so that, even at 20% distance from the saturation, the count distribution has tails in the saturated region; an intensity evaluated at 10% from the saturation limit gives an error of 13%, but even an apparently safe intensity of 40% [5] from the limit signal produces quite a low but still appreciable 0.8% error. This is therefore potentially a much larger source of error in HAADF quantitative analysis rather than the effect of a non-constant angular efficiency evaluation.

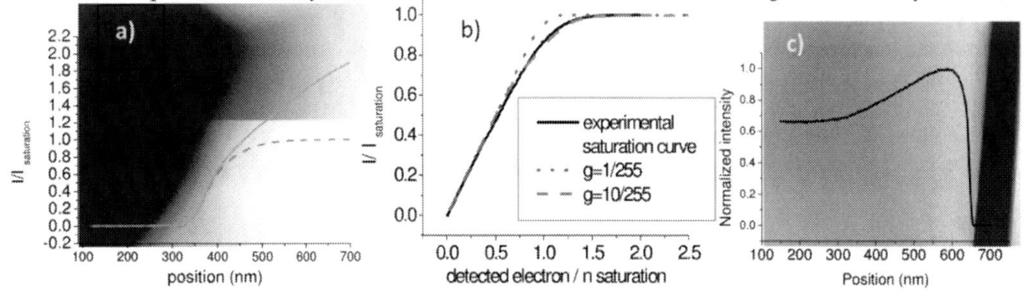

Figure 3. a) Split ADF image of the GaAs sample where two gain settings were used. The intensity profiles from the same area are also shown as solid and dash lines. The two curves are normalized in order to be coincident in the low thickness region. b) The saturation curve deduced from profiles in (a) and compared with simulated curves according to eq. 2 for two values of the gain, *g*.

Finally, an experiment of comparative thickness measurement is sketched in fig. 3c where an ADF image (46-128 mrad) of a GaAs sample is shown. An intensity profile normalized to its maximum is also shown superimposed. Both show that, after reaching a maximum, the intensity decreases further from the edge. HAADF measurements (115-322 mrad) demonstrate that the thickness is continuously increasing so this maximum should corresponds to a thickness of about 450 nm as discussed in fig. 2c. This appears consistent with preliminary thickness measurements by EELS.

5. Conclusions

In conclusion, I have demonstrated, with simulations and experiments, that relative methods based on the comparison of intensities in different regions can be used for quantitative chemical analysis and thickness determination even if the detailed angular efficiency of the detector is unknown. The error involved in the non-flat response of the detector is potentially lower than the non-linearity errors caused by high gain settings. This is particularly the case for very high angle ADF without a Cs corrector on the probe where the number of electrons collected is lower and fluctuations are large. Care should be taken therefore to avoid subtle non-linearity even when saturation is not evident.

References

[1] Grillo V, Carlino E and Glas F 2008 *Phys. Rev. B* **77** 054103
[2] LeBeau J M and Stemmer S 2008 *Ultramicroscopy* **108** 1653–1658
[3] http ://www.gatan.com
[4] Rosenauer A et al. 2009 *Ultramicroscopy* **109** 1171.
[5] E H, Nellist PD, Lozano-Perez S and Ozkaya D 2010 *J. Phys.: Conference Series* **241** 012067
[6] Carlino E and Grillo V 2005 *Phys. Rev. B* **71** 235303.
[7] Carlino E, Grillo V and Palazzari P 2007 in: *Microscopy of Semiconducting Materials* (edited by A. Cullis and P. Midgley, IOP Conf. Proc. Institute of Physics, Bristol)
[8] Walther T and Humphreys CJ 1997 *Inst. Phys. Conf. Ser.* **153** 303

Quantification of series of X-ray spectra taken at different tilts in analytical transmission electron microscopy

MC Parri and T Walther

Department of Electronic and Electrical Engineering, University of Sheffield, Mappin Street, Sheffield S1 3JD, UK

m.parri@sheffield.ac.uk

Abstract. In 2007, it was suggested that by varying the take-off angle between repeated measurements of the same region in energy-dispersive X-ray spectroscopy (EDXS) in a transmission electron microscope (TEM) a significant improvement of the accuracy of X-ray quantification could be achieved. Simultaneously, an analytical expression of the depth within the thin foil specimen of the feature analysed was derived, which in turn could be used to better quantify the material of interest. Simulated results for the absorption of a variety of X-rays in different semiconductor substrates using different Monte Carlo programs will be presented, specifically regarding a model which corrects for tilting the specimen toward or away from the detector: a common method of varying the take-off angle. A set of experimental EDX spectra of capped InAs quantum wells or dots taken in plan-view geometry by tilting the specimen holder will be used for comparison of experimental with the simulated data.

1. Explanation of principle

An accurate quantification of a sample from EDX data can be heavily dependent on absorption correction, particularly for the softer (i.e. lower energy) X-rays. This correction is typically achieved by estimating the thickness and density of the material between the source of these X-rays and the detector. By comparing the detected intensities of various X-ray lines to the expected intensity, one can attempt to deduce the reduction in intensity caused by absorption and hence correct for it. However, this approach is not very accurate and often leads to substantial error margins.

A method for calculating the depth of a layer was presented in 2007[1]: one takes multiple readings of the same area under the same conditions except that the take-off angle is varied. This introduces an artificial change in the effective depth of the layer, allowing one to deduce the thickness of the intervening material and therefore quantify the material with potentially greater accuracy.

The correct relationship for calculating this depth, d, is given as equation (1). λ is the attenuation wavelength, I is the detected intensity and θ is the take-off angle: λ is a constant value that is a reflection of the photoabsorption cross-section at the X-ray energy of the corresponding overlayer material at the relevant energy. Since the decay length λ is a constant in an exponential function, it is immediately apparent that accurate knowledge of its value is necessary, as a small deviation from the true value will give a considerably greater percentage difference in the value of d. Therefore, a great portion of this report will be dedicated to its calculation.

$$d = \frac{\lambda \ln\left(\frac{I_1}{I_2}\right)\left(\cos(\theta_1 - \theta_2) - \cos(\theta_1 + \theta_2)\right)}{4\sin\left(\frac{\theta_1 - \theta_2}{2}\right)\cos\left(\frac{\theta_1 + \theta_2}{2}\right)} \quad (1)$$

2. Methods of calculating the attenuation wavelength

Briefly, the characteristic absorption wavelength is calculated by acquiring or simulating a logarithmic plot of intensity vs. depth of overlayer and applying linear regression to the result. However, creating a sample for each data point, each with different overlayer thickness, is resource intensive and hence not feasible experimentally.

This technique, 'depth variation', as sketched in figure 1, must also take into account beam straggling, as this causes the quantity of generated X-rays to increase the deeper the layer of interest. An accurate result requires either correction for this changing initial intensity (which is feasible in simulations but impossible in a practical experiment) or a high enough beam energy in practice such that the decrease in beam energy in the specimen foil is inconsequential.

By considering the conditions under which characteristic absorption wavelength can be calculated (i.e. change in depth and ergo X-ray intensity) it is clear that by moving the detector (and changing the effective depth) an entire series can be acquired from a single sample. This method, 'detector tilt', figure 2, is simply the realisation of the theoretical approach, and as such does not require a stopping power correction.

Unfortunately, many microscopes do not allow for the X-ray detector to be moved: in these changing the take-off angle is done by tilting the specimen, which itself presents a problem. As the sample tilts from the horizontal, the path taken by the downwards-travelling electrons will pass through a greater quantity of the material, thereby exciting more X-rays. Given that X-rays are produced in roughly linear proportion to the thickness of the material travelled through, trigonometry dictates that the intensity can be reduced to the equivalent of the horizontal set-up while retaining the increased effective depth presented by the specimen tilt. Hence, multiplying the detected intensity by the cosine of the tilt angle of the specimen from the horizontal causes the system to appear as the 'detector tilt' example, allowing the same equation to be used. This can be seen in figure 3.

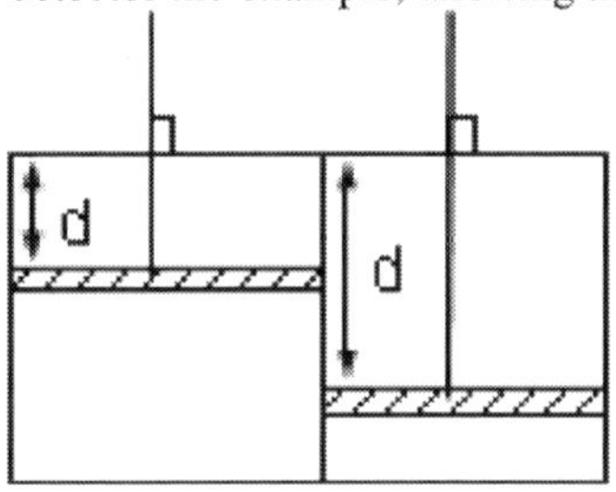

Figure 1: Depth variation method. Notice $\theta = 90°$ take-off angle.

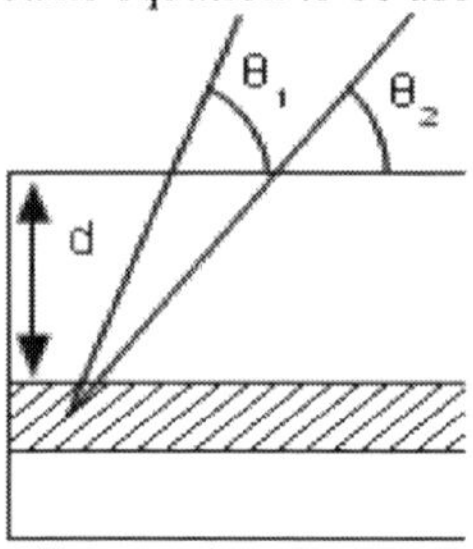

Figure 2: Detector tilt method

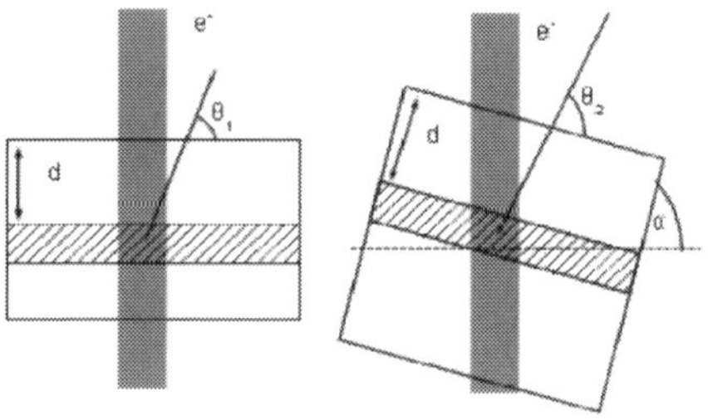

Figure 3: Specimen tilt method

To demonstrate the facility of the methods described in section 2, a series of simulations were run in NISTMonte[2] where the geometry was manipulated so as to model the three situations. All three simulations were designed such that they were functionally identical. In summary, the samples consisted of 140nm of GaAs (density 5.316 g/cm^{-3}), within which was embedded 5nm of InAs (density 5.67 g/cm^{-3}), with a beam energy of 200keV. In the first ('depth variation') the take- off angle was held at 90° and the depth of the layer varied through d=125.6, 95.0, 76.9, 65, 56.7, 50.6, 46.0 and 42.4 nm. The detector tilt method used the InAs layer at d=30nm deep and the detector was moved to 15, 20, 25, 30, 35, 40, 45 and 50°: at 30nm and 45°, this gives the same effective depth as the first method. For the specimen tilt, the layer was again held at 30nm depth but the specimen tilted through θ=−10, −5, 0, 5, 10, 15, 20 and 25°, where negative sign means away from the detector: given a typical take-off angle of θ=25° this is again equivalent to the first method.

The results of these simulations for the In L-line in GaAs are shown in table 1. It is clear that the depth variable method is far superior in terms of statistical spread of the simulation. Unfortunately, as already mentioned, this method is impractical for experimental work, since neglecting the beam straggling correction would, in this case, throw off the result by up to 9%.

	λ_{InL} [nm]
variable depth	2768.5±1.1
detector tilt	2784.7±11.0
specimen tilt	2735.8±44.9

Table 1: Method comparison

It is worth noting that applying the depth variable correction to the detector tilt method (i.e. dividing detected intensity by generated intensity) improves the error by an order of magnitude, to a similar value as that of the variable depth method.

The specimen tilt method has the worst error, likely due to the correction for the cosine of the tilt angle. Normalising the detected intensity with respect to the generated intensity does not improve the error here. For the specimen tilt, the linear fit was significantly worse than for the other two methods, implying that either the situation causes more data scatter or that the simulation had some minor flaw.

3. Comparison of programs

Table 2 shows the comparison of λ between different simulation programs, for increasing energy of the X-ray lines. The 'depth variation' technique was used, with I/I_0 correction applied, since (as seen above) this has the greatest accuracy of the three methods. It is immediately obvious that the differences are minor for the lower energies but become large for the higher energies, well beyond the stated error. This is due to systematic differences in the models used by the various programs.

Edax Electron Flight Simulator (EEFS) has a low scatter, implying good self-consistency, and yet displays lower calculated values for most lines than the other programs. NISTMonte displays even lower scatter (due to the quality of the linear fit and rounding errors), but shows higher values for all lines. The average value for the In (L) line, 2526nm, was used for the calculations in section 4, since this was the most reasonable value available, although the standard deviation of ±179nm would suggest an error in In_L intensity of approximately 2% at 200nm GaAs thickness.

Line	NISTMonte	EEFS[3]	Hurricane[4]	Casino[5]
N (K)	144.2±0.1	134.43±0.1	140±3	138±4
Al (K)	365.4±0.1	344.62±0.2	342±1	354±1
Si (K)	502.5±0.1	418.95±0.3	468±10	468±1
P (K)	725.7±0.1	625.88±0.5	667±10	668±4
In (L)	2770.2±0.2	2392.64±8.7	2390±32	2551±2
Sb (L)	3627.3±0.3	3064.53±11.7	3007±42	3334±4

Table 2: Decay lengths λ (nm) for GaAs coverage. Error bars give consistency with exponential function using numerical output from the four different programs.

4. Simulated Data

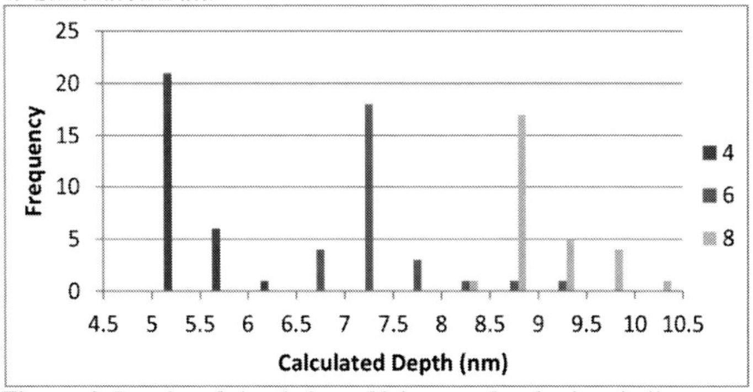

Figure 4: Results of simulations. Values on the abscissa indicate lower and upper limits of the histogram channels. Note the correct results for depths are the nominal thicknesses of the cap plus ~2nm.

In Figure 4 the results of three NISTMonte simulations are shown. 4nm of InAs was simulated with a GaAs capping layer of 4nm, 6nm and 8nm thickness. These simulations ignore bremsstrahlung radiation, detector sensitivity and noise. In a real sample this thin, one would expect considerable counting noise.

The data of Figure 4 serve to illustrate that having different thicknesses can have strong effects, even with such thin material. This conclusion does not hold well for experimental data, as overcoming counting errors must either be achieved through considerable tilts, limiting the number of available data points, or long lifetimes, which might cause beam damage during analysis.

5. Experimental data

In this section data from an experimental sample is shown. The sample is an InAs quantum dot layer on GaAs and has been capped by nominally 6nm GaAs.

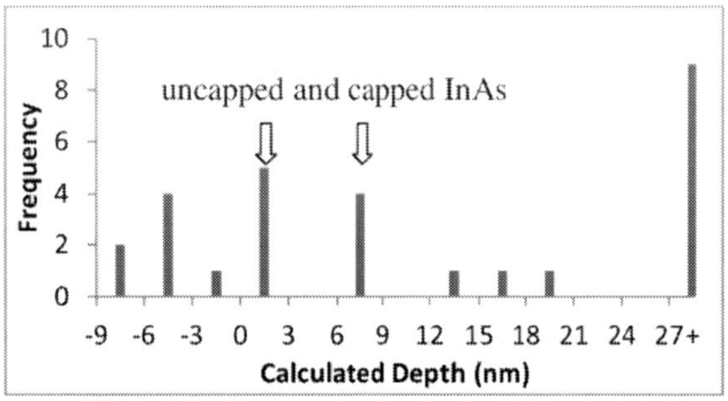

Figure 5: Histogram of calculated depths.

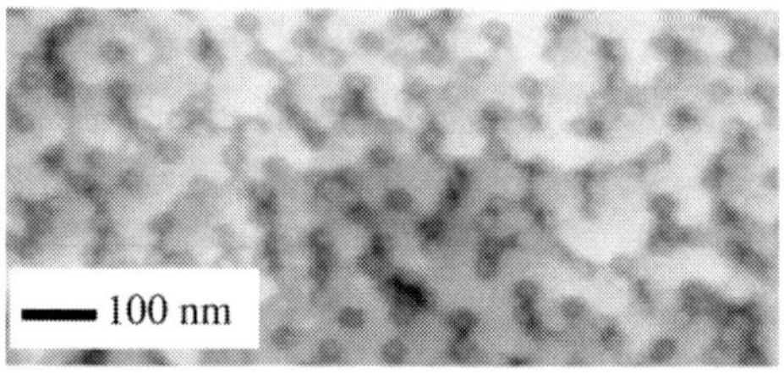

Figure 6: AFM topograph [6]. Notice that some of the islands are not capped and the irregular thickness of the cap layer.

There was considerable scatter in the intensity, possibly caused by holes in the capping layer. Despite this, applying equation (1) to each data point with respect to every other data point gives some results that appear accurate, in addition to many other results which do not, as seen in figure 5. In this sample, it was discovered via atomic force microscopy (AFM) that islands had formed, and that the depth beneath the cap was not a fixed value in this case, however, the mean value may be approximated by the nominal value of the cap (~6nm) plus half of the InAs thickness (~2nm), i.e. ~8nm. This explains the two peaks indicated by arrows in figure 5.

Those depths that are calculated as negative are the result of I_1 being greater than I_2 in equation (1), i.e. the intensity decreased as the sample was tilted towards the detector. This result is inconsistent with theory and may be due to parallax effects by the sample not being at the correct eucentric height in the pole-piece anymore when the holder is tilted towards the maximum tilt angles. Negative results are unphysical and result either from sample inconsistency or large relative error in I (in this case, low signal to noise ratio).

There are four results which sit within the correct channel (i.e. 6 to 9), with none on either side. Data at channel 27+ cover a large range (37.5 to 409.4) without any apparent clustering and seem to be artifacts due to noise that has a particularly strong effect at low take-off angles.

6. Summary

Three different methods of calculating the attenuation wavelength were described and simulated data used to support their similarity. Four different Monte Carlo programs were used to calculate the attenuation wavelength and their systematic differences compared. An experimental data set was used and compared with simulations to test if the method could work experimentally.

References

[1] Walther T, 2008, Proc. EMAG 2007, Glasgow, *J. Phys: Conf. Ser.* **126**, 012090
[2] Ritchie N W M, 2005, *Surf. Interface Anal.* **37**, 1006
[3] SmallWorld, 2011, LLC: http://www.small-world.net/efs.htm
[4] Rickerby DG and Thiot JF, 1994 *Mikrochmica* Acta **114**, 421
[5] Hovington P, Drouin D, Gauvin R., 1997 *Scanning* **19**, 1
[6] Liew SL, Walther T, Irsen S, Hopkinson M, Skolnick MS and Cullis AG, 2008, Proc. Microscopy of Semiconducting Materials 2007, Cambridge, *Springer Proc. in Physics* **120**, 259

Investigation of optical and concentration profile changes of InGaNAs/GaAs heterostructures induced by thermal annealing

Robert Imlau[1], Knut Müller[1], Oleg Rubel[2], Rafael Fritz[2], Marco Schowalter[1], Kerstin Volz[2] and Andreas Rosenauer[1]

[1] University Bremen, Otto-Hahn-Allee 1, 28359 Bremen, Germany
[2] Philipps University Marburg, Hans-Meerwein-Straße, 35032 Marburg, Germany

E-mail: rimlau@ifp.uni-bremen.de

Abstract. In this contribution we compare optical and structural properties of $In_{0.2}Ga_{0.8}N_{0.024}As_{0.976}$ quantum wells before and after annealing at 550 °C and 600 °C in an N_2 atmosphere. We measure strain and chemically sensitive contrast to determine local indium and nitrogen concentrations using a TEM 3-beam image formed by the 000, 220 and 200 beams. For quantification of the concentrations Bloch wave simulations are used, which include bonding and static atomic displacements. The samples were grown by metal-organic vapour phase epitaxy, prepared with focused ion beams (FIB), thinned with low energy ion milling and investigated in a Cs-corrected Titan 80/300 microscope using an L-shaped objective aperture. Imaging conditions with the Laue circle centre at (0 4.2 0) are used as they show a weak dependence on the lamella thickness.

Absorption measurements show a blueshift of the band gap of 19 ± 7 meV (550 °C) and 36 ± 7 meV (600 °C) after annealing. The average nitrogen concentration was found to be 2 ± 1 % and is unaffected by the annealing temperature. In contrast, the mean indium concentration appears to decrease from 18.5 ± 2 % before to 15 ± 1 % after annealing. Together with the blueshift, this observation is discussed in terms of a modification of electron structure factors, caused by preferred coordination of N to In atoms.

1. Introduction

Introducing even small amounts of nitrogen into an InGaAs crystal reduces band gap and lattice parameter, allowing to tune both of them [1, 2]. Hence, $In_xGa_{1-x}N_yAs_{1-y}$ is an interesting material for optoelectronic devices emitting at the telecommunication wavelengths of 1.3 μm and 1.55 μm [3, 4]. Post-growth annealing leads to an increase of the originally low luminescence intensity of these devices but also to a band gap blueshift [5, 6]. The blueshift can reach up to 100 meV depending on indium content, growth mode and the annealing conditions temperature, time and atmosphere. To explain the blueshift of $In_xGa_{1-x}N_yAs_{1-y}$ during annealing, structural properties were investigated. Some authors found evidence for an increase of preferred indium-nitrogen bonding [5, 7, 8], while others found evidence for long-range redistribution of nitrogen or indium [6, 9]. So there might be different mechanisms contributing to the band gap increase.

To investigate these structural changes, atomic-scale nitrogen and indium content has to be detected with highest possible accuracy. To this end, two sources of information are needed.

Grillo et al. developed a TEM-method based on measuring strain and the amplitude of the chemically sensitive (200) reflection of a single well, using a zone axis high-resolution and an off-zone dark field micrograph [10] under kinematic approximation, whereby imaging of exactly the same specimen area is difficult.

Another method was presented by Müller et al. [9] recently, using the 200, 220 and 000 beams to determine strain and 200 reflection amplitude simultaneously from a single TEM image. Strain and 200 amplitude provide the two pieces of information needed to determine the concentration of indium and nitrogen. This method is used here to investigate the reason for the blueshift observed after annealing a metal-organic vapour phase epitaxy (MOVPE) grown $In_{0.2}Ga_{0.8}N_{0.024}As_{0.976}$ quantum well.

2. Experimental Setup

A 15 nm thick quantum well with nominal composition $In_{0.2}Ga_{0.8}N_{0.024}As_{0.976}$ was grown with MOVPE at 520 °C. The sample was annealed under N_2 atmosphere at 550 °C for 10 minutes and afterwards for another 10 minutes at 600 °C. For each annealing step, optical experiments were performed using a JASCO V-670 spectrophotometer. The cross-sectional TEM specimens have been prepared in zone axis [001] using an FEI focused ion-beam machine Nova Nanolab 200. The specimens were thinned by argon ion milling at 400 eV using a Technoorg Linda gentle mill in order to remove the amorphous layer and to reduce the thickness. Before thinning the specimen thickness was measured with an accuracy of ± 15 nm using STEM as described in [11]. The thinning time was chosen in order to reach the same thickness for each lamella of about 20 nm. All (S)TEM measurements were performed with an FEI Titan 80/300 equipped with a corrector for the spherical aberration of the objective lens, operated at 300 kV. The TEM images were recorded on a 2048 × 2048 pixel CCD camera.

3. Optical properties before and after annealing

In order to determine the band gap of the quantum well, absorption spectra are evaluated in the range of 0.75 eV to 1.5 eV photon energy. The absorption $A(E)$ is calculated from the measured transmission $T(E)$ and the determined reflectance $R(E)$, via $A(E) = 1 - T(E) - R(E)$. In the investigated range, the reflectance depends almost linearly on the photon energies [12], giving $R(E) = m \cdot E + b$. Below the band gap of the investigated wells, no absorption occurs and only transmission and reflection take place. Here, the assumed linear reflectance is determined.

Figure 1 shows the obtained absorption spectra for each annealing step. Energies below 0.85 eV do not show absorption. Towards higher energies the absorption increases exponentially, caused by the impurity band. For higher energies the absorption shows a square-root-like behavior caused by the combined density of states of the valence and conduction bands. The square root functions fitted to these energies give the band gaps. The band gap is 944 ± 5 meV before annealing. After annealing at 550 °C and 600 °C blueshifts of 19 ± 7 meV and 36 ± 7 meV occur.

4. Structural properties before and after annealing

Before discussing the structural properties, we outline the used method shortly. See [9] for more details. An L-shaped objective aperture is used to select only the 200, 220 and 000 beams. Fourier filtering the 3-beam image gives the 220 fringe and the 200 amplitude image as depicted in figure 2, which are used to measure the strain and the 200 amplitude locally. To determine the indium and nitrogen concentration maps from the locally measured strain and the 200 fringe amplitude, reference tables were used as depicted in figure 3. To calculate these, elasticity theory was applied and Bloch wave simulations were performed [13]. The structure factors used account for static atomic displacements (SADs) via empirical potentials [14]. Furthermore, bonding is included in form of modified atomic scattering amplitudes (MASAs) [15]. In order to achieve a

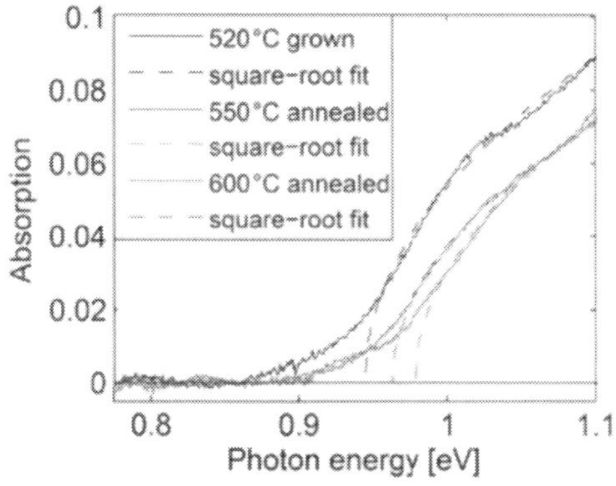

Figure 1. Absorption spectra before and after thermal annealing at 550 °C and 600 °C.

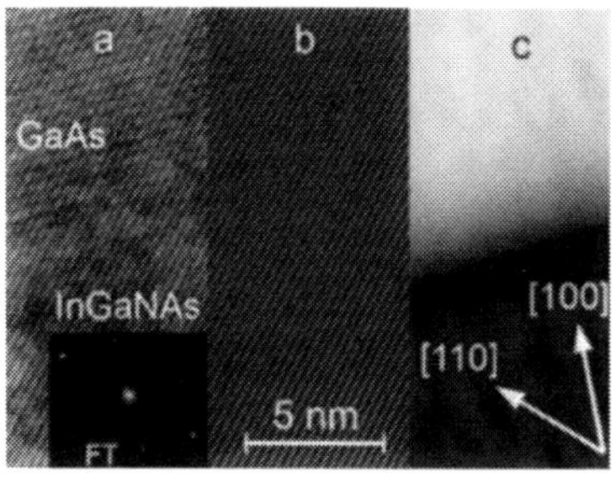

Figure 2. (a) Experimental 3-beam image of the $In_{0.2}GaN_{0.024}As/GaAs$ well before annealing acquired near zone axis [001] with a Laue circle center at (0 4.2 0). Filtering the diffractogramm shown in (a) as insert gives (b) 220 fringe image and (c) the chemically sensitive amplitude of the 200 beam.

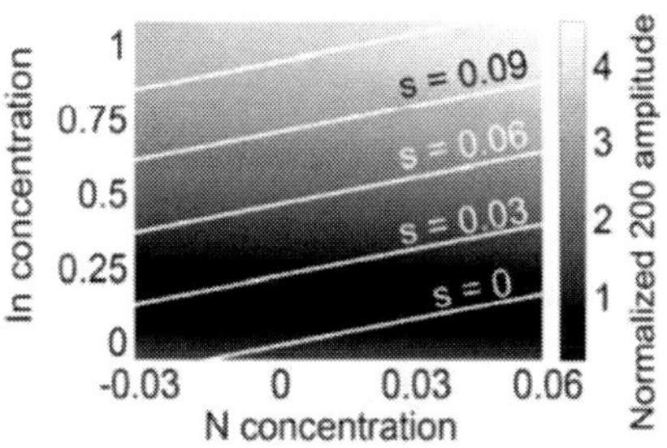

Figure 3. Reference data for the normalized 200 fringe amplitude and the normalized [220] fringe distance in [100] direction (white lines) as a function of indium and nitrogen concentration for a Laue circle centre of (0 4.2 0) and a thickness of 20 nm [13].

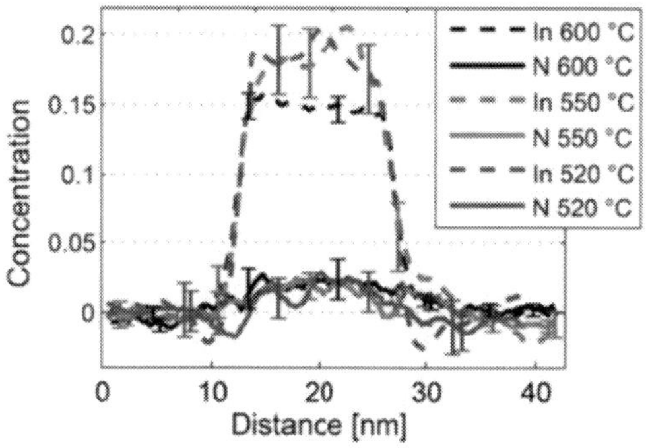

Figure 4. Indium and nitrogen concentration profiles before (520 °C) and after annealing at 550 °C and 600 °C.

weak dependency on the lamella thickness, imaging conditions were optimized [13]. By the use of a Laue circle center of (0 4.2 0) this dependency is minimized.

Figure 4 shows indium and nitrogen concentration profiles before and after annealing. Each profile is obtained by averaging the concentration maps parallelly to the well over a distance of 10 nm. With increasing annealing temperature, the average indium content decreases from $18.5 \pm 2\,\%$ to $15 \pm 1\,\%$ in the well, while the nitrogen content of $2 \pm 1\,\%$ is not measurably affected by the thermal annealing. The given precision accounts for inaccurately known specimen thickness, unknown relaxation in electron-beam direction and concentration fluctuations.

5. Discussion

Upon annealing a band gap blueshift of $36 \pm 7\,\text{meV}$ is observed, whereas the indium concentration seems to decrease by $3.5 \pm 2.5\,\%$. The blueshift could be explained by the diffusion of indium out of the InGaNAs quantum well. In case of a diffusion into the GaAs layers one would expect a broadening of the indium concentration which was not observed (see fig. 4). As there are no indications of indium diffusion after annealing, this does not seem to occur. The indium concentration decrease can be explained by an increasing number of In-N bonds, leading to a change of the 200 structure factors [8], and hence to a misassignment of the concentrations. A growing number of In-N bonds is also plausible because it minimizes the strain energy of the crystal [7] and it is discussed as a reason for the blueshift after thermal annealing [5].

To investigate the dependence of the 200 structure factor F on the number of In-N bonds, calculations were performed accounting for SADs as well as bonding via MASA. To account for SADs, super-cells with a size of 30 × 30 × 30 unit cells have been relaxed [14], where every nitrogen atom is surrounded only by gallium atoms (N-4 Ga), indium atoms (N-4 In), or by a random distribution according to the gallium/indium ratio. For the observed concentrations before annealing, we find a decrease of the 200 structure factor with increasing frequency of N-In bonds. Compared to F_{random}, $F_{\mathrm{N-4\,Ga}}$ and $F_{\mathrm{N-4\,In}}$ show an increase of 5 % and a decrease of 18 %, respectively. A redistribution towards preferred N-In coordination leads to an underestimation of the indium content, which was observed after annealing. Hence, the observed band gap blueshift is attributed to an increase of the N-In bonds.

6. Conclusion

The seeming decrease of indium concentration and the observed blueshift after thermal annealing are explained by preferred coordination of nitrogen and indium. If long range redistributions of indium or nitrogen had been present here, they would have been observed with the used 3-beam method [9].

7. Acknowledgements

This work was supported by the DFG under contracts RO2057/4 SCHO 1196/3 and VO 805/4,5.

References

[1] Kondow M, Kitatani T, Larson M C, Nakahara K, Uomi K and Inoue H 1998 *J. Cryst. Growth* **188** 255

[2] Bellaiche L 1999 *Applied Physics Letters* **75** 2578–2580

[3] Nakahara K, Kondow M, Kitatani T, Larson M C and K U 1998 *IEEE Phot. Technol. Lett.* **10** 487

[4] Fischer M, Reinhardt M and Forchel A 2000 *Electr. Lett.* **36** 1208

[5] Klar P J, Grünning H, Koch J, Schäfer S, Volz K, Heimbrodt W, Karmal Saadi A M, Lindsay A and O'Reilly E P 2001 *Phys. Rev. B* **64** 121203

[6] Albrecht M, Grillo V and Remmele T *et al. 2002 Appl. Phys. Lett.* **81** 2719

[7] Rubel O, Volz K, Torunski T, Baranovskii S D, Grosse F and Stolz W 2004 *Appl. Phys. Lett.* **85** 5908

[8] Volz K, Torunski T, Rubel O and Stolz W 2008 *J. Appl. Phys.* **104** 053504

[9] Müller K 2011 *J. Phys.: Conf. Ser. Submitted*

[10] Grillo V, Albrecht M, Remmele T, Struck H P, Egorov A Y and Reichert H 2001 *J. Appl. Phys.* **90** 3792

[11] Rosenauer A, Gries K, Mueller K, Pretorius A, Schowalter M, Avramescu A, Engl K and Lutgen S 2009 *Ultramicroscopy* **109** 1171

[12] Adachi S 1992 *Physical Properties of III-V Semiconductor compounds* (J. Wiley and S.)

[13] Müller K, Schowalter M, Rosenauer A, Rubel O and Volz K 2010 *Phys. Rev. B* **81** 075315

[14] Rubel O, Nemeth I, Stolz W and Volz K 2008 *Phys. Rev. B* **78** 075207

[15] Rosenauer A, Schowalter M, Glas F and Lamoen 2005 *Phys. Rev. B* **72** 085326

Nanoscale EELS analysis of elemental distribution and band-gap properties in AlGaN epitaxial layers

H Amari[1], HY Zhang[1], L Geelhaar[2], C Chèze[2], M J Kappers[3] and T Walther[1]

[1] Department of Electronic & Electrical Engineering, University of Sheffield, Sir Frederick Mappin Building, Mappin Street, Sheffield S1 3JD, UK
[2] Paul-Drude-Institut für Festkörperelektronik, 10117 Berlin, Germany
[3] Department of Materials Science & Metallurgy, University of Cambridge, Cambridge, CB2 3QZ, UK

E-mail: h.amari@sheffield.ac.uk

Abstract. Aluminium gallium nitride (AlGaN) is a strong candidate for high-power and high-temperature electronic devices and short-wavelength (visible and ultraviolet) optoelectronic devices. For band-gap engineering of nitride layers, it is essential to be able to perform an accurate local measurement of their optical properties. In this work, core-loss electron energy loss spectroscopy (EELS), plasmon spectroscopy and valence EELS (VEELS) are compared for the investigation of the local chemistry and band-gap of AlGaN.

1. Introduction

EELS is often applied to characterize the optical and chemical properties of wide band-gap nitride compounds. While the core-loss region of an EEL spectrum provides information about the elemental distribution and the crystal structure of the material, the low-loss region contains information about the chemical composition and electronic structure as well as the optical properties of the specimen. In this work, core-loss EELS, plasmon and valence EELS are compared for the investigation of AlGaN layer samples that cover a wide compositional range from Ga-rich ($Al_{0.23}Ga_{0.77}N$) to Al-rich ($Al_{0.87}Ga_{0.13}N$).

2. Sample description

Different nitride heterostructures were used in our experiments. In the Ga-rich sample ($Al_{0.23}Ga_{0.77}N$), a 140 nm GaN buffer layer was grown at 700 °C on top of (0001) Al_2O_3 by molecular beam epitaxy (MBE) followed by the $Al_{0.23}Ga_{0.77}N$ at 800 °C [1]. For the Al-rich samples a thick AlN buffer layer (~ 4 µm) was deposited on (0001) Al_2O_3 substrate by metal-organic chemical vapour deposition (MOCVD) followed by 2 µm $Al_xGa_{1-x}N$ layers (x = 0.44, 0.72, and 0.87) deposited at 1090 °C [2]. The Al and Ga ratios were investigated by Rutherford Back Scattering (RBS) or X-ray diffraction (XRD).

3. Experimental

TEM specimens were prepared by focused ion beam (FIB) thinning which has been used to obtain sample thicknesses in the range 50 – 700 nm. Analytical electron microscopy was carried out in a JEOL 2010F field-emission TEM/STEM microscope operated at 197 keV and equipped with annular dark field (ADF) and bright field (BF) detectors coupled with a Gatan Digiscan system. For electron energy loss spectroscopy (EELS) a Gatan imaging filter (GIF 2000) was used.

4. Results and discussion

4.1. STEM and EFTEM analyses

Fig. 1 shows BF images of the Ga-rich and one of the three Al-rich structures studied in this work. The FIB milling was employed for obtaining window regions of different thicknesses (t) as shown. In the Ga-rich sample, the 134 nm GaN and the 139 nm AlGaN layer appear dark because of their higher mean atomic number with respect to sapphire (substrate), and both show columnar grain growth. We also note the formation of an oxide on top of the AlGaN epilayer in the Ga-rich sample. Using energy-filtered transmission electron microscopy (EFTEM), the sample thickness (t) was measured locally in different positions along the whole structure, as shown in Table 1, from relative thickness maps using the inelastic mean free paths (λ) of the materials for a collection angle of β = 40 mrad, via the EELS routines of Gatan Digital Micrograph, employing Egerton's approximations for effective atomic number and mean energy loss [3]. Points presented along the growth direction in figure 1 indicate positions of thickness measurements obtained by EFTEM and reported in Table 1.

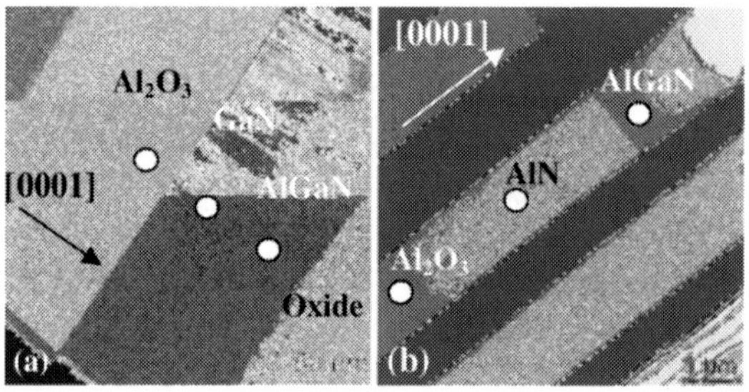

Figure 1. BF image of (a) $Al_{0.23}Ga_{0.77}N$/GaN/sapphire, and (b) $Al_{0.44}Ga_{0.56}N$/AlN/sapphire. Points indicate positions of thickness measurements by EFTEM reported in Table 1.

Table 1: Thicknesses measured by EFTEM.

sample	λ (nm)	t (nm)
$Al_{0.23}Ga_{0.77}N$	90	86
$Al_{0.44}Ga_{0.56}N$	93	64
$Al_{0.72}Ga_{0.28}N$	99	105
$Al_{0.87}Ga_{0.13}N$	103	103

4.2. Determination of local composition by EELS

To investigate the Al and Ga distribution in AlGaN layers, a series of EELS analyses in TEM mode were applied. All core-loss EEL spectra were acquired with an energy resolution of 0.9 eV (measured from full width at half maximum (FWHM) of the zero loss peak (ZLP)), exposure times of 1s, dispersion of 1eV/channel, in diffraction mode with a convergence and collection semi-angle of about 0.8 mrad and 12.5 mrad, respectively. Typical core loss spectra including Al-K and Ga-L edges of GaN, $Al_{0.44}Ga_{0.56}N$, and AlN layers are presented in figure 2. The atomic ratio of Al to Ga (x_{Al}/x_{Ga}) was calculated using the cross-sections, σ, from either Hartree-Slater or hydrogenic (plus white line) models in the Digital Micrograph quantification routine, based on the following formula:

$$\frac{x_{Al}}{x_{Ga}} = \frac{\sigma_{Ga}(\Delta,\beta)}{\sigma_{Al}(\Delta,\beta)} \times \frac{I_{Al}(\Delta,\beta)}{I_{Ga}(\Delta,\beta)} \quad (1)$$

where $\sigma_i(\Delta, \beta)$ are the cross-sections under the corresponding ionization edges for integration window size Δ and semi-collection angle β, $I_i(\Delta, \beta)$ are the intensities (i = Al, Ga).

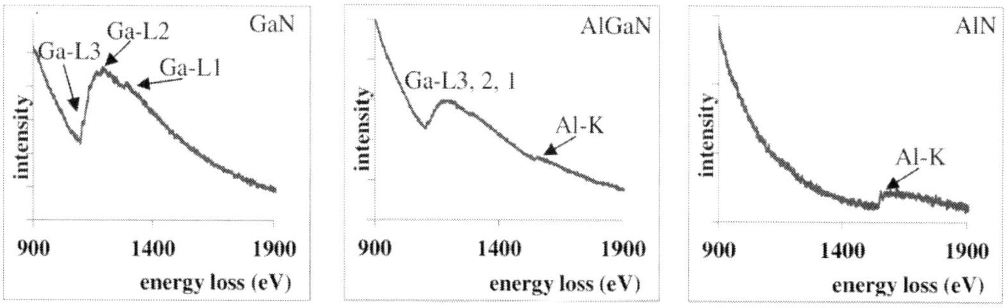

Figure 2. Core- loss spectra of GaN (left), $Al_{0.44}Ga_{0.56}N$ (middle) and AlN (right).

To investigate the influence of the integration window on the quantification, we quantified our layers using different integration windows covering the range 50-250 eV, using the same settings for the offset and the width of the background for each element (see Table 2). Because we didn't apply any de-convolution (the thickness of the samples varied from about t/λ ~0.6 to ~1), it was essential to calculate all Al/Ga ratios by using the same integration window for both edges [4]. As we can see from Fig. 3 (a) which presents results from the Hartree-Slater model, the apparent Al composition ratio from EELS ($x_{Al}/[x_{Ga} + x_{Al}]$; with $x_{Al} + x_{Ga} = 1$) depends strongly on the integration window size for lower Al-content while it appears stable at higher Al-content.

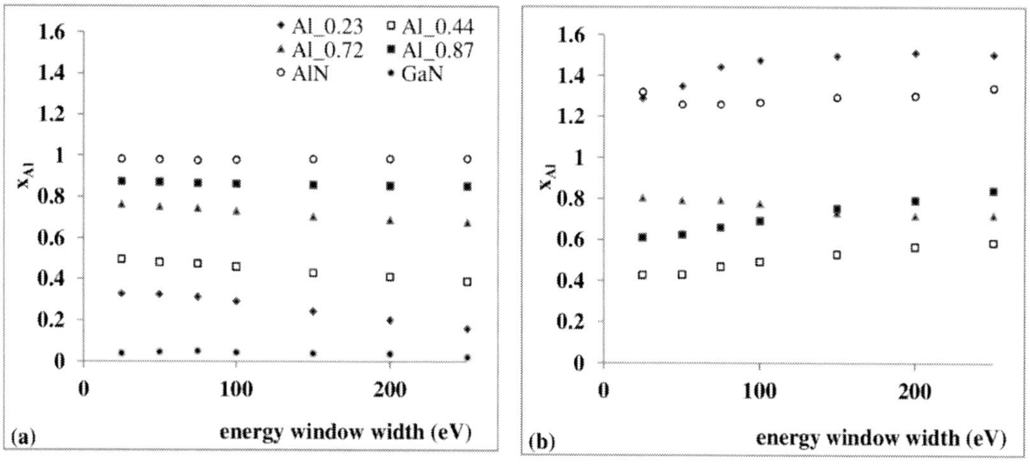

Figure 3. (a) Al and Ga ratio quantification using Hartree-Slater and (b) Al and N ratio using hydrogenic plus white line models, as a function of the width of the energy windows used for integrating the intensity after background subtraction. The latter quantification using the N K-edge produces apparent values > 1.

Table 2: Energy window settings used for elemental quantifications of the core-loss spectra.

	background fit		integration window	
element	offset (eV)	width (eV)	start (eV)	width (eV)
Al	-120	98	1560	50-250
Ga	-169	115	1115	50-250
N	-48	35	400	50-250

For the N edge, using a larger window for fitting the background gave inferior fits that crossed the spectrum at high energies, which we attribute to weak intensity oscillations in the loss region 300-350 eV due to carbon.

4.3. Determination of plasmon excitations

Low-loss spectra were collected using an energy dispersion of nominally 0.05eV/channel, assumed to be correct within a few %, and a typical acquisition time of 0.5s in order to limit the electron beam damage to the sample and also to limit energy drift. Fig. 4 depicts typical low-loss spectra from all the AlGaN samples showing two main characteristics: the main plasmon excitations at slightly different energies (E_p) and a peak superimposed at ~ 23 eV on the high-energy tail of the plasmon excitations for Ga-rich samples which is related probably to the transition from Ga 3d [5]. The experimental results shown in Fig. 4 demonstrate that within the compositional range of interest, E_p is sensitively depending on both features.

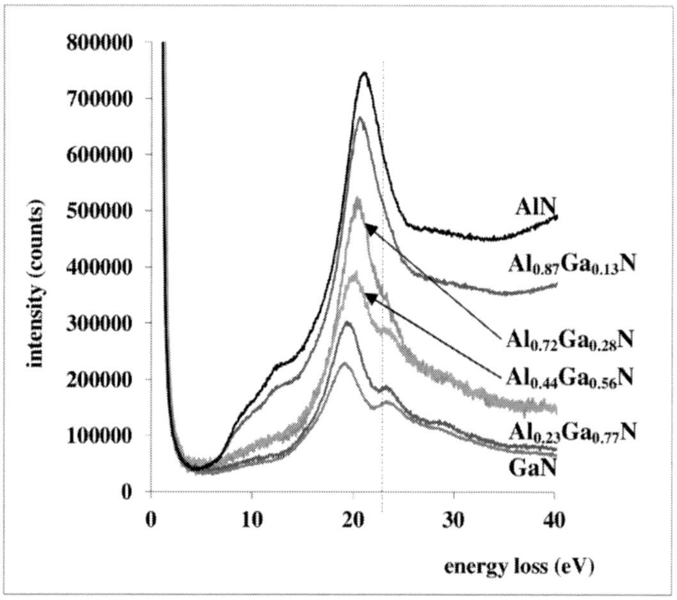

Figure 4. Comparison of experimental low-loss spectra of AlGaN.

Table 3: plasmon peak energies.

sample	E_p (eV)
GaN	19.15
$Al_{0.23}Ga_{0.77}N$	19.3
$Al_{0.44}Ga_{0.56}N$	20.25
$Al_{0.72}Ga_{0.28}N$	20.35
$Al_{0.87}Ga_{0.13}N$	20.55
AlN	21

Plotting plasmon excitation as a function of x_{Al} and performing a linear fit yields a 1.8eV shift with x_{Al}: E_p (eV) = (19.12 ± 0.16) + x (1.80 ± 0.25); with a linear correlation coefficient (R^2) = 0.9245.

4.4. Determination of band-gap

In addition to giving us information on plasmon excitations, the low-loss region of the spectra provides us with transitions across the band-gap. The main problems in looking at the band-gap directly are to separate the signal onset from the tail of the zero-loss peak and to take into consideration possible Cerenkov radiation. For the first problem, we believe that using a sufficiently high energy resolution (about 0.8 eV) and using the microscope with suitable experimental conditions (high energy dispersion of about 0.05eV/channel and an acquisition time of 0.5s) separating the onset of the band-gap from the zero loss background is possible. For the second problem, because of the small angular range of Cerenkov radiations (for the collection of significant Cerenkov contribution a small collection aperture of the order of ~0.1mrad is needed) [6], we believe that using a collection angle of about 12 mrad will limit the influence of these radiations, which for AlGaN with a refractive index of n = 2.15 (AlN)...2.3 (GaN) could occur for acceleration voltages above 60 kV, but have actually not been observed by us. Also, nitrides cover a band-gap energy from 3.4 eV (for GaN) to 6.2 eV (for AlN), and above ~ 3 eV no radiative effects have been reported to influence the spectrum for thick samples [7].

To reveal directly the onset of the band-gap, we refer to one of the optical principles of direct transitions in semiconductors, according to which the intensity (I) has a square-root dependence on the band gap (E_g) [8], so accordingly, I^2 would have a linear dependence on E_g. Thus, by plotting I^2 vs. energy-loss (E), as shown in Fig. 5, and extrapolating the linear region to zero, the value of the direct band-gap transition energy is obtained. We performed linear fits to the experimental spectra over energy ranges that started just above the maximum band-gap of AlN (E_g = 6.20 eV) [9] and varied the upper range of the fitting interval until we obtained the best linearity (i.e. 6.25 to 13, 14, 15 eV etc...) (see Table 4).

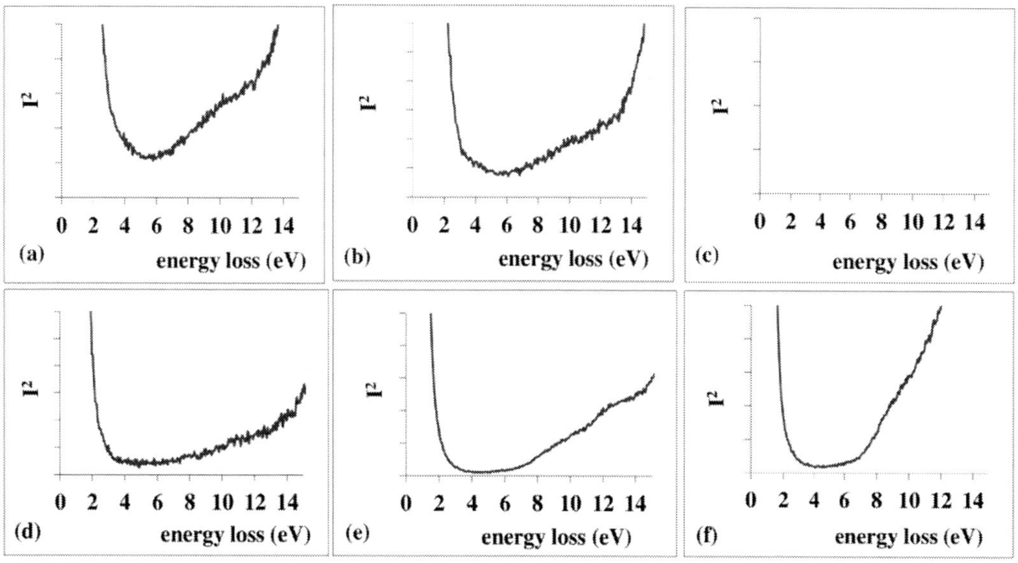

Figure 5. Comparison of the square of the intensity versus energy of (a) GaN, (b) $Al_{0.23}Ga_{0.77}N$, (c) $Al_{0.44}Ga_{0.56}N$, (d) $Al_{0.72}Ga_{0.28}N$, (e) $Al_{0.87}Ga_{0.13}N$, and (f) AlN.

As we can see from Table 4, there is a strong dependence of the band-gap on Al-content, which is in excellent agreement with theory. The best linearity was obtained almost for the same intervals (6.25 to ~14 eV) for most alloy compositions. Also, we note that the linear regression coefficient (R^2) gives a direct measure of the uncertainty in the extrapolated band-gap, ΔE_g, which varied from ± 0.1 eV (if R^2>0.99) to ± 0.3 eV (if $R^2 \approx 0.93$) (see Fig 6).

Table 4: Band-gap measurements of all AlGaN layers.

sample	fit interval	R^2	band-gap (eV)
GaN	[6.25-12]	0.9820	3.26 ± 0.15
$Al_{0.23}Ga_{0.77}N$	[6.25-13.5]	0.9709	3.67 ± 0.19
$Al_{0.44}Ga_{0.56}N$	[6.25-14]	0.9335	4.79 ± 0.32
$Al_{0.72}Ga_{0.28}N$	[6.25-14]	0.9744	5.05 ± 0.20
$Al_{0.87}Ga_{0.13}N$	[6.25-15]	0.9934	6.09 ± 0.10
AlN	[6.25-15]	0.9917	6.40 ± 0.11

Linear fit: E_g (eV) = (3.14 ± 0.20) + x (3.18 ± 0.31); with a linear correlation coefficient (R^2) = 0.9618.

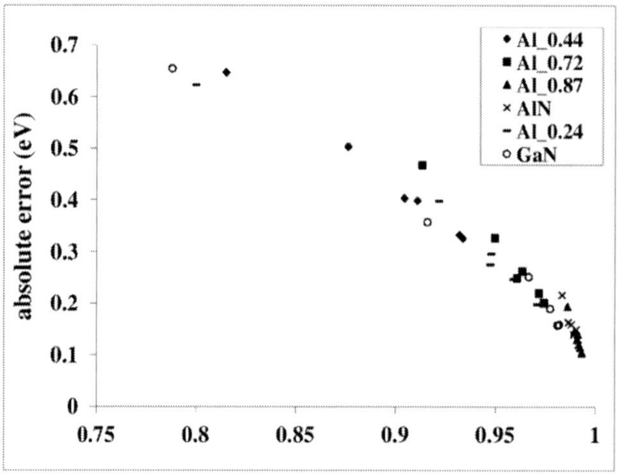

Figure 6. Uncertainty of extrapolated band-gap vs. fitting quality expressed by R^2.

5. Conclusion

It can be concluded that EELS, at least under the experimental conditions described in this paper, gives direct access to the optical properties of AlGaN layers. Calculations of the band-gap using plots of the square of the intensity vs. energy-loss have shown good agreement with theory. Reproducible systematic changes have been observed in the plasmon excitations and the band-gap from $Al_xGa_{1-x}N$ as a function of the Al-content. H. Amari acknowledges funding by EPSRC under grant EP/F02374X/1.

References

[1] Riechert H *et al.* 1997, *Mater. Res. Soc. Symp. Proc.* **449**, pp. 149
[2] Chang TY *et al. 2010*, *J. Appl. Phys.* **108**, 123522
[3] Egerton RF: *EELS in the Electron Microscope*, Plenum (1996, New York) p. 304
[4] Walther T and Mader W 1999, Microsc. Semicond. Mater. Conf., Oxford, *Inst. Phys. Conf. Ser.* **164**, 121
[5] Gass M *et al.*, 2004, *Appl. Phys. Lett.* **84**, 1453
[6] Gu L *et al.*, 2007, *Phys. Rev. B* **75**, 195214
[7] Festenberg CV, 1969, *Z. Physik* **227**, 453
[8] Schubert EF, *Light-Emitting Diodes*, 2nd edition: Cambridge University Press, 2006
[9] Zhang X *et al.* 1995, *Appl. Phys. Lett.* **67** 1745

Journal of Physics: Conference Series **326** (2011) 012040 doi:10.1088/1742-6596/326/1/012040

2D-composition mapping in InGaN without electron beam induced clustering of indium by STEM HAADF Z-contrast imaging

Andreas Rosenauer[1], Thorsten Mehrtens[1], Knut Müller[1], Katharina Gries[1], Marco Schowalter[1], Stephanie Bley[1], Parlapalli Venkata Satyam[1], Adrian Avramescu[2], Karl Engl[2], Stephan Lutgen[2]

[1] University of Bremen, Otto-Hahn-Allee 1, D-28359 Bremen, Germany
[2] OSRAM Opto Semiconductors GmbH, D-93055 Regensburg, Germany

E-mail: `rosenauer@ifp.uni-bremen.de`

Abstract. Investigation of composition in InGaN quantum wells and quantum dots by TEM is hampered by formation of electron beam induced agglomeration of indium, which occurs if the specimen is exposed to the electron beam for a few minutes. In this contribution we demonstrate that compositional analysis of InGaN nanostructures is possible without this artifact if STEM Z-contrast imaging is applied instead of parallel beam illumination. The suggested method for composition analysis in InGaN is based on a comparison of intensity normalized with respect to the incident electron beam with simulated image intensity. Simulations are performed with the STEMsim program using the frozen lattice multislice approximation for which static atomic displacements were taken into account.

1. Introduction

In scanning transmission electron microscopy (STEM) using a high angle annular dark field (HAADF) detector, image intensity depends on atomic numbers of scattering atoms, which allows for simple qualitative interpretation of composition distribution. Quantitative measurement of composition is possible if the measured image intensity is compared with accurate simulations computed with the frozen lattice approximation, as shown by Grillo et al. [1]. In reference [1], relative image intensity was compared with simulation and the specimen thickness was obtained by the projection method. These authors also pointed out that taking into account the static atomic displacements that occur if atoms with different covalent radii occupy the same sublattice of a sphalerite type crystal, improves the accuracy of composition determination significantly. Recently, LeBeau and Stemmer suggested use of image intensity normalized with respect to the incident electron beam for quantitative comparison with simulation [2] using external signal amplifiers. Conditions, for which the internal signal amplifiers of a TITAN 80/300 TEM equipped with a Fischione Model 300 HAADF detector can be used, were investigated in reference [3]. In this publication it was also shown that comparison of experimental and simulated image intensities allows measurement of specimen thickness in GaN and evaluation of composition in AlGaN/GaN using line scans of images obtained at low magnification. Here we report on 2D-composition mapping in InGaN. This material is sensitive to electron beam irradiation as it leads to clustering of indium after a few minutes in the TEM using the parallel beam illumination mode [4, 5]. In contrast, we show here that InGaN is much more stable during STEM investigation under usual imaging conditions.

2. Experimental conditions

Our test sample was grown by metal organic vapour phase epitaxy (MOVPE) at temperatures in a range of 700 °C to 1100 °C. The process parameters were carefully chosen in order to minimize the indium fluctuations. The sample contains InGaN layers with a maximum In concentration of approximately 25 %. STEM investigations were performed with a TITAN 80/300 TEM equipped with a Fischione Model 300 HAADF detector. The imaging parameters were as follows: Acceleration voltage 300 kV, spherical aberration constant of the probe forming lens $C_S = 1.2$ mm, radius of the condenser aperture 9 mrad, camera length 196 mm, HAADF detector angular range 33 to 200 mrad, gun lens 6, spot size 9, extraction voltage 4.5 kV, emission current 47 μA and 9.5×10^{-11}A probe current. Brightness and contrast were adjusted as described in reference [3]. For these settings of the illumination, the contrast transfer of the amplifiers is sufficiently linear [3]. Using a detector scan, we normalize intensities of STEM images with respect to the incident electron beam as suggested in [2]. The images consist of 2048×2048 scan points taken with a dwell time of 19 μs, resulting in a scan time of 80 s per image.

3. Image simulation

We used the STEMsim program [7] for simulation of the data needed for comparison with experiment. The simulations were performed for biaxially strained InGaN layers with In concentrations < 50 % and specimen thicknesses < 200 nm with the frozen lattice approximation taking into account the nonhomogeneous sensitivity of our HAADF detector. Static atomic displacements were computed with the LAMMPS program [8] using the Stillinger-Weber parametrization published in reference [9]. Each computed STEM image was averaged over the area of one image unit cell and the results were tabulated in dependence of In concentration and specimen thickness.

4. Evaluation procedure

Figure 1a shows a small part of a STEM image that will be used to demonstrate the evaluation steps. The intensity of this image is normalized with respect to the incident electron beam [2, 3]. Figure 1b shows the same part of the image after noise filtering using a Wiener filter. To estimate the noise level, a circular area is chosen in the Fourier transformed image in such a way that it only contains noise. The noise level used for the Wiener filtering was chosen as 0.3 times the maximum absolute value found in the circular area. This Wiener filter is applied to facilitate detection of atomic column positions. For each bright spot in the image we search for the pixel with highest intensity. The position of this pixel is used as estimated position of the atomic column. Then, a local mean intensity is computed for each atomic column. To this end, the area of the image is segmented into two dimensional Wigner-Seitz cells. The highlighted area in Fig. 1c shows one of these cells as an example. Using the original non-Wiener filtered image, intensities of all pixels corresponding to one Wigner-Seitz cell are averaged which defines the local mean intensity (Fig. 1d). Then, the specimen thickness is evaluated for each atomic column within the GaN (blue region in Fig. 1d) by a comparison of the local mean intensity with image simulation. A polynomial is fitted to specimen thicknesses obtained in the GaN regions from which interpolated thickness values are obtained for the InGaN region of the image. Figure 1e gives the thickness map obtained for the example image. Finally, a comparison of local mean intensity (Fig. 1d) and evaluated specimen thickness (Fig. 1e) with the simulated data is used to obtain the In concentration evaluated for each atomic column. The resulting In concentration map is given in Fig. 1f.

5. Results and Discussion

Figure 2a and b show images 1 and 32 of a series of 32 images taken from the same region. Although the electron beam had scanned the same specimen area for 42 minutes after exposure

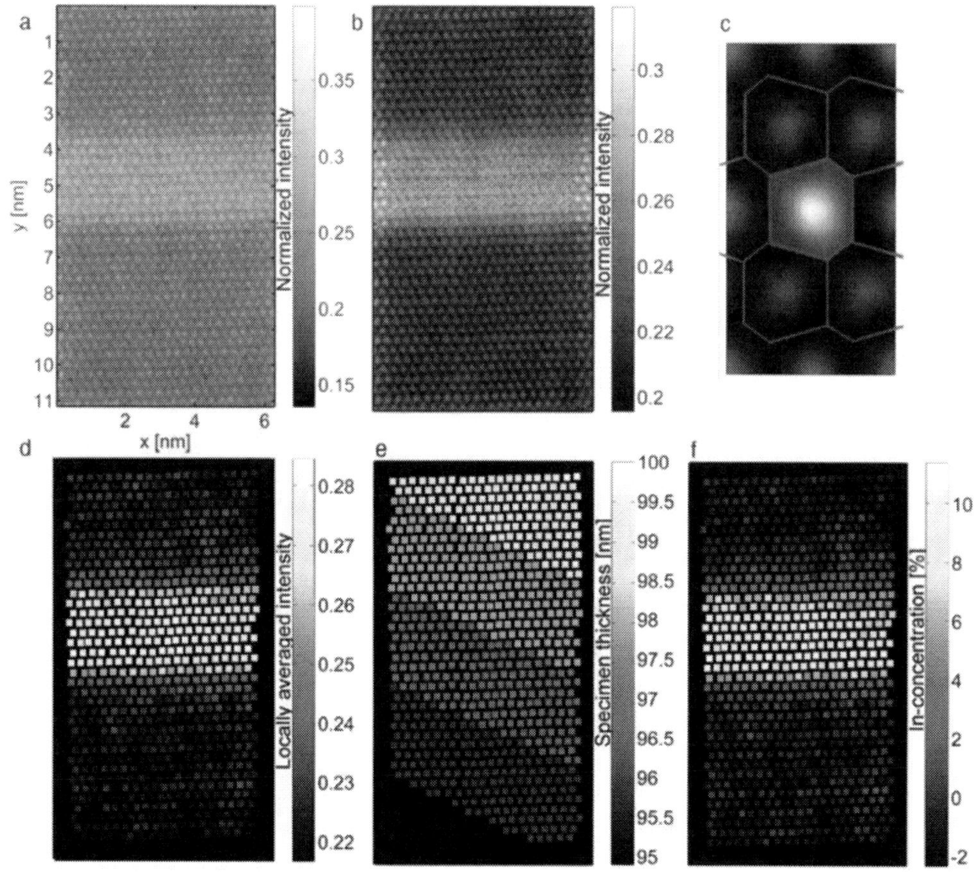

Figure 1. a) Small part of a STEM image, b) Wiener filtered image with detected positions of atomic columns, c) drawing explaining the segmentation of the STEM image into Wigner-Seitz cells. d,e,f) are color coded maps showing d) the local mean intensity, e) the evaluated specimen thickness and f) the evaluated indium concentration.

of image 32, there is no significant change of the contrast pattern. Fig. 2c displays a map of the In concentrations evaluated from image 1. Comparing Fig. 2c with d, it becomes obvious that no clustering of In can be observed after taking 32 images from the same area. An overview image that contained the illuminated area revealed a slightly increased background intensity most probably caused by carbon contamination. This leads to a decrease of the intensity contrast between GaN and InGaN, and thus to the reduction of the measured maximum In concentration from approximately 18 % in Fig. 2c to 16 % in Fig. 2d.

The small electron beam irradiation damage of InGaN we observed in our high resolution STEM experiments is in contrast to high resolution TEM, where In clustering is encountered after a few minutes [5]. A possible reason are different electron doses used in these experiments. In our STEM experiment, we used a probe current of 9.5×10^{-11} A. For 80 s, this leads to a total dose on the investigated 50×50 nm^2 large area of 7.6 nC, resulting in a dose of 0.003 nC/nm^2. In reference [5], the electron beam current was reduced substantially below the maximum attainable from the microscope and was 35 A cm^{-2}. This led to In clustering after 220 s, corresponding to a dose of 0.08 nC/nm^2, which is significantly larger than the dose used in our STEM experiments.

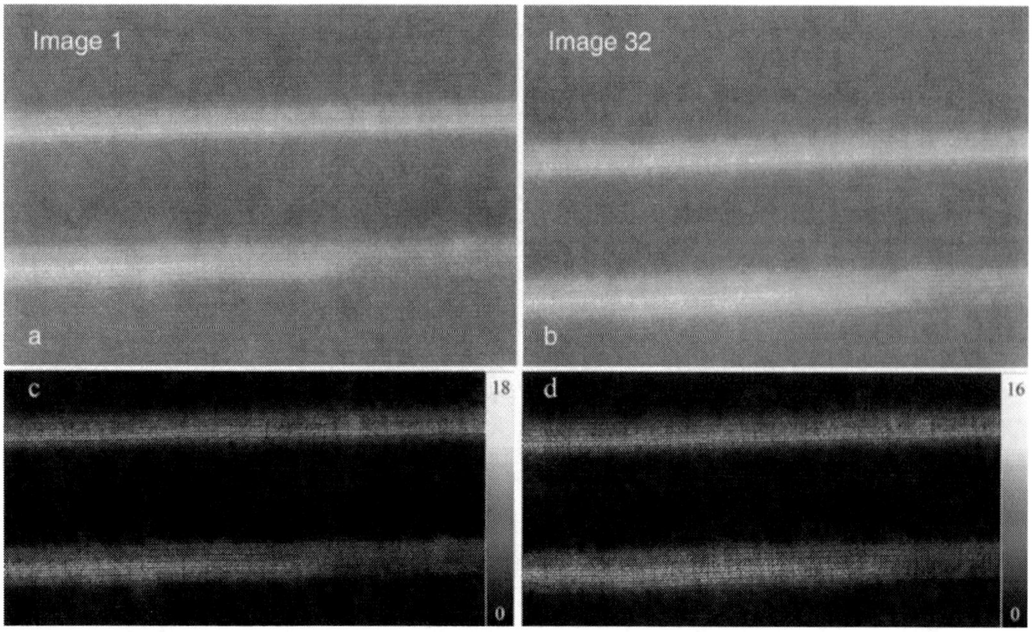

Figure 2. a,b) Images 1 and 32 of a series of 32 STEM images obtained from the same specimen area. c,d) Color coded maps of the In concentration evaluated from the STEM images shown in the upper row. The legend shows the In concentration in percent.

However, after exposure of 32 images, the dose was 0.1 nC/nm^2 in our experiment, and no In clustering was observed. This indicates that the amount of In clustering not only depends on the electron dose, but can also be influenced by other parameters such as the growth method and the In-concentration [6].

6. Conclusion

We showed a procedure to measure 2D indium concentration maps from high resolution STEM images. We did not observe In clustering after scanning the same specimen area for 42 minutes. These findings are an important step towards investigation of indium clusters in as-grown samples, whose existence is a controversial issue.

Acknowledgements

This work was supported by the Deutsche Forschungsgemeinschaft (DFG) under contract numbers RO 2057/4 and RO 2057/8.

References

[1] Grillo V, Carlino E and Glas F 2008 *Phys. Rev. B* **77**, 054103.
[2] LeBeau JM and Stemmer S 2008 *Ultramicroscopy* **108**, 1653
[3] Rosenauer A et al. 2009 *Ultramicroscopy* **109**, 1171
[4] O'Neill JP et al. 2003 *Appl. Phys. Lett.* **83**, 1965
[5] Smeeton TM, Kappers MJ, Barnard JS, Vickers ME and Humphreys CJ 2003 *Appl. Phys. Lett.* **83**, 541
[6] Albrecht M et al. 2001, MSM-2001, Oxford, *IOP conference series* **169**, 267
[7] Rosenauer A, and Schowalter M 2007 *Springer Proceedings in Physics* **120**, 169
[8] Plimpton SJ 1995 *J. Comp. Phys.* **117**, 1.
[9] Lei HP, Chen J, Petit S, Ruterana P, Jiang XY and Nouet G 2006 *Superlattices and Microstructures* **40**, 464.

Comparison of the contrast in conventional and lattice resolved ADF STEM images of InGaAs/GaAs structures using different camera lengths

Y Qiu, L Lari, I M Ross and T Walther

Kroto Centre for High-Resolution Imaging and Analysis, Dept. Electronic and Electrical Engineering, University of Sheffield, Mappin Street, Sheffield, S1 3 JD, UK

E-mail: Y.Qiu@shef.ac.uk

Abstract. A procedure to quantify annular dark field (ADF) images in scanning transmission electron microscopy (STEM) has been applied to two 200kV transmission electron microscopes (TEMs), a JEOL 2010F and a double aberration-corrected JEOL 2200FSC. A series of ADF images is acquired as a function of the camera length (i.e. inner detection angle). Then the intensity ratio of InGaAs and GaAs is plotted vs. camera length and extrapolated to zero, at which point the contrast behaves exactly as predicted by Rutherford's scattering. The linearity of ADF intensity ratio vs. camera length improves significantly by using the JEOL 2200FSC compared to the JEOL 2010F at medium resolution. A high-resolution ADF image at 2MX nominal magnification acquired in the JEOL 2200FSC shows the same linearity of intensity ratio vs. camera length, independent of whether the ratios of the average background intensities or the fringe amplitudes are used for the analysis. This is explained by both group III and group V atoms contributing to the {111} fringes observed, similar to low resolution data.

1. Introduction

Annular dark-field (ADF) scanning transmission electron microscopy (STEM) has been very successful because of the strong atomic number (Z) dependence. However, for crystalline specimens low angle Bragg beams complicate the contrast interpretation of the images. The image cannot be interpreted simply in terms of thickness and atomic number contrast. Therefore a high-angle annular detector (≥ 40 mrad) was proposed by Howie [1] with the aim to avoid the coherent Bragg scattering.

In ADF STEM the image intensity decreases while the contrast increases with increasing detection angle (decreasing camera length) [2]. For infinite scattering angle (camera length equal to zero), which cannot be achieved in any experiment directly, Rutherford scattering by the atomic nuclei is dominant and the intensity is proportional to the square of the atomic number (Z^2) of the atoms at constant specimen thickness [3]. Therefore, a method was proposed to quantify the chemical composition in binary systems such as SiGe/Si, where only Si vs Ge atomic exchanges occur, by extrapolating the ADF image contrast as a function of camera length to zero [4]. The method was used in this work to evaluate the peak indium concentration in an annealed InGaAs/GaAs heterostructure, where only In vs Ga atomic exchanges occur.

2. Experimental

Eight layers of InAs/GaAs QDs covered by 5 nm InGaAs were deposited on a Si-doped GaAs (100) substrate using molecular beam epitaxy (MBE). These dot-in-a-well (DWELL) structures were

Journal of Physics: Conference Series **326** (2011) 012041 doi:10.1088/1742-6596/326/1/012041

separated by 33 nm GaAs deposition. Then thermal treatment was applied to the sample in a N_2 ambient for 5 minutes at 750ºC with the aim to minimize the compositional fluctuation in the DWELL structures. ADF STEM images were taken by using both the JEOL JEM 2010F TEM and the aberration-corrected JEOL JEM 2200FSC TEM while varying the camera length (i.e. inner detection angle) to investigate the peak In concentration in the DWELL structure.

For each acquisition, the detector gain was set at the initial stage of each experiment to ensure the highest intensity level (acquired at longest camera length) was just below detector saturation level and an associated dark current signal (intensity recorded with gun valve closed) was acquired as a reference (background intensity, which was subtracted before calculating the intensity ratios).

The JEM 2010F TEM is equipped with a Schottky-type field-emission gun (FEG) operating at 197 kV. In STEM mode, it normally provides an electron beam ~ 0.25 nm in size at ~9.5 mrad semi-angle of beam convergence with 0.5 mm spherical aberration constant.

The aberration-corrected JEM 2200FSC TEM is equipped with a FEG operating at 200 kV and two aberration correctors. The upper aberration corrector for similar STEM eliminates the opening error of the probe-forming condenser lens system. An electron probe with ~ 0.1 nm diameter can be formed and used to perform STEM. By applying a larger aperture, the electron beam can also be focused to a similar size provided by the JEM 2010F TEM, but with much more current in it. As a result the signal-to-noise ratio is improved.

3. Results

Figure 1 presents an ADF STEM image at 120 kX nominal magnification of the sample after annealing acquired in the JEOL JEM 2010F, with camera length decreasing from top to bottom during the acquisition. The bottom stripe was acquired while the gun valve was closed. The dark current signal, I_0, for each experiment was directly read from this kind of image stripe section using the Gatan Digital Micrograph software.

Figure 2 shows a high resolution ADF STEM image at 2 MX with the camera length decreasing from top to bottom during acquisition in the JEOL 2200FSC. Weak lattice fringes were resolved and the {111} lattice fringes are used to calculate the average intensity ratio (R) and the ratio of the fringe amplitudes (Q) in InGaAs and GaAs for each camera length.

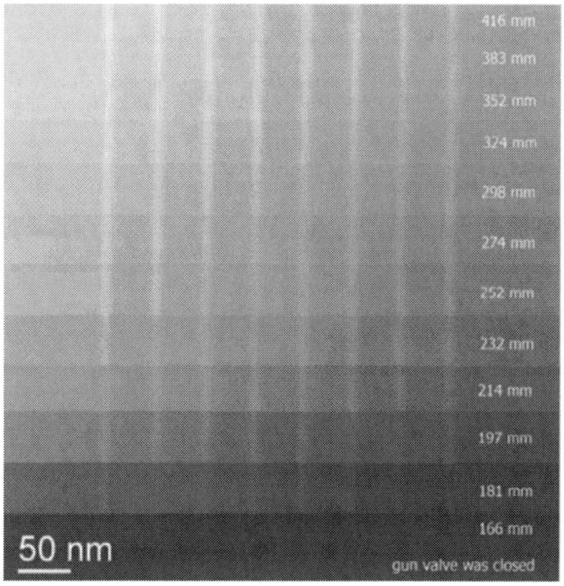

Figure 1. ADF STEM image at 120 kX magnification with camera length decreasing from top to bottom. The corresponding collection angle, β, from top to bottom are 18.0 mrad, 19.6 mrad, 21.3 mrad, 23.2 mrad, 25.2 mrad, 27.4 mrad, 29.8 mrad, 32.3 mrad, 35.1 mrad, 38.1 mrad, 41.4 mrad, and 45.2 mrad. The last stripe was acquired while the gun valve was closed.

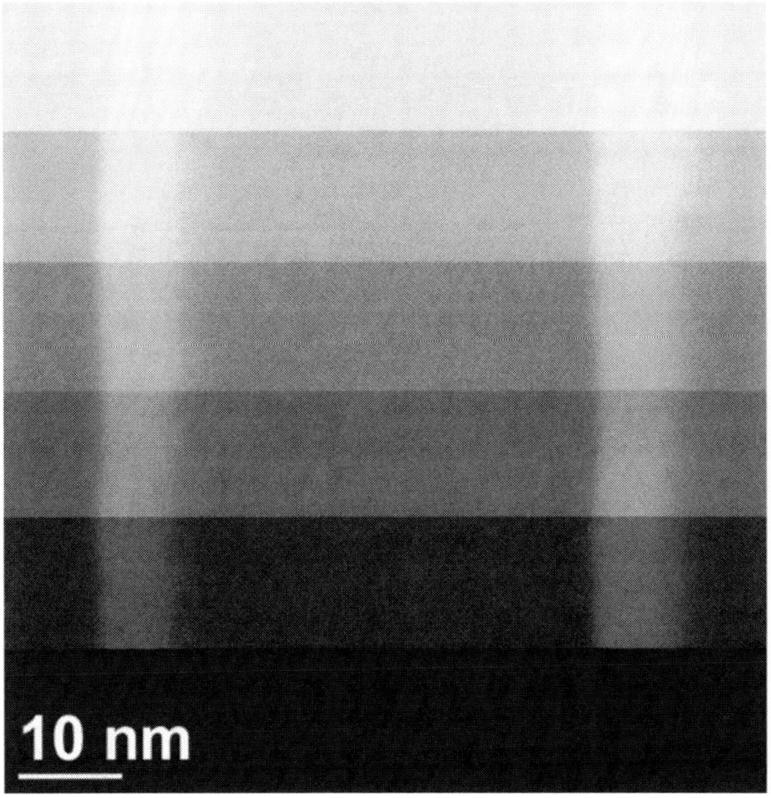

Figure 2. High resolution ADF STEM image at 2 MX magnification with the camera length decreasing from top to bottom during the acquisition in the JEOL JEM-2200FSC. The corresponding camera lengths for each stripe (collection angle, β) from top to bottom are 362 mm (20.7±0.8 mrad), 290 mm (26.2±0.9 mrad), 217 mm (39.3±1.2 mrad), 181 mm (47.1±1.6 mrad) and 145 mm (57.3±2.3 mrad). The last stripe was acquired while the gun valve was closed.

4. Discussion

The average intensity ratio (R) and the ratio of {111} fringe amplitudes of $In_xGa_{1-x}As$ to GaAs vs. camera length are plotted in figure 3. It shows that for the JEOL 2010F at the shortest camera lengths the intensity ratios calculated scattered dramatically, probably due to the poor signal-to-noise ratio. There is also a noted curvature around 250 mm camera length. This may relate to non-linearities in the electro-magnetic field of the inter-mediate lens used to alter the camera length [5]. Linearity has been improved significantly by using the JEOL 2200FSC at both medium and high resolution as the regressional coefficients now approach unity (cf. Table 1). That's because the aberration corrected instrument is able to provide a smaller probe with higher current, giving a higher signal-to-noise ratio. For large camera lengths the ratio of {111} fringe intensity is generally low. It increases fast with decreasing camera length and finally yields a similar intercept for zero camera length. However, the relative compositional error evaluated by using fringe amplitude ratios is larger than that obtained by using an average intensity ratio, which again is due to the contribution of noise to the lattice fringes.

Assuming a Gaussian distribution of standard deviation σ of the measurements, the 2σ-interval gives 95.4% confidence that the In peak concentration from the three separate measurements of R is 0.291±0.062, 0.248±0.008 or 0.266±0.014 from the ADF images taken in the JEOL 2010F at 120 kX, the JEOL 2200FSC at 250 kX or the JEOL 2200FSC at 2MX, respectively. Consequently, the peak indium concentration in the $In_xGa_{1-x}As$ lies between 0.25 and 0.26.

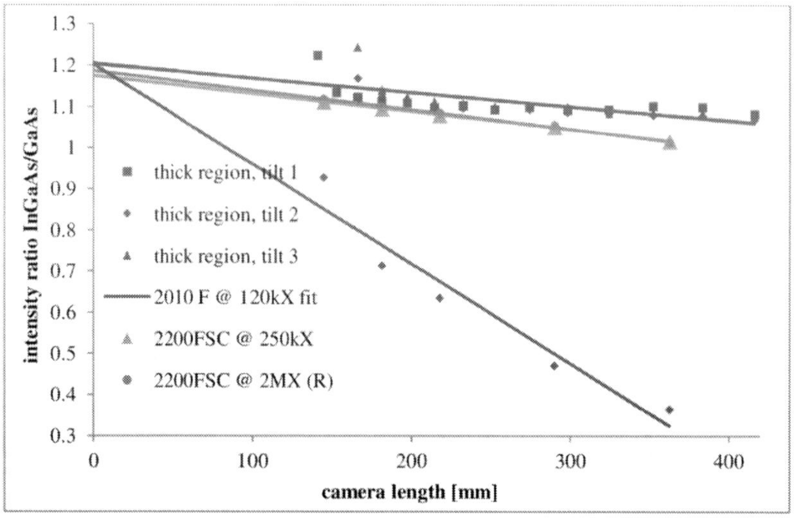

Figure 3. Plots of the average intensity ratio (*R*) and the ratio of {111} fringe amplitudes (*Q*) of InGaAs to GaAs vs. camera length

Table 1. Results of linear regressional analysis of the plots in figure 3

Condition	2010F TEM	2200 FSC TEM		
nominal magnification	120 kX	250 kX	2 MX	
intensity ratio	average	average	average	fringe amplitude
intercept *b* from lin. reg. fit	1.204±0.024	1.174±0.003	1.187±0.005	1.202±0.093
lin. regression coefficient	0.752	0.998	0.995	0.935
In peak concentration, $x \pm \Delta x$ (ε=2)	0.291±0.031	0.248±0.004	0.266±0.007	0.288±0.132
relative error, $\Delta x/x$	10.7%	1.6%	2.6%	45.8%

5. Conclusion

Aberration corrected STEM improves linearity of the data, mainly due to improving the signal-to-noise ratio by forming a smaller probe with higher current compared to conventional STEM. This reduces the relative error of the evaluated chemical composition by using average intensity ratio significantly. This work also shows that the result can be confirmed by using the fringe amplitude but with a larger relative error due to the higher contribution of noise to the lattice fringes.

References

[1] Howie A 1979 *J. Microscopy* **177** 11
[2] Treacy MMJ, Howie A and Pennycook SJ 1980 *Proc. Electron Microscopy and Analysis Group Conference 1979, Inst. Phys. Conf. Ser.* **52** 261
[3] Treacy MMJ, Howie A and Wilson CJ 1978 *Philos. Mag. A* **38** 569
[4] Walther T 2006 *J. Microscopy* **221** 137
[5] Walther T, Qiu Y and Cullis AG 2010 Proc. Electron Microscopy and Analysis Group Conference 2009, Sheffield, *J. Phys. Conf. Ser.* **241** 012068

GaAs-based Nanowires Studied by Low-Temperature Cathodoluminescence

Anders Gustafsson[1], Jessica Bolinsson[1], Martin Ek[2] and Lars Samuelson[1]

[1] Solid State Physics and the Nanometer Structure Consortium (nmC), Lund University, Box 118, S-221 00 Lund Sweden

[2] Polymer and Materials Chemistry, Lund Universtiy, Box 124, S-221 00 Lund Sweden

Anders.Gustafsson@ftf.lth.se

Abstract. We present cathodoluminescence data of nanowires (NWs) grown using size-selected gold particles as seeds. The NWs have a GaAs core with a diameter of 50 nm and a length of several μm. The NWs in this study were generally covered with a shell of AlGaAs. With increasing growth temperature, the emission intensity increases significantly. From a variety of growth conditions, we conclude that the exposed sides of the NWs during growth play an important role in the emission intensity. The diffusion of carriers was studied by inserting a segment of GaInAs in GaAs NWs. By capping the NWs with an AlGaAs shell, we observe a tenfold increase in the diffusion length along the core.

1. Introduction

Over the past decades, there has been a quest to fabricate high-quality nanowires (NWs). These structures are interesting from a fundamental point of view. They also have potential for uses in future devices [1]. Semiconductor NWs seeded by metal particles have attracted much attention lately. In this study we present cathodoluminescence (CL) data from NWs seeded by size selected gold aerosol particles [2]. One important figure of merit for NWs is the intensity of the emission. We have studied how the emission intensity is influenced by the growth temperature of the core. We have also studied how the diffusion of carriers along the core is influenced by the addition of a shell.

2. Experimental details

The GaAs NWs were grown using metal-organic vapour phase epitaxy (MOVPE), using 50 nm aerosol gold particles [2] to seed the growth on (111)B-oriented GaAs substrates [3]. The size of the particle determines the diameter or the resulting NWs, and the majority of the NWs grow in a <111>B direction. In order to study the effect of the growth temperature on the emission, a series of samples were grown in the temperature range of 380 to 500°C, in steps of 20°C [4]. The NWs were grown to a length of 3 - 4 μm, and for samples intended for studies of diffusion lengths (L_D), a short GaInAs segment was inserted. Most of the NWs were covered by an AlGaAs shell grown at a higher temperature (630°C), resulting in a shell thickness of about 20-30 nm, as determined by transmission electron microscopy. While growing the AlGaAs shell, some growth also takes place underneath the seed particles [9]. This results in the growth of a short AlGaAs NW core region at the top of the NWs.

Capping the NW cores with a higher band gap material (e.g. AlGaAs on GaAs) significantly improves the emission from the NWs.

The CL studies were performed in a scanning electron microscope (SEM) with a lHe cold stage. The emission was detected by a dedicated light-collection system [5, 6]. The NWs were studied in two modes: 1) as grown on the substrate in side view, and 2) broken off and transferred to Si substrates to avoid background emission from the GaAs substrate. The temperature was typically 6-10 K and monochromatic images as well as average spectra were recorded. An acceleration voltage of 5 keV and a probe current of 50 pA were used. Under these conditions, a spatial resolution along the NWs of about 50 nm can be achieved. We have used the inclusion of a low-bandgap segment (GaInAs, ≈5% In) in the core (GaAs) of a NW as a detector for diffusion into the segment. There is an intensity increase of the GaInAs emission when the electron beam is moved along the GaAs core towards the segment caused by diffusion from the core to the segment. Intensity profiles were extracted from monochromatic images and a single 1D exponential was fitted to each side of the intensity profile [7].

3. Result and Discussion

CL measurements were carried out on NWs grown under the different growth conditions described above. Spectra were recorded in side view from NWs grown at the different temperatures between 380 and 500°C. The scanned area typically contains 20-30 NWs. For each growth temperature, a reference spectrum was recorded by moving the sample to scan over that same area size on the side of the substrate instead. Two of the spectra (380 and 500°C) are shown in figure 1a. All spectra recorded for the NWs grown at different temperatures are dominated by a single peak at around 1.475 eV. The origin of the peak is not clear, but it appears to be related to twin boundaries in the nanowires. The main difference is that the intensity relative to the substrate increases with growth temperature of the core. The increase is about 30 times as the growth temperature is increased from 380 to 500°C. This increase has an exponential dependence on the growth temperature. This is shown in figure 1b, where the intensity ratio between the NW and the substrate emission is plotted on a logarithmic scale.

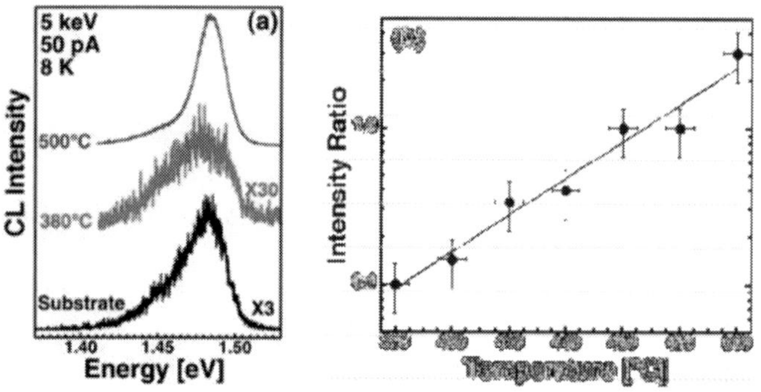

Figure 1. a) comparison between the spectral features of the NWs grown at 380°C and at 500°C, and of the substrate. b) The ratio between the NW and substrate intensity as a function of growth temperature.

One source of the enhanced emission can be found in SEM images of the NWs. With increasing temperature, the NWs have a more tapered shape, as the higher temperature allows for some growth on the sides of the NWs during core growth, as can be seen in figure 2. This results in a slightly larger core diameter and thereby larger volume. We have observed that growth from larger-diameter particles leads to a higher emission intensity for NWs in the same growth run. However, this volume effect is not enough to explain the 30-fold increase in the intensity. Figure 2 also shows monochromatic CL images recorded at the peak of the NW emission. For the NWs grown at 380°C, the individual NWs are barely distinguishable in the image, whereas they are clearly visible at 480°C. For the low-temperature growth, the emission from the substrate is stronger than from the NWs. The ratio is reversed for the high temperature growth.

To further study the temperature effects, we have made a sample with a mixed-temperature core, started at 480°C and then switched to 380°C half way through the core growth. The emission from the entire core is as weak as for the 380°C core. We have also grown samples with a GaAs core and a

GaAs shell, the latter grown at the normal shell growth temperature of 630°C. When covered by an additional AlGaAs shell, these NWs show weaker emission from the low-temperature core than from the corresponding NWs without the intermediate GaAs shell. However, the part of the GaAs core grown during the GaAs shell growth shows very strong emission intensity.

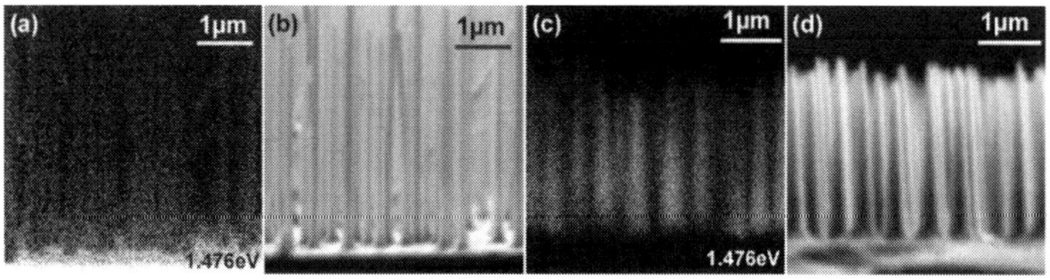

Figure 2. Two sets of images, a monochromatic CL image and the corresponding SEM image, from NWs grown at a) and b): 380°C and c) and d): 500°C.

The intensity of the emission depends strongly on competing non-radiative recombination. It is therefore essential to reduce the number of non-ratiative recombination centers. For NWs, these are mainly associated with surface states and traps [8]. Non-radiative channels will reduce the total life time of the carriers and thereby also their diffusion length. We have investigated the effect of the shell on the diffusion length of carriers in the GaAs core, measured as described above on NWs transferred to Si substrates. The result is presented in figure 3. The monochromatic CL and corresponding SEM images are shown for a capped and an uncapped NW. The figure also shows two intensity traces along the NWs. By fitting exponentials to the intensity profiles (int$\propto$exp($-x/L_D$), it is possible to extract the diffusion lengths along the GaAs parts of the cores. In both cases, the diffusion length below the segment (left in the images and line traces) show similar values of about 100nm, which is much lower than expected for high-quality wires, e.g. about 3 μm for V-grooved wires [7]. Above the segment, there is a dramatic increase in the diffusion length for the capped core ($\approx$1μm) but there is almost no difference for the uncapped core.

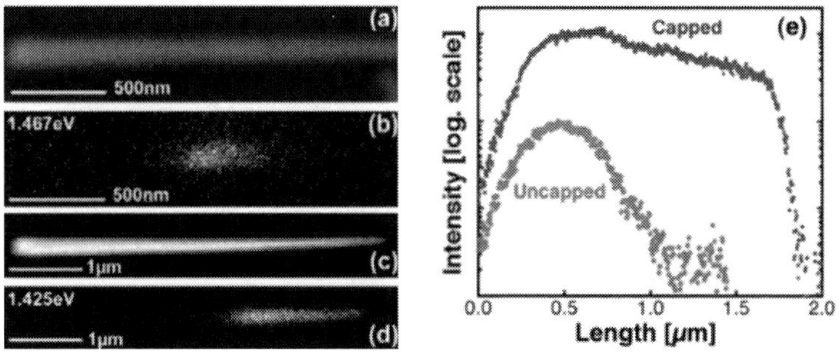

Figure 3. a) SEM and b) CL image of an uncapped segment of GaInAs in a GaAs core. s) SEM and d) CL image of an AlGaAs-capped segment of GaInAs in a GaAs core. e) line traces of the intensity of the CL images of (b) and (c).

To explain the difference in the diffusion length below and above the segment for the capped core, we originally speculated that the carriers were lost to a radial GaInAs layer, a result of radial growth when growing the segment [9]. We observed weak emission at an even lower energy than the segment emission, which we attributed to a radial quantum well (RQW) with a higher In content than in the segment itself. However, this was on the edge of the range of the detector used so it was not clear. We have recently started to use a detector with a more suitable energy range and we have now been able to, not only confirm this but to show that the RQW emission is very strong. This is shown in figure 4 for NWs on the substrate, where the emission from the RQW is significantly stronger than the emission from the core and segment. Though not shown here, the emission pattern varies when the detection energy is varied, but it stays below the spatial location of the segment.

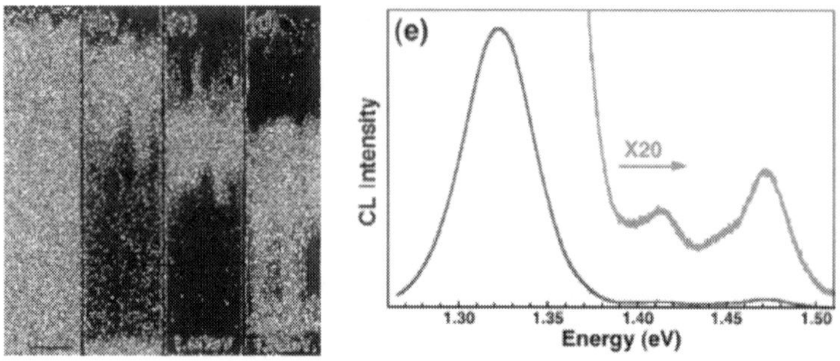

Figure 4. a) SEM image and monchromatic CL emission of b) GaAs core, c) GaInAs segment and d) radial GaInAs layer. e) spectrum of the same area.

From the experimental results, the importance of the GaAs surface is clear. The lack of emission intensity is most likely related to the non-radiative recombination of traps on the GaAs surface [8]. It is evident from the measurements of the diffusion lengths that capping the wires with a passivating shell improves the diffusion length significantly. For the capped NWs, the improvement is related to a reduction in the surface recombination rate, related to surface traps [8], where fewer carriers are drained from the core. The same effect can be observed when the core is covered with a shell of lower bandgap material, a RQW. The difference is that the additional recombination path is radiative and can be observed, like in figure 4d. The draining of carriers by the RQW results in a similar diffusion length as for the uncapped core. As from the temperature dependence, the fact that low-temperature core growth on a high temperature-grown core results in emission intensity similar to the low-temperature core is significant. This means that the reduced emission intensity is probably a surface effect rather than a bulk effect, as the surface states become trapped and embedded in the structure. This leads us to believe that a higher growth temperature leads a lower density of surface states on the core, which in turn leads to stronger emission intensity.

Acknowledgements
The work has been performed with funding from, mainly, nmC@LU, with support from national (e.g. VR, TFR, NFR and NUTEK), KAW and EU programs. The authors would like to thank Kilian Mergenthaler, Mats-Erik Pistol and Niklas Sköld for fruitful discussions and experimental support. The TEM study was performed at nCHREM.

References

[1] Yang P, Yan R, and Fardy M 2010 Nano Letters **10** 1529
[2] Ohlsson BJ, Björk MT, Magnusson MH, Deppert K and Samuelson L and Wallenberg LR 2001 Appl. Phys. Lett. **79** 3335
[3] Seifert W, Borgström M, Deppert K, Dick KA, Johansson J, Larsson MW, Mårtensson T, Sköld N, Svensson CPT, Wacaser BA, Wallenberg LR and Samuelson L 2004 J. Crystal Growth **272** 211
[4] Bolinsson J, Gustafsson A, Ek M, Mergenthaler K, Jacobsson D, Pistol M-E, Samuelson L (unpublished)
[5] Gustafsson A, Pistol M-E, Samuelson L and Montelius L 1998 J. Applied Physics **84** 1715
[6] The emission was collected by a parabolic mirror and dispersed through a monochromator with a 1μm blaze grating. The intensity was recorded by a GaAs (Hamamatsu R943-02) or a GaInAsP (Hamamtsu R5509-43) photomultiplier tube in current mode.
[7] Gustafsson A, Bolinsson J, Sköld N, and Samuelson L 2010 Appl. Phys. Lett. **97**, 072114
[8] Demichel O, Heiss M, Bleuse J, Mariette H and Fontcuberta i Morral A 2011 Appl. Phys. Lett. **97**, 201907
[9] Bolinsson J, Mergenthaler K, Samuelson L and Gustafsson A 2011 J Crystal Growth **315** 138

Correlated micro-photoluminescence and electron microscopy study of a heterostructured semiconductor nanowire

J Todorovic[1], A F Moses[2], D L Dheeraj[2], P Olk[2], B O Fimland[2], H Weman[2] and A T J van Helvoort[1]

[1] Department of Physics, [2]Department of Electronics and Telecommunications, Norwegian University of Science and Technology, NO-7491 Trondheim, Norway

a.helvoort@ntnu.no

Abstract. Correlation between the optical and the structural properties of an individual heterostructured nanowire (NW) is crucial for optimising the NW synthesis and device design. In this work, low temperature (10 K) micro-photoluminescence (μ-PL), low and high voltage scanning transmission electron microscopy and conventional transmission electron microscopy are applied to the same individual NWs. The studied NWs, grown by Au-assisted molecular beam epitaxy, have a wurtzite GaAs core with a zincblende GaAsSb axial insert, enclosed with an AlGaAs radial shell and a GaAs capping layer. By subsequent analysis of one and the same NW by μ-PL and several electron microscopy techniques in different microscopes, we can relate the spectral features of a single NW to the structural features, such as different crystal phases, lattice defects and composition.

1. Introduction

III-V semiconductor nanowires (NWs) are attractive for a broad range of applications as for example solar cells, lasers, single photon sources and sensors. The electro-optical properties of III-V NWs are directly related to the crystal phase (zincblende (ZB) or wurtzite (WZ)), lattice defects and the composition. In practice, NWs from the same batch can differ for example by dimensions or defect configuration. Therefore structural and electro-optical properties of the same NW have to be investigated. Direct correlation between the optical properties and the structural characteristics of an individual NW is crucial for understanding band alignments and quantum confinement effects as well as for optimizing the NW synthesis and device design for future applications. Transmission electron microscopy (TEM) based techniques are commonly used to fully characterize structural elements of NWs. However, conventional TEM specimens are not well suited for micro-photoluminescence (μ-PL) studies. So far, only a few correlated PL-TEM studies of the same NW have been reported [1,2].

In this work, low temperature (~10 K) μ-PL, scanning (transmission) electron microscopy (S(T)EM) at 30 kV and 200 kV and conventional transmission electron microscopy (TEM) at 200 kV are applied on the same WZ GaAs/AlGaAs core-shell NWs with ZB GaAsSb core inserts. Due to the complexity of this particular heterostructure, a range of different electron microscopy techniques were required to understand which structural features actually affected the optical properties. By detailed correlated study of one NW, particular structural characteristics are directly connected with the observed PL features.

2. Experimental details

The NWs were grown by Au-assisted molecular beam epitaxy [3,4]. The heterostructure consists of WZ GaAs core with ZB GaAsSb axial insert, AlGaAs radial shell, and GaAs cap. In addition, an axial WZ AlGaAs segment forms on top of the WZ GaAs core during AlGaAs radial shell growth [4]. NWs were ultrasonically dispersed in isopropanol and drops of the solution were transferred onto a Si TEM grid with a 50 nm thick amorphous silicon nitride support film (ThinWindows Inc.). For safe handling and good heat conduction in the μ-PL experiments, the TEM grid was sealed between two Mo washers with central holes of 2 mm [5].

μ-PL was performed in a Janis ST-500 cryostat at ~ 10 K, using continuous laser excitation at the wavelength of 532 nm. The power density at the specimen was varied from 1 W/cm^2 to about 50 W/cm^2. The detection system included a Horiba Jobin Yvon iHR500 spectrometer with a 300 grooves/mm grating and an Andor Newton^{EM} Si electron multiplying charge-coupled device (EMCCD). The spectral resolution of the system is about 200 μeV.

Structural characterization was performed on a Hitachi S-5500 S(T)EM (30kV, cold field emission gun) for low magnification bright field (BF) STEM and on a Philips CM30 TEM (200 kV, LaB_6) for BF TEM, dark field (DF) TEM, and selected area electron diffraction (SAED). A JEOL 2010F TEM/STEM (200 kV, Schottky field emission gun) was used for lattice imaging (HREM) and high angle annular DF (HAADF) STEM (probe size approx. 0.5 nm, 49-120 mrad collection angle range).

3. Results and discussion

A NW, containing all structural elements of the studied heterostructure after breaking off from the substrate, was selected from the PL-TEM dataset of about 30 NWs for a more detailed TEM and STEM characterization. Figure 1 shows that the PL spectra of the NW contain three bands, designated as A (1.35-1.45 eV), B (1.50-1.55 eV) and C (1.58-1.65 eV), which are characteristic for this NW heterostructure. The BF STEM (30 kV) image in the inset of figure 1 presents an overview of the NW divided into regions I (containing ZB GaAsSb insert surrounded by the WZ GaAs core and the AlGaAs shell, full line box), CS (WZ GaAs/AlGaAs core-shell, dashed line box) and AS (WZ AlGaAs axial segment, dotted line box) which are the origins of the characteristic PL emissions A, B and C respectively [2,4,5]. Low voltage STEM provides quick insight into morphology and size of the NW and can be performed before μ-PL as well [5].

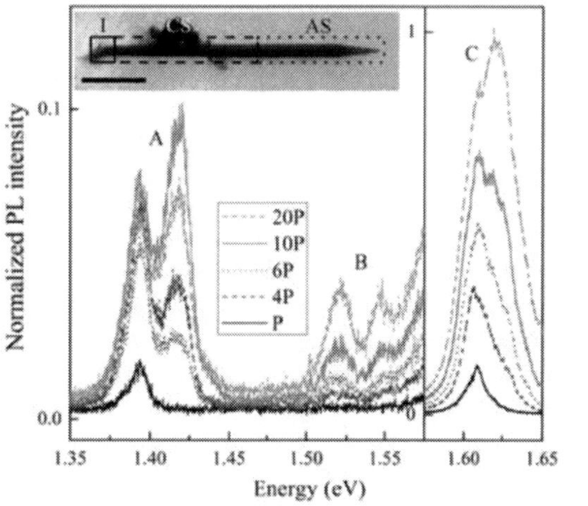

Figure 1. Low temperature (10 K), power dependent PL spectra (P = 2.55 W/cm^2) of a NW together with 30 kV BF STEM image (inset). For this NW heterostructure, A, B and C designate the three characteristic PL bands which originate respectively from the regions I, CS and AS of the NW (inset). The PL spectra were taken before the electron microscopy studies. The scale bar in the inset is 500 nm. Note that the NW is anti-tapering towards the tip.

The ground state of GaAsSb related PL emission (band A) is observed at 1.39 eV (figure 1) for this NW. No blue shift of this PL band can be observed with increase of excitation power. Such behavior indicates that both electrons and holes are confined within the GaAsSb insert (i.e. type I band alignment between ZB GaAsSb insert and WZ GaAs core) [6]. At higher excitation powers a PL peak at 1.42 eV appears (figure 1). The origin of this PL transition is not clear at present.

Region I of the NW is shown in detail in figure 2. DF TEM (figure 2(a)) visualizes the different crystal phases present with the highest contrast: the ZB insert (dark) in the WZ NW (bright) with some stacking faults (SFs). The NW has a diameter of 76 nm in region I, just above the fracture, and a 38 nm long ZB GaAsSb insert. There are two adjacent SFs in the WZ GaAs core starting at about 21 nm above the insert (figure 2(a,b)). The crystal phases were confirmed by SAED (not shown). Note that the crystal phases and the SFs can be identified by lattice imaging (HREM) on the 50 nm thick SiN support film. Type I behavior of GaAsSb related PL emission (figure 1) indicates that the two SFs (figure 2(a,b)) are far enough away from the insert not to affect carrier recombination within the insert.

Figure 2 shows that the fracture of the NW occurred within ZB and WZ sections in region I, which might have impact on the PL emission from the ZB GaAsSb insert. HAADF STEM was used to visualize the difference between the WZ GaAs core and the WZ AlGaAs shell, or the ZB GaAsSb insert and the ZB AlGaAs shell. The HAADF STEM image in figure 2(d) and area intensity profile in the inset show the core-shell structure of regions I and CS with a core diameter of 27 nm. The shell is about 25 nm thick, whereas the width of the NW at the lower ZB-WZ interface is 49 nm as viewed in a <110> projection. This indicates that the fracture of the ZB section occurred mostly within the ZB AlGaAs shell. Since the insert is mostly intact and almost completely passivated by ZB AlGaAs shell and WZ GaAs core, the GaAsSb related PL emission (band A in figure 1) remains observable.

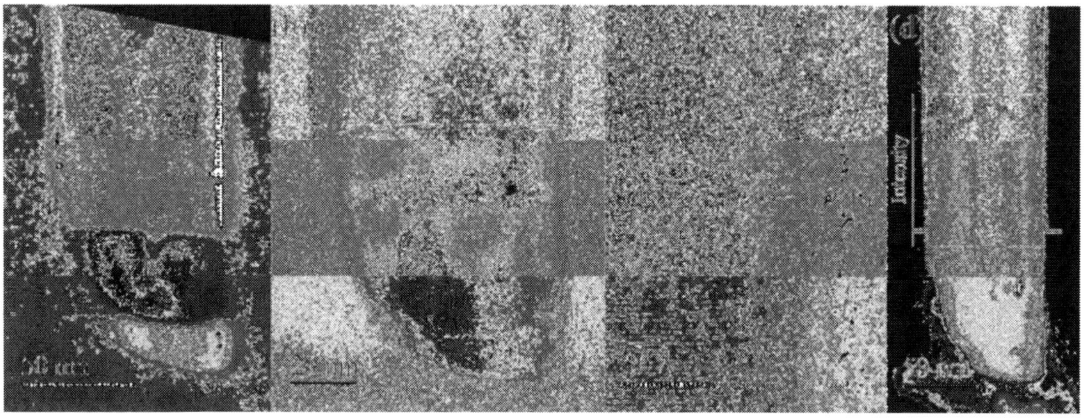

Figure 2. Detailed TEM characterization of the region I: (a) DF TEM image on a <110> zone. (b) HREM image. (c) HREM image of the upper ZB-WZ interface. (d) HAADF STEM image few degrees off a <110> zone with 100 nm area intensity profile (inset). Diameter of the core is 27 nm.

The upper part of the NW (regions CS and AS) is shown in figure 3. HAADF STEM shows where the transition between the core-shell structure and the axial WZ AlGaAs segment occurs (figure 3(a)). The WZ GaAs/AlGaAs core-shell segment above the two SFs in region I (figure 2(a,b,d)) is SF free for about 0.7 µm, which is the estimated core length above the insert based on the growth parameters [3]. The WZ region with high SF density (about 0.3 µm long) follows as observed in HAADF STEM and BF TEM (figure 3(a,b)). The gradual transition from WZ GaAs/AlGaAs core-shell to axial WZ AlGaAs segment starts in this region. In the axial AlGaAs segment, the SF density is reduced (figure 3(a,c)) compared to the core-shell transition region. The area intensity profiles (figure 3(d)) of the HAADF STEM image (figure 3(a)) show the core-shell structure in area 1 and the axial AlGaAs segment (no core-shell structure) in area 2. The anti-tapering shape of the NW (inset of figure 1) explains the higher intensity of the axial AlGaAs area profile (profile 2, figure 3(d)).

In the PL spectra (figure 1), band B (GaAs related PL emission) is broad and weak compared to bands A and C. The B-band is related to a mixture of various type II recombinations due to the SFs in the WZ GaAs core [5]. The peak at 1.55 eV that appears at higher excitation powers is not completely understood at present.

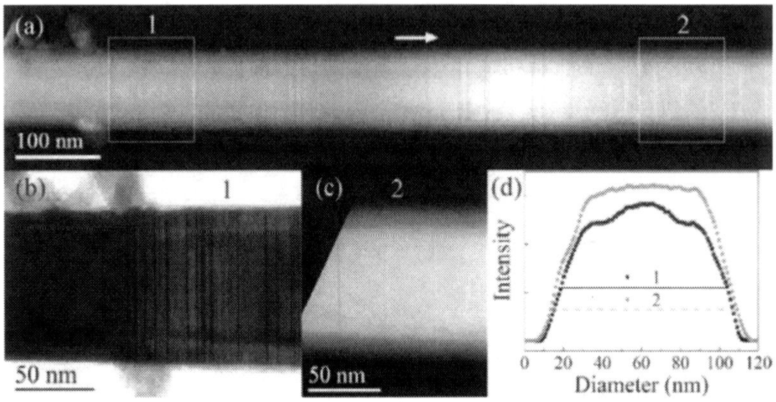

Figure 3. (a) DF STEM image few degrees off zone with two 100 nm long areas selected. The arrow shows the growth direction of the NW. (b) BF TEM image on a <110> zone including area 1. (c) DF STEM image of area 2. (d) Area intensity profiles of areas 1 and 2 in DF STEM image in (a).

The AlGaAs related PL emission is coming from the axial AlGaAs segment and not from radial AlGaAs shell [4,5]. The significant volume of the axial AlGaAs segment (40 % of the total NW length of 2.2 μm), the presence of SFs in the axial AlGaAs and the high SF density in the WZ GaAs/AlGaAs core-shell transition region can partly account for the strong AlGaAs PL emission compared to GaAsSb and GaAs PL emissions since SFs act as carrier traps. SFs here seem to prevent carrier diffusion from their origin, within the axial AlGaAs segment, through the WZ GaAs core towards the ZB GaAsSb insert.

4. Conclusions

One single NW has been studied by μ-PL and in three different electron microscopes to directly correlate the structural features to the optical properties. ZB GaAsSb related PL emission exhibits type I band alignment when the SFs in surrounding WZ GaAs core are far away (about 21 nm) from the ZB GaAsSb insert. By combining HREM and HAADF STEM, it was determined that the fracture within the ZB phase occurred mainly within the AlGaAs shell. Such fracture still allows PL emission from the ZB GaAsSb insert. The GaAs related PL emission, at about 1.52 eV, originates from type II recombination between SFs and the WZ GaAs core with varying distribution of SFs in the region of transition from core-shell to axial AlGaAs segment. The AlGaAs related PL emission originates from the WZ AlGaAs axial segment and is most prominent in the PL spectra of this NW due to the presence and distribution of SFs in the axial segment and in the transition region from core-shell to the axial segment.

Acknowledgment

This work was supported by "RENERGI" program (Grant No. 190871) of the Research Council of Norway.

References

[1] Bao J, Bell D C, Capasso F, Wagner J B, Mårtensson T, Trägårdh J and Samuelson L 2008 *Nano Letters* **8** 836
[2] Heiss M et al. 2011 *Phys. Rev. B* **83** 045303
[3] Dheeraj D L et al. 2008 *Nano Letters* **8** 4459
[4] Zhou H L et al. 2009 *Nanotechnology* **20** 415701
[5] Todorovic J, Moses A F, Karlberg T, Olk P, Dheeraj D L, Fimland B O, Weman H and van Helvoort A T J 2011 *Nanotechnology* **22** 325707
[6] Hoang T B, Moses A F, Ahtapodov L, Zhou H, Dheeraj D L, van Helvoort A T J, Fimland B O and Weman H 2010 *Nano Letters* **10** 2927

Polarity determination in ZnSe nanowires by HAADF STEM

M Den Hertog[1], M Elouneg-Jamroz[1], E Bellet-Amalric[2], S Bounouar[2], C Bougerol[1], R André[1], Y Genuist[1], JP Poizat[1], K Kheng[2] and S Tatarenko[1]

CEA-CNRS group "Nanophysique et Semiconducteurs"

[1] Institut Néel, CNRS et Université Joseph Fourier, 25 rue des Martyrs, BP 166, 38042 Grenoble Cedex 9, France

[2] CEA-Grenoble INAC-SP2M, 17 rue des Martyrs, 38054 Grenoble Cedex 9, France

Martien.den-hertog@grenoble.cnrs.fr

Abstract. High angle annular dark field scanning transmission electron microscopy is used to analyze the polarity of ZnSe nanowires grown, by molecular beam epitaxy, on GaAs substrates. The experimental results are compared to simulated images in order to verify possible experimental artefacts. In this work we show that for this type of nano-objects, a residual tilt of the specimen below 15 mrad, away from the crystallographic zone axis, does not impair the interpretation of the experimental images.

1. Introduction

ZnSe nanowires (NWs) are considered interesting structures for novel electronic and opto-electronic device functionalities. In ZnSe NWs both the hexagonal wurtzite (WZ) and cubic zinc blende (ZB) structure can occur [1]. These crystallographic structures for binary compounds are non-centrosymmetrical. Typically, along <111> in ZB and <0001> in WZ, one can define a polarity of the structure based on whether the II to VI (Zn to Se) bonds are aligned in the growth direction or opposite to it. It's important to investigate the polarity of such materials as it can affect their physical properties, in particular their optical properties. Indeed the crystal polarity is at the origin of electrical polarization along the growth direction: piezoelectric polarization in ZB or spontaneous polarization in WZ structures. In this paper we investigate the direction of the Zn to Se bond, referred to as the polarity, of ZnSe NWs epitaxially grown on GaAs substrates using High Angle Annular Dark Field Scanning Transmission Electron Microscopy (HAADF STEM). We do not study a possible effect of the direction of the ZnSe bond or the presence of polarization fields. HAADF STEM is an efficient technique to probe polarity along nanowires as it is sensitive to chemical contrast, namely sensitive to the atomic number Z. The experimental results are compared to simulated HAADF images to evaluate how the experimental conditions could affect the reliability of the polarity determination.

2. Experimental details

The NWs are grown by molecular beam epitaxy (MBE) using gold as a catalyst. Samples used in this study are all grown in a solid source MBE system. The source materials for the MBE system are elemental Zn (6N) and Se (6N). Commercial GaAs (111)B substrates are first deoxidized by annealing at 640°C under As pressure in a connected III-V MBE growth chamber. We then grow a 70 nm GaAs buffer layer. In order to avoid Ga incorporation in the NWs and to improve the surface quality and the epitaxial relation between the NWs and the substrate, a thin ZnSe buffer layer (about 50 nm thick) is grown at 280°C as described in [2]. Gold is deposited from an effusion cell on the ZnSe buffer layer, at room temperature, in a dedicated metal deposition chamber connected by UHV to the II-VI and III-V growth chambers. The nominal thickness of gold is on the order of 0.1 nm. The sample was heated at 500°C to cause dewetting of the gold layer and create nanocatalyst particles. ZnSe NWs are grown between 300°C and 450°C with a Zn:Se beam equivalent pressure (BEP) ratio of 1:4 and pressures in the 10^{-7} Torr range.

HAADF STEM images were obtained on a probe Cs corrected FEI TITAN TEM (300 kV). TEM samples were prepared by classical mechanical polishing followed by Ar ion milling using a Gatan PIPS. HAADF STEM multi slice simulations were realized using the code developed by Kirkland [3], using a 6x6 zinc blende (ZB) ZnSe unit cell with a sample thickness of 8 nm, corresponding to 20 (110) planes and comparable to the NW diameter (Fig. 1B). The input potential and wavefunction were sampled with 512 px and the final image or linescan was made on a length of 1.2 nm sampled with 64 px. The electron optical parameters of the Titan, that were included in the simulation, were 300 kV acceleration voltage, a spherical aberration of 10 µm, a convergence angle of a 15 mrad, and defocus of 0 nm. The defocus was not known exactly, however, according to simulation a small defocus (for instance 10 nm) decreases the contrast and the contrast difference between Zn and Se atoms but does not affect the visibility of the polarity. The contributions of the chromatic aberration are not included in these calculations. Images were calculated using 50 and 200 mrad inner and outer angles for the HAADF detector.

3. Results and Discussion

An SEM image of the ZnSe NW sample is shown in Fig. 1A. It can be observed that most of the NWs are vertical, parallel to the <111> normal of the substrate, i.e. very likely in epitaxial relation with the ZnSe buffer layers grown on the GaAs (111)B substrate. A few of them are inclined following the other <111> directions of the substrate. The typical diameter of the wires is 10 nm (Fig. 1B) and their length is ~150 nm.

An HAADF STEM image of a NW from this sample is shown in Fig. 1B. Many stacking faults are present along the growth direction giving rise to ZB and WZ regions, growing either [111]-oriented or [0001]-oriented. An averaged zoom of the NW of Fig. 1B is shown in Fig. 1C, where three atomic planes can be observed: the two lower planes have the WZ structure and the two upper planes have the ZB structure respectively, alternatively we could describe Fig. 1C by the ZB structure with a rotational twin or stacking fault between the bottom and the middle plane. The signal to noise ratio in this image is improved by averaging over several regions in the NW around and including the boxed region in Fig. 1B [4]. Thanks to the Cs probe correction of the Titan TEM, the spatial resolution enables us to clearly resolve the dumbbells in the crystal lattice and a sufficient signal to noise ratio allows determining the polarity of the NWs. On most dumbbells a brighter contrast is observed in the top part of the dumbbell. Since growth was carried out on GaAs (111)B, where As (Z=33), the heavier atom, is up in the dumbbell, we find correspondingly that in the NW the heavier atom, being Se (Z=34) is up in the dumbbell. Using an atomic model of the structure we can see that the Zn to Se bond lying in the image plane is pointing down the NW. It seems that irrespective of defects present along the NW, the polarity in the wire follows the polarity imposed by the substrate. In the following these results are compared to simulated HAADF STEM images to investigate possible experimental artifacts.

HAADF STEM experimental images have to be compared to simulations in order to evaluate how much this approach is sensitive to misalignment of the samples regarding polarity determination. Evaluation of the effects of a small specimen tilt on the image is necessary as it is difficult to align the specimen exactly on the crystallographic axis. The diffraction pattern (and Ronchigram in STEM) is easily used to orient the specimen on a crystallographic axis, however, small residual specimen tilts below <1 deg (17 mrad) are difficult to observe but can obviously influence the image contrast. In [5] the effect of a small tilt away from the zone axis orientation was studied using multislice simulation for the example of silicon on different zone axis orientations. It was shown that in the case of a specimen tilt on the order of 1°, a reduction by 2 of the high-resolution image contrast should be expected. More importantly it was shown that the effect of a small crystal tilt can cause an asymmetry in the contrast on dumbbells in a silicon crystal along the [011] zone axis [5,6]. In the simulation, tilting of the specimen is equivalent to shifting the probe function between slices (and is only valid for small tilts of no more than ~17 mrad). Shifting of the probe function between slices will cause asymmetric intensity on the two atoms in the dumbbell. Therefore extreme care should be taken in the determination of the polarity. In practice such small residual tilts can be measured by comparison of micro diffraction patterns with simulations [6].

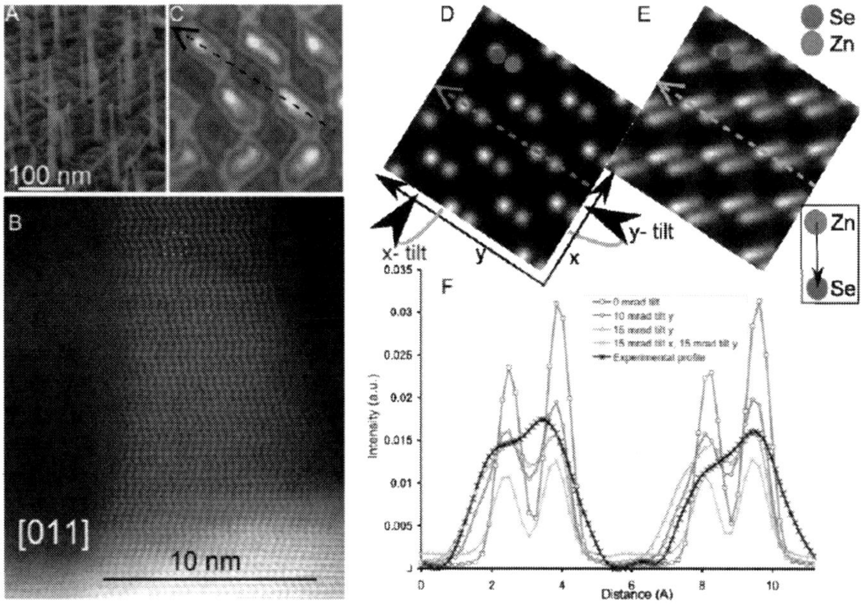

Figure 1. Determination of the nanowire polarity and comparison with simulations. (A) SEM image of NWs grown on GaAs(111)B substrate at 350°C (B) Experimental HAADF STEM image (filtered with a median filter using a 7 px square window) of the NW sample shown in (A). The wire grows in a <111> direction and contains many rotational twin defects/stacking faults on the growth plane. (C) Zoom averaged over several regions close to and including the squared region marked in (B); systematically a brighter contrast is observed on the upper atom in the dumbbell. (D) HAADF STEM simulation along the [011] direction for an 8 nm thick ZB ZnSe unit cell. The Zn and Se atoms are indicated respectively as well as the x and y tilting directions. (E) HAADF simulation with 15 mrad tilt in the x and y direction with respect to the previous simulation and phonon contributions at room temperature. The direction of the Zn to Se bond in the image plane is shown in the inset. (F) Simulated line profiles made along the arrows marked in (D,E) for different y and (x,y)-tilt values off the [011] direction compared with a normalized experimental profile made in (C) along the black arrow, integrated over 0.05 nm.

In the case of epitaxially grown NWs, the specimen is aligned on the crystallographic axis of the substrate. However, as the NWs are very thin and embedded in glue for sample preparation, they can be slightly bent and are therefore not necessarily oriented exactly the same as the substrate. Moreover, as the crystal volume of ZnSe NWs is extremely small, and these samples damage rapidly under the influence of the electron beam, micro diffraction measurements were not possible on these samples. Instead we chose to simulate images and linescans in ZnSe with different specimen tilts up to 15 mrad to see if such tilts would influence the contrast on Zn and Se atoms.

Simulated HAADF STEM images are shown of ZB ZnSe for the [011] axis without specimen tilt (Fig. 1D) and with 15 mrad tilt in the x and y direction (Fig. 1E). The polarity is Se, meaning that the (111) plane is terminated by Se. Clearly the polarity remains visible (meaning that the brightest contrast is observed on the heavier atom being Se) even at relatively important specimen tilts for this sample thickness. In Fig. 1F line scans are shown made along the dumbbells (dotted arrows in Figs. 1D,E) at different x and y tilts. It should be noticed that both x or y tilts weaken the intensity on the dumbbell but a tilt in the x direction preserves the intensity ratio of the two atoms while a tilt in the y direction lowers this ratio. We found that the visibility of the polarity in the case of ZnSe NWs is rather independent of the specimen tilt because the specimen is thin, only 8 nm as compared to 25 and 35 nm considered in [5,6]. At our specimen thickness the difference in atomic number (ΔZ=4) is sufficient to observe the polarity even if a specimen tilt close to 1° is present.

For comparison an experimental profile is shown obtained in Fig 1C. The experimental line profile was qualitatively normalized to allow comparison with the simulated profiles. The minimum intensity in the profile was set to zero and the maximum intensity was set to a value that allowed the best superposition of simulated and experimental profiles. The comparison of simulated and experimental profiles indicates that indeed residual specimen tilt is present. Furthermore the experimental profile is broader than the simulated profiles even at high tilt. Probably a broadening of the intensity is due to the amorphous material present around the NW, mainly glue used for the specimen preparation. This amorphous shell will cause some background noise in the image and loss of spatial resolution, but will not affect relative intensities on atomic columns since the amorphous shell is relatively homogeneous. Also mechanical vibrations can broaden the signal.

4. Conclusion

We have shown that the polarity in ZnSe NWs can be determined using HAADF STEM. In the case of epitaxially grown NWs the polarity is Se, as imposed by the GaAs starting substrate. Comparison with simulated images shows that even if a relatively large residual tilt of the specimen away from the crystallographic zone axis would be present, the polarity remains visible.

Acknowledgements

This work is supported by the French National Research Agency (ANR) through Nanoscience and Nanotechnology Program (Project BONAFO n°ANR-08-NANO-031-01) who provided a research fellowship for MdH. MEJ acknowledges financial support from the Nanosciences Foundation (RTRA). We acknowledge Jean-Luc Rouviere for use of the script to average over several regions in the STEM image.

References

[1] Xia D. et al, *Chin. Phys. Lett.* **23** (2006) 1317.
[2] Robin IC, Aichele T, Bougerol C, André R et al., *Nanotechnology* **18** (2007) 265701.
[3] Kirkland E J, Roane R F and Silcox J, *Utramicroscopy* **23** (1987) 77.
[4] Rouviere J-L, Bougerol C et al., *Appl. Phys. Lett.* **92** (2008) 201904
[5] Maccagnano-Zacher SE, Mkhoyan KA, Kirkland EJ et al., *Ultramicroscopy* **108** (2008) 718-726.
[6] Yamazaki T, Kawasaki M, Watanabe K, Hashimoto I and Shiojiri M, *Ultramiscroscopy* **92** (2002) 181-189.

Reconstruction of an InAs nanowire using geometric and algebraic tomography

R S Pennington[1], S König[2], A Alpers[2], C B Boothroyd[1] and R E Dunin-Borkowski[3,1]

[1] Center for Electron Nanoscopy, Technical University of Denmark, DK-2800 Kongens Lyngby, Denmark
[2] Zentrum Mathematik, Technische Universität München, Boltzmannstr. 3, D-85747 Garching bei München, Germany
[3] Ernst Ruska-Centre for Microscopy and Spectroscopy with Electrons, Forschungszentrum Jülich, D-52425 Jülich, Germany

E-mail: rp@cen.dtu.dk

Abstract. Geometric tomography and conventional algebraic tomography algorithms are used to reconstruct cross-sections of an InAs nanowire from a tilt series of experimental annular dark-field images. Both algorithms are also applied to a test object to assess what factors affect the reconstruction quality. When using the present algorithms, geometric tomography is faster, but artifacts in the reconstruction may be difficult to recognize.

1. Introduction

Conventional algebraic tomography reconstruction algorithms such as the simultaneous iterative reconstruction technique (SIRT) [1] reconstruct a three-dimensional object of varying density from a tilt series of projected images. In contrast, geometric tomography (GT) [2] reconstructs only the shape of the object, using additional information such as e.g. convexity, if applicable.

Here, we reconstruct the cross-section of an InAs nanowire from a tilt series of experimental annular dark-field images using a GT algorithm [3], and compare to the results from a conventional SIRT reconstruction. Additional comparisons using a simulated test object are used to assess the fidelity of each reconstruction and the ease with which artifacts can be identified.

2. Conventional algebraic reconstruction

A high-angle annular dark-field scanning transmission electron microscopy (HAADF STEM) image series of an InAs nanowire was acquired using a probe-aberration-corrected FEI Titan 80-300 operated at 300 kV with a probe convergence semi-angle of 18 mrad and an inner detector semi-angle of 100 mrad. The images were acquired over a total angular range of 139° with a 1° tilt increment. The original 2048×2048 pixel images were then binned by a factor of 4 to reduce noise in the reconstruction. A bright-field TEM image of the nanowire and two representative HAADF STEM images and line profiles from the tilt series are shown in Fig. 1.

SIRT reconstruction was performed using the TomoJ software with 15 iterations [4, 5] after aligning the original tilt series using cross-correlation and a Canny-Deriche edge-detection filter. Cross-sections of the wire determined from reconstructions using 14 and 140 projections are shown in Fig. 2 and Fig. 3, respectively, for both of the sections of the nanowire marked in Fig. 1.

The SIRT reconstruction is blurred in the direction of the "missing wedge", contains artifacts in the region outside the wire, and requires a subjective choice of threshold to find the nanowire boundary.

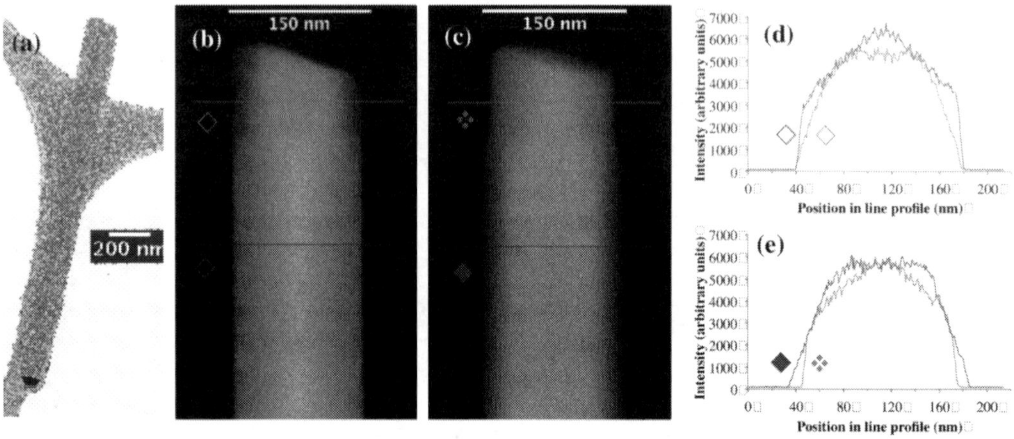

Figure 1. (a) is a bright-field TEM image of the InAs nanowire specimen used for tomography; (b) and (c) aligned binned HAADF STEM images taken from the tilt series of the wire at angles of 7° and 37°, respectively, with 8-pixel-wide line profiles from the marked regions shown in (d) and (e), exhibiting the roughly-hexagonal faceting expected from the reconstruction, but not noise levels. There is a growth defect between the red and blue line profile, leading to rotation in the blue profile relative to the red.

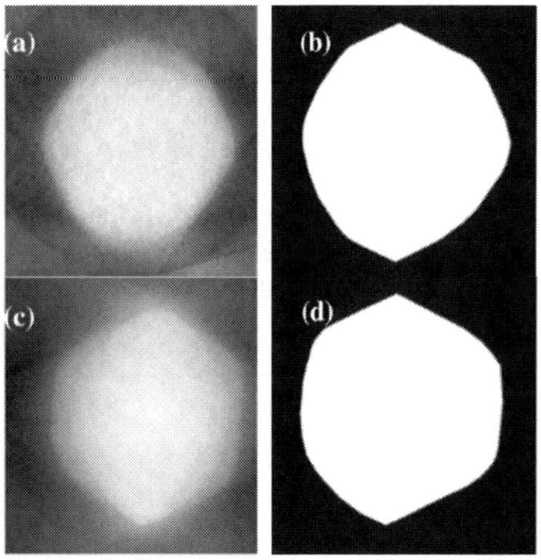

Figure 2. Reconstructed cross-sections from HAADF STEM tilt series using (a,c) SIRT and (b,d) GT algorithms from 14 equally-spaced projections over a tilt range of 139°. (a,b) are from the pale red line in Fig. 1, (c,d) from the dark blue line. Nanowire width is 140 nm measured from profiles in Fig. 1.

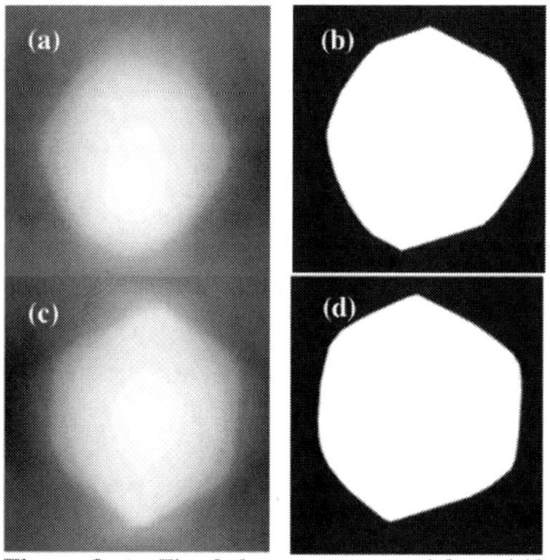

Figure 3. As Fig. 2, but reconstructed from 140 equally-spaced projections over a tilt range of 139°.

3. Geometric tomography from shadow outlines

The GT algorithm used in the present study, introduced by Gardner and Kiderlen [3], reconstructs only from the shadow outlines in each projection, neglecting any density variation. It is computationally simple: after thresholding, it fits a polygon to be consistent with the shadow outline from all projections. The algorithm used here converges as the number of projections approaches infinity [6]. Our implementation is more than an order of magnitude faster than SIRT [4,5]. Previous work has applied the GT concept to TEM images, but the back-projection algorithm used in [7] needs both post-reconstruction thresholding and is not proved to be convergent; the algorithm used in [8] is also not proved to be convergent. Our GT implementation [3] was applied to the same nanowire cross-sections (Fig. 1) that were used in section 2 after thresholding the input images at a value of 300 counts. For the present sample, our GT reconstructions from 14 and 140 projections are nearly identical, but the SIRT reconstructions improve with increased sampling. Our GT implementation and SIRT are in reasonable agreement outside the areas affected by the missing-wedge at the top and the bottom of the wire (Figs. 2, 3), where the SIRT reconstruction is blurred and our GT reconstruction is fully uninterpretable.

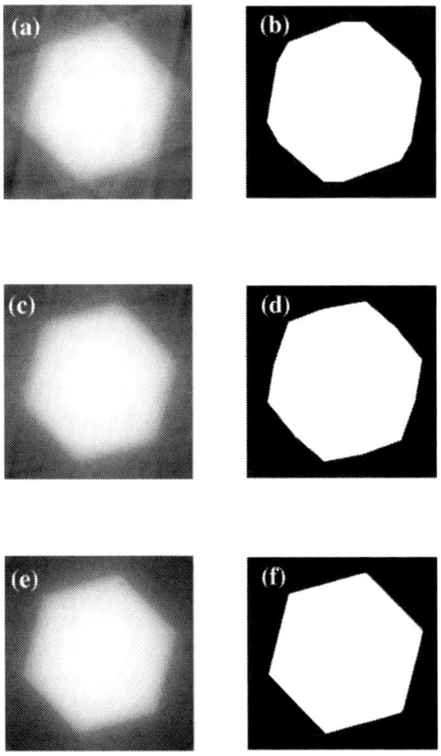

Figure 4. The effect of angular spacing on test data reconstruction for (a,c,e) SIRT and (b,d,f) geometric tomography for (a,b) 20°, (c,d) 10°, and (e,f) 1° with no noise and 180° tilt range.

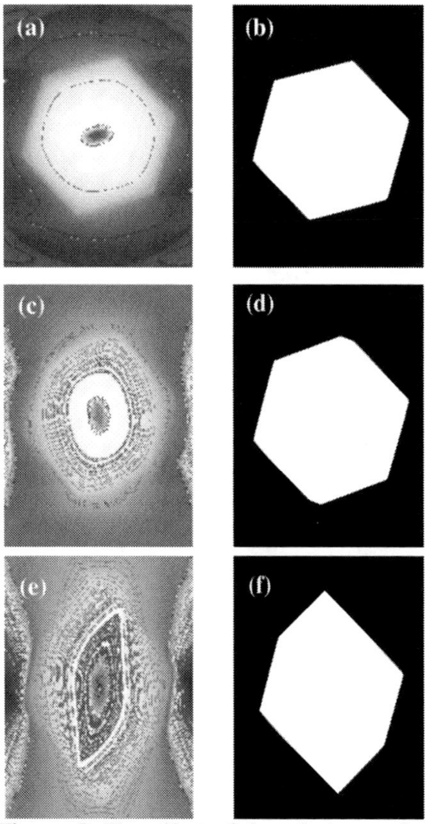

Figure 5. The effect of the missing wedge on test data reconstruction for (a,c,e) SIRT and (b,d,f) geometric tomography for (a,b) 160°, (c,d) 140°, and (e,f) 90° tilt range with no noise and 1° angular spacing.

4. Comparison of both geometric and algebraic algorithms using test data

To compare how individual factors affect this GT implementation and conventional SIRT, a simulated tilt series was generated from a regular, binary hexagon. The angular spacing between images (i.e. number of projections), tilt range and noise level were all individually varied, and image intensity fluctuations were used to simulate diffraction contrast.

Increasing the angular spacing creates artifacts outside the wire for the SIRT reconstruction, but a decrease in corner sharpness using GT (Fig. 4). Limiting the tilt range leads to missing-wedge effects in both algorithms, but the GT missing-wedge is more similar to the rest of the reconstruction (Fig. 5). The GT reconstruction can be sensitive to noise if it makes thresholding difficult, and this can lead to e.g. apparent facet artifacts. (Fig. 6), which may be a problem for the experimental GT reconstruction in section 3. For diffraction contrast, the SIRT reconstruction shows minimal change, but the GT reconstruction shows no change, because the model used here does not affect the outlines (Fig. 7).

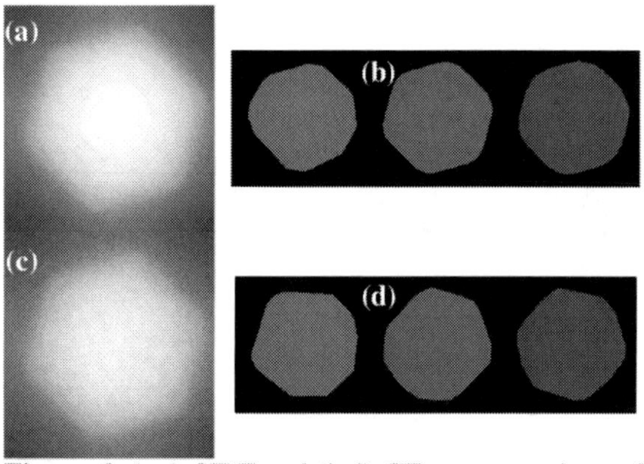

Figure 6. (a,c) SIRT and (b,d) GT reconstructions of the test hexagon with Gaussian noise level 10× (top) and 80× (bottom) the original intensity per object pixel. Color goes from red to green to blue (left to right) in (b,d) as the pre-reconstruction threshold T increases (b: T=300,500,700) (d: T=2000,2700,2900).

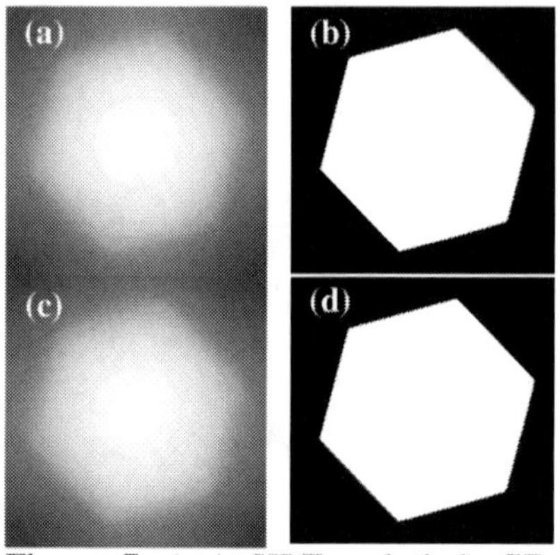

Figure 7. (a,c) SIRT and (b,d) GT reconstructions of the test hexagon for random diffraction contrast (50% chance of ≤25% intensity increase (a,b) or ≤75% intensity increase (c,d)).

5. Conclusions

A geometric tomography algorithm based on image outlines has been compared with algebraic tomography for an InAs nanowire with a convex cross-section. The missing wedge is a problem for both algorithms. Artifacts in the reconstruction may be more difficult to recognize for geometric tomography. Other geometric tomography algorithms can use the image intensity (e.g. [9]), instead of only the shadow outlines, potentially yielding a better reconstruction, especially in the missing-wedge region.

References

[1] Herman G T 2009 *Fundamentals of Computerized Tomography*, (London: Springer)
[2] Gardner R J 2006 *Geometric Tomography* (New York: Cambridge University Press)
[3] Gardner R J and Kiderlen M 2009 *IEEE Trans. Pattern Anal. Machine Intell.* **31** 556-562
[4] Abràmoff M D, Magalhães P J and Ram S J 2004 *Biophotonics International* **11** 36-43
[5] Messaoudil C, Boudier T, Sanchez Sorzano CO and Marco S 2007 *BMC Bioinformatics* **8** 288
[6] Gardner R J, Kiderlen M and Milanfar P 2006 *Ann. Statist.* **34** 1331-1374
[7] Saghi Z, Xu X and Möbus G 2008 *Journal of Microscopy* **232** 186-195
[8] Petersen T C and Ringer S P 2009 *J. Appl. Phys.* **105**, 103518
[9] Gardner R J and Milanfar P 2003 *Discrete and Computational Geometry* **29** 279-303

Towards quantitative three-dimensional characterisation of buried InAs quantum dots

S Kadkhodazadeh[1,2], E S Semenova[2], M Schubert[2], M Thuvander[3], K M Stiller[3], K Yvind[2] and R E Dunin-Borkowski[1]

[1]Center for Electron Nanoscopy, Technical University of Denmark, DK-2800, Kgs. Lyngby, Denmark

[2]Department of Photonic Engineering, Technical University of Denmark, DK-2800 Kgs. Lyngby, Denmark

[3]Chalmers University of Technology, Applied Physics, 412 96 Gothenburg, Sweden

shka@cen.dtu.dk

Abstract. InAs quantum dots grown on InP or InGaAsP are used for optical communication applications operating in the 1.3 – 1.55 μm wavelength range. It is generally understood that the optical properties of such dots are highly dependent on their structural and chemical profiles. However, morphological and compositional measurements of quantum dots using transmission electron microscopy can be ambiguous because the recorded signal is usually a projection through the thickness of the specimen. Here, we discuss the application of scanning transmission electron microscopy tomography to the morphological and chemical characterisation of surface and buried quantum dots. We highlight some of the challenges involved and introduce a new specimen preparation method for creating needle-shaped specimens that each contain multiple dots and are suitable for both scanning transmission electron microscopy tomography and atom probe tomography.

1. Introduction

Self-assembled III-V semiconductor quantum dots (QDs) are of interest for applications that include optical amplifiers, temperature-stable low-noise lasers and mode-locked lasers [*e.g.* 1-3]. Important electronic properties such as density of states and internal carrier dynamics, which determine the optical characteristics of the QDs, are largely affected by the morphologies and chemical compositions of both the QDs and the material surrounding them. A thorough understanding of these properties is therefore essential for successful applications. The majority of structural characterisation carried out on InAs QDs has involved the study of either uncapped QDs using atomic force microscopy (AFM), or projections through the thickness of specimens in transmission electron microscopy (TEM). However, the spatial resolution of AFM is often insufficient to accurately determine the exact morphologies of QDs. Furthermore, most device applications require QDs to be overgrown with a material that has a different bandgap and lattice parameter. This step, known as capping, can introduce significant changes to the structure and chemistry of the QDs. While conventional TEM techniques can be used to study capped QDs in plan-view or cross-sectional geometries, the resulting images can give ambiguous information about their three-dimensional properties. Here, we discuss the applicability of high angle annular dark field (HAADF) scanning transmission electron microscopy (STEM) tomography to the study of InAs/InGaAsP QDs. We discuss the challenges of three-dimensional characterisation of buried QDs and present a new method of preparing specimens for combined experiments using both electron tomography and atom probe tomography (APT).

2. Results and discussion

InAs/InGaAsP samples comprising 1.65 ML of InAs deposited on (001) $In_{0.85}Ga_{0.15}As_{0.35}P_{0.65}$ (lattice matched to InP) were grown by metal organic vapour phase epitaxy (MOVPE). Buried QDs were overgrown with 100 nm of InGaAsP. Plan-view and cross-sectional TEM specimens were prepared using mechanical polishing, dimpling and argon ion milling and examined at 300 kV using an FEI Titan TEM equipped with a field emission electron source and a spherical aberration corrector on the condenser lens system. Inner and outer annular dark-field detector collection semi-angles of 48 and 240 mrad, respectively, were used with an electron probe convergence semi-angle of 17.5 mrad. Alignment and reconstruction of tomographic data was carried out using the simultaneous iterative reconstruction technique (SIRT) algorithm in Inspect3D software (typically 15 iterations were used). Visualisation of reconstructions was performed using Aviso v.6 software.

2.1. Morphological characterisation using electron tomography

HAADF STEM tomography was carried out on plan-view specimens to study the morphologies of surface QDs. Figure 1(a) shows a HAADF STEM image of a representative surface QD. Consistently, elongated hexagonal shapes were observed for the bases of the surface QDs. The elongation direction was determined to be [110] using selected area electron diffraction, high-resolution HAADF STEM and AFM. The HAADF STEM images also suggest a double-terraced geometry for the surface QDs, with steeper facets around their bases and shallower facets towards their tops. This observed geometry is consistent with a theoretical model of InAs QDs on an InGaAs substrate lattice matched to InP [4] shown in figure 1(b). A tomographic reconstruction of a QD produced using a series of 67 images obtained at 2° intervals over a tilt range of ± 66° is displayed in figure 1(c). The elongated hexagonal base is visible in the three-dimensional reconstruction.

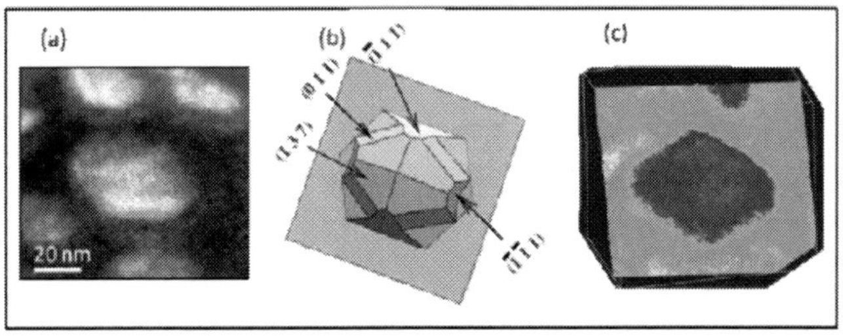

Figure 1. (a) HAADF STEM image of a surface InAs QD on an InGaAsP substrate that is lattice matched to InP. (b) Theoretical energetically favourable shape of InAs QDs on an InGaAs substrate lattice matched to InP, taken from [4]. (c) Tomographic reconstruction of a surface QD from 67 HAADF STEM images taken at 2° intervals over the tilt range ±66° using the SIRT algorithm

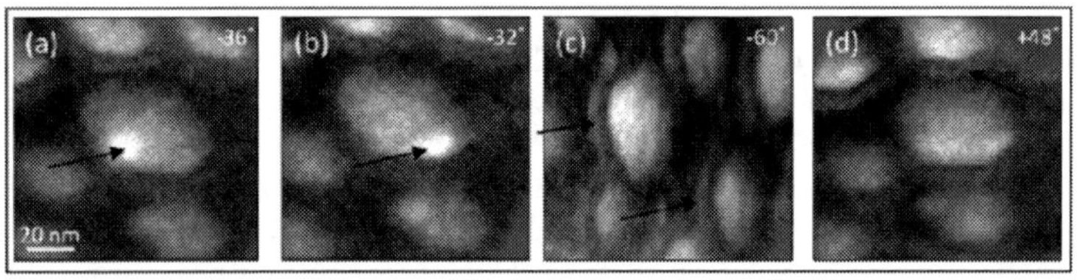

Figure 2. HAADF STEM images of a surface InAs/InGaAsP QD at different tilt angles, showing different diffraction effects appearing at different tilt angles

Despite the large inner detector angle used, interesting diffraction effects appear in the tilt series of HAADF STEM images. Figures 2(a) and (b) show regions of localised bright contrast that change rapidly with specimen tilt and may be associated with surface strain. Furthermore, at large tilt angles bright bands appear around the bases of QDs, as shown in figures 2(c) and (d). These bands may also be associated with diffraction effects. These features are interesting in their own right. However, they are inconsistent with the projection requirement, which states that the recorded intensity should be a monotonic function of a property of the object [5], and may therefore lead to artefacts in the tomographic reconstruction.

HAADF STEM tomography of capped QDs was attempted in both cross-sectional and plan-view specimen geometries. In both cases, a limitation was found to be the available tilt range before the specimen became too thick for imaging. In the case of cross-sectional specimens, it is also possible that only a section of a QD may be contained within the specimen. The tilt axis chosen for tomography of cross-sectional specimens was found to be particularly important, as the use of tilt axes other than [001] (the growth direction) resulted in the signal from the wetting layer becoming very weak after only a few degrees of tilt. Figure 3 shows an attempt at three-dimensional reconstruction of a capped InAs/InGaAsP QD reconstructed from 37 HAADF STEM images acquired at 2° intervals over a tilt range of only ± 36°. Unfortunately, the images became too blurred and noisy beyond this tilt range. Elongation due to the large missing wedge in Fourier space resulted in artefacts in the reconstruction, which make complete morphological characterisation impossible. Moreover, the contrast arising from strong electron channelling at low order zone-axis orientations results in significant departures from the projection requirement for tomography. HAADF STEM tomography analysis of a buried InAs/InGaAsP prepared in plan-view geometry is shown in figure 4. A series of 61 images acquired at 2° intervals over a ± 60° tilt range was used for reconstruction. The QD appears to have an elongated shape along the <110> direction. Although evidence of faceting of the QD is present in both the original HAADF STEM images and the final reconstruction, the facets are not as clearly visible as those of the surface QD in figure 1. This difference may result from chemical intermixing between the QD and the capping material during overgrowth, as well as from artefacts in the tomographic reconstruction.

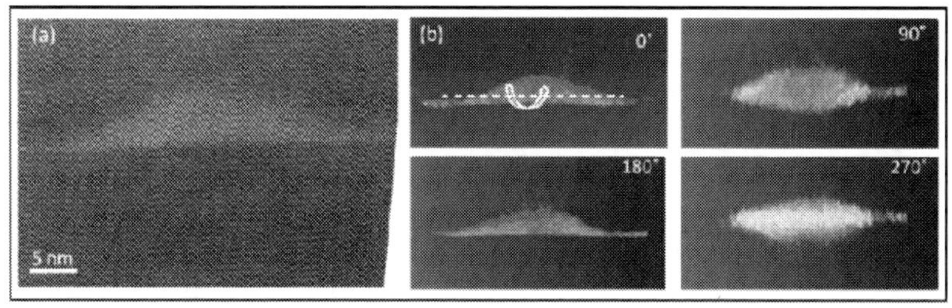

Figure 3. (a) HAADF STEM image of a buried InAs/InGaAsP QD viewed along the [110] direction in cross-sectional geometry. (b) Reconstructed image of the same QD from 37 images acquired at 2° intervals over a ±36° tilt range. The reconstruction was carried out using the SIRT algorithm (15 iterations). The arrow indicates the direction of rotation.

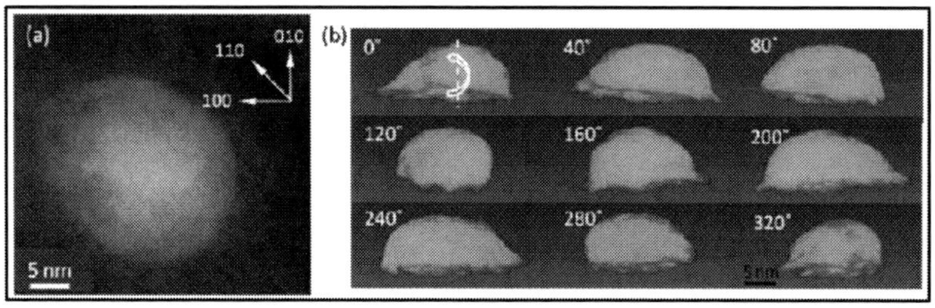

Figure 4. (a) HAADF STEM image of a buried InAs/InGaAsP QD in plan-view geometry viewed along the [001] orientation. (b) Tomographic reconstruction of the QD from 61 images acquired at 2° intervals over a ±60° tilt range using the SIRT algorithm (15 iterations). The arrow indicates the direction of rotation.

2.2. Fabrication of needle-shaped specimens

The specimen thickness limitation in electron tomography can, in principle, be overcome by fabricating needle-shaped specimens using focused ion beam (FIB) milling, to allow tilting by up to 180° without significant increase in projected specimen thickness. However, the preparation of a needle-shaped specimen containing a single QD is highly challenging, as most QDs are too small to be visible during FIB preparation. Moreover, the FIB can introduce considerable damage in III-V semiconductors, including amorphisation and gallium ion implantation [6]. We have fabricated needle-shaped specimens that are 100 nm in diameter, using reactive ion etching, selective wet etching and critical point drying in plan-view geometry (figure 5). The choice of a plan-view geometry for the needles means that each specimen contains several QDs. The needles can be detached from the substrate by cleaving (see figure 5(b)) or lifted out and mounted onto suitable grids using a micro-manipulator in the FIB with minimal additional damage (figure 5(c)). Besides electron tomography, these needle-shaped specimens can be used for APT, for which needle-shaped specimens with sharp tips (narrower than 100 nm) are required. Our ongoing experiments involve the application of both HAADF STEM tomography and APT to the same QDs in order to better understand their morphologies and compositions. A comparison between the reconstructions obtained using these two techniques will also assist in the evaluation and mitigation of potential artefacts in each case.

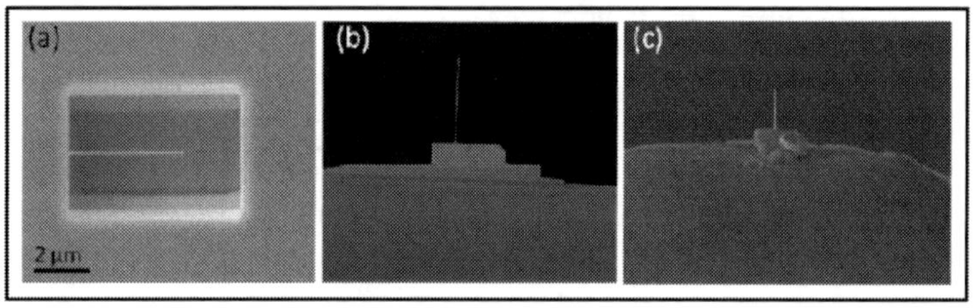

Figure 5. (a) Scanning electron microscopy images of needle-shaped specimens fabricated using reactive ion etching, selective wet etching and critical point drying. The needle specimens can be (b) detached from substrates by cleaving or (c) lifted out and mounted onto a suitable grid.

3. Conclusions

HAADF STEM tomography of surface and buried InAs/InGaAsP QDs has been presented and some of the challenges of the application this technique for the morphological and chemical characterisation of buried QDs have been discussed. A new method of specimen preparation suitable for both electron tomography and APT of buried QDs has been introduced.

References

[1] Akiyama T and Arakawa Y 2007 *Proc. IEEE* **95** 1757
[2] Sugwara M and Usami M 2009 *Nat. Photonics* **3** 30
[3] Thompson M G, Rae A R, Xia M, Penty R V and White I H 2009 *IEEE J. Sel. Top. Quantum Electron.* **15** 661
[4] Kratzer P, Chakrabarti A, Liu K K Q, Scheffler M 2009 *New J. Phys.* **11** 73018
[5] Hawkes P W 1992 *The electron microscope as a structure projector: three-dimensional imaging with the transmission electron microscope* (New York: Plenum Press) pp 17-38
[6] Rubanov S, Munroe P R 2005 *Microsc. Microanal.* **11** 446

Measuring two dimensional strain state of AlN quantum dots in GaN nanowires by nanobeam electron diffraction

A Béché[1], C Bougerol[2], D Cooper[3], B Daudin[3] and JL Rouvière[3]

[1] FEI Company, Achtseweg Noord 5, 5600 GG Eindhoven, The Netherlands
[2] Institut Néel, CNRS, BP 166, 25 rue des Martyrs, 38042 Grenoble, France
[3] CEA-Grenoble, 38054 Grenoble, France

E-mail: armand.beche@fei.com

Abstract. Nanobeam electron diffraction (NBED or NBD) is a transmission electron microscope (TEM) based technique able to measure strain with nanometre spatial resolution. This technique is especially well designed for nanostructures as reference diffraction pattern can be acquired anywhere on the sample. NBED is here applied on AlN/GaN superlattices grown on SiC and results compared to finite element simulations in order to assess the accuracy of the method. Strain profiles were then acquired with success on GaN nanowires presenting 3 nm GaN inclusions separated by 3 nm thick AlN.

1. Introduction

Measuring strain at the nanometer scale became of heavy need these last ten to fifteen years due to the introduction of strain in devices on the ten nanometers scale in the microelectronic industry [1]. Presently, the only instrument able to achieve strain measurement at this scale with sufficient strain sensitivity is the transmission electron microscope (TEM). This versatile instrument offers different ways of measuring strain, in real space (high resolution TEM [2,3], dark field electron holography (DFEH) [3,4]) or reciprocal space (convergent beam electron diffraction [5], nanobeam electron diffraction [6]).

Nanobeam electron diffraction (NBED or NBD) consists of forming an almost parallel sided electron probe with nanometer probe diameter that will be scanned over the sample surface. Compared to the other techniques, NBED is well designed for measuring strain on nano-objects as it can be applied to thin samples, in contrast to CBED, it does not require a reference close to the region of interest unlikely DFEH and offers a very large field of view compared to HRTEM. Moreover, if the structure is aligned along a zone axis, strain can be measured in the two directions normal to the electron beam.

These abilities make NBED an excellent technique for measuring strain in nano-objects and it is applied to GaN/AlN nanowires in the present study. In order to realize reliable strain measurement, NBED profiles were first measured for an AlN/GaN superlattice grown on SiC and compared to finite element simulations.

2. Experimental conditions

NBED strain profiles were realized on an FEI Titan microscope equipped with a Cs probe correction module which was acting as pure transfer lens during the experiment. The microscope high tension was set to 200kV and a high angle annular dark field detector (HAADF) was used to record an image of the sample and precisely locate the probe on the region of interest. Diffraction patterns were acquired on a 2k x 2k Gatan Ultrascan charge-coupled device (CCD) camera located before the Gatan Imaging Filter.

As in most TEM based techniques, strain measurement was made relative to a reference region in a relaxed strain state. In the following, the strain ε will then be defined using the *Lagrange* formalism $\varepsilon=(a_{roi}-a_{ref})/a_{ref}$ where a_{ref} is the lattice parameter of the reference region and a_{roi} the lattice parameter of the region of interest. Moreover, all axes directions have been defined such that x is the growth direction, y the in-plane direction and z the direction of electron propagation. The processing of the diffraction patterns was done using a specially designed program based on an automated diffraction spot fitting algorithm.

The first sample is a superlattice consisting of an arrangement of 6 layers of GaN(16 nm)/AlN(13 nm) deposited on top of a 60 nm thick AlN layer grown on a (1-100) SiC substrate. NBED was mainly used on this sample to determine if relaxation was occurring in one specific direction. Two samples were prepared in two different crystallographic orientations, [0001] and [11-20], by FIB milling at low energy with extra care to prevent modification of the strain field.

The second sample consists of GaN nanowires of 50 nm in diameter grown on <111> Si substrates. The top parts are composed of 11 periods of 3 nm GaN separated by 3 nm AlN barriers. The sample, made by simple cleavage of the substrate and the wires, was studied in the cross-section direction.

3. AlN/GaN superlaticce on SiC

The first NBED experiment was conducted on the AlN/GaN superlattice and results compared to finite element simulations. This comparison allows checking the reliability of strain measurement on such materials.

For this experiment, an electron probe of 6 nm in diameter was created using a 20 µm C2 aperture, resulting in 0.4 mrad semi-convergence angle. These conditions were specifically chosen for this sample as they allowed us to obtain good spatial resolution with respect to the geometry and excellent strain sensitivity. A picture of the probe and a corresponding diffraction pattern taken through a crystal oriented along the [11-20] zone axis are displayed in Figure 1.

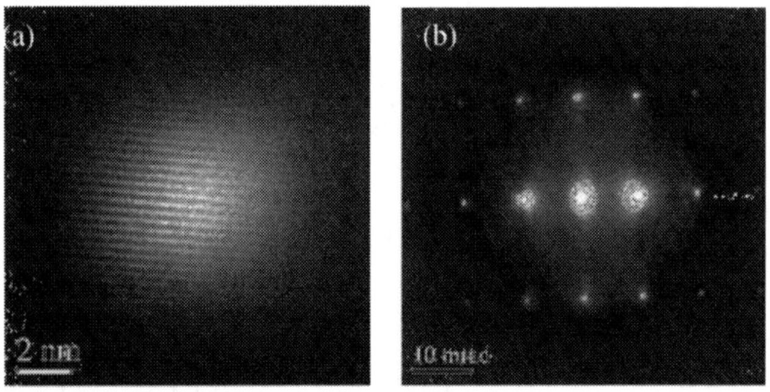

Figure 1. (a) Image of the electron probe through a [11-20] SiC crystal using a 20 µm C2 aperture. (b) Diffraction pattern in the [11-20] zone axis acquired using the probe displayed in (a).

Figure 2a displays a HAADF image of the sample acquired with the previously cited NBED probe. The first profile in Figure 2b, acquired along the [11-20] direction, reveals almost perfect epitaxy between the substrate and the superlattice, the strain in the y direction being equal to zero. In the x direction, the difference in strain between AlN and GaN layers is clearly visible and each layer can be precisely defined. The first 60 nm thick AlN layer appears in a strain state of ~1% compared to the SiC substrate. The GaN layers are also in tensile strain, but with much higher strain value: 4.3%. It is interesting to notice that the noise in the unstrained region of the sample is very low (±0.05%).

Journal of Physics: Conference Series **326** (2011) 012047 doi:10.1088/1742-6596/326/1/012047

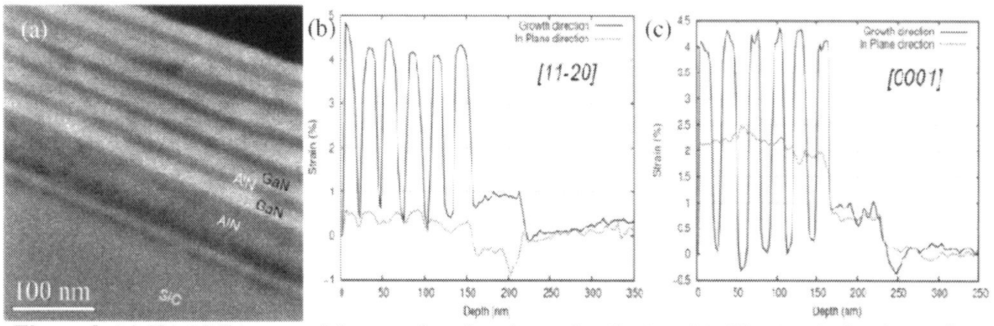

Figure 2. (a) HAADF image of the sample using the probe displayed in Fig. 1a. (b) Strain profiles acquired along the [11-20] zone axis. (c) Strain profiles acquired along the [0001] zone axis.

The same experiment was then realized on a sample oriented along the [0001] zone axis, as displayed in Figure 2c. In the x direction, a behaviour similar to the one observed in the [11-20] direction occurred: the first AlN layer is 1% tensely strained and the GaN layers on top are 4.3% strained. However, the y direction now also reveals a strain relaxation of the thick AlN layer as its strain state is just under 1%. This reveals the presence of defects (no preservation of the epitaxial conditions) in the in-plane direction (11-20), the zone axis direction in the previous sample. The GaN/AlN superlattice presents an even higher relaxation with 2% difference in lattice parameter compared to the SiC substrate.

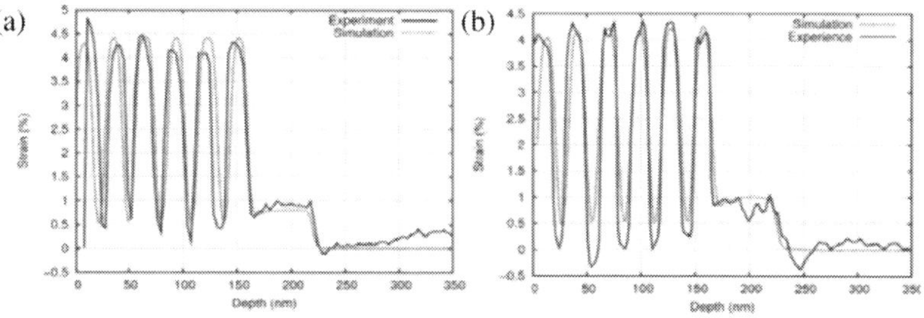

Figure 3. (a) Comparison between experimental and simulated strain profile for the [11-20] direction. (b) Same comparison for the [0001] direction.

Numerical modelling of the AlN/GaN superlattice sample has been performed on both previous samples in order to assess the reliability of the NBED measurement. The Comsol Multiphysics software was used to model the strain relaxation which appears in the thin TEM lamella. Figures 3a and 3b display the comparison between the experimental and the simulated strain profiles in the x direction for both [11-20] and [0001] orientations. The relaxation effect in the y direction for the second sample in the (11-20) orientation was taken into account, as well as the convolution imposed by the probe size [6].

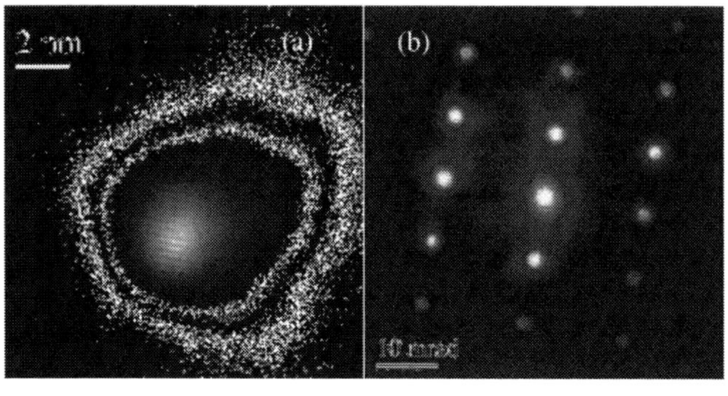

Figure 4. (a) Image of the electron probe through a (11-20) GaN crystal using a 50 μm C2 aperture. (b) Diffraction pattern in the (11-20) zone axis acquired using the probe displayed in (a). The difference in convergence compared to Figure 1b is clearly visible in the fainter diffraction spots.

The agreement between experimental and simulated curves is really impressive and the small differences can be attributed to the lack of homogeneity of the layer thicknesses and orientations. These results prove that NBED is a reliable technique for measuring strain that can be used on more complicated devices.

4. GaN nanowires with AlN/GaN superlattice

As the thickness of the layers deposited on top of the GaN nanowires are much smaller than the one described in the previous section, the same NBED settings cannot be used. Indeed, a 6 nm probe diameter would have led to a blurring of the AlN/GaN layers and a lack of details in the strain profiles. In order to decrease the probe diameter, a bigger C2 aperture was selected (50 µm) which results in the formation of a 2 nm diameter probe (Figure 4a). As the probe diameter reduces, the Rayleigh criterion implies an increase of the semi-convergence angle to 1 mrad. The diffraction pattern, displayed in Figure 4b, reveals wider diffraction spots which will result in a decrease of the strain sensitivity of the NBED measurement.

This NBED probe was then employed on the nanowires to obtain the image displayed in Figure 5a. A strain profile was then acquired along the dashed line and is reported in Figure 5b. For this measurement, the nanowire was oriented along the (11-20) zone axis and the first GaN part taken as a reference. The noise level in the unstrained part of the wire (± 0.15%) is much higher than in section 3 due to the increase in semi-convergence angle. The relaxation of a part of the strain (-1.5 to -2%) in the *y* direction is clearly visible in the superlattice but does not correspond to a fully relaxed state (-4%), contrary to the result for the *x* direction. This relaxation in the growth direction was expected as this upper surface is a free surface.

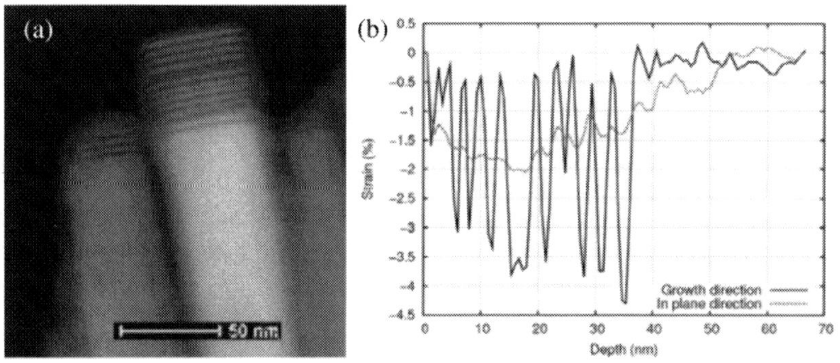

Figure 5. (a) HAADF image of the GaN/AlN nanowire acquired with the probe displayed in Fig. 4a. (b) Strain profile acquired in the middle along the nanowire displayed in (a).

5. Conclusion

The evolution of the transmission electron microscopes these last five years allows obtaining nanometer scaled probe with almost parallel beams and strain profiles with very good strain sensitivity (5×10^{-4}). The nanobeam electron diffraction technique (NBED or NBD) is then a well suited method for measuring strain at the nanometre scale in nano-objects such as nanowires for several reasons: it does not required a reference region close to the region of interest, it can be applied on a sample with a wide range of thicknesses, and it is easy to process.

References

[1] Auth C et al.2008 *Intel Technology Journal* **12**:02, p77-85.
[2] Hÿtch MJ, Snoeck E and Kilaas R 1998 *Ultramicroscopy* **74**:3, p131-146.
[3] Sarigiannidou E et al.2005 *Applied Physics Letters* **87**:20, p203112.
[4] Hÿtch M, Houdellier F, Hüe F and Snoeck E 2008, *Nature* **453**, p1086-1089.
[5] Béché A, Rouvière JL, Barnes JP and Cooper D 2011 *Ultramicroscopy* **111**:3, p227-238.
[6] Béché et al. 2009 *Applied Physics Letters* **95**:12, p123114.

Three-dimensional measurement of composition changes in InAs/GaAs quantum dots

A Wahra[1], C-H Tan[2], J Xie[2], P Vines[2] and R Beanland[1]

[1] Department of Physics, University of Warwick, Coventry CV4 7AL, United Kingdom
[2] Department of Electronic and Electrical Engineering, University of Sheffield, Mappin St, Sheffield S1 3JD United Kingdom

E-mail: a.wahra@warwick.ac.uk

Abstract. We present a transmission electron microscope (TEM) analysis of InAs/GaAs quantum dots (QDs) which have been subject to thermal annealing. Annealing produces a significant shift in infrared detector response to longer wavelengths, indicating changes in QD size, composition, or both. A three-dimensional reconstruction of an 'average' QD is obtained by combining compositionally-sensitive dark field 002 TEM images of many QDs, followed by fitting to a structural model assuming cylindrical symmetry. Errors in compositional measurements are estimated, and the validity of the model is assessed by TEM image simulations using 2-beam dynamical electron diffraction theory. As-grown material exhibits an increasing concentration of In towards the top of the QDs, whereas annealed material shows diffusion of indium from the upper part of the QD towards the sides and base. There is no measurable change in the position of the outer interface of the QD.

1. Introduction

The properties of InAs/GaAs quantum-dot (QD) structures are of great interest for many optoelectronic devices such as lasers for telecommunications [1], optical amplifiers and detectors [2,3]. Indium incorporation during growth, and redistribution during thermal treatment, plays a significant role in determining their optical and electronic properties. Annealing leads to a blue shift and a narrowing in optical emission, and has been shown to alter the size, shape and the chemical composition of QDs [4-6]. A variety of transmission electron microscopy (TEM) techniques have been used to evaluate the composition in QDs (e.g. [7-9]); however, a simple method of three-dimensional compositional analysis does not yet exist since all TEM images are a 2D projection of the 3D structure [10]. Significant changes in QD shape, size and composition are known to occur during encapsulation by overlying material [11]; uncapped dots are likely to be significantly different to those in bulk material, leading to the requirement for characterisation of QDs fully inside a matrix of different material. Here, we assess the possibility of using 002 dark-field (DF) diffraction contrast images to obtain quantitative 3D composition measurements using a simple structural model. Quantum dot infrared photodetector (QDIP) structures were grown on (001) GaAs by molecular beam epitaxy; 50nm thick upper and lower cladding layers of $Al_{0.1}Ga_{0.9}As$ surrounded a 30-repeat QD structure, in which nominally 2.4ML InAs QDs were deposited at 510°C between 1nm $In_{0.15}Ga_{0.85}$ layers separated by 40nm GaAs spacers. Modulation doping was employed in the spacer layer at an equivalent value of 2 electrons/dot. QD densities were 3 x 10^{10} cm^{-2} and exhibited room temperature

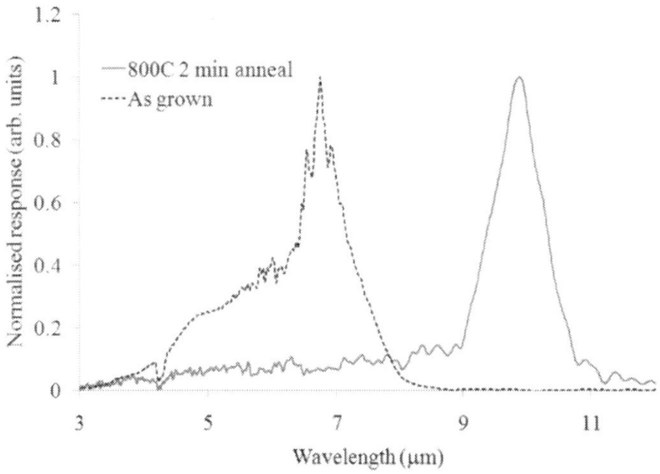

Figure 1. Photoresponse of as-grown and annealed QDIP structures.

photoluminescence at 1260 nm. Post-growth annealing was carried out at 800°C for 2 minutes. As-grown samples exhibited peak photoresponse at 6.75 μm with an applied bias of -14V which red shifted to almost 10 μm after annealing (Fig. 1), caused by a large change in bandgap (i.e. intermixing). Cross-section specimens were prepared using standard techniques (mechanical thinning followed by ion milling using Ar^+ at 6 and 3kV) and examined in a JEOL 2000FX TEM at 200kV.

2. Results

Figure 2 shows dark field 002 TEM images of (a) as-grown and (b) annealed QDs. These images are an average of several QDs with imposed mirror symmetry in order to reduce noise in the data. The InAs QD appears as bright central region (high In content) surrounded by a dark halo. Although the annealed QD appears larger, the position of the outer dark line is essentially unchanged; a more accurate description is that In and Ga are redistributed inside the dot during annealing to give a more uniform internal composition.

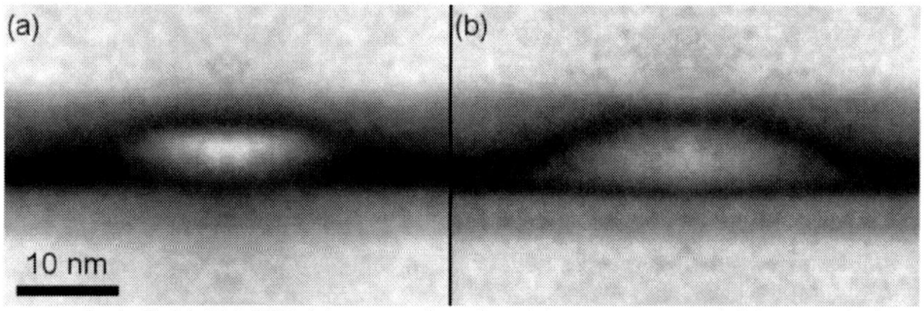

Figure 2. Average dark field 002 TEM images showing (a) as-grown and (b) annealed InAs QDs.

3. Analysis

Previous studies of InAs QDs have shown that they appear almost identical when viewed from any viewpoint perpendicular to the growth direction [12], i.e. their symmetry is close to cylindrical. This suggests that a structural model with cylindrical symmetry can be used, which greatly reduces the computational effort required to extract compositional data from images such as Fig. 2. Furthermore, since the 002 diffracted beam is relatively weak, close to kinematic electron diffraction conditions apply and the image may be regarded as a simple 2D projection I_{2D} of a 3D structure I_{3D}. Under the condition of cylindrical symmetry, we may represent any smoothly varying 3D structure using a series of n Gaussian functions, i.e.

$$I_{2D} = \int_{-t/2}^{t/2} I_{3D}\, dz \quad , \quad I_{3D} = \sum_{i=1}^{n} a_i \exp\left[-\frac{(x-x_0)^2+z^2}{2\sigma_i^2}\right] + C \tag{1}$$

where the projection direction is along the z direction, a_i, and σ_i are parameters describing the ith Gaussian function and C is a constant (background) value. In this analytical approach, the 3D model can be obtained directly from the 2D image since the relationship between I_{2D} and I_{3D} is exact. The specimen thickness is an unknown parameter and thus must be supplied by other measurements or by obtaining the best fit between the model and experiment. Thus, by fitting each horizontal line in the image to a series of Gaussian functions + constant background it is possible to produce a '3D intensity model' for a specimen of thickness t. In practice, we find that the series can be truncated at $n = 2$ while still giving a satisfactory fit to the data. A 3D composition distribution X_{3D} may be obtained from I_{3D} using the kinematic relationship between intensity and composition as shown in Fig. 3. For the $In_xGa_{1-x}As$ system this is double valued for compositions below $x \approx 0.47$; a choice must thus be made between the two solutions. Here, since the QD has a GaAs substrate and capping layer we may propagate the lower composition value (from outside the QD towards its centre) to the point at which the gradient of intensity reverses. This results in an artificially sharp boundary between the QD and surrounding material in the composition maps.

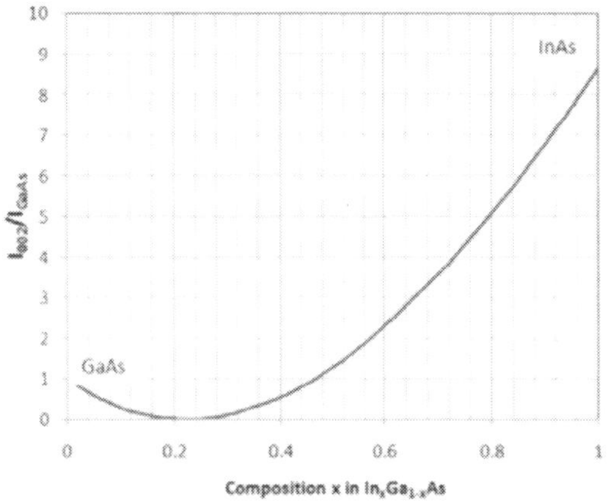

Figure 3. Relationship between intensity I_{002}/I_{GaAs} and In content x.

In order to evaluate this procedure, the 3D composition distribution X_{3D} was used as the input to a fully dynamical image simulation as shown in Fig. 4 for both the as-grown and annealed structures for a range of specimen thicknesses. It can be seen that, for a given image, as the specimen thickness increases the amount of indium in the dot must also increase to maintain the bright core at the same level relative to the GaAs. The central row of images in Fig. 4 shows the difference between the dynamical simulation and the experimental image (Fig. 2). Although all of the simulated images are in reasonable agreement with the experiment, it can be seen that the model tends to overestimate In content in very thin specimens (the dot is too bright, giving a dark region in the difference image) and underestimate the In content in thicker specimens (the dot is too dark, giving a bright region in the difference image). The actual specimen thickness t is unknown for the images of Fig. 2 but is probably $50 < t < 100$ nm.

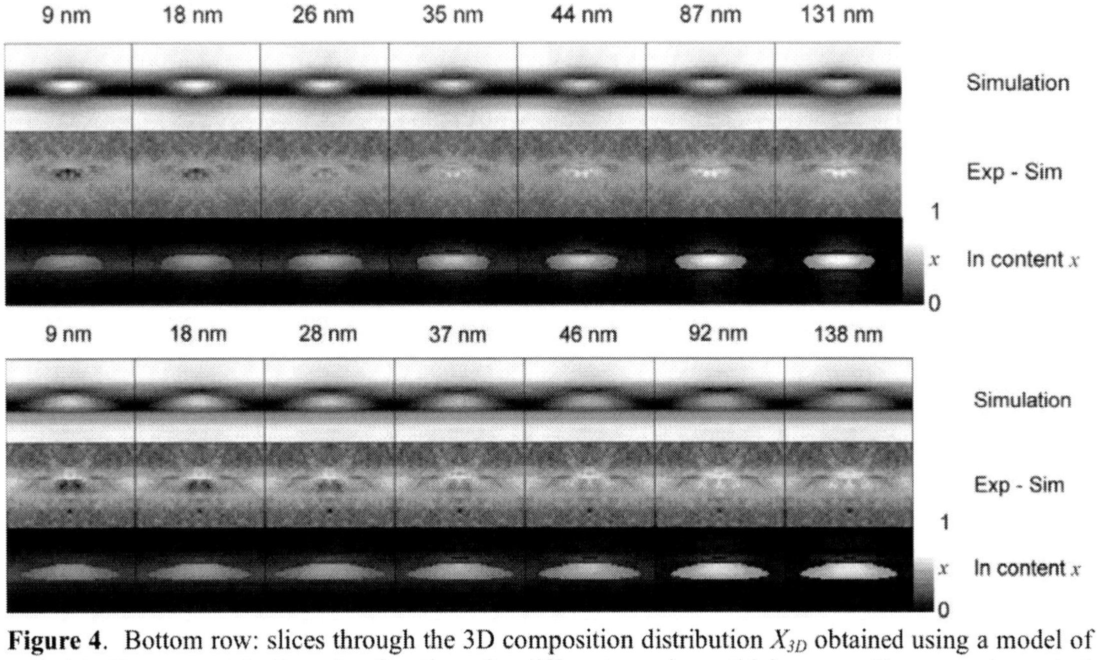

Figure 4. Bottom row: slices through the 3D composition distribution X_{3D} obtained using a model of cylindrically symmetric Gaussian functions for different specimen thicknesses. Top row: dynamical simulations of dark field 002 images using X_{3D}. Changes in lattice parameter as a function of

composition were included but shear strains were ignored. Centre row: difference between the experimental and simulated images.

4. Discussion and Summary

These results imply that it may be possible to use a simple model of QD structure to obtain accurate 3D compositional data. However, some problems still remain. First, the need for an accurate measurement of specimen thickness is essential, since it is possible to fit the experimental data to a wide composition range while maintaining a good fit for all reasonable specimen thicknesses. Second, although the contrast passes through a minimum between the bright QD centre and the surrounding GaAs, the intensity does not reach zero. This is presumably due to the effects of finite resolution, internal strain and/or inelastic scattering; the use of an 'average' image, obtained by summing images from several QDs, will also increase this effect. This leads to an artificially sharp boundary between the QD and the surrounding material, and a poor estimate of In content close to the minimum of Fig. 3 ($0.15 < x < 0.3$). Finally, the effects of strain contrast are also visible in the compositional model; the darker area above the QD is interpreted as a change in In content outside the QD.

Despite these limitations, the 3D compositional data in the as-grown sample clearly shows a concentration of In towards the top of the dot and the V-shaped region of high In content [1]. The maximum In concentration in the QD for a specimen thickness of 87nm is $x = 0.7$. The annealed sample shows a redistribution downwards and outwards from the region of high In content, leading to a cone-shaped region of high In content and a lowered maximum of $x = 0.6$ for a specimen thickness of 87nm.

Acknowledgements

AW acknowledges support from the Libyan Department of Cultural Affairs. The Gatan Orius digital TEM camera used in this research was funded by Birmingham Science City: Creating and Characterising Next Generation Advanced Materials, with support from Advantage West Midlands and part funded by the European Regional Development Fund.

References

[1] Stangl J, Holý V and Bauer G, 2004 *Reviews of Modern Physics*, **76** 725
[2] Horiguchi N, Futatsugi T, Nakata Y, Yokoyama N, Mankad T and Petroff PM 1999 *Jap. J. Appl. Phys.* **38** 2559
[3] Bosi M and Pelosi C, 2007 *Prog. Photovolt: Res. Appl.* **15** 51
[4] Kosogov AO, Werner P, Gösele U, Ledentsov NN, Bimberg D, Ustinov VM, Egorov A Yu, Zhukov AE, Kop'ev PS, Bert NA and Alferov Zh I, 1996 *Appl. Phys. Lett.* **69** 3072
[5] Qiu Y, Zhang ZY, Hogg RA, Cullis AG and Walther T, 2010, *J. Phys: Conf. Ser.* **209** 012036
[6] Babinski A, Jasinski J, Bozek R, Szepielow A and Baranowski JM, *Appl. Phys. Lett.* **79** 2576
[7] Scheerschmidt K and Werner P 2002, in "Nanoscience and Nanotechnology", ed. M. Grundmann, Springer
[8] Lemaître A, Patriarche G and Glas F, 2004 *Appl. Phys. Lett.* **85** 3717
[9] Blank H, Litvinov D, Schneider R, Gerthsen D, Passow T and Scheerschmidt K, 2009, *Cryst. Res. Technol.* **44** 1083
[10] Beanland R, 2005 *Ultramicroscopy* **102** 115
[11] Howe P, Ru ECL, Clarke E, Murray R and Jones TS, 2005 *J. Appl. Phys.* **98** 113511
[12] Beanland R, Sanchez AM, Hernandez-Garrido JC, Wolf D and Midgley PA, 2010 *J. Microscopy* **237** 148

Evaluation of the In desorption during the capping process of diluted nitride In(Ga)As quantum dots

DF Reyes[1], DL Sales[1], R Gargallo-Caballero[3], JM Ulloa[2], A Hierro[2], A Guzmán[2], R García[1] and D González[1]

[1]Departamento de Ciencia de los Materiales e I. M. y Q. I., Facultad de Ciencias, Universidad de Cádiz, 11510 Puerto Real, Cádiz, Spain

[2]ISOM – Departamento Ingeniería Electrónica E.T.S.I. Telecomunicación, Universidad Politécnica de Madrid, Ciudad Universitaria s/n, 28040 Madrid, Spain

[3]Paul Drude Institut für Festkörperelektronik, Hausvogteiplatz 5-7, 10117 Berlin, Germany

E-email: daniel.fernandez@uca.es

Abstract. Diluted nitride self-assembled In(Ga)AsN quantum dots (QDs) grown on GaAs substrates are potential candidates to emit in the windows of maximum transmittance for optical fibres (1.3-1.55 μm). In this paper, we analyse the effect of nitrogen addition on the indium desorption occurring during the capping process of $In_xGa_{1-x}As$ QDs (x=1 and 0.7). The samples have been grown by molecular beam epitaxy and studied through transmission electron microscopy (TEM) and photoluminescence techniques. The composition distribution inside the dots was determined by statistical moiré analysis and measured by energy dispersive X-ray spectroscopy. First, the addition of nitrogen in In(Ga)As QDs gave rise to a strong redshift in the emission peak, together with a large loss of intensity and monochromaticity. Moreover, these samples showed changes in the QDs morphology as well as an increase in the density of defects. The statistical compositional analysis displayed a normal distribution in InAs QDs with an average In content of 0.7. Nevertheless, the addition of Ga and/or N leads to a bimodal distribution of the Indium content with two separated QD populations. We suggest that the nitrogen incorporation enhances the indium fixation inside the QDs where the indium/gallium ratio plays an important role in this process. The strong redshift observed in the PL should be explained not only by the N incorporation but also by the higher In content inside the QDs.

1. Introduction

In the past two decades, alloys based on diluted nitrogen III-V semiconductors have aroused great interest, especially, since Kondow proposed them as substitutes for GaInAsP alloys for use in telecommunications devices [1]. The smaller size and bigger electronegativity of N compared to other V-elements should produce simultaneously a reduction of lattice constant and bandgap in III-V alloys. In this sense, the incorporation of small amounts of N in In(Ga)As Stranski-Krastanov QDs grown on GaAs substrate by Molecular Beam Epitaxy (MBE) has deserved more attention, due to the fact that it allows to extend the emission wavelength to the windows of maximum transmittance for optical fibres

(1.3-1.55μm) [1-4]. However, obtaining the desired emission with the addition of N faces some difficulties, mainly due to two effects: (i) the appearance of defects that degrade the optical properties of the material and (ii) interdiffusion phenomena produced during the QDs capping process, which is necessary for any application. During the coating process, an exchange between the In/Ga is produced, causing morphological and compositional changes in QDs, making a controlled tuning to the desired wavelength difficult [5]. In this work, the influence of nitrogen on the In/Ga exchange and the effect of incorporation of Ga during growth of InAsN QDs is characterized by different TEM techniques.

2. Experimental details

Four samples of In(Ga)As(N) QDs have been grown by MBE with different Ga and N contents, as shown in Table 1. The structure of samples consists of 200 nm GaAs buffer layer grown on a (001) GaAs substrate at 590 °C, 4 or 6 (Ga)InAs(N) monolayers (ML) at 440 °C, a 200 nm GaAs capping layer, the firsts 27 nm being grown at the same temperature as QDs (to reduce segregation) and the rest of the layer at 590 °C. Finally, a surface layer of (Ga)InAs(N) QDs is grown under the same conditions as the buried one. For the cases of samples AN and BN, N was supplied by an Oxford Applied Research radio frequency (RF) plasma source using 0.1 sccm flow of pure N_2 (6N) and 100W power, which corresponds to a nominal N incorporation of ~1%. Further details of the growth and photoluminescence (PL) of these samples can be found in Ref.[6]. The samples were studied using a JEOL JEM 2011 working at 200 kV in conventional transmission electron microscopy (CTEM) mode.

Table 1. Nominal composition (atomic %) and number of deposited monolayers in the active layer (QDs) of studied samples.

Sample	Ga%	In%	N%	ML
A	0	100	0	4
AN	0	100	1	4
B	30	70	0	6
BN	30	70	1	6

3. Results and discussions

Moiré fringe patterns are registered inside the QDs of every sample when imaging in diffraction contrast bright field mode using $\boldsymbol{g}$=(022) (**Figure 1**.a). These moiré fringes come from the interference pattern between two diffracted beams $\boldsymbol{g}_1$ and $\boldsymbol{g}_2$ when transmitted through the superposition of two lattices. In this case, it corresponds to the interference produced when the electron beam passes through the GaAs capping layer and the QD, the last being plastically relaxed. As a result, fringe distance in moiré patterns depends on the lattice mismatch between the (220) planes of the substrate and the relaxed QD structure, and therefore, on the QDs composition and the degree of plastic relaxation. In this framework, a statistical study of the composition of In and Ga within the QDs is carried out [4, 7].

The average distance between fringes was measured at QDs half height (**Figure 1**.b), allowing the calculation of (220) planes distance of QDs (details of this study can be found in Ref. [7]). No correlation between QD height and distance between fringes are observed in any of the studied cases. This result suggests that the lattice parameter of the plastically relaxed QDs does not depend on their size, indicating the same degree of plastic relaxation for all QDs. Therefore, the interplanar spacing mainly depends on the QDs composition. If totally relaxed QDs are assumed for all samples, the average composition inside QDs can be estimated from the interplanar spacings. Accordingly, **Figure 2** shows the histograms of In content of all the samples. The QDs of sample A present a normal distribution with a median of 0.73±0.03 of In content, while the sample AN show a bimodal distribution centred around 0.78±0.01 and 0.93±0.01, having higher In content than QDs in sample A.

These results suggest a stronger retention of In during the capping process of InAsN QDs compared to InAs ones, reaching a virtually total blocking of In/Ga intermixing in some of the analyzed QDs. On the other hand, the histograms of samples B and BN present a bimodal distribution of the In content, centred around 0.52±0.03 and 0.64±0.05 for both samples. However, population inversion is found in both samples. The integrated area of the second (higher In content) peak is 1.76 times larger than the first peak of the BN sample, while for sample B the integrated area of the second peak is 0.44 times the area of the first peak, thereby producing a larger number of QDs with higher In content by the addition of N. It should be noted that we have disregarded the nitrogen content in the lattice parameter calculation. Even though the nitrogen reduces the lattice constant, a consideration of nitrogen at percentage levels would imply an even higher amount of estimated In-content in the InAsN QDs.

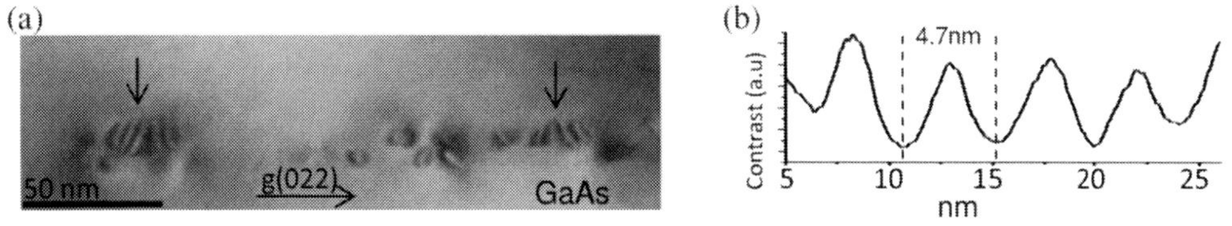

Figure 1. (a) BF (220) CTEM image taken in the [110] zone axis showing buried QDs. Arrows point to moiré fringe patterns found at QDs sites. (b) Typical image intensity contrast profile acquired at QDs half height used to measure fringe distances.

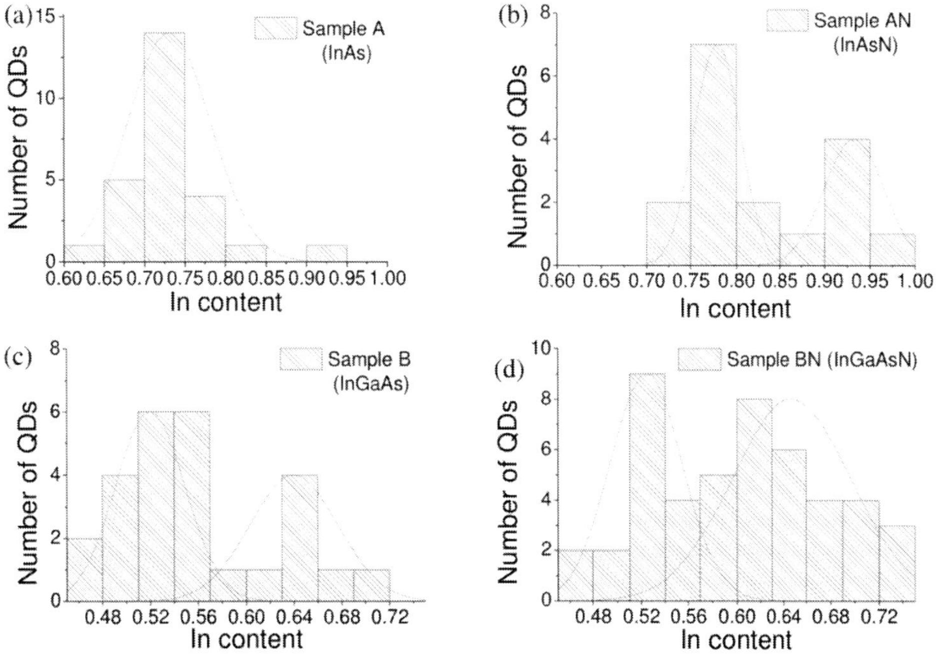

Figure 2. Histograms of determined In content of A (a), AN (b), B (c) and BN (d) QD samples respectively.

These results are in agreement with PL data (not shown in this article), where it is appreciated that a small incorporation of N produces a strong redshift in the emission peak, together with a large loss of intensity and monochromaticity compared to the PL emission of samples grown without N. Thus, the redshift in the sample AN (from 1076 nm to 1301 nm), and BN (from 1046 nm to 1550 nm) is explained both by the addition of N together with the higher In content observed within the QDs.

However, the largest displacement is observed in the BN sample, despite of having a lower content of In, which could be due to an increased content of N, as well as to the greater height of QDs. Previous report regarding the dependence of N incorporation on the Ga content into GaInAsN QDs demonstrated that the maximum N incorporation into GaInAsN QDs occurred at 30% Ga content [6]. Such behaviour was attributed to the competition between different mechanisms, such as the strong stability of the Ga-N bonds and the surface strain during the QDs growth. The increase of the number of gallium atoms bonded to nitrogen in the as-grown condition would explain the huge redshift in the bandgap of BN sample [8]. Further experiments are being performed to clarify the effects of the nitrogen in the atomic distribution of III atoms during the QDs growth and capping processes.

4. Conclusions

All the QDs studied showed moiré patterns due to a full plastic relaxation, allowing the statistical calculation of the composition within the QDs. A small amount of nitrogen diluted in InAs quantum dots during growth leads to an increased In content, as well as a bimodal distribution. As a consequence, there is a higher In content (up to 20%) in QDs of InAsN compared to InAs ones. Otherwise, the addition of Ga or Ga and N simultaneously during the growth of InAs QDs produces a bimodal distribution with two populations centred at the same positions in both samples (B and BN). In this case, the retention of In content is smaller, increasing only the number of QDs with a higher In content. The main retention of indium inside quantum dots suggests that nitrogen gives rise to a blockage of the exchange processes In/Ga during the capping step of the QDs, obtaining a enrichment of indium inside quantum dots and contributing to the redshift observed in the PL. The strong redshift of InGaAsN can be explained by an increase of Ga-N bonds compared to In-N ones in the as-grown condition.

Acknowledgment

Financial support from CICYT (project MAT2010-15206), the Spanish MCI (Consolider-Ingenio 2010 IMAGINE CSD2009-00013), Junta de Andalucía (projects TEP-383 and P09-TEP-5403), Comunidad de Madrid (project P2009/ESP-1503) and European Science Foundation (COST Action MP0805) are gratefully acknowledged.

References

[1] Kondow, M., et al., 1997, GaInNAs: a novel material for long-wavelength semiconductor lasers. *IEEE Journal of Selected Topics in Quantum Electronics* **3**, 719-730.

[2] Kuroda, M., et al., 2005, Growth and characterization of InAsN alloy films and quantum wells. *Journal of Crystal Growth* **278**, 254-258.

[3] Shih, D.K., et al., 2003, Band gap reduction in InAsN alloys. *Japanese Journal of Applied Physics* Part 1 **42**(2A), 375-383.

[4] Schumann, O., et al., 2004, Morphology and optical properties of InAs(N) quantum dots. *Journal of Applied Physics*, **96**, 2832-2840.

[5] Costantini, G., et al., 2006, Interplay between thermodynamics and kinetics in the capping of InAs/GaAs(001) quantum dots. *Physical Review Letters* **96**(22).

[6] Gargallo-Caballero, R., et al., *Dependence of N incorporation into (Ga)InAsN QDs on Ga content probed by rapid thermal annealing*, in *Physica Status Solidi C: Current Topics in Solid State Physics, Vol 6, No 6*, M. Walther, Editor 2009, Wiley-V C H Verlag Gmbh: Weinheim. p. 1441-1444.

[7] Reyes, D.F., et al., 2011 Inhibition of In desorption in diluted nitride InAsN quantum dots. *Applied Physics Letters* **98**(7).

[8] Klar, P.J., et al., 2001, (Ga, In)(N, As)-fine structure of the band gap due to nearest-neighbor configurations of the isovalent nitrogen. *Physical Review B* **64**, 121203.

Analysis of partially oxidised epitaxial silicon mono-layers on germanium virtual substrates using aberration corrected scanning transmission electron microscopy

I M Ross[1], D J Norris[1], B De Jaeger[2], W Lee[2], M Meuris[2] and T Walther[1]

[1] Kroto Centre for High Resolution Imaging and Analysis, Department of Electronic & Electrical Engineering, University of Sheffield, Sheffield S1 3JD, UK
[2] IMEC, Kapeldreef 75, B-3001 Leuven, Belgium.

Abstract: We report the application of aberration corrected STEM imaging to study the interface structure of a partially oxidised epitaxial-Si passivation layer, nominally 2 mono-layers thick, deposited on a Si(001)/Ge virtual substrate. HAADF imaging revealed evidence of 1-2 mono-layers of crystalline Si between the Ge virtual substrate and overlying SiO_2. An interface roughness of 1-2 mono-layer steps was also apparent between the epitaxial-Si and Ge virtual substrate. The total thickness of the residual crystalline Si and overlying SiO_2 suggests that the initial epitaxial-Si layer may have been 1-2 mono-layers thicker than the nominal value.

1. Introduction

Germanium on silicon based metal oxide semiconductor (MOS) field effect transistors offer potentially improved performance due to an enhanced carrier mobility compared to conventional silicon based technology [1]. One drawback of the application of pure germanium in MOS devices is the poor processing characteristcs of GeO_2 compared to SiO_2 [1,2]. Hence, a critical step in the development of such devices is the deposition of a few mono-layers (ML) of partially oxidised epitaxial silicon (epi-Si) passivation between the germanium virtual substrate (Ge-VS) and a high-k hafnium oxide dielectric cap. The thickness and quality of this Si layer has a significant impact on the ultimate device performance [3] therefore knowledge of its structure and nature of the interface with the Ge-VS is desirable. In previous studies [1,4] the spatial resolution offered by conventional transmission electron microscopy (TEM) and scanning transmission electron microscopy (STEM) has been insufficient to clearly elucidate the detailed structure of the interface. Hence in the current study, we have applied aberration corrected STEM imaging to investigate the microstructure of a partially oxidised epi-Si layer with a nominal thickness of half a unit cell deposited on pure Ge-VS. Analysis was performed in a 300kV aberration corrected TEM/STEM equipped with a cold-field emission gun and both probe and image correction [5]. The high angle annular dark-field (HAADF) and bright-field (BF) STEM resolution provided by this instrument is routinely better than 1Å. Moreover, the chemical composition sensitivity of the atomic number (Z) contrast offered by HAADF imaging can in principle reveal information about the nature of the Ge-VS/Si interface and the degree of oxidation of the epi-Si layer.

2. Experimental

Device layers were prepared on a ~2μm relaxed germanium virtual substrate deposited on (001) Si wafers using conventional silicon processing techniques at 650°C. Following a short H_2 anneal, an epi-Si layer was grown at 500°C using SiH_4. By adjustment of the process time and temperature the thickness of the epi-Si layers can be controlled, in this instance to nominal ~2ML. Partial oxidation was achieved by subsequent immersion in 1ppm ozone for 60 seconds before deposition of an 8nm thick HfO_2 high-k dielectric cap by atomic layer deposition. Full details of the device processing route are provided elsewhere [1].

Cross-sections for examination in the TEM/STEM were prepared along the [110] direction by conventional mechanical polishing and thinning to electron transparency using Ar^+ ion beam milling. Imaging was performed in a 300kV cold field emission gun (C-FEG) double aberration corrected TEM/STEM (JEOL Z3100-R005). STEM imaging was performed with both the imaging and probe forming corrector optics optimised to provide a Cs value in each case less than ±1μm. Under such conditions transfer beyond 0.1nm is routinely achieved [5]. High angle annular dark-field (HAADF) STEM images were recorded with a probe current of 9.4pA, a convergence semi-angle of 21mrads and a STEM dark-field inner and outer detection angles of 62mrad and 165mrad respectively. Bright-field (BF) STEM images were collected using a Gatan BF detector at a nominal camera length of 12cm. Specimen thickness was estimated by analysis of the zero energy-loss peak using the standard routine within the Gatan *Digital Micrograph (DM)* software.

To assess the relative atomic column intensities in the HAADF images subtraction of the underlying background signal was performed. Firstly, the image intensity recorded in vacuum was subtracted from the raw image followed by a light 3x3 band-pass filter to reduce detector noise. The underlying background signal was then removed by assuming the background intensity between each row of atomic columns within the HAADF image was the same as that under the atomic columns. A series of contrast intensity profiles were subsequently generated for each row of atomic columns using the Gatan *DM* software each integrated over three pixels [6]. The contrast profiles were finally normalised with respect to the Ge columns to give a series of contrast ratio profiles. Similar analysis was extended to regions of the specimen of known composition, i.e. the interface between the pure Si (wafer substrate) and pure Ge (virtual substrate) to provide an internal calibration of the effective HAADF contrast.

3. Results and Discussion

A low magnification BF-STEM image, Fig. 1a, provides an overview of the device structure with the main regions of interest labelled. Fig. 1b shows a BF-STEM image of the interface between the (001) Si substrate and the overlying Ge-VS. A periodic modulation in the image contrast is observed at the interface with a period of ~10nm. This periodicity corresponds to complete relaxation of the lattice mismatch of the Si and Ge (a_{Si}=0.5431 nm and a_{Ge}=0.56575 nm) where 25x $d\mathrm{Ge}_{(110)}$ ~ = 26x $d\mathrm{Si}_{(110)}$ = 9.97 nm. Closer examination of the regions indicated in Fig. 1b reveals a series of stacking faults just above the Si/Ge-VS interface, an example of which is shown in the HAADF image, Fig. 1c.

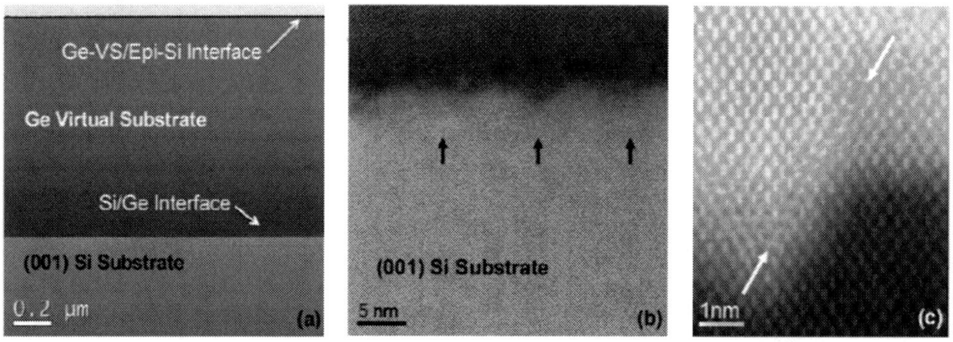

Fig. 1: (a) BF-STEM overview of the bulk device architecture, (b) BF-STEM image of the interface between the (001) Si substrate and the Ge-VS illustrating the periodic contrast variations at the interface (arrowed), c) HAADF image of one such region indicated in Fig. 1b revealing stacking faults (bound by white arrows).

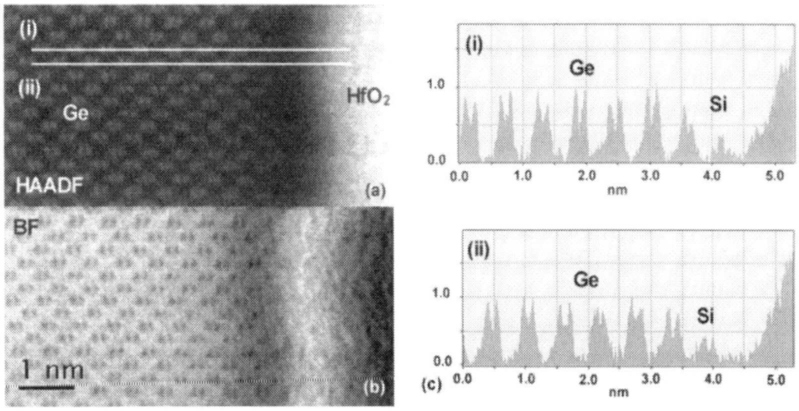

Fig. 2: (a) HAADF STEM image of the Ge-VS/Epi-Si interface (growth direction left to right) the 0.14nm germanium dumbbells are clearly resolved within the Ge-VS region, (b) the corresponding BF STEM image and (c) contrast intensity profiles across adjacent rows (i) and (ii) of atomic columns defined in 2(a).

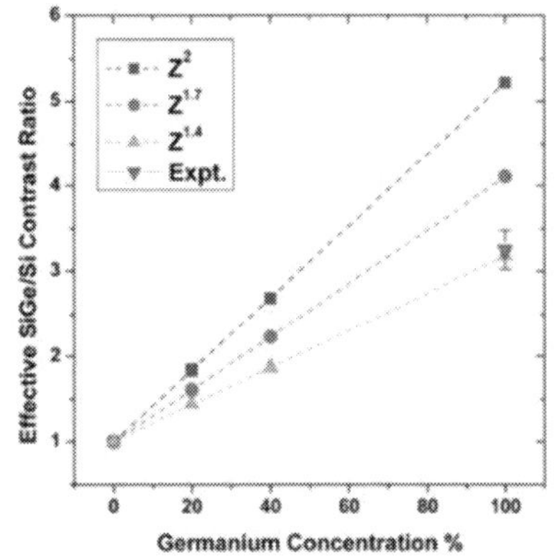

Fig. 3: Plot of theoretical HAADF SiGe/Si Z-contrast dependency for a range of Si-Ge binary alloys (Z^2, $Z^{1.7}$ and $Z^{1.4}$) and experimentally derived effective contrast intensity ratio for pure Ge and Si (Ge/Si=3.25±0.23).

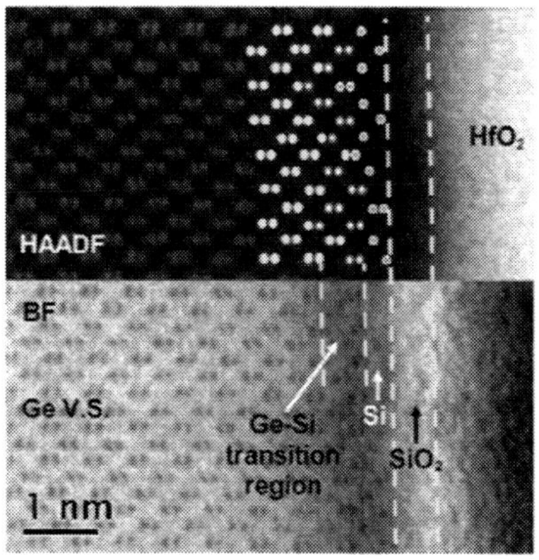

Fig. 4. Composite HAADF and BF image with the atomic columns designated as either Ge or Si based on the relative I_{Ge}/I_{Si} contrast ratio measured from each row of atomic columns as illustrated in Fig. 2c.

Fig. 2a and 2b show a HAADF and corresponding BF STEM image respectively of the surface Ge-VS/Si/SiO_2/HfO_2 interface region (from left to right), the Ge (004) dumbbells (0.14nm) are clearly resolved in the virtual substrate and the HfO_2 appears with bright contrast to the right hand side of Fig. 2a. The atomic number contrast offered by HAADF imaging provides a means of extracting information about the atomic species through measurement of the relative image contrast. Fig. 2c shows two contrast intensity profiles, (i) and (ii), from adjacent rows of atomic columns after back ground subtraction and normalisation. Both profiles indicate a reduction in the relative intensity of the atomic column contrast at the interface between the Ge-VS and the amorphous oxide speculatively labelled as crystalline Si (c-Si) in Fig. 2c. The mean experimental contrast intensity ratio derived from Fig. 2a (I_{Ge}/I_x) (where x corresponds to the intensity of the suspected Si atomic columns) corresponds to 2.75±0.30 which is significantly lower than the expected of ~5.2 assuming a Z^2 dependency predicted by pure Rutherford scattering. A plot of the theoretical contrast intensity ratios for the binary Si:Ge system for a range of Z dependencies is presented in Fig. 3. In order to assess the dependency observed in our current data, we have performed an internal calibration for regions of know composition (pure Si substrate and pure Ge-VS). The experimental results suggest an I_{Ge}/I_{Si} contrast ratio of 3.25±0.23, i.e. closer to a $Z^{1.4}$ dependency as indicated in Fig. 3. Within the experimental error this value approaches that observed for the profiles obtained in Fig. 2c and provides some confidence that 1-2 ML coverage of c-Si remains after partial oxidation. Subsequently, by similar analysis of the (I_{Ge}/I_x) profiles for each row of atomic columns in Fig. 2a we are able to assign atomic columns with a (I_{Ge}/I_x) ratio between 2.4-3.5 as Si and those with ratios approaching unity as Ge as illustrated in Fig. 4. A transition region with values between these two extremes is evident suggesting apparent atomic column intermixing within the transition region and/or an interface roughness of 1-ML extending along the electron beam direction through the sample. Alternatively, delocalisation of the probe intensity to neighbouring atomic columns through the relatively thick sample, determined as ~33nm by energy-loss spectroscopy may also be significant [7]. Factors including detection angle and specimen thickness can all potentially influence the relative HAADF contrast while the presence of interface strain in similar structures has been shown previously to modulate the Z-contrast under near zone axis conditions [8]. The thickness of the overlying SiO_2 layer was estimated at 0.60-0.68nm. The interface between the SiO_2 and HfO_2 was taken as the half maximum of the contrast intensity gradient however the exact nature of this interface was difficult to assess. Future spectroscopic profiling and image simulation are therefore required to support the experimental STEM data before further conclusions can be drawn.

4. Conclusions

Partially oxidised epitaxial-Si passivation layers, nominally 2 ML thick, deposited on pure Ge-VS have been investigated using aberration corrected STEM BF and HAADF imaging. Studies of the (001)-Si/Ge-VS interface reveal the presence of periodic stacking faults originating several ML above the interface. Analysis of the HAADF Z-contrast at the Ge-VS/epi-Si interface indicates that 1-2 ML of c-Si remains after surface oxidation. In addition, an apparent projected interface roughness between the Ge VS and c-Si region of 1-2 ML is observed. The total thickness of the remaining c-Si and overlying SiO_2 suggests that the initial epi-Si layer must have been 1-2 ML thicker than the nominal value. The nature of the interface between the amorphous SiO_2 and the HfO_2 dielectric cap however is less distinct and requires further clarification.

Acknowledgements

The authors would like to thank the Engineering and Physical Sciences Research Council for their financial support (EP/F033893/1 – "Renaissance Germanium").

References

[1] De Jaeger B, Bonzom R, Leys F, Richard O, Van Steenbergen J, Winderickx G, Van Moorhem E, Raskin G, Letertre F, Billon T, Meuris M and Heyns M 2005 *Microelectron. Eng.* **80**, 26
[2] Wu N, Zhang Q, Zhu C, Chan D S H, Li M F, Balasubramanian N, Chin A, Kwong D-L 2004 *Appl. Phys. Letts.* **85**, 4127
[3] Vincent B, Loo R, Vandervorst W, Delmotte J, Douhard B, Valev V K, Vanbel M, Verbiest T, Rip J, Brijs B, Conard T, Claypool C, Takeuchi S, Zaima S, Mitard J, De Jaeger B, Dekoster J, Caymax M 2011 *Solid State Electron.* doi:10.1016/j.sse.2011.01.049
[4] Norris D J, Ross I M, Cullis A G, Walther T, Myronov M, Dobbie A, Whall T, Parker E H C, Leadley D R, De Jaeger B, Lee W and Meuris M 2010 *Proc. EMAG 2009, Sheffield, J. Phys: Conf. Ser.* **241** 012044
[5] Sawada H, Tanishiro Y, Ohashi N, Tomita T, Hosokawa F, Kaneyama T, Kondo Y, Takayanagi K 2009 *J. Electron Microsc.* **58**, 357
[6] Robb P D and Craven A J 2008 *Ultramicroscopy* **109**, 61
[7] Grillo V 2009 *Ultramicroscopy* **109**, 1453
[8] Walther T and Humphreys C J 2007 *Proc. EMAG97,Cambridge, Inst. Phys. Conf. Ser.* **153**, 303

Silicide formation process in ultra-thin Ni-silicide film for advanced semiconductor devices: mechanism of $NiSi_2$ formation at low temperature

N Ikarashi and K Masuzaki

LSI Research Laboratory, Renesas Electronics Corporation
1120 Shimokuzawa, Chuo-ku, Sagamihara, 252-5298, Japan

E-mail: nobuyuki.ikarashi.xz@renesas.com

Abstract. Atomic-resolution high-angle annular dark-field scanning-transmission electron microscopy and ab-initio calculations were used to reveal the reaction involved in the formation of ultra-thin Ni silicide film at 300°C. We found that a Ni-adamantane structure, in which Ni atoms occupy the tetrahedral interstitial voids of Si, forms at the initial stage of the reaction. We also found that the adamantane structure is under considerable compressive stress due to lattice-mismatch at the adamantane structure-Si interface (5.6%). Then, $NiSi_2$ forms just beneath the Ni-adamantane structure at a much lower temperature than the $NiSi_2$ formation temperature reported for the reaction between a Ni layer and Si substrate (800°C). Our analyses strongly suggest that the Ni-adamantane structure acts as a precursor in the formation of $NiSi_2$.

1. Introduction

In advanced metal-oxide-semiconductor field effect transistors (MOSFETs), thin NiSi films (< 20 nm) are used as contacts in the source and drain regions [1]. A serious problem is the unexpected formation of $NiSi_2$ during the silicidation process: $NiSi_2$ raises the resistivity and roughens the silicide/Si interface, degrading device performance.

Previous studies have indicated that in the reaction between a Ni film and Si substrate, Ni_2Si, NiSi, and $NiSi_2$ are sequentially formed as the sintering temperature increases [2]. The formation temperature of $NiSi_2$ is about 800°C. However, the $NiSi_2$ phase is reportedly formed at a much lower temperature in some cases [3-5]. The mechanism of $NiSi_2$ formation at low temperatures, which is used in the silicidation process (< 400°C), has yet to be clarified.

We examined the atomic structure of a very thin Ni-silicide layers formed by depositing monolayers of Ni on a Si (001) surface at 300°C, using atomic-resolution high-angle annular dark-field scanning-transmission electron microscopy (HAADF-STEM). We also examined the stability of these thin Ni-silicide layers using first-principle calculations. From these results, we infer that the surface layer acts as a precursor in forming $NiSi_2$ at low temperature.

2. Experiment

We prepared two samples in which 0.5-nm-thick (sample 1) and 1.0-nm-thick (sample 2) Ni films were deposited on Si (001) substrates by chemical vapor deposition. Deposition rate was 0.2 nm/min. The samples were then formed into <110> cross-sectional specimens using mechanical thinning and Ar-ion milling. In HAADF-STEM observations, the electron probe convergence angle was 20 mrad, and the annular detector was set to collect the electrons scattered at angles between 45 and 110 mrad. First-principles calculations were carried out using the WIEN2k package [6]. STEM image simulation software xHREM was used in the STEM image analysis.

3. Results

Cross-sectional views over wide areas of the specimens are shown in Fig. 1. Thin bright surface layers were observed in both samples. In sample 2, bright triangular areas were observed just beneath the surface layer. The facets of the triangular areas were parallel to the (111) of the Si substrate. The faceted shape (Fig. 1(b)) as well as its atomic arrangement, shown in the micrograph in Fig. 3, indicated that the triangular areas were $NiSi_2$.

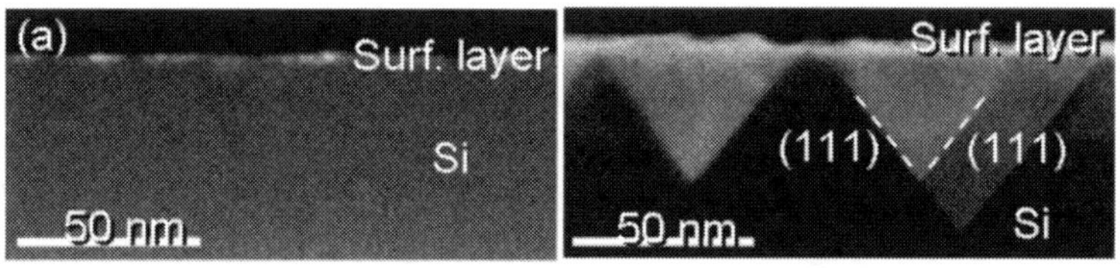

Figure 1. HAADF-STEM of (a) sample 1 and (b) sample 2. Bright areas are reaction products.

High-resolution observations of the surface layer revealed that Ni atoms in the surface layers occupy tetrahedral interstitial voids of the Si substrate. Figure 2(a) shows a representative image of the surface layer of sample 1. Bright dots are evident at the tetrahedral interstitial voids of Si as well as at the Si lattice sites. The image intensity distribution measured along the (110) lattice plane of Si, which is indicated by the arrow in Fig. 2(a), is shown in Fig. 2(b). Peaks are evident at some interstitial sites in the surface area (the left side of the graph), strongly suggesting that the sites are occupied by Ni atoms. Moreover, in some Si atomic column pairs, the peak intensities of individual Si atomic columns significantly differed. This difference indicates that Si atoms dissociated themselves from the lattice sites that showed smaller peak intensities.

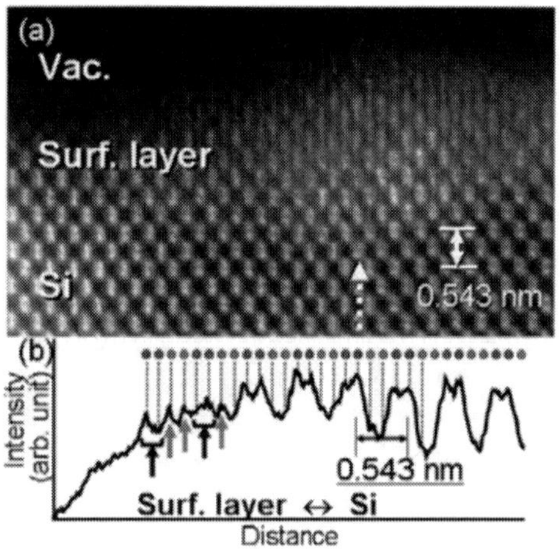

Figure 2. (a) Close-up of surface layer in sample 1 and (b) intensity distribution measured along a lattice plane denoted by arrow in (a). Blue and red balls in (b) represent Si lattice sites and interstitial voids, respectively. Peaks at blue ball positions (denoted by red arrows) indicate that interstitial sites are occupied. Intensities of Si atomic column pairs (denoted by black arrows) significantly differ.

A close-up view of the interface between the surface layer and the bright triangular areas in sample 2 is shown in Fig. 3(a). To determine the atomic arrangement in the surface layer, we first performed STEM image simulations (Fig. 3(b), (c)). In the simulations of the surface layer, we assumed that Ni atoms occupy tetrahedral interstitial voids at a certain occupancy factor on the basis of the findings in Fig. 2 (the structure is referred to as adamantane structure [7, 8]). The simulated images showed that the intensities at the Si lattice sites and interstitial voids (i.e., Ni sites) were similar at the Ni occupancy factor of 0.55 (Fig. 3(d)), reproducing the observed image. In addition, the simulated image of a CaF_2-type $NiSi_2$ (Fig. 3(c)) well reproduced the observed image of the triangular area in Fig. 3(a).

Moreover, despite the different atomic arrangement in the surface layer and that in the triangular areas, no crystal lattice discontinuities, such as dislocations, were noticeable at the interface. That is, $NiSi_2$ was formed epitaxially beneath the surface layer, strongly suggesting that the surface layer plays a crucial role in $NiSi_2$ formation.

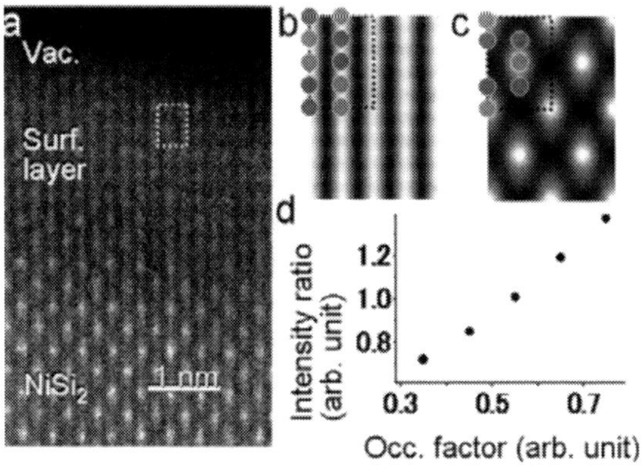

Figure 3. Close-up of interface between surface layer and bright triangular areas ($NiSi_2$) in sample 2: (a) observed image, (b) and (c) simulated images. Dotted rectangle shows two-dimensional unit cell in the projection. Blue and red circles indicate the Si- and Ni atomic-column positions, respectively. (d) Ratio of Ni atomic-column intensity to Si atomic-column intensity in surface layer was plotted against Ni occupancy factor. Intensities were almost equal when Ni occupancy factor was set to 0.55.

To examine the stability of the Ni-adamantane structure formed on a Si substrate, we calculated the total energies of its unit cell as a function of their lattice constants. In the calculation, we set the Ni occupancy factor to 1.0, for simplicity. The calculated total energy curves are shown in Fig. 4 together with that of Si. The reference energy of each curve, E_{min}, was set to the calculated minimum energy of each structure. The calculated Si equilibrium lattice constant, at which the total energy was minimized, was 0.534 nm, which agreed well with the measured one [9]. The calculated equilibrium lattice constant of the Ni-adamantane structure was 0.565 nm, which was 5.8% larger than that of Si. Hence, the Ni-adamantane structure should be under compressive stress when it is formed epitaxially on a Si

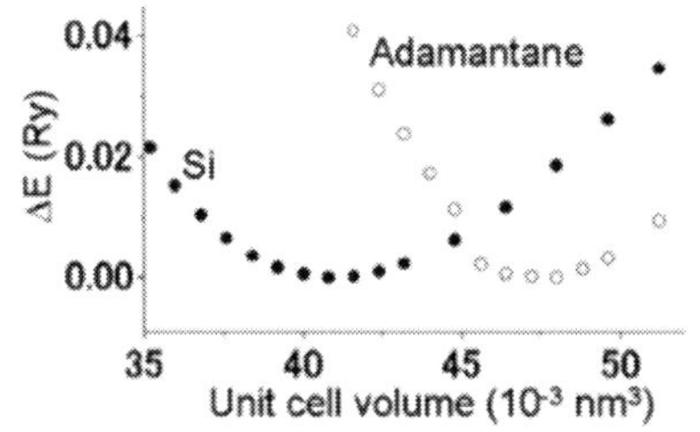

Figure 4. Calculated total-energy curves as function of unit cell volumes. Vertical axis, ΔE, shows difference between total energy and minimum total energy.

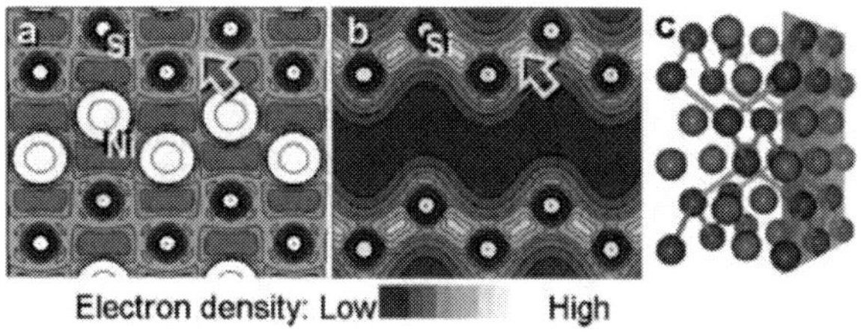

Figure 5. Calculated electron density distribution in (a) adamantane structure and in (b) Si. (c) Lattice plane along which electron density is displayed. Electron densities at centers of Si-Si bonds (indicated by arrows) are 0.37×10^3 electrons/nm^3 in (a) and 0.57×10^3 electrons/nm^3 in (b).

substrate. This considerable lattice mismatch and resulting interfacial stress possibly prevents the surface layer from becoming thicker.

4. Discussion

According to the first-phase rule [10], the Ni_2Si phase is the first silicide phase in the reaction between a Ni film and Si substrate: an atomic diffusion at the Ni/Si interface generates an interface layer with a eutectic composition (Ni:Si = 2:1) and a silicide crystal whose composition is congruent with the interface layer nucleates. In our samples, the Ni diffusion from the thin Ni layer may be too small to form a Ni-rich (Ni_2Si) interface layer, but the penetration of a small amount of Ni atoms into Si forms a Ni-adamantane structure instead.

Hence, the thin Ni film (~ 1 nm thick) and/or the small Ni deposition rate (0.2 nm/min) used in this experiment would be necessary for adamantane structure formation and following low temperature $NiSi_2$ formation. This is because large Ni film thickness and/or large Ni deposition rate would cause a large Ni diffusion and result in Ni-rich interface layer formation.

Considerable stress (5.8%) of the Ni-adamantane structure may be the cause of the low-temperature $NiSi_2$ formation. This is because the stress of the Ni-silicide layer should be reduced by $NiSi_2$ formation, which is more lattice matched to Si (the lattice constant is 0.5395, 0.6% smaller than Si [9]). Thus, we observed a $NiSi_2$ region below an initial adamantane surface layer in sample 2 (1.0-nm-thick Ni sample), in which sufficient Ni diffused to form $NiSi_2$, while only a surface adamantane layer was observed in sample 1 (0.5-nm-thick Ni sample) . In addition, the decrease in the covalency of the Si-Si bonds in the surface layer possibly allows the changing of atomic arrangement at the temperatures as low as 300°C.

5. Summary

The mechanism of the $NiSi_2$ formation at a low temperature (300°C) was investigated using HAADF-STEM and ab-initio calculations. STEM analysis showed that a Ni-adamantane structure was formed at the initial stage of the reaction. In addition, $NiSi_2$ was formed just beneath the surface layer with the Ni-adamantane structure. Our computational results indicated that the equilibrium lattice constant of the Ni-adamantane structure was larger than that of Si, indicating that the stress at the Ni-adamantane structure/Si interface should be considerably larger than that at the $NiSi_2$/Si interface. We thus infer that the Ni-adamantane structure may act as a precursor in $NiSi_2$ formation.

References

[1] *International Technology Roadmap for Semiconductors* (Semiconductor Industry Association, San Jose, CA, 2009), 2009 edition
[2] *Properties Of Metal Silicides* (IEE, London) ed. M. Karen 1995
[3] Tung R T, Gibson J M and Poate J M 1983 *Phys. Rev. Lett.* **50** 429.
[4] Gibson J M and Batstone J L 1989 *Surf. Sci.* **208** 317.
[5] Hesse D, Werner P, Heydenreich J and Mattheis R, 1994 *Proc. Mat. Res. Soc. Symp.* **320** 221.
[6] Blaha P, Schwarz K, Madsen G K H, Kvasnicka D and Luitz J 2001 *WIEN2k, An Augmented Plane Wave Plus Local Orbitals Program for Calculating Crystal Properties*, Vienna University of Technology, Austria
[7] Cheung N W and Mayer J W 1981 *Phys. Rev. Lett.* **46** 671
[8] Chang Y -J and Erskine J L 1982 *Phys. Rev. B* **26** 4766
[9] Wyckoff R W G 1963 *Crystal Structures* (John Wiley & Sons, New York, London).
[10] Walser R M and Bene R W, 1976 *Appl. Phys. Lett.* **28** 624.

Mapping boron in silicon solar cells using electron energy-loss spectroscopy

M Duchamp[1], C B Boothroyd[1], A Kovács[1], S Kadkhodazadeh[1], T. Kasama[1], M S Moreno[2], B B Van Aken[3], J-P Barnes[4], M Veillerot[4], S B Newcomb[5], R E Dunin-Borkowski[6,1]

[1] Center for Electron Nanoscopy, Technical University of Denmark, DK-2800 Kongens Lyngby, Denmark
[2] Centro Atómico Bariloche, AR-8400 S. C. de Bariloche, Argentina
[3] ECN Solar Energy, P.O. Box 1, NL-1755 ZG Petten, The Netherlands
[4] CEA-Leti, MINATEC Campus, 17 rue des Martyrs, FR-38054 Grenoble Cedex 9, France
[5] Glebe Laboratories, Newport, Co. Tipperary, Ireland
[6] Institute for Microstructure Research, Forschungszentrum Jülich, D-52425 Jülich, Germany

E-mail: martial.duchamp@cen.dtu.dk

Abstract. Electron energy-loss spectroscopy (EELS) is used to study the B distribution in a p-i-n layered solar cell structure. The boron concentration in the p-doped Si layer is expected to be ~10^{21} cm^{-3} and should not exceed 10^{17} cm^{-3} in the neighbouring intrinsic layer. We show that B concentrations as low as 3×10^{20} cm^{-3} (0.6 at. %) can be measured using EELS. Our measurements are in close agreement with real space *ab-initio* multiple scattering calculations and secondary ion mass spectrometry measurements.

1. Introduction

Thin films of amorphous silicon (a-Si:H) are used as active absorber layers in solar cells deposited on steel foil produced by roll-to-roll processing [1]. In order to improve the efficiency of such solar cells, the nature of the defects that act as charge recombination centres and decrease the internal electric field in the active Si layer needs to be understood. In previous work, B contamination of the active Si layer has been measured using secondary ion mass spectrometry (SIMS) and the efficiency has been shown to improve when the B concentration was reduced [2]. However, the lateral and depth resolution of SIMS are limited to about 100 μm and 5 nm, respectively.

Electron energy-loss spectroscopy (EELS) in the transmission electron microscope (TEM) has nanometre spatial resolution, but the detection of low B concentrations is difficult, especially in Si, because energy-loss near-edge structure (ELNES) from the Si L-edge, caused by scattering of inner-shell electrons to the conduction band, extends from 99 eV to >300 eV and interferes with the B K-edge at 188 eV [3]. By using C_S-corrected scanning (S)TEM EELS, B concentrations of 7×10^{20} cm^{-3} have been measured with a 3 nm electron beam [4], while Boothroyd *et al.* detected 0.5 at. % B in Ni_3Al by using an iterative averaging process to reduce channel-to-channel variations in the gain [5].

Here, we measure B concentrations in a-Si:H and a-SiC:H layered samples that are of interest for the development of solar cells. In order to determine the minimum detectable B concentration, we

Published under licence by IOP Publishing Ltd

characterize samples that contain both no B and two different concentrations of B. We examine the Si ELNES and B K-edge structures as a function of B concentration. Our results are compared with SIMS measurements of B content in the same samples.

2. Experimental details

Three Si-based layers were grown using plasma-enhanced chemical vapour deposition on Ag and ZnO layers. Test samples contained 200-nm-thick a-Si:H and a-SiC:H layers that were (i) undoped, (ii) lightly B-doped and (iii) highly B-doped. For comparison, typical solar cell stacking grown on a steel substrate involves the growth of (i) 20 nm p-doped Si, (ii) 300 nm of undoped Si and (iii) 20 nm of lightly B-doped SiC layers. A bright-field (BF) TEM image of a real solar cell sample is shown in Fig 1(a).

Thin film silicon solar cells are typically made on plastic, glass or metallic substrates, making sample preparation for TEM difficult. Here, 100nm-thick-TEM specimens were prepared using focused-ion-beam (FIB) milling and finished with a 5 keV Ga^{2+} beam. EEL spectra were acquired in TEM diffraction mode on an FEI Tecnai T-20 equipped with a LaB_6 electron source and an FEI Titan 80-300 equipped using a field emission electron source, with Gatan Tridiem imaging filters. All measurements were carried out at 120 keV to limit electron beam damage.

At each region of interest, 6 energy-loss spectra that were shifted in energy were acquired. Channel-to-channel gain variations were reduced by using an iterative averaging process that is described in detail elsewhere [5]. In order to minimise shot noise, the exposure time was kept as high as possible without damaging the specimen. For large area measurements, each spectrum was acquired for one minute using a beam diameter of approximately 200 nm in TEM diffraction mode. For higher spatial resolution measurements, line scans were acquired using a 5 nm spot size in TEM diffraction mode. Each spectrum recorded for 1.5s.

Real space multiple scattering calculations were performed using FEFF9 9.05 software, taking into account experimental parameters such as beam energy and collection angle. In order to include core hole effects, the final-state-rule approximation was used. The Hedin-Lundqvist self-energy was used to take inelastic losses into account. Calculations were performed for crystalline SiC and Si. Static disorder of approximately 8 times the Debye-Waller factor was added to simulate amorphous materials. Variations in B concentration were modelled by changing the cluster size (with a small cluster size corresponding to a high B concentration) and/or with different numbers of B atoms replacing Si in different coordination shells. The B K-edge was calculated for concentrations of approximately 0.7 at. % and 2.1 at. % B in Si and SiC.

3. Si L_I edge ELNES: simulations and experimental data

Ab initio calculations of ELNES from the Si L_I-edge are shown in Fig. 1(b) for crystalline and disordered Si and SiC. Both materials have significant ELNES close to the B K-edge position, which is observed to depend on C concentration. The simulations predict a decrease in ELNES when disorder is included, for both Si and SiC. Fortunately, there is no Si L_I-edge peak exactly at the B K-edge position, making the detection of the B K-edge on top of the Si and SiC L_I ELNES easier. Figure 1(b) shows that a reasonable match is obtained between the experimental and simulated Si L_I-edge positions, even though the L_{23} contribution is difficult to extract and has an influence on the experimental data. Unfortunately, a simple power-low background-subtraction is difficult due the presence of the strong Si L_I-edge ELNES below 185 eV.

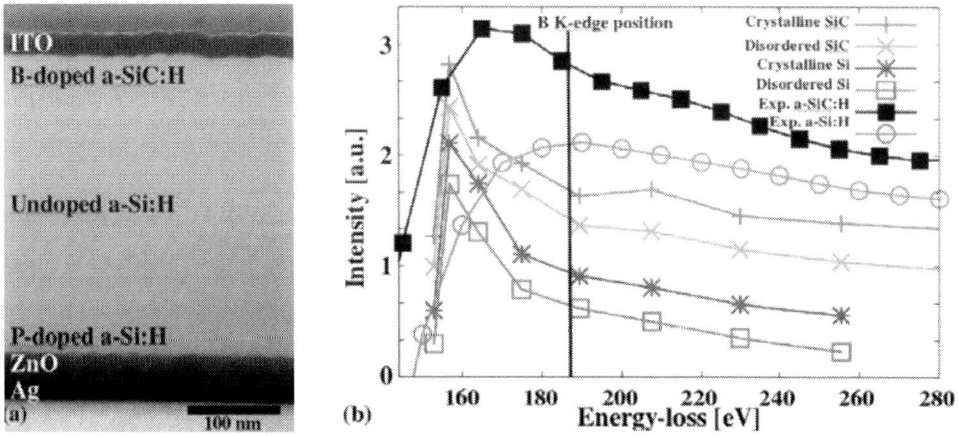

Figure 1. (a) BF TEM image of an a-Si solar cell prepared using FIB milling. (b) Si L_1-edge ELNES calculations for Si and SiC for both crystalline and disordered atomic configurations. Experimental EEL spectra for a-Si:H and a-SiC:H are also shown. The spectra have been shifted vertically for clarity.

4. B K-edge: quantitative analysis

Experimental background-subtracted B K-edge EEL spectra measured from samples containing different B concentrations are shown in Figs 2 (a) and (b). The signals were therefore separated from their backgrounds using a combination of a power law and first order affine functions. The B peaks from highly-doped a-SiC:H and a-Si:H layers are clearly recognisable in Fig 2 and match the simulated B peaks well after refining the crystallinity and cluster size in the simulations. The increase in the intensity of the SiC spectrum at energies higher than 195 eV is due to the Si L_1 ELNES and is consistent with the peak centred at 210 eV found in the calculations (see Fig 1(b)). An energy shift of the lightly doped samples is observed which is not yet understood.

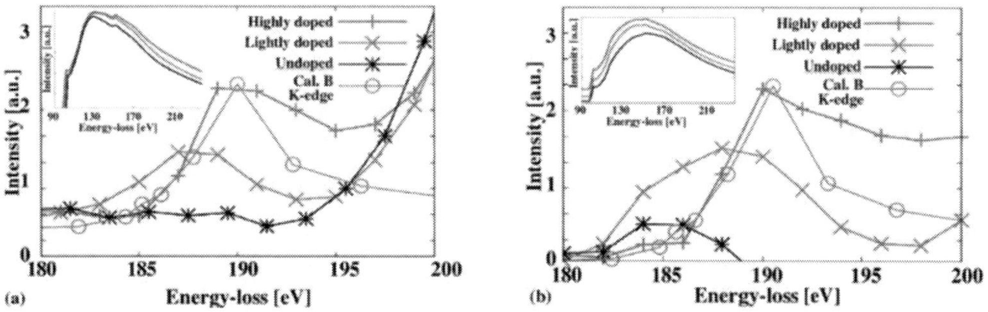

Figure 2. EEL spectra acquired from (a) a-SiC:H and (b) a-Si:H for three different B concentrations. Backgrounds have been subtracted using power law and first order affine functions fitted to the range 175 to 185 eV. The inserts show the spectra before background subtraction of the Si ELNES.

B compositions were determined from the spectra using Hartree-Fock cross-section [6] and are given in Table 1. The relative B concentration obtained by SIMS was estimated by assuming the total number of atoms to be 5×10^{22} cm^{-3}. For the undoped sample, no measurable B was detected using EELS. The lowest B concentration that was measured by EELS with a 200 nm spot size was 0.6 at.%.

Table 1. B and C concentrations in test samples measured by EELS and SIMS

	Si		SiC		
	B_{EELS} [%]	B_{SIMS} [%]	C_{EELS} [%]	B_{EELS} [%]	B_{SIMS} [%]
Highly doped	4.74	6	15	1.96	3.04
Lightly doped	1.85	1.4	16.29	0.62	0.6
undoped	0	3.10^{-3}	14.42	0.06	3.10^{-3}

Figure 3 shows the result of an initial attempt to measure B profile with EELS using a 5 nm spot size in an a-SiC:H sample close to the ITO interface, as would be present in a real solar cell. The B concentration measured using EELS is higher than that from SIMS. The discrepancy may be explained by the fact that the SIMS measurements is averaged over large regions.

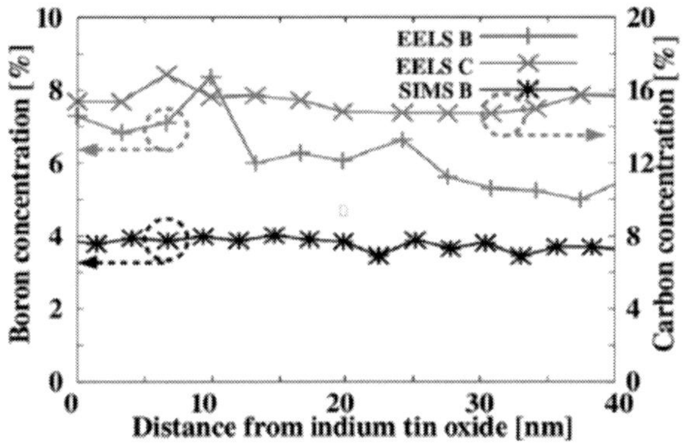

Figure 3. B and C concentrations measured using EELS with a 5 nm spot size. B concentrations measured using SIMS are also shown.

5. Conclusions

B concentrations in a-Si:H and a-SiC:H test structures have been measured using EELS in combination with iterative averaging. Reasonable agreement has been found between SIMS and EELS measurements, within a factor better than 2. B concentrations as low as 0.6 at.% have been measured in a-SiC:H and a detection limit for B of approximately $3\times10^{20}cm^{-3}$ has been estimated for the present experimental parameters. An attempt to measure B concentration profiles by EELS with a 5 nm spot size shows promise for the characterization of a real solar cell sample. Results obtained using other TEM techniques, such as low-loss EELS and off-axis electron holography will be compared with the present approach elsewhere.

References

[1] Van Aken B B, Dorenkamper M, Devilee C, Heijna M C R, Loffler J and Soppe W J 2009 *34th IEEE photovoltaic specialists conference* **1-3** 426

[2] Kroll U, Bucher C, Benagli S, Schönbächler I, Meier J, Shah A, Ballutaudb J, Howling A, Hollensteinb Ch, Buchelc A and Poppellercet M 2004 *Thin solid films* **451-452** 525

[3] Muto S, Sugiyama H, Kimura T and Tanabe T 2004 *Jap. J. Appl. Phys.* **43** 1076

[4] Asayama K, Hashikawa N, Kajiwara K, Yaguchi T, Konno M and Mori M. 2008 *Appl. Phys. Exp.* **1** 074001

[5] Boothroyd C B, Sato K and Yamada K 1990 *Proc. XIIth Int. Congress for Electron Microscopy, Seattle, Washington, USA* 80

[6] Egerton R F 1999 *J. Electron Microscopy* **48** 711

Investigation of boron implantation into silicon by quantitative energy-filtered transmission electron microscopy

T Walther[1] **and J England**[2]

[1] Department of Electronic & Electrical Engineering, University of Sheffield, Sir Frederick Mappin Building, Mappin Street, Sheffield S1 3JD, UK
[2] Varian Semiconductor Equipment Associates, 35 Dory Road, Gloucester MA01930, USA

t.walther@sheffield.ac.uk

Abstract. High-resolution electron microscopy (HREM), chemical mapping by energy-filtered transmission electron microscopy (EFTEM), electron energy-loss spectroscopy (EELS) and energy-loss spectroscopic profiling (ELSP), also called spectroscopic transverse image profiling by EFTEM (stripeTEM) have been combined to study the distribution of boron after implantation into silicon at low energy (0.5keV) and rapid thermal annealing. The key to the experiments is that stripeTEM combines both ~1nm spatial resolution and sub-at% sensitivity. It is shown that the near-surface spike of boron visible in secondary ion mass spectrometry (SIMS) profiles is an artefact of surface profiling. The real distribution of boron is several nm deeper into the specimen, in agreement with both Monte Carlo simulations and shoulders apparent in the SIMS profile. Quantification of the experimental stripeTEM data shows that the B concentration is approximately ~2at% in the SiO_2 and ~4at% in the Si, with statistical errors of about ±0.2at% due to noise and background fitting and systematic errors of ±0.6at% due to uncertainties in the scattering cross-sections.

1. Introduction

For electronic devices doping of layers at specific depths in silicon (Si) wafers is commonly established by ion implantation, and a suitable p-dopant is boron (B). A subsequent anneal is then used to ensure that, firstly, the implanted ions move from interstitial to lattice sites so they become electrically active, and, secondly, other point and also more extended lattice defects produced during the implantation are effectively healed out [1].

2. Implantation, anneal, SIMS measurements and Monte Carlo simulations

The Si wafer investigated was first pre-amorphised using 12 keV Ge, co-implanted with 7keV C (at a dose of $1\times10^{15}cm^{-2}$) and then doped using carborane implantation [2] to give an implant equivalent to 0.5keV B at a dose of $1\times10^{15}cm^{-2}$. A rapid thermal anneal at 1025°C was performed for a few seconds to re-grow the crystal and activate the boron. Si interstitials released during the crystal re-growth can lead to transient enhanced diffusion and clustering of B interstitials [2].

The post anneal boron depth distribution was measured by secondary ion mass spectroscopy (SIMS) profiling. Fig. 1 shows the SIMS result, with an apparent peak at just ~1nm below the surface and two broader peaks at ~4nm and ~12nm depths. Fig. 2 depicts the predicted profile for the as-implanted boron from Monte Carlo simulations using the SRIM code [3], which predicts an ion range of ~4nm, in agreement with an extrapolation from values tabulated for higher ion energies [4]. This suggests that the 4nm and 12nm peaks are due to the as-implanted profile and transient enhanced diffusion during the anneal. A previous transmission electron microscopy (TEM) study on a higher dose and energy boron only implant [5] suggested that the surface peak in the post anneal profile could be an artefact of the SIMS technique. The aim of this study is to investigate this further for a case that included amorphisation and co-implants, and see if any evidence of B clustering could be observed.

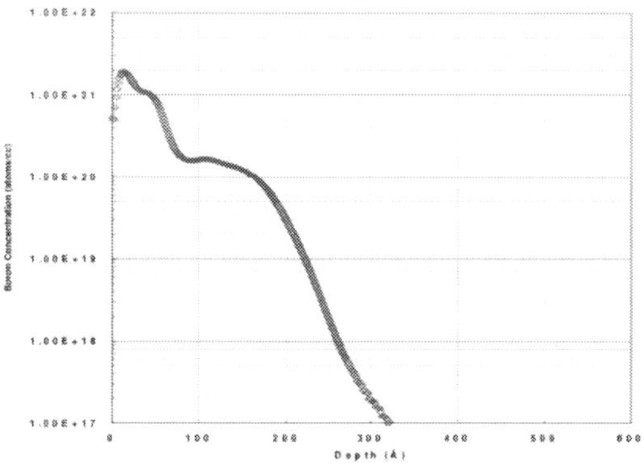

Figure 1. SIMS depth profile of B distribution

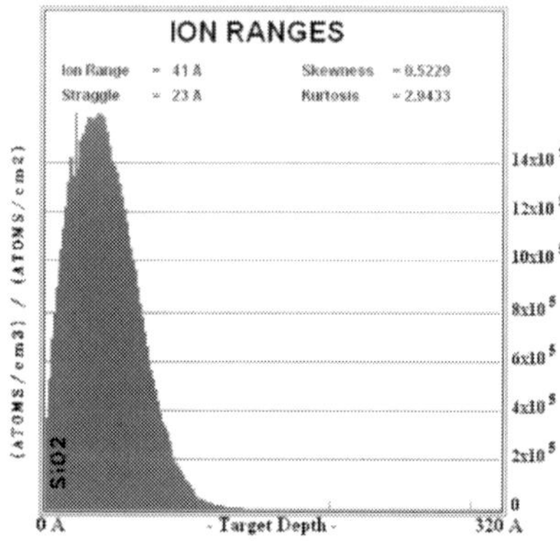

Figure 2. Monte Carlo simulation of B implantation into Si/SiO_2 [3].

3. TEM Experiments

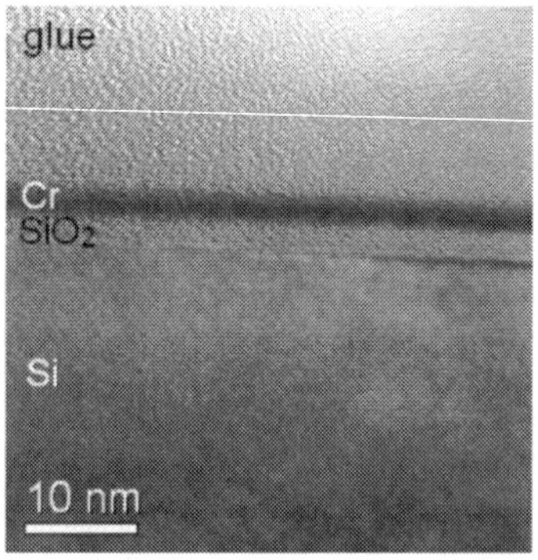

Figure 3. Cross-sectional elastic bright field (BF) image formed by using a slit aperture in the energy-dispersive plane to exclude all inelastic scattering. The top of the implanted Si shows diffraction contrast due to strain.

Figure 4. High-resolution electron microscopy along [100] zone axis with {022} lattice fringes of 0.19nm spacing in Si. The top left depicts the amorphous glue, then follow the Cr layer (dark), native SiO_2 (speckle contrast) and the Si substrate (grey) at the bottom right.

A stripe of 30mm x 3mm was cleaved from the wafer piece provided and a ~3nm Cr layer was evaporated on top of it to seal and define the surface for later analysis. Then a piece was glued together with a SiGe multi-layer sample as reference for both magnification and analytical signals, the stack was backed by more pieces of Si, cut along <100> directions using a wire saw, ground, mechanically polished and finally milled until perforation by argon ions at 3keV energy, 3mA combined current from top and bottom surface, ~15° angle of incidence and 2kV sample bias.

The sample was then investigated near [100] zone axis orientation in a JEOL 2010F field-emission gun transmission electron microscope (FEG-TEM) equipped with a Gatan Imaging Filter (GIF) operated at 197-200kV acceleration voltage. The methods applied were bright-field imaging (BF), high-resolution imaging (HREM), energy-filtered imaging (EFTEM) and energy-loss spectroscopic profiling (ELSP). Figures 3 and 4 show the structure in cross-section at medium and high resolution, respectively. Below the 2nm SiO_2 layer there is a clear indication for a 2-3nm wide band of darker contrast in the top of the crystalline Si.

The chromium layer is 2.6±0.1 nm wide and delineates the former free surface which has been ion implanted. The Cr layer has actually been partially oxidized in air, as shown in Figs. 5e and f. The native oxide layer thickness has been determined to be 2.1±0.2 nm. A ~0.8nm thick layer of strongly varying contrast has been observed within the top 2nm of silicon, indicating strain contrast, presumably due to numerous point defects. Depending on orientation and focus settings, this strained layer may appear bright or dark (see figure 3). No extended lattice defects or precipitates have been observed, however, and the silicon lattice seems intact everywhere else.

Energy-filtered imaging and electron energy-loss spectroscopic profiling indicate the presence of a sub-nm thin layer buried ~2nm under the Si/SiO_2 interface which is completely invisible in HREM. It only shows up as a faint dark line in EFTEM at energy losses between the Si L-edge and the B K-edge and ratio images calculated from those (shown in Fig. 6c), but these are impossible to quantify. While ratio and three-window imaging techniques consistently indicate a B rich layer in the Si, multiple window background fitting to an extended series of energy filtered images acquired with 5eV slit width from 150eV to 225eV reveals the presence of small B-rich clusters ~0.3nm thick and up to 1nm in length, positioned ~4nm below the original wafer surface which has been marked by a line in Fig. 6d, however, the analysis is now complicated by the oxide layers becoming unstable under the electron beam and changing structure as a function of time, thereby producing spurious artefacts in the form of larger patches of alternating bright and contrast at the position of the oxide layers in Fig. 6d.

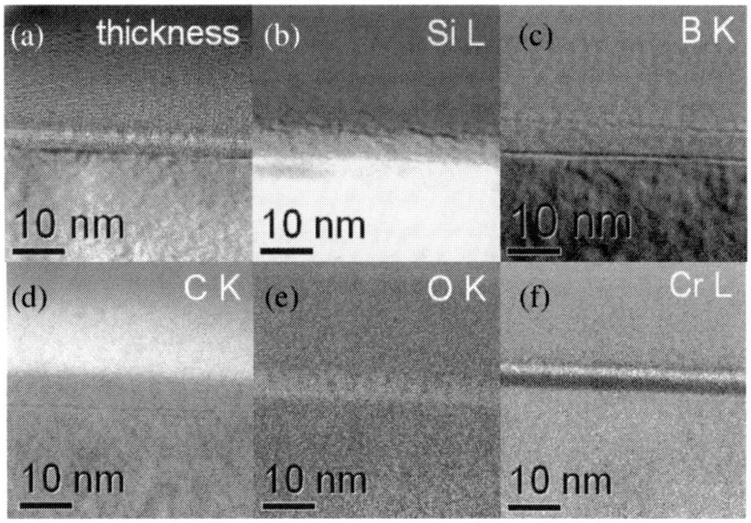

Figure 5. Results from energy-filtered imaging. (a) is a relative thickness map obtained from the ratio of inelastic to elastic scattering. The max. thickness near the image top is ~0.65 times the inelastic mean free path in the Si, which is ~125nm at the imaging conditions chosen, while the sample thickness is only 0.13 times the mean free path near to the Si/SiO_2 interface. The elemental distribution maps (b-f) have been calculated for the corresponding energy-loss core-loss edges using the standard 3-window-method and inverse power-law background fitting and subtraction. These maps show the presence of the corresponding elements (the more the lighter the image).

Energy-loss spectroscopic profiling allows us to extract electron energy-loss spectra (EELS) of the Si L-edge commencing at 99eV and the B K-edge commencing at nominally 188eV, at the positions of interest with ~1nm spatial resolution in the growth direction. Quantification of these is rather difficult. For the Si this is due to the general uncertainty in the scattering cross-sections of L-edges and for the B it is due to the weakness of the signal (i.e. smaller inelastic scattering cross-section plus lower concentration). The quantification procedure was as described in [6, 7]:

- vertical interpolation of spectroscopic images by factor 2 (vertical stretching) to 0.23nm/line
- cross-correlation of images at specimen edge to eliminate relative drift along the vertical
- line-by-line EELS data extraction from spectroscopic images
- EELS background fit (inverse power law)
- background subtraction
- signal integration over 70eV (Si), 50eV (B, C) or 30eV (O) from edge onset, note: no de-convolution has been applied; B K-edge commences at ~195eV, higher than literature values (188eV) suggest
- normalisation for different exposure times and scattering cross-sections
- plot of profiles
- ratio of boron signal to sum over signals from all major elements yields atomic concentration

An evaluation of the relative edge intensities taking into account the different cross-sections consistently yields values between 2at% and 4at% B if the integration range chosen is at least 30eV wide (integration windows only 20eV wide or smaller positioned directly after the edge onset yield erroneous results due to the delayed nature of the Si L-edge and possibly a slight energy drift). Fig. 7 is a patchwork of such electron energy-loss spectroscopic profiles that display intensity as a function of growth direction upwards and as energy loss horizontally. The dispersion along the horizontal direction is 0.103eV/channel, the original sampling along the vertical direction is 0.46nm/pixel. The Cr signal defines the surface, and EEL spectra shown in Figs. 8 and 9 have been extracted from lines corresponding to the positions of the SiO_2 and the Si near the clusters observed. Taking into account the different scattering cross-sections B concentrations of 2.6 and 4.0at% are calculated, respectively, with statistical errors of about ±0.2at% due to noise and background fitting and systematic errors of 0.6at% due to uncertainties in the scattering cross-sections necessary for scaling the electron intensities measured. These values are close to the detection limit of the technique.

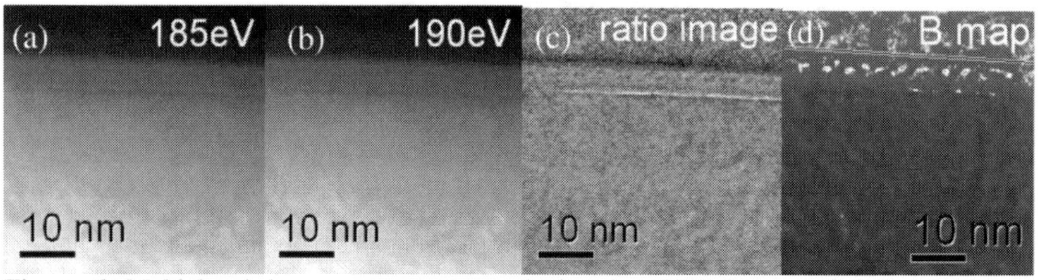

Figure 6. Multiple-window energy spectroscopic imaging (ESI) has revealed B-rich clusters ~6.5nm below the surface of the Cr layer, i.e. ~4nm deep under the surface the original (oxidised) wafer (d). They are located where in energy-filtered imaging just before the B K-edge at 188eV a dark layer is observed within the Si (a, b), and ratio imaging suggests additional contrast due this edge (c) but cannot be quantified. Above the line drawn in (d) the B map only shows noise, just below it the SiO_2 is unstable and yields artefacts, but the bright regions that are also bright in (c) indicate a few B clusters in the form of thin platelets.

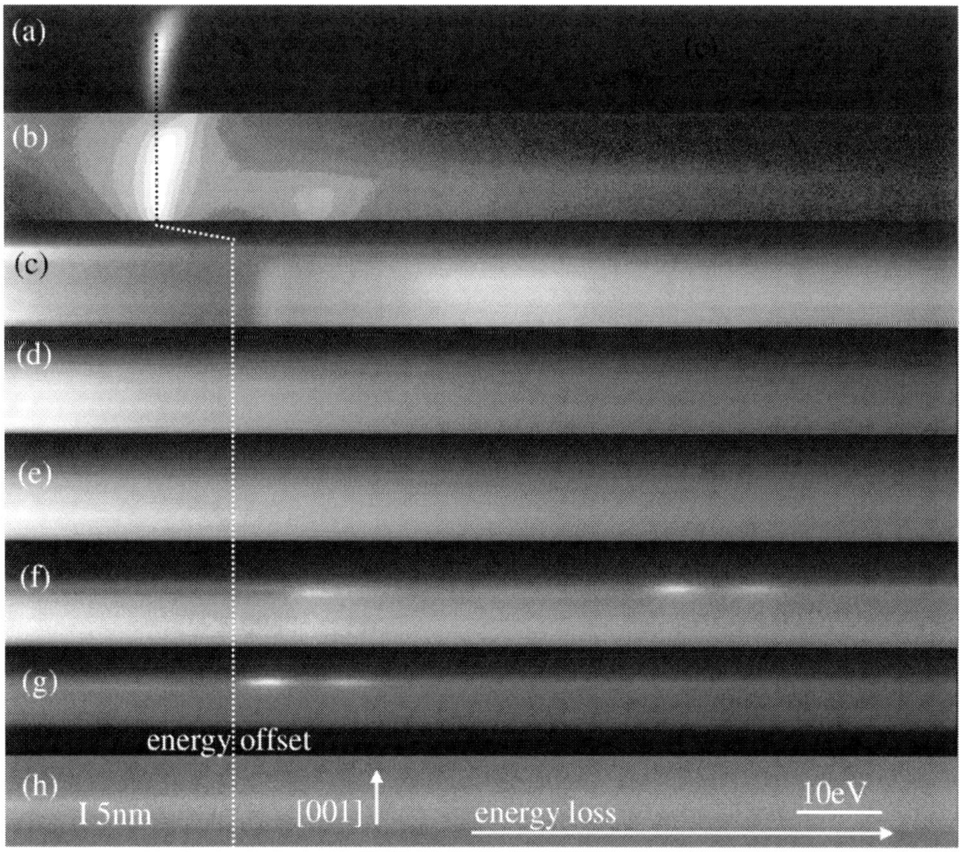

Figure 7. Patchwork of energy-loss spectroscopic profiles recorded with TEM in image and GIF in spectroscopy mode. The energy offsets chosen are 0eV (zero loss peak in (a) and the same on a logarithmic scale in (b), showing the plasmon loss within the thicker Si part), 99eV for the Si L-edge (c), 188eV for the B K-edge (d), 284eV for the C K-edge (e), 532eV for the O K-edge (f), 575 eV for the Cr $L_{2,3}$-edge (g) and 1217eV for the Ge $L_{2,3}$-edge (h).

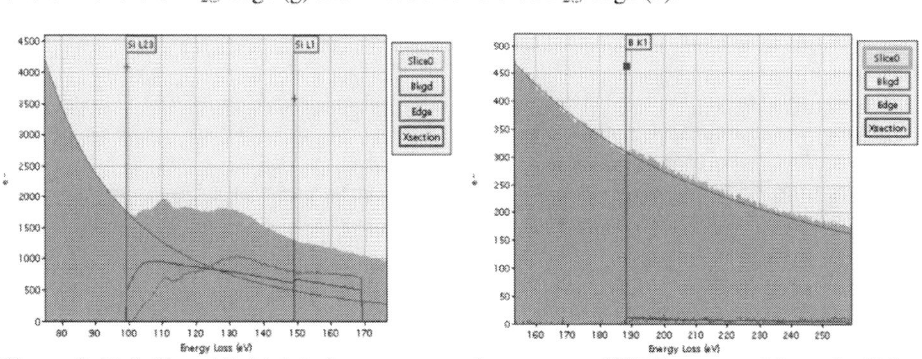

Figure 8. Si (left) and B (right) electron energy-loss spectra (EELS) extracted from the SiO_2 region (10s exposures). The ratio of Si/B=11.8±0.9 implies a boron concentration of 2.6±0.2at% within the SiO_2. Turquoise/grey are experimental spectra, red the background fits, dark blue the theoretical cross-sections from Hartree-Slater theory and green the extracted edges. No de-convolution was applied because the sample was sufficiently thin (t/λ=0.10) to neglect plural inelastic scattering.

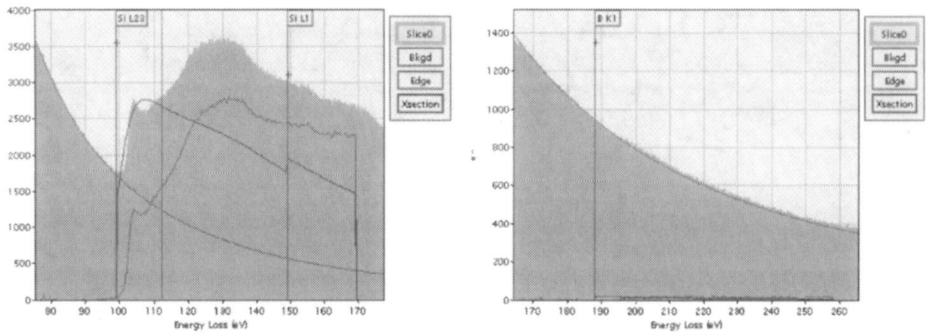

Figure 9. Si and B EELS from the Si near the dark region in Fig. 2 where the B clusters were observed (10s). The ratio of Si/B=23.9±1.1 implies a boron concentration of 4.0±0.2at%. t/λ=0.13

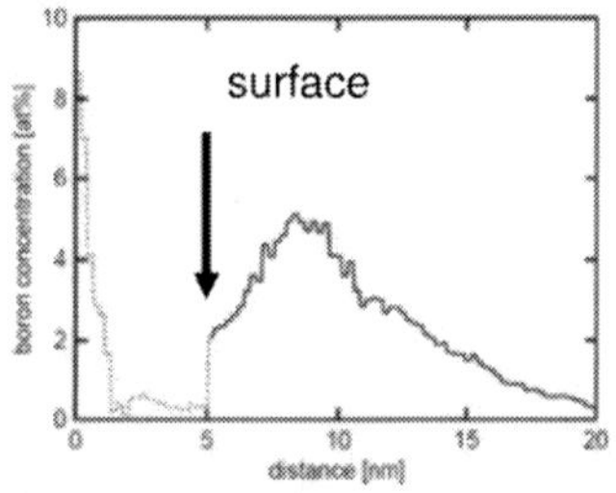

Figure 10. Quantitative B profile extracted from the stripeTEM data in Figs. 7c and 7d

Finally, the boron concentration profile along the growth direction shown in Fig. 10 was extracted from the B and Si spectra. Values that correspond to positions in the glue or Cr layer are very noise and shown dotted; only positions with significant counts in both edges are marked by the solid line. Interestingly, the maximum B peak coincides in depth position roughly with where the lower row of bright patches in Fig. 6d occurs, suggesting that these are indeed due to small B-rich precipitates.

4. Summary

In summary, boron was detected by energy-filtered imaging in the form of sub-nanometre tiny platelets clustered around 4nm depth below the original wafer surface (now SiO_2/Cr interface). Energy-loss spectroscopy of the silicon and boron signals has yielded an effective B concentration in the SiO_2 of ~3at% and in the Si of a maximum of 4at%, which corresponds to the clusters observed. These values agree with the experimental shoulder observed in the SIMS profile at about 4nm sputter depth, but disagree with the B-rich spike reported by SIMS nearer the surface. No evidence for B concentrations above 5at% was found, neither could we find extended lattice defects or precipitates.

References

[1] Nastasi M, Mayer JW and Hirvonen JK 1996 *Ion-Solid Interactions: Fundamentals and Applications* (Cambrige Unversity Press, Cambridge)
[2] Kawski J and Merrill M 2009 *Sol. State Technol.* **52**, 20
[3] Ziegler JF, Ziegler MD and Biersack JP 2010 *Nucl. Instrum. Methods Physics Res.* B **268**, 1818
[4] Carter G and Grant WA 1976 *Ion Implantation of Semiconductors* (Edward Arnold, London) p 57
[5] Wang TS, Cullis AG, Collart EJH, Murrell AJ and Foad MA 2000 *Appl. Phys. Lett.* **77**, 3586
[6] Walther T and Mader W 1999 Proc. Microsc. Semicond. Mater. Conf., Oxford (ed. AG Cullis and R Beanland) *Inst. Phys. Conf. Ser.* **164**, 121
[7] Walther T 2003 *Ultramicroscopy* **96**, 401

Nano-field mapping for the semiconductor industry

D Cooper[1], *A Béché[2], G Servanton[3], R Pantel[3], P Morin[3] and J-L Rouviere[2]

[1] CEA-LETI, Minatec, 17 rue des Martyrs, F38054 Grenoble, France
[2] CEA-INAC, Minatec, 17 rue des Martyrs, F38054 Grenoble, France
[3] ST Microelectronics, 850 Rue Jean Monnet, F-38926 Crolles, France

Email: david.cooper@cea.fr

Abstract. There is a need to measure the dopant potentials and strain fields in semiconductor materials with nm-scale resolution. Here we show that off-axis electron holography is a powerful technique that can be used to measure the fields present in a high-k metal gate 28-nm node nMOS device with a contact etch stop liner stressor. Off-axis electron holography has been used to map the positions of the active dopants with a spatial resolution of 1 nm. The experimental results have been compared to electron energy loss spectroscopy maps. Finally, dark field electron holography has also been used to provide strain maps and the experimental results have been verified using nanobeam electron diffraction.

1. Introduction

Off-axis electron holography is a transmission electron microscopy based technique that can be used to recover the phase on an electron. An electron biprism is used to interfere an electron wave that has passed through a specimen with an electron wave that has passed through only vacuum to provide an interference pattern, known as the hologram. From the electron hologram, phase maps of the specimen can be reconstructed. As the phase of an electron is sensitive to magnetic fields, strain fields and electrostatic potentials, such as from the presence of dopants, electron holography can be used to map these properties with nm-scale resolution. In the absence of magnetic fields, the specimen can be tilted to a weakly diffracting condition to remove the contribution from strain and the phase of the electrons $\Delta\phi$, that have passed through a specimen of thickness t can be given as,

$$\Delta\phi = C_E \int_0^t V(x,y,z)dz$$

where C_E is a constant dependent on the energy of the electron wave (7.29×10^{-3} rads V^{-1} nm^{-1} for 200 kV electrons), V is the electrostatic potential and z is the electron beam direction [1]. Although in principle, the relationship between the measured phase and the electrostatic potential is straightforward, problems that arise from the effects of focused ion beam (FIB) specimen preparation and charging of the specimens during examination cause what is known as the electrically inactive thickness on the specimens. However, systematic studies of electrical junctions suggest that these problems are significantly reduced for the very high dopant concentrations that are present in state-of-the-art semiconductor devices [2].

2. Experimental

Here a 28-nm-technology high-k metal gate nMOS device was prepared for examination by back side FIB-milling. A 176-nm-thick specimen was finished at an operating voltage of 8 kV as this provides a compromise between reducing the surface damage and preserving the parallel sides that are required for electron holography. In order to reduce the charging of the oxide regions above the semiconductor devices, it was necessary to carbon coat the specimens. For dopant profiling, electron holograms were acquired using an FEI Titan operated at 200 kV. To obtain a large field of view for electron holography, the Lorentz lens was used with the conventional objective lens switched off. To improve the signal to noise ratio in the reconstructed phase images, it was possible to acquire the electron holograms for 64 seconds due to the electrical and mechanical stability of the TEM [3]. Figure 1(a) shows an electron hologram of the specimen with a fringe spacing of 2.2 nm and a field of view of 360 nm. The corresponding potential map shown in Figure 1(b) has been reconstructed with a spatial resolution of 6.6 nm. Although the signal to noise ratio in the reconstructed phase image map was $2\pi/200$, the spatial resolution in the final potential map is insufficient for the determination the exact position of the electrical junction. Therefore it is necessary to improve the spatial resolution of the technique.

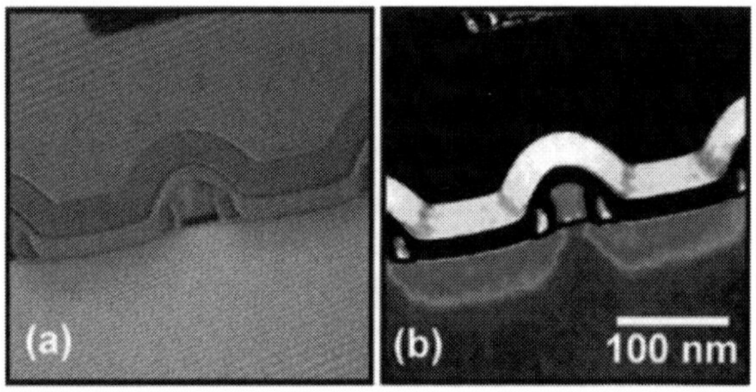

Figure 1. (a) Electron hologram of a 28-nm-gate technology nMOS device. The fringe spacing is 2.2 nm and the field of view is 360 nm. (b) Potential map showing the presence of active dopants in the source and drain regions of the device.

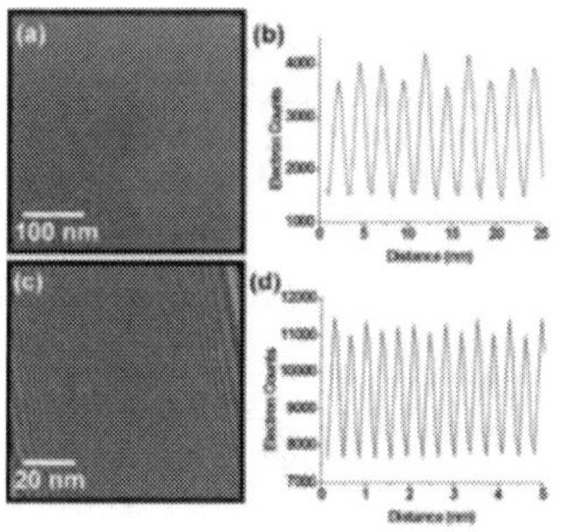

Figure 2. (a) Electron hologram acquired in Lorentz mode. (b) Profile showing 2.2 nm fringe spacing. (c) Electron hologram acquired in free lens control. (d) Profile showing 0.3 nm fringe spacing.

Electron holography performed in "Lorentz Mode" provides a field of view of between 1 micron and 250 nm. These large fields of view are required as even the latest generation semiconductor devices are larger than the typical fields of view that are used in TEM. However, these large fields of view also restrict the achievable spatial resolution. The spatial resolution is normally three times the fringe spacing, however, due to the limited coherence of the electron beam and the finite number of CCD pixels that are used to record the hologram it is not possible to improve the spatial resolution simply by increasing the biprism voltage to reduce the hologram fringe spacing. In conventional high-resolution mode, the field of view of the electron hologram is around 30 nm, which is too small for the examination of most semiconductor specimens, especially as a reference region is required to be in the field of view of the electron hologram. In order to improve the spatial resolution for dopant profiling, the microscope has been operated in objective lens mode using free lens control [4]. Figure 2(a) shows an empty hologram with a field of view of 360 nm that has been acquired in Lorentz mode for 64 seconds. Figure 2(b) shows a profile of the hologram revealing a fringe spacing of 2.2 nm which will lead to a reconstructed phase image with a spatial resolution of 6.6 nm. Figure 2(c) shows an empty electron hologram that has been acquired for 64 seconds using free lens control with a field of view of 75 nm. The corresponding profile can be seen in Figure 2(d), the fringe spacing is 0.33 nm which will lead to a spatial resolution of 1 nm in the reconstructed phase image. In the past it was not possible to obtain 1 nm spatial resolution, as typical semiconductor specimens were too large, however, the miniaturisation of state-of-the-art semiconductor devices means that the region of interest and the reference can now be included in the small field of views that are necessary to sample the holographic fringes. Figure 3(a) shows a potential map with 1 nm spatial resolution of the nMOS device that has been acquired using off-axis electron holography operated in free lens control. The electron hologram was acquired for 64 seconds to improve the signal to noise ratio. It was not possible to have a vacuum region close enough to the specimen for a field of view of 75 nm so the substrate was used as a reference. Although the substrate is lightly doped, simulations suggest that the potential is constant in the reference region. From the potential map, we have determined that the electrical junction has is to 4 nm under each side of the gate due to the diffusion of the dopant atoms. In order to verify these results, the same sample was examined using electron energy loss spectroscopy (EELS) using an FEI Osiris operated at 120 kV [4]. In order to sample the dopant diffusion under the gate, several devices were examined. Spectra were acquired over a specimen area of 380 by 140 nm and sampled by 180 by 65 pixels, leading to a spatial resolution in the As map of 2.1 nm. As a consequence 12,000 spectra were acquired at a speed of 0.3 seconds each. Figure 3(b) shows a 32 by 30 pixel detail from the Arsenic map for the specimen and it can be seen that the As diffusion under the gate is around 4 nm which is consistent with the electrical potential map.

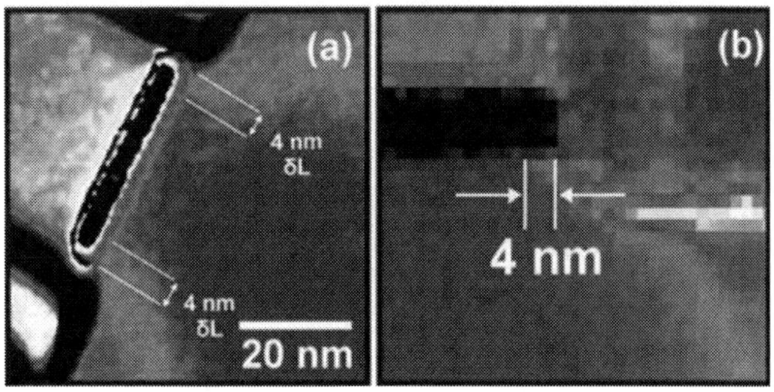

Figure 3. (a) Potential map of the nMOS device specimen with a spatial resolution of 1 nm. (b) EELS map of the As dopant atoms in the specimen with a spatial resolution of 2.1 nm.

Dark field electron holography is a recently developed technique that can be used to provide quantitative strain maps [6]. The difference between dark field holography and conventional holography is that the specimen is intentionally tilted to a strongly diffracting condition and the information for the required set of lattice places can be selected by choosing a diffracted beam by using an objective aperture. The strain can then be calculated by using a geometrical phase analysis (GPA) algorithm. The same physical specimen that was used for dopant profiling was also used to provide the strain maps. Dark field holography was performed using the Titan operated at 200 kV, again electron holograms were acquired for 64 seconds with a fringe spacing of 2.2 nm. To measure the strain in the growth direction, the specimen was tilted close to a 2 beam condition with the (002) diffraction spot selected, for the in-plane direction, the (020) spot was selected. Figure 4(a) shows a dark field electron hologram for the device for the (020) in plane direction, (b) shows a strain map and (c) shows a strain profile. The measured value of 0.2 % tensile strain is close to what is expected from simulations. The results were verified by using nanobeam electron diffraction (NBED) on the same sample during the same session on the TEM [7]. NBED uses a near-parallel electron beam which is scanned across the specimen in STEM mode. By using the three condenser lens system in the FEI Titan, an electron beam with a convergence angle of 0.2 mrads and a FWHM of 6 nm can be obtained. The shifts in the diffraction spots allow the local strain to be assessed using peak finding software that has been developed at CEA. Figure 4(d) shows a dark field electron hologram for the (002) growth direction, (e) shows the strain map and (f) profiles. Again the measured value of 0.2 % compressive strain was verified by NBED. The values of strain that are measured are extremely low which demonstrates the excellent sensitivity of the two different techniques. However, it was extremely straightforward to obtain these strain maps by dark field electron holography, whereas, significant efforts were required to measure the strain by NBED.

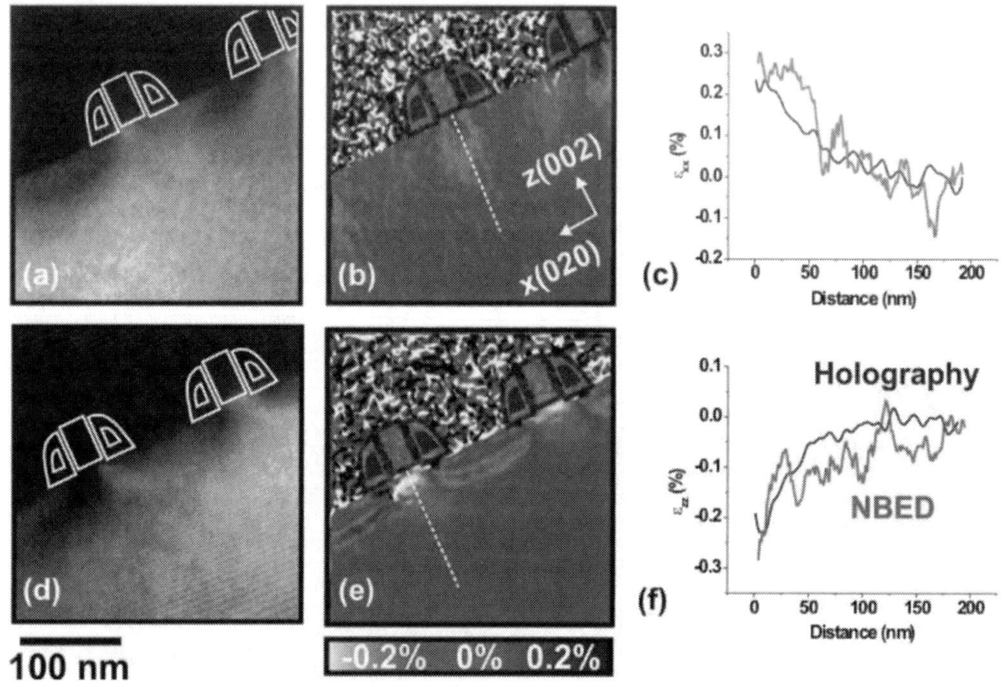

Figure 4. (a) Dark field electron hologram for the (020) in-plane direction, (b) corresponding strain map and (c) strain profiles obtained by dark field holography and NBED. (d) Dark field hologram for the (002) growth direction, (e) corresponding strain map and (f) profiles acquired by both dark field holography and NBED.

3. Conclusion

In conclusion, we have shown that off-axis electron holography is a powerful technique that can be used to map the active dopants and strain fields in semiconductor devices with nm-scale resolution. The active dopant maps have been compared to EELS maps and the dopant diffusion distance under the gate has been measured. In addition the strain fields in the conduction channel have been quantitatively measured by dark field electron holography and the results are consistent with those measured by NBED.

Acknowledgements

This work was partially supported through the Recherché Technologie de Base (RTB) Program and performed within the IBM – ST Microelectronics – LETI alliance.

References

[1] Rau WD, Schwander P, Baumann F, Hoppner W and Ourmazd A, 1999 *Phys. Rev. Lett.* **82**, 2614

[2] Cooper D, Ailliot C, Barnes JP, Hartmann JM, Salles P, Benassayag G and Dunin-Borkowski RE, 2010 *Ultramicroscopy* **110**, 383

[3] Cooper D, Truche R, Rivallin P, Hartmann JM, Laugier F, Bertin F and Chabli A, 2007 *Appl. Phys. Lett.* **91** 143501

[4] Sickmann J, Formanek P, Linck M, Muehle U and Lichte H, 2010 *Ultramicroscopy* **111** 290

[5] Servanton G, Pantel R, Juhel M and Bertin F, 2010 *Micron* 41, 118

[6] Hytch M, Houdellier F, Hue F and Snoeck E, 2008 *Nature.* **453** 1086

[7] Béché A, Rouviere JL, Clement L and Hartmann JM, 2009 *Appl. Phys. Lett.*, **95** 123114

TEM of Nano-LEDs made by laser writing

M W Fay[1], A Nasir[2], O Makarovsky[2], A Rastelli[3], L Eaves[2], R Campion[2], C T Foxon[2] and A Patanè[2]

[1] Nottingham Nanotechnology and Nanoscience Centre, University of Nottingham, Nottingham NG7 2RD, UK

[2] School of Physics & Astronomy, University of Nottingham, Nottingham NG7 2RD, UK

[3] Institute for Integrative Nanosciences, IFW Dresden, D-01069 Dresden, Germany

Michael.fay@nottingham.ac.uk

Abstract. Using a focused laser beam, nanoscale current channels and nanoscale light emitting diodes (LEDs) have been made by the controlled thermal diffusion of manganese interstitial ions (Mn_i) out of a semiconductor heterostructure containing a layer of p-type ferromagnetic (GaMn)As. Here we use EDX, EELS and dark field imaging to probe directly the diffusion of Mn interstitials in the vicinity of the nano-LED region.

1. Introduction

Recently, it has been demonstrated how the controlled thermal diffusion of manganese interstitial ions (Mn_i) out of a semiconductor heterostructure containing a layer of p-type ferromagnetic (GaMn)As can be used to create nanoscale current channels [1-2] and nanoscale light emitting diodes (LEDs) [3]. Our devices are p-i-n resonant tunnelling LEDs with a two-dimensional AlAs/GaAs/AlAs quantum well embedded in the intrinsic (i) layer. This is sandwiched between a bottom n-type GaAs contact layer and a top p-type layer of $Ga_{1-x}Mn_xAs$ with x=5%. By annealing a small spot with a focused laser beam, we are able to create a preferential path for charged carriers when a forward bias is applied to the p-i-n diode, thus activating a nanoscale region of the device to emit light at a bias well below the threshold voltage for emission from the non-annealed regions. The technique also provides real-time in situ characterisation and control of the nanoscale current channels during their formation and allows us to produce an ordered array of LEDs or else a shaped, light-emitting region with submicron precision [3].

2. Experimental

For TEM analysis, we selected a region annealed with a focused laser beam of diameter <1000nm, wavelength λ=515nm and power P= 40mW. These laser annealing conditions produced visible surface effects, allowing accurate location of the annealing region in the FIB-SEM, while producing enhanced photoluminescence and electroluminescence in close proximity to the annealed region. This allowed us to probe how the material composition changes with increasing distance from the laser spot, i.e. with decreasing annealing temperatures, which we estimate to be >500°C inside the laser spot and decreasing to T_A<100°C at distances d>5000nm.

In preparing the samples for TEM analysis, it is important to minimize amorphisation of the sample and further diffusion of the Mn interstitials. Lift-out samples were initially prepared for TEM analysis from a specific microscopic area containing a nano-LED region in an FEI Quanta 200 3D. These were attached at one side to an Omniprobe support and were then further FIB milled, finishing at 10kV, 16pA. The samples were plasma cleaned for ten minutes in an inductively couple plasma of Ar/25% O_2. The sample was then further milled in a Fischione 1010 argon ion miller, finishing at 500V, 5mA. Samples were further plasma cleaned for five minutes immediately prior to TEM analysis.

The Mn-distribution within and outside the active region of the LED was analysed using EDX, EELS and 002 dark field imaging in a JEOL 2100F equipped with an Oxford Instruments INCA energy TEM 250 EDX system and a Gatan Tridium image filter and Gatan Orius camera.

3. Results and discussion

TEM analysis reveals a significantly altered region comprising a complex structure of amorphous and polycrystalline material directly under the annealing point, of the order of ~400nm wide and ~200nm deep in the case of a 40mW laser anneal (Figure 1a). The transition from heavily restructured to apparently intact GaAs structure appears to be fairly abrupt, with the quantum well structure still clearly visible within ~50nm of the restructured material (Figure 1b).

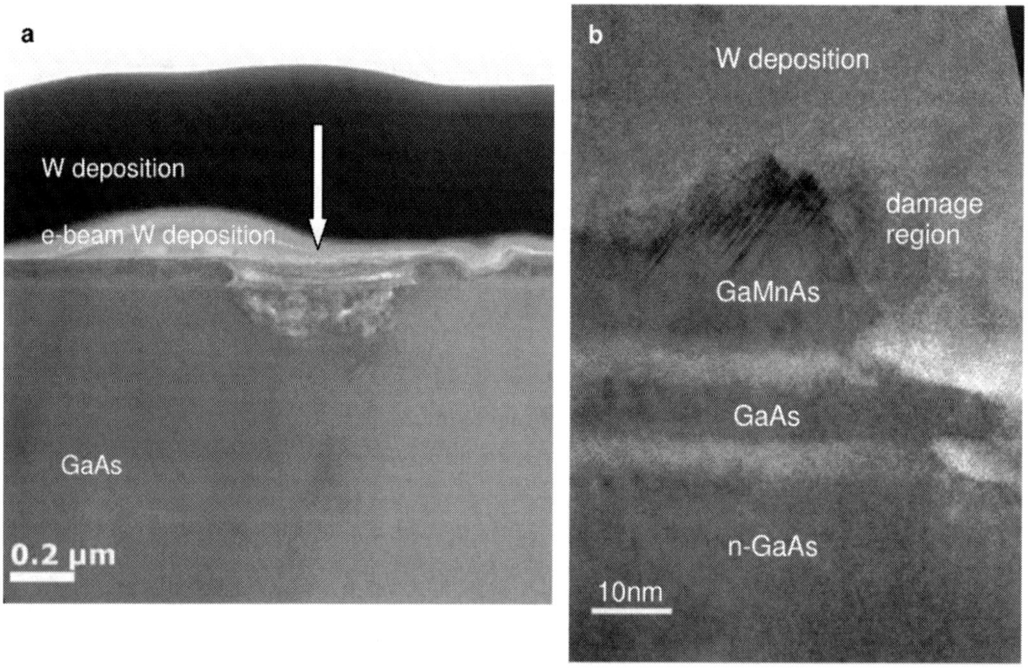

Fig. 1. a) Bright field TEM image showing the extent of the structurally altered region (arrowed) directly under the laser during annealing. b) HRTEM image showing that the device structure is largely preserved immediately adjacent to the region structurally altered by the laser annealing.

Elemental mapping using EDX reveals that there has been significant evaporation of As from the structure at the immediate point of the laser anneal. The Ga and Mn signals in this heavily restructured region correspond to a strong O signal, indicating a high level of oxidation. However, an Al-rich layer is still clearly visible, albeit shifted slightly towards the substrate, and appears to correspond to a region with a low level of oxidation (see Figure. 2).

In a region of ~50nm at the edge of the highly damaged region, there has been strong diffusion of Mn towards the surface, where EDX indicates that it has been oxidised. There is no evidence of significant diffusion of other elements in this region, with the AlAs/GaAs/AlAs interfaces remaining clearly defined. However, within ~100nm of the edge of the significantly amorphised region, the overall Mn distribution is comparable to that of regions several hundred nm from the annealed region (see Figure 3). Below the AlAs barrier layers, the level of Mn appears to be generally below the detectable limits in this EDX system, both near to the anneal and at some distance.

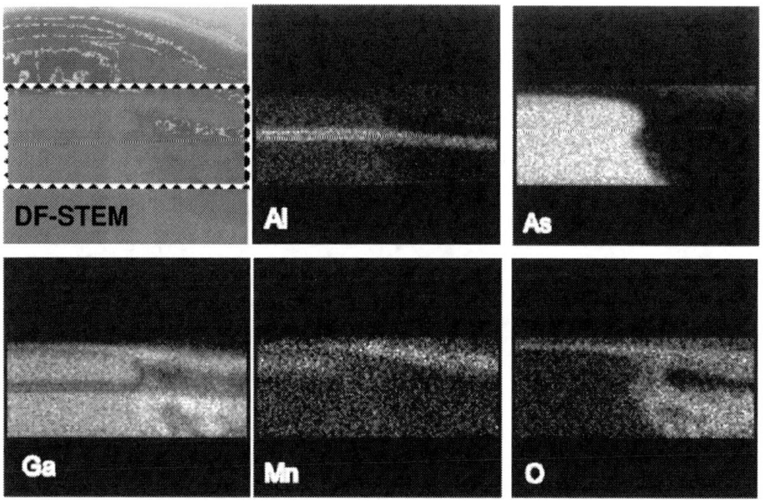

Fig. 2. EDX maps of the interface between the damaged region and the intact, active device structure. DF-STEM image is 365 nm across.

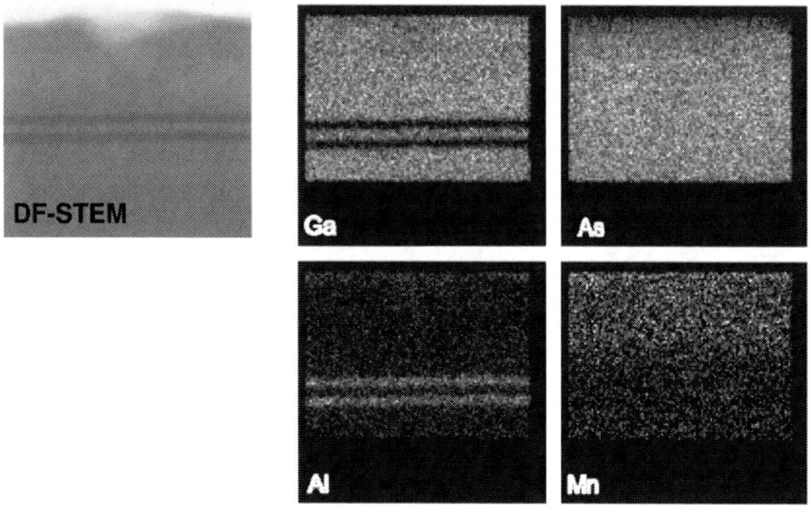

Fig. 3. EDX maps acquired at ~500nm from the edge of the annealed region. Gradation can be seen in the Mn composition, increasing to the surface of the sample. DF-STEM image is 125nm across.

002 dark field imaging has the potential to probe the distribution of Mn interstitials specifically, as the structure factor is not significantly affected by substitutional Mn, but is strongly reduced by the energetically favoured type 2 interstitials (four As nearest neighbours)[4]. However, for analysis over

a larger area this technique requires a flat sample with little buckling, meaning that it can be sensitive to sample preparation. 002 dark field analysis of a ~250nm thick lift out sample shows a darker intensity region both above and below the quantum well structure, extending ~500nm from the edge of the amorphised region. Noting the relatively high temperature (T_A >500°C) required to create the amorphised region of GaAs and the characteristic temperature for the Mn_i diffusion (T_A <200°C [1]), the extended dark regions in Figure 3 could be associated with the diffusion of interstitial-Mn from the top (GaMn)As layer into the underlying structure. Further corroborative evidence is required before this can be firmly concluded.

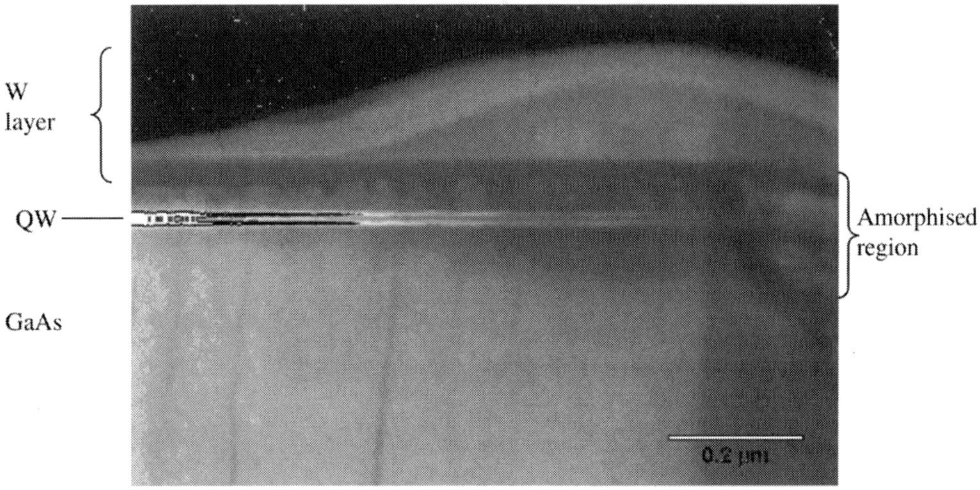

Fig. 3. 002 Dark Field image, with the annealed region to the right. The darker region to the right of the image is a thickness artefact. Some surface damage to the sample may have occurred during ion beam milling.

4. Conclusions

The microstructure and elemental compositions of a nanoLED created by laser-induced thermal annealing have been appraised. At 40mW, the laser has produced a central largely amorphised region depleted of As, surrounded by a 50nm wide region where Mn has diffused towards the surface without otherwise affecting the structure of the material. 002 Dark Field shows darker intensity below the QW structure and over a few hundred nm from the amorphised region, which may indicate the presence of interstitial Mn, but requires further corroborative evidence.

References

[1] Makarovsky O, Balanov A G, Eaves L, Patanè A, Campion R P, Foxon C T and Airey R J, 2010 *Phys. Rev. B* **81**, 035323.

[2] Wijnheijmer A P, Makarovsky O, Garleff J K, Eaves L, Campion R P, Gallagher B L and Koenraad P M 2010 *Nano Letters* DOI: 10.1021/nl102739y.

[3] Makarovsky O, Kumar S, Rastelli A, Patanè A, Eaves L, Balanov A G, Schmidt O G, Campion R and Foxon C T 2010 *Advanced Materials* **22**, 3176.

[4] Glas F, Patriarche G, Largeau L and Lemaître A 2004 *Phys. Rev. Lett.* **93**, 086107-1.

Direct observation of carrier distribution in laser diode using off-axis electron holography and Lorentz microscopy

H Sasaki[1,2], K Yamamoto[2] and T Hirayama[2]

[1] Yokohama R&D Laboratories, Furukawa Electric Ltd., 2-4-3 Okano, Nishi-ku, Yokohama, 220-0073, Japan
[2] Japan Fine Ceramics Center, 2-4-1 Mutsuno, Atsuta-ku, Nagoya, 456-8587 Japan

E-mail: hsasa@yokoken.furukawa.co.jp

Abstract. The carrier distribution of a buried laser diode structure was examined using off-axis electron holography and Lorentz microscopy, and a two-dimensional structure of p-n junctions was clearly observed. The sensitivity of off-axis electron holography and Lorentz microscopy to different carrier concentrations of GaAs specimen was compared.

1. Introduction

The measurement of carrier distributions in semiconductor devices is very important in the development of new devices and for the analysis of device failure. The application of off-axis electron holography in the silicon semiconductor industry has attracted considerable attention in recent years [1-3]. The first Lorentz microscopy observation of Si p-n junction was carried out by Merli et al. [4-6]. Twitchett et al. quantitatively examined the Si p-n junction using both off-axis electron holography and Lorentz microscopy [7]. We have been concerned with compound semiconductors such as GaAs [8] in observation using off-axis electron holography, and found that the process of removing the damaged layer formed by a focused ion beam (FIB) is unavoidable [9,10] when the TEM specimen is fabricated using a FIB.

We compared the sensitivity of off-axis electron holography and Lorentz microscopy to different carrier concentrations in a GaAs specimen. In addition, a two-dimensional carrier distribution of a buried laser diode structure was observed using these methods.

2. Experimental Details

2.1. GaAs test sample with step-like carrier concentration

We examined p-n thin films grown on a GaAs substrate by metal-organic chemical vapour deposition (MOCVD). The structure of the TEM specimen is shown in Fig. 1. Silicon and zinc were used as n-type and p-type dopants, and the composition of the samples was for testing the sensitivity of electron holography and Lorentz microscopy to different carrier concentrations.

The specimen was prepared using an SIINT SMI3050TB (Triple Beam) FIB system, which combines a Ga ion beam, a scanning electron microscope for process monitoring, and an Ar ion beam to remove the FIB damaged layers [11]. The detailed procedure for preparing TEM specimens was reported in ref. [10]. The specimen suitable for TEM observation was thinned to 300 nm in the FIB system. Finally, the FIB damaged layer was removed by milling the sample surface for 5 minutes with a 1.0-kV Ar ion beam in the FIB system.

Electron holographic and Lorentz microscopy observations were made with a JEOL-3100F TEM operating at 300 kV, and equipped with a Schottky field-emission electron gun and an electron biprism. Digital holograms with 1024 × 1024 pixels were recorded using a slow-scan charge-coupled device (CCD) camera system. We used 13 holograms with different initial phases for phase reconstruction. Phase-shifting electron holography [12] was used to reconstruct the phase images because it produces a phase map that has high spatial resolution and phase-measurement sensitivity.

2.2. Semiconductor Laser Diode
An InP-based buried laser diode structure was produced by using MOCVD and etching process. The FIB system was used to thin down a TEM specimen to 300 nm. Surfaces of the specimen were milled for 15 minutes using an Ar ion beam at an accelerating voltage of 1.0 kV at liquid N_2 cooling temperature and an incident angle of 10°.

A Hitachi HF-3300 TEM operated at 300 kV equipped with a cold field-emission electron gun and an electron biprism was used for the electron holography and Lorentz microscopy experiment. Digital 4096 × 4096 holograms were recorded using a slow-scan CCD camera system. The hologram was reconstructed using a conventional Fourier transform method.

3. Results and Discussions

3.1. GaAs with step-like carrier concentration
The reconstructed phase image is shown in Fig. 1(a). The p- and n-type regions are clearly seen as areas of dark and bright contrast, and some differences of changing dopant concentrations can also be seen. The interface between the 1×10^{19} cm^{-3} region and 1×10^{18} cm^{-3} regions is clearly seen in the p-type area, and the 1×10^{17} cm^{-3} and 1×10^{16} cm^{-3} regions can be distinguished. A detailed analysis of the phase image has been presented in ref. [10].

The defocus images are shown in Fig. 1(b)-(g). The p-n junctions were clearly seen at both 0.6 mm

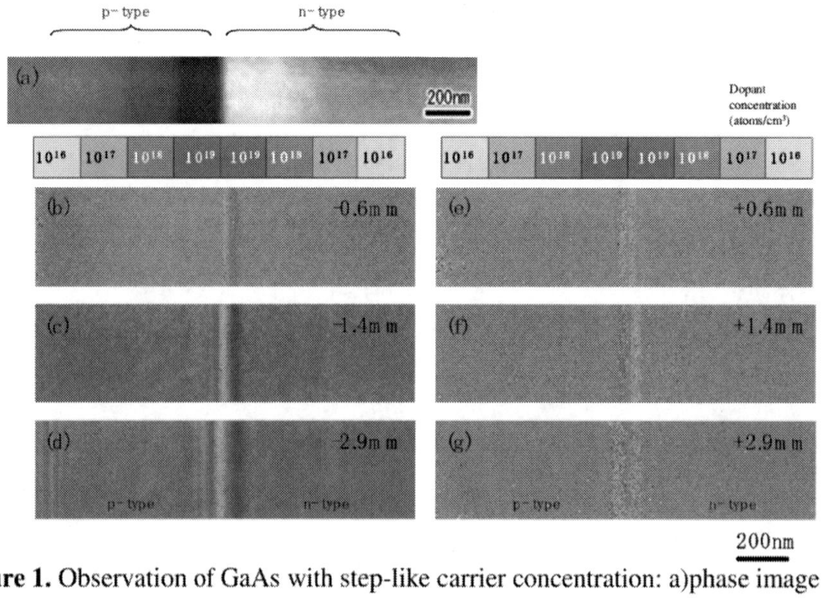

Figure 1. Observation of GaAs with step-like carrier concentration: a)phase image b-g) defocus images.

under-focus and over-focus, but interfaces of different carrier concentration are hardly observed in each image. In the 1.4 and 2.9 mm defocus images, the interface between 1×10^{19} cm^{-3} and 1×10^{18} cm^{-3} regions in the p-type area are slightly seen. Compared with the phase image of the off-axis electron holography, the defocus image of Lorentz microscopy has lower sensitivity to different dopant concentrations.

From a practical point of view, Lorentz microscopy is not useful in the detection of differences of only one order of magnitude in carrier concentrations. However, if the location of an abrupt p-n junction needs to be checked, Lorentz microscopy is very practical because it does not requires an electron biprism and is an easier measurement to carry out than off-axis electron holography. In addition, the vacuum region which a reference wave comes through is unnecessary. For observations using off-axis electron holography, the vacuum region has to be made close to the region of interest, so preparing a TEM sample for a semiconductor device can be very difficult.

3.2. Semiconductor Laser Diode

To evaluate how well commercial devices perform, they need to be evaluated using Lorentz microscopy and off-axis electron holography. We used these methods to observe a buried laser diode structure. An in-focus image of a laser diode is shown in Fig. 2(a), where only the active layer is observable. A defocus image at -6.7 mm under-focus is shown in Fig. 2(b). The two-dimensional structure is clearly seen by black and white double contour lines. A defocus image at +6.7 mm over-focus is shown in Fig. 2(c). At double contour lines, black and white lines are inverted from the under-focus image. Since these inverted black and white lines appear in both images, these lines indicate p-n junctions.

A reconstructed phase image of the laser diode device which had poor operating characteristics is shown in Fig. 3(a). In this hologram, the interference region and spacings between the interference fringes are 5 μm and 30 nm. The p-n junction of the laser device can be also clearly seen. As discussed previously, a significant advantage of off-axis electron holography is its high sensitivity. The highly n+ doped area, which cannot be observed in defocus images, is visible in the phase image. Figure 3(b) shows an enlarged phase image from the dashed frame in Fig. 3(a). The interference region and spacings between the interference fringes are 1.5 μm and 5 nm at this hologram condition. The phase image shows clear fine structure of the laser device. The designed location of the p-n junction was the dashed line in Fig. 3(b), but the observed p-n junction is in a different location. The arrow indicates that both sides of n-regions are connected. This is the cause of high leakage current in the failed laser device.

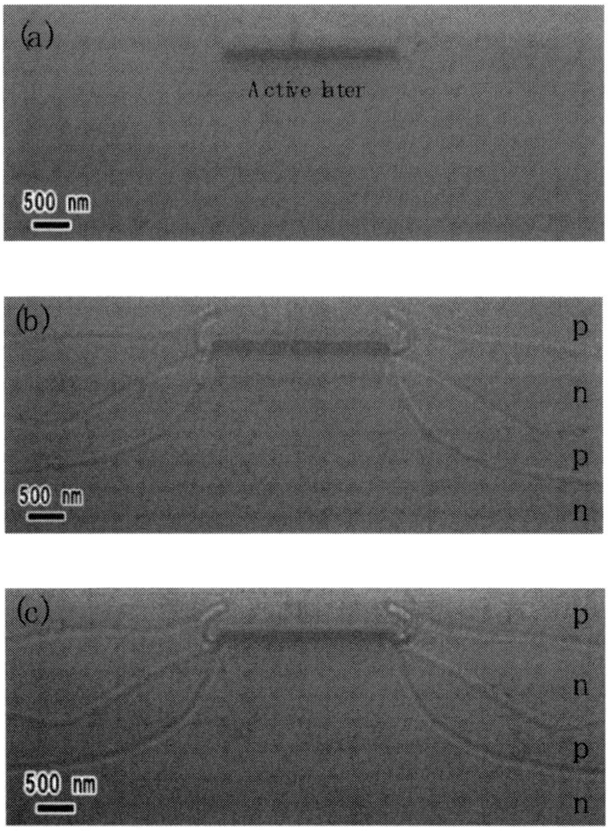

Figure 2. Observation of semiconductor laser diode using Lorentz microscopy a) in focus, b) at defocus of -6.7 mm under-focus, c) at defocus of +6.7 mm over-focus.

4. Conclusions

We clearly observed the GaAs p-n junction using off-axis electron holography and Lorentz microscopy. The interface between 1×10^{19} cm^{-3} and 1×10^{18} cm^{-3} doped regions was clearly seen using electron holography, but was hardly visible with Lorentz microscopy. Off-axis electron holography is more sensitive to different dopant concentrations. We also observed the buried laser diode structure using these two techniques. The two-dimensional structure of the p-n junctions was clearly observed, which demonstrated an example of device failure analysis.

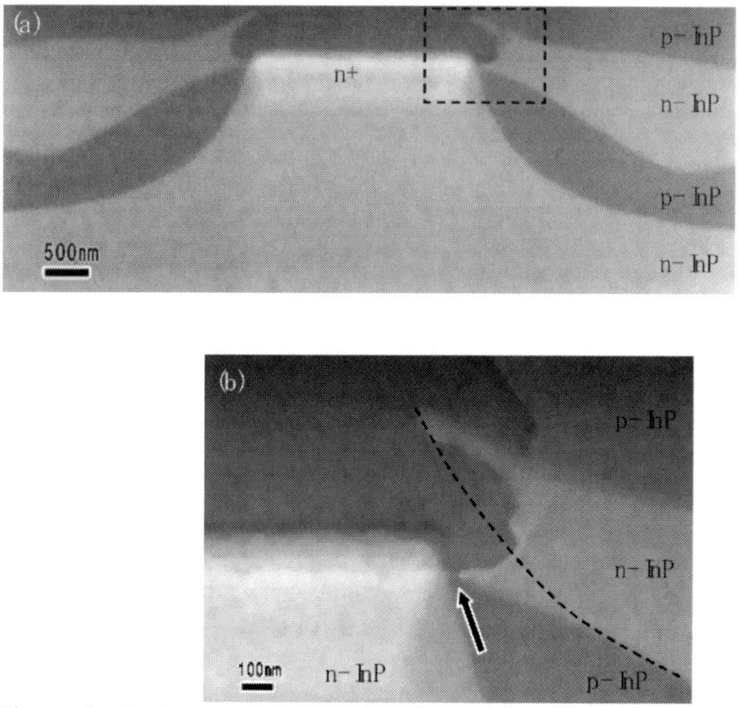

Figure 3. a) phase image of semiconductor laser diode b) enlarged phase image from dashed frame in Fig. 3(a). Designed location of p-n junction was at dashed line.

References

[1] Rau WD, Schwander P, Baumann FH, Hoppner W and Ourmazd A 1999 *Phys. Rev. Lett.* **82** 2614
[2] Wang Z, Hirayama T, Sasaki K, Saka H and Kato N 2002 *Appl. Phys. Lett.* **80** 246
[3] Twitchett AC, Dunin-Borkowski RE and Midgley PA 2002 *Phys. Rev. Lett.* **88** 238
[4] Merli PG, Missiroli GF and Pozzi G Phys. 1973 *Phys. Status Solidi (a)* **16** K89
[5] Merli PG, Missiroli GF and Pozzi G Phys. 1973 *Phys. Status Solidi (a)* **20** K87
[6] Merli PG, Missiroli GF and Pozzi G Phys. 1975 *Phys. Status Solidi (a)* **30** 699
[7] Twitchett AC, Dunin-Borkowski RE and Midgley PA 2006 *Philos. Mag. A* **86** 5805
[8] Sasaki H, Yamamoto K, Hirayama T, Ootomo S, Matsuda T, Iwase F, Nakasaki R and Ishii H 2006 *Appl. Phys. Lett.* **89** 244101
[9] Sasaki H, Ootomo S, Matsuda T, Yamamoto K and Hirayama T 2007 *Proc. MSM2007* 395
[10] Sasaki H, Yamamoto K and Hirayama T 2010 *J. Phys.Conf. Ser.* **209** 012028
[11] Takahashi H, Sugiyama Y, Yamamoto Y, Sadayama S, Iwasaki K and Fujii T 2006 *Proc.16th Int Microscopy Congress*, (Sapporo: Sapporo Convention Center) 1104
[12] Yamamoto K, Kawajiri I, Tanji T, Hibino M and Hirayama T 2000 *J. Electron Microsc.* **49** 31

Conventional and 360 degree electron tomography of a micro-crystalline silicon solar cell

M Duchamp[1], A Ramar[1], A Kovács[1], T Kasama[1], F-J Haug[2], SB Newcomb[3], C Ballif[2] and RE Dunin-Borkowski[4,1]

[1] Center for Electron Nanoscopy, Technical University of Denmark, DK-2800 Kongens Lyngby, Denmark

[2] Photovoltaics and Thin Film Electronics Laboratory, Ecole Polytechnique Fédérale de Lausanne, Rue Breguet 2, CH-2000 Neuchâtel, Switzerland

[3] Glebe Laboratories, Newport, Co. Tipperary, Ireland

[4] Institute for Microstructure Research, Forschungszentrum Jülich, D-52425 Jülich, Germany

E-mail: martial.duchamp@cen.dtu.dk

Abstract. Bright-field (BF) and annular dark-field (ADF) electron tomography in the transmission electron microscope (TEM) are used to characterize elongated porous regions or cracks (simply referred to as cracks thereafter) in micro-crystalline silicon (μc-Si:H) solar cell. The limitations of inferring the 3D geometry of a crack from a tilt series of images acquired from 100-nm-thick focused ion beam (FIB) milled TEM specimen are discussed. In an attempt to maximize the specimen tilt range and to reduce the effects of diffraction and phase contrast on the reconstruction, both BF and ADF electron tomography are used to acquire 360° tilt series of images from a FIB-prepared needle-shaped μc-Si:H specimen.

1. Introduction

Thin film microcrystalline silicon (μc-Si:H), prepared by plasma-enhanced chemical vapour deposition, is a promising material for use as the active absorber layer in low cost solar cells [1], for which efficiencies higher than 8% have been obtained on flexible substrates [2] and over 10% on glass substrates [3], with higher efficiencies when used in tandem solar cells based on a stack of a-Si:H and μc-Si:H cells. The use of rough substrates has been demonstrated to increase light trapping efficiency and overall solar cell efficiency [4]. However, the presence elongated porous regions (called cracks thereafter) between adjacent agglomerations of nanocrystals is thought to decrease electrical performance, as they can introduce shunts between the top and bottom contacts, thereby decreasing the open-circuit voltage and the fill-factor of the solar cell [5].

Transmission electron microscopy (TEM) combined with electron tomography is a suitable technique for the analysis of the redistribution of different materials in a three-dimensional volume. However, the detection of a region of lower density within a crystalline matrix using electron tomography is still challenging [6], even if the requirement that the recorded image intensity should be a monotonic function of a desired property proportional, e.g. specimen thickness or chemical composition, is approximately justified by acquiring annular dark-field (ADF) images, which are less sensitive to diffraction contrast than conventional bright-field (BF) images.

Here, we assess the applicability of both BF and ADF electron tomography reconstruction of (i) a crack in a μc-Si:H matrix and (ii) fine-grained Si nanocrystals in a needle-shaped sample in which missing wedge artifacts are reducing by allowing images to be acquired as the specimen is tilted through 360° in the TEM [7].

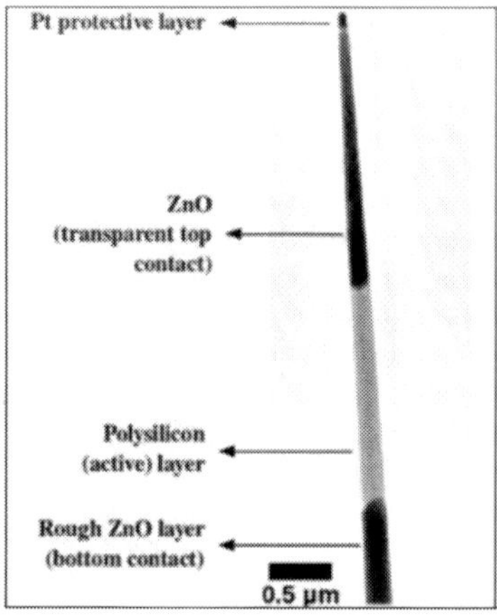

Figure 1. Bright-field TEM image of a needle-shaped μc-Si:H solar cell specimen prepared for 360° electron tomography using focused ion beam milling. Diffraction contrast in the Si layer is averaged other the CCD camera pixels due to the low magnification mode.

2. Experimental details

The solar cell samples examined here were grown onto pyramidal ZnO surface textures that had been deposited by low-pressure chemical vapour-deposition processes onto glass substrates [8]. Single-junction thin-film μc-Si:H layers were grown in the form of n-i-p sequences using plasma-enhanced chemical vapour deposition under standard conditions of plasma power, temperature and silane flux (4 W, 200 °C and 7.3 sccm, respectively).

TEM specimen preparation of thin film silicon solar cells can be highly challenging since they are made on plastic, glass or metallic substrates. Here, both conventional 100-nm-thick lamellar and 200-nm-diameter needle-shaped specimens (Figure 1) were prepared using focused ion beam (FIB) milling in a dual-beam FEI Helios microscope. Tilt series of BF TEM images were acquired in an FEI Tecnai T20 TEM, while tilt series of ADF scanning (S)TEM images were acquired in an FEI Titan TEM, using 47.4 and 10.2 mrad inner ADF detector semi-angles for crack and polysilicon characterization, respectively. All measurements were carried out at 120 keV to minimize electron beam damage. Each tilt series of images was aligned manually due to the limited number of recognizable specimen figures. Tomographic reconstructions were performed using the simultaneous iterative reconstruction technique (SIRT) algorithm with 30 iterations using FEI Inspect3D software.

3. Results and discussion

3.1. Crack characterization in a conventional TEM lamella

Figure 2 (a) shows a tomographic reconstruction of a crack in μc-Si:H obtained from (a) BF TEM image series acquired from a conventional 100-nm-thick FIB-prepared TEM lamella using a tilt range of ±60°. The growth direction of the μc-Si:H layer is the *y*-direction. The crack is visible in the form of a region of reduced density close to the centre of the volume. Similar contrast at the edge of the reconstructed volume is an artefact that may have resulted from the limited tilt range. In Fig. 2 (a), the approximate shape of the crack is recovered, but its precise lateral dimensions are likely to be in error as a result of the effect of local variation in diffraction contrast in the μc-Si:H matrix on the reconstruction.

Figures 2 (b) and (c) show corresponding results obtained by applying ADF TEM tomography with a tilt range of ±60°. Fig. 2 (b) shows a section through the reconstructed volume in which the crack is shown in grey, the μc-Si:H matrix in red and residual artefacts that may arise from the limited tilt angle in yellow. Figure 2 (c) shows a 3D iso-surface representation of the volume occupied by the crack. The bounding box represents the size of the reconstruction volume in the *y*- and *x*-directions, and a region in the *z*-direction that was selected to remove missing wedge artefacts. The use of a needle-shaped sample that allows a 360° tilt range to remove such artefacts is described below.

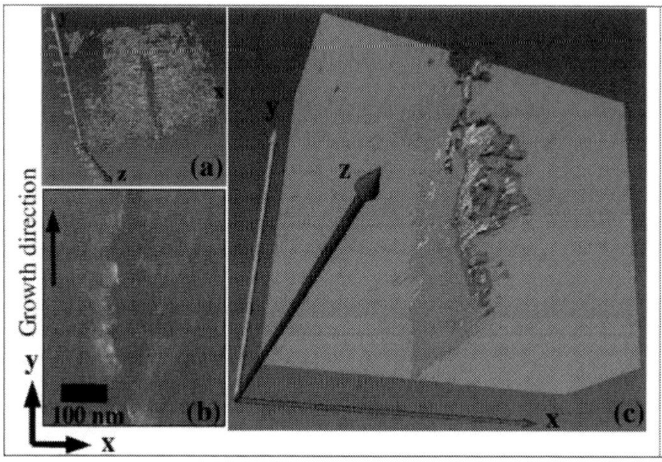

Figure 2. (a) Electron tomographic reconstruction of a crack in μc-Si:H obtained from bright-field TEM tilt series of images. (b-c) Corresponding annular dark-field tomographic reconstruction showing (b) the section through reconstructed volume and (c) an iso-surface representation of the crack volume in a region of the reconstruction that excludes many of the most serious missing wedge artefacts.

3.2. μc-Si:H characterization in a needle-shaped specimen

Figure 3 (a) shows a BF TEM picture of a needle-shaped specimen, taken from a tilt series of images of μc-Si:H acquired over a full 360° tilt range. In Fig. 3 (a), individual crystals are in strongly diffracting orientation and the local growth direction of the μc-Si:H is perpendicular to the ZnO. Fig. 3 (b) shows a section through a reconstructed volume created from the tilt series of images that include Fig. 3 (a). The reconstruction reveals the presence of a 20-nm-thick amorphous layer at the edge of the needle resulting from amorphization of the edge of the μc-Si:H specimen during ion milling. Fig. 3 (c) shows a vortex projection of the reconstructed volume in which the contrast has been affected by diffraction contrast from Si nanocrystals visible in the original BF TEM images in the tilt series.

Figure 3 (d) shows an ADF STEM image of the same needle-shaped sample acquired using an inner semi-angle detector of 47.4 mrad. This image is much less sensitive to diffraction contrast than the BF image shown in Fig. 3 (a) and much more directly interpretable in terms of variations in projected density. Fig. 3 (e) shows a section through the reconstructed volume, in which the darker contrast at the edge of the needle is due to Ga^{2+} implantation during FIB milling. The contrast from the Si nanocrystals in the needle is less pronounced than in Fig. 3 (b) but is still visible. Fig. 3 (f) shows two orthogonal sections through the reconstructed volume.

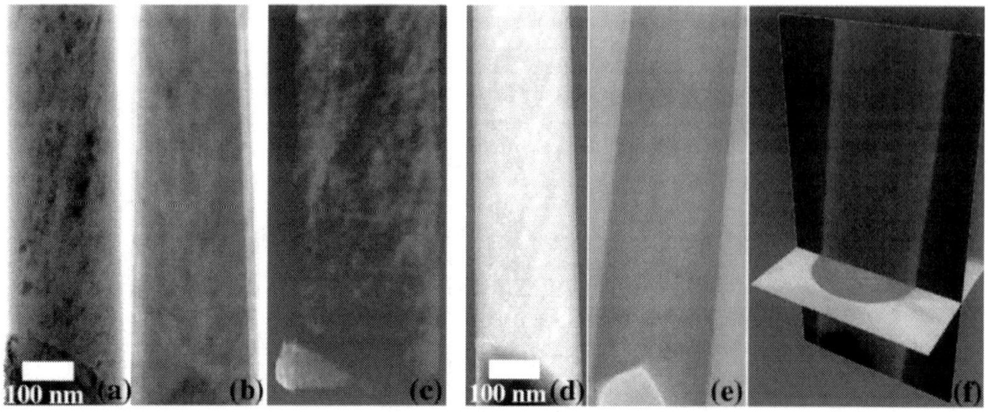

Figure 3. (a) BF TEM image of a 200 nm diameter needle-shaped specimen of μc-Si:H on rough ZnO; from a tomographic tilt series acquired with a tilt range of 360°. (b) shows a section through the reconstructed volume, while (c) shows a vortex projection. (d) ADF image taken from a tilt series of images of the same needle using an inner detector semi-angle of 47.4 mrad. (e) shows a central section while (f) shows two orthogonal sections of the reconstructed volume.

4. Conclusions

Tomographic tilt series of BF and ADF images have been used to obtain three-dimensional reconstructions of μc-Si:H in a thin film solar cell. A reconstruction of a crack in a conventional FIB-prepared lamella shows artefacts resulting from the effect of the limited tilt range used. A needle-shaped sample is used to achieve a 360° tilt range to overcome this problem, although two effects of diffraction contrast and difficulties with image alignment are not eliminated completely. Ongoing work involves addressing these problems and preparing a needle-shaped specimen that contains a single crack within its volume.

References

[1] Shah A, Torres P, Tscharner R, Wyrsch N and Keppner H 1999 *Science* **285** 692
[2] Haug F-J, Soderstrom T, Python M, Terrazzoni-Daudrix V, Niquille X and Ballif C 2009 *Solar Energy Materials and Solar Cells* **93** 884
[3] Yamamoto K, Toshimi M, Suzuki T, Tawada Y, Okamoto T, Nakajima A 1998 *MRS Spring Meeting, San Francisco, April 1998*
[4] Čampa A, Krč J and Topič M 2009 *J. Appl. Phys* **105** 083107
[5] Python M, Vallat-Sauvain E, Bailat J, Domine D, Fesquet L, Shah A and Ballif C 2008 *Journal of Non-Crystalline Solids* **354** 2258
[6] Kübel C, Voigt A, Schoenmakers R, Otten M, Su D, Lee T-C, Carlsson A and Bradley J 2005 *Microscopy and Microanalysis* **11** 378
[7] Midgley P A and Dunin-Borkowski R E 2009 *Nature Materials* **8** 271
[8] Faÿ S, Kroll U, Bucher C, Vallat-Sauvain E and Shah A 2005 *Solar Energy Materials and Solar Cells* **86** 385

Characterization of stress transmission from silicon nitride layer to transistor channel

R Thomas[1,2], D Benoit[1], L Clement[1], P Morin[1], D Cooper[2] and F Bertin[2]

[1] STMicroelectronics, 850 rue Jean Monnet, F-38926 Crolles, France

[2] CEA-LETI, Minatec Campus, 17 rue des Martyrs, F-38054 Grenoble, France

E-mail : remi.thomas-cro@st.com

Abstract. In this paper, we characterize contact etch stop layer (CESL) stress transmission to transistor conduction channel with a new observation technique using transmission electron microscopy in dark-field mode, called dark-field electron holography. First, we explain CESL elaboration by presenting deposition mode and post-deposition film treatment. Then, we describe in detail dark-field electron holography experiments and we establish a link between experimental results and simulation.

1. Introduction

Contact etch stop is a silicon nitride film that has been introduced in MOS stacks at 0.25 μm node technology. Its role consists of stopping oxide etching during contact elaboration. Indeed, etching is performed in two steps: first oxide etching and then nitride etching. This highly selective etching is to avoid shallow trench isolation (STI) etching and generate short-circuits. Ito & al. [1] have shown that CESL possesses an intrinsic mechanical stress that, transmitted to the conduction channel, improves carrier mobility and consequently electrical performance of the transistor. In the present work, we characterize stress in the conduction channel by dark-field electron holography by establishing strain mapping of crystalline silicon.

2. Contact etch stop layer elaboration

CESL is a silicon nitride film deposited by plasma enhanced chemical vapour deposition (PECVD) process. The tool used at STMicroelectronics is Producer from AMAT. Precursor species are SiH_4 and NH_3, diluted in a neutral gas which is in this case N_2. A plasma is produced, which breaks precursor molecules to form very reactive radicals. Then, an amorphous film composed of silicon, nitride and hydrogen (a-$Si_xN_yH_z$) is formed on devices. This film possesses an intrinsic stress which can be set from compressive (-3 GPa) to tensile (+1.2 GPa) according to deposition conditions. By a stress transmission phenomenon through the structure, carrier mobility increases in the conduction channel, which consequently improves electrical performance of transistors.

Post-deposition film treatment, for example a UV-cure process, can then be applied. This type of treatment gives sufficient energy to break Si-H and N-H chemical bonds, which implies hydrogen desorption and lattice reticulation (creation of Si-N and Si-Si chemical bonds, N-N bonds having a too low bonding energy to be stable). UV-cure treatment improves tensile stress but is harmful for CESL under compressive stress.

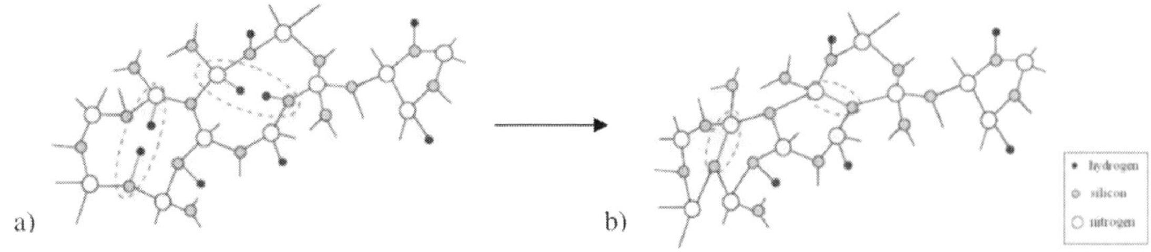

Figure 1. CESL microstructure (a) as deposited (b) after post-treatment.

As a consequence, it can be useful to visualize strain in silicon and more precisely in the conduction channel, in order to understand the mechanisms which play a role in stress transmission.

3. Dark-field electron holography characterization technique

Dark-field electron holography is a characterization technique which allows evaluating local stress in a crystal with a large field of view [2]. Dark-field mode consists of tilting the electron beam. This manipulation allows us to work with a diffracted beam on-axis and in consequence to study sample properties along one family of crystalline planes. The principle of dark-field electron holography relies on the difference of electron beam trajectory between a strained and an unstrained zone of the crystal. Electrons that are crossing the strained part have their phase changed while the ones that pass through strain free matrix retain their phase. Then a positively charged biprism recombines the two beams to form a hologram of the structure in the microscope image plane, which then contains information on strained and unstrained lattice. As a consequence it is possible from this hologram to characterize crystal strain.

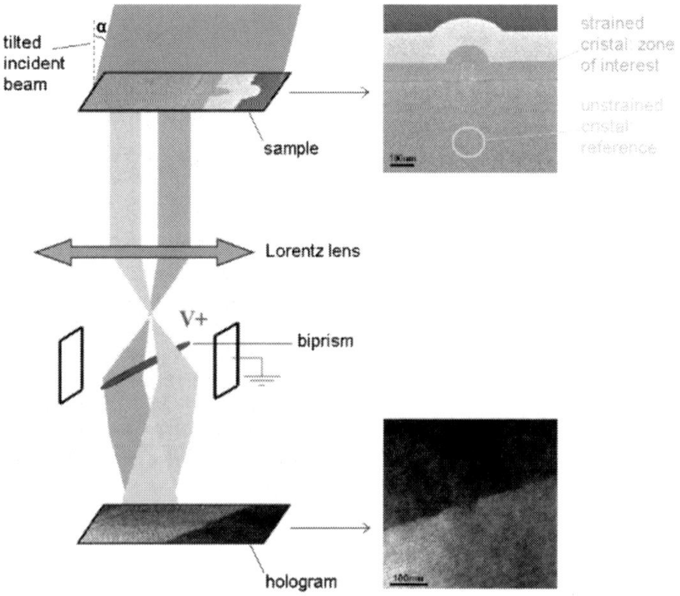

Figure 2. Dark-field electron holography working principle.

4. Experimental conditions

Dark-field electron holography is performed on an FEI TECNAI F20 transmission electron microscope equipped with a field emission gun (FEG TEM) working at an acceleration voltage of 200keV. Extraction voltage is set to 3800V with a small spot size and a large gun lens. To realize those experiments, the TEM column must be equipped with a biprism and a Lorenz lens. Dark-field holograms are acquired on a 1k x 1k multiscan CCD camera behind a Gatan imaging filter (GIF).

Samples are extracted from bulk and thinned with an FEI HELIOS dual beam focused ion beam (FIB) and scanning electron microscope (SEM) tool using an Omniprobe system. Refining is performed at a current of 8pA and a final cleaning at low energy (5kV) in order to reduce surface amorphization. Lamella thickness is calculated using energy filtered TEM (EFTEM) imaging, with the formula $\frac{t}{\lambda} = \ln\left(\frac{I_{total}}{I_{elastic}}\right)$, where t is the thickness and λ the inelastic mean free path.

Holograms are acquired along {220} crystalline family planes, which correspond to transistor carrier migration direction between source and drain through the silicon channel.

5. Lamella preparation

This step is determining the quality of dark-field holography experiments. Indeed, a lamella too thick will stop too many electrons and prevent us to get an exploitable signal; on the contrary, with a lamella too thin, there is a risk of stress relaxation which will imply lamella bending. Good compromise is a lamella around 100 nm ± 20 nm thin.

Classical lamella preparation employed at STMicroelectronics is to etch a window by focused ion beam of some micrometers square. This technique is interesting because it is time-effective but it presents the major drawback of having to generate a transfer of structure morphology, which induces a thickness gradient in silicon. As a consequence, dark-field electron holography lamellas are prepared by backside milling, which consists of etching the lamella rotated by an angle of 180°. This necessitates longer preparation time but does not introduce variations in silicon thickness.

a) 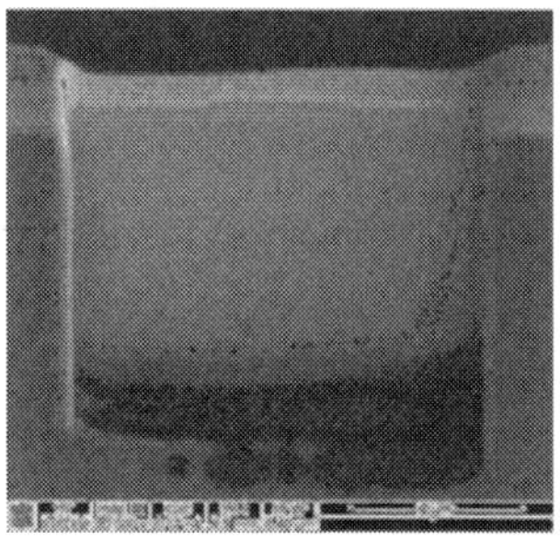 b) 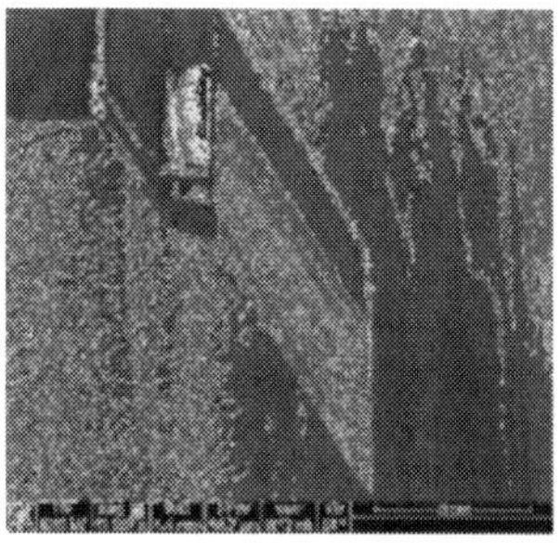

Figure 3. Lamella preparation by (a) window milling (b) backside milling.

6. Silicon stress characterized by dark-field electron holography

To observe stress in the conduction channel, we have performed extremely simplified structures compared to classical MOS. Indeed, it allows reducing delay associated to transistor making and suppress other factors that could contribute to stress generation in the conduction channel. The structure is composed of a polysilicon gate (45 nm width, 60nm height), CESL stressor (100 nm thickness, 1.6 GPa intrinsic stress) and oxide capping (150 nm thickness). This last element is inserted in the stack in order to protect the structure during milling. The silicon wafer is oriented along <110> crystalline direction.

Structures described above are characterized by dark-field electron holography. Samples consist of nine customized MOS separated by a distance of 120 nm (pitch: 165 nm). As only crystalline materials are visible on the resulting strain map, gate positions are represented by white squares.

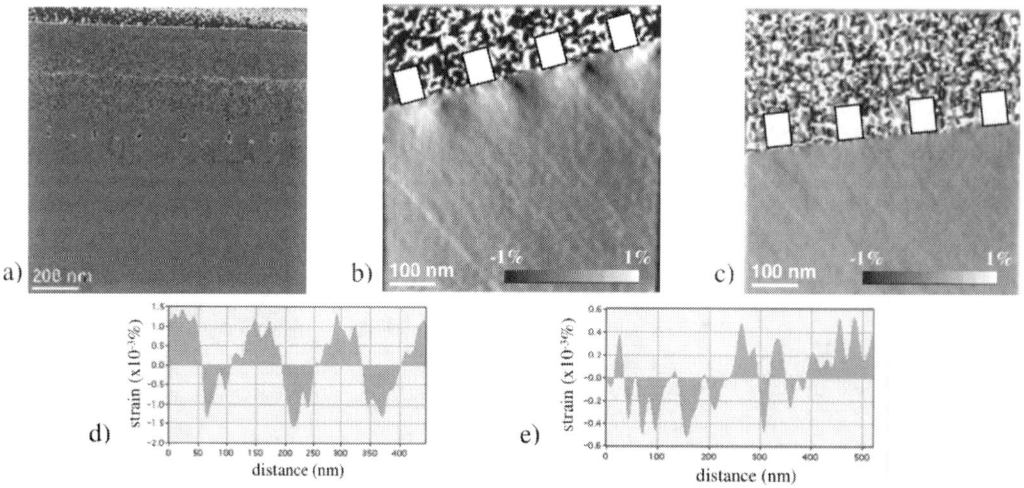

Figure 4. (a) TEM image of the structure. Strain mapping along carrier displacement direction of (b) CESL stressed structure (c) structure without CESL. Strain profile in conduction channel along carrier mobility direction (parallel to crystal surface) of the device (d) with CESL (e) without CESL.

By observing the strain map in Fig. 4(b), two distinct deformation zones appear: tensile strain under the gates and compressive strain in-between. As a consequence, tensile strain allows improving carrier mobility in MOS conduction channel. By simulating this device using COMSOL Multiphysics, these results are confirmed. In both cases, the strain in the conduction channel is around 0.10-0.15%.

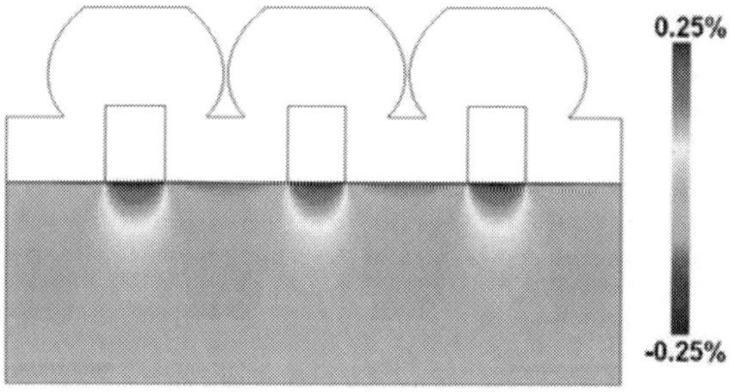

Figure 5. Simulated strain map of the CESL stressed structure.

On the strain map in Fig. 4(c), we clearly see that this stress is small compared to CESL because it doesn't appear on the map. When we study the deformation profile in the conduction channel under polysilicon gates, strain is quantified to be around 0.05%, which is minimal compared to the CESL stressed device.

7. Conclusion

Dark-field electron holography is a very interesting technique to observe and quantify stress transmission from a stressor like CESL to conduction channel. As a consequence, this opens a new way of quantification of stress induced by CESL in conduction channel as a function of different parameters like film thickness, film stress, distance between gates, etc.

References

[1] Ito S, Namba H, Yamagushi K, Hirata T, Ando K, Koyama S, Kuroki S, Ikezawa N, Suzuki T, Saitoh T, Horiushi T, 2000, Mechanical Stress Effect of Etch-Stop Nitride and its Impact on Deep Submicron Transistor Design, *Proc IEDM*, 247-250

[2] Hytch M, Houdellier F, Hue F, Snoeck E, 2008, *Nature* ***453***, 1086-1090

Point field emitters based on rod-like ZnO nanocrystals

V V Artemov[1], **L N Demianets and N A Kiselev**
Institute of crystallography RAS
119333 Moscow Leninskiy pr.59, Russia

Email: artemov@ns.crys.ras.ru

Abstract. New semiconducting anisotropic materials for field emitters, such as ZnO, are of interest from practical and scientific standpoints. Usually, rod-like nanocrystals have a length of few hundred microns and a radius of curvature less than hundred nm. In the present work, rod-like nanocrystals ZnO with high aspect ratio are investigated. They are obtained by high-temperature pyrolysis of Zn compounds. FIB Quanta 200 3D is used for preparation, structure characterization and field emission experiments. With the help of a Kleindiek micromanipulator, the most suitable crystal rods are transferred to the tungsten needle tip and fixed using tungsten gaseous injection system (GIS). As a result, ZnO point emitters are formed with a tip radius of curvature less than 50 nm. Emission characteristics revealed a field amplification coefficient (β) of about 1500 for an anode-cathode distance of 50 μm.

1. Introduction

Rod-like ZnO crystals with high aspect ratio are promising as semiconducting, high melting temperature materials for field emission. There are few investigations dedicated to the field emission of ZnO individual crystals. Current-voltage ultra low threshold field emission characteristics were obtained from a single multipod [1]. Anode-cathode distance dependence of the enhancement coefficient was investigated by transmission electron microscopy in situ [2]. In [3], the correlation between resistance and emission of individual ZnO nanostructures has been measured as an anode with manipulator was brought directly to the ZnO tip.

2. Experimental

Rod-like crystals were obtained by high temperature pyrolysis of organic Zn compounds in presence of Al as a catalyst. Rods that are most suitable for the emission experiments are hundred microns long and from 40 to 100 nm in diameter. The shape of crystals has been characterized by scanning electron microscopy (SEM) with a JEOL JSM- 7401F and the crystal structure has been investigated with an FEI Titan 80-300 (Figure 1,2,3). From lattice imaging (HREM) it is possible to say that the growth direction is [0001] and the surface of the rod-like crystals is clean. Both SEM and HREM revealed narrowing endings and round tips ($r \approx 50$ nm).

[1] To whom any correspondence should be addressed.

Figure 1. SEM micrograph of massive formation of rod-like ZnO crystals.

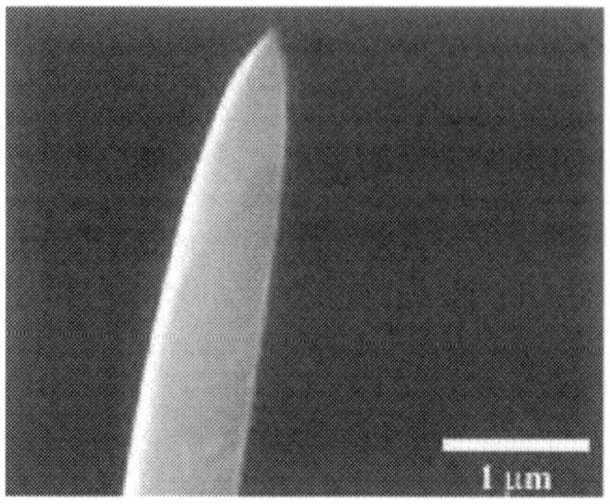

Figure 2. SEM micrograph of the ZnO rod ending.

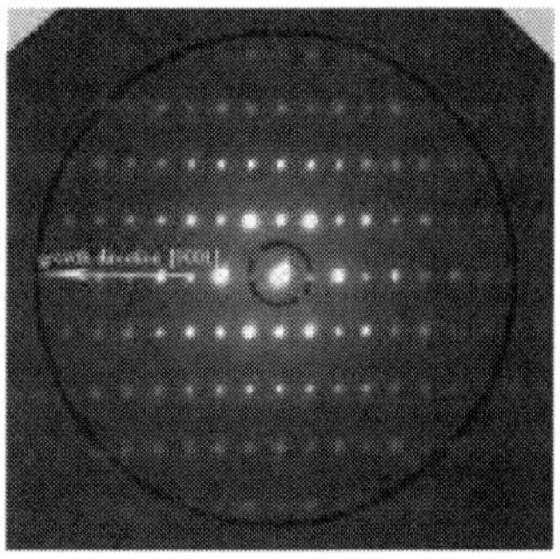

a

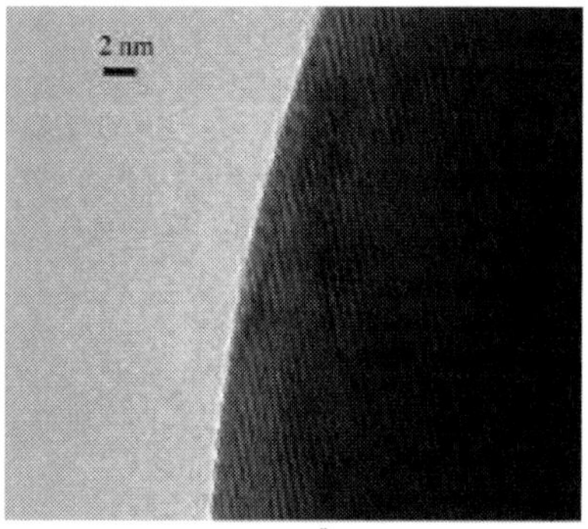

b

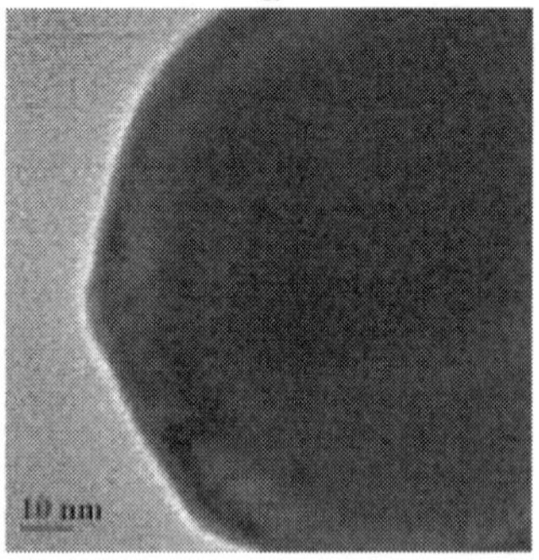

c

Figure 3. HREM of rod-like ZnO crystals
(a) electron diffraction along [11-20], growth direction is [0001]
(b) profile image of the crystal
(c) tip of the crystal

Emission experiments have been performed in situ using a focused ion beam (FIB) Quanta 200 3D with Kleindiek micromanipulator and tungsten gaseous injection system (GIS). The anode was mounted on the end of micromanipulator to enable control of the anode-cathode distance in the range 5-50 μm. The anode itself was prepared by tungsten evaporation on manipulator needle. The anode diameter was about 5 μm (figure 4). The emission measurements have been performed at pressure about 10^{-4} Pa. The field amplification coefficient β can be calculated using equation (1) [4]:

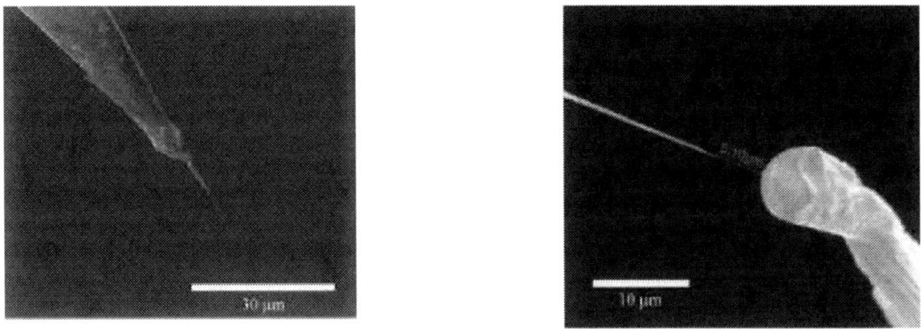

a b

Figure 4. Field emission experiment:
(a) rod –like ZnO mounted on the tungsten needle tip,
(b) anode is moved close to the ZnO tip.

$$\beta = 1.2\left(\frac{l}{r} + 2.15\right)^{0.9} \quad (1),$$

where l is the length of the rod and r is its apex radius. So for a 100 µm rod-like emitter with apex radius 50 nm, the amplification coefficient is β=1123.

The current-voltage characteristics have been recorded with a Keithley 6487 combined unit, which contains the current meter and the controlled source with voltages up to 500V in steps of 5 V. When the anode-cathode distance is 2 µm at 500 V then the ZnO crystal is destroyed (corresponding to an electric field of 250 MV/m).

One can model the emission properties of cold field emitters using Fowler-Nordheim equation:

$$\ln(\frac{J}{E_l^2}) = \ln(\frac{A}{\varphi}) - \frac{B\varphi^{3/2}}{E_l} \quad (2),$$

where J is the current density, E_l the intensity of the local electric field, $A = 1.56 * 10^{-10}$ A* V^{-2} *eV, $B = 6.83*10^3$ $eV^{-3/2}$ *V*m and φ the work function for ZnO. Local electric field is usually calculated using the formula:

$$E_l = \beta V / d \quad (3),$$

where V is the applied voltage, and d is the distance between the anode and cathode. Accordingly, Eq. (2) and $J=I/2\pi r^2$ (relationship between current I and current density J for the spherical cathode) in our case yield:

$$\ln(\frac{I}{E^2}) = \ln(\frac{2\pi r^2 A\beta^2}{\varphi}) - \frac{B\varphi^{3/2}}{E\beta} \quad (4)$$

From the slope K of this dependence it is possible to calculate the coefficient β:

$$\beta = -\frac{B\varphi^{3/2}}{K} \quad (5)$$

Using the micromanipulator some thin and long rods have been picked out and mounted on the end of tungsten needle (obtained by chemical etching) by local tungsten evaporation. As a result, a point emitter has been formed (Figure 4). Table 1 shows the results of the calculation of the coefficient β for four different distances d for the same sample on the basis of the curves in Figure 5, 6.

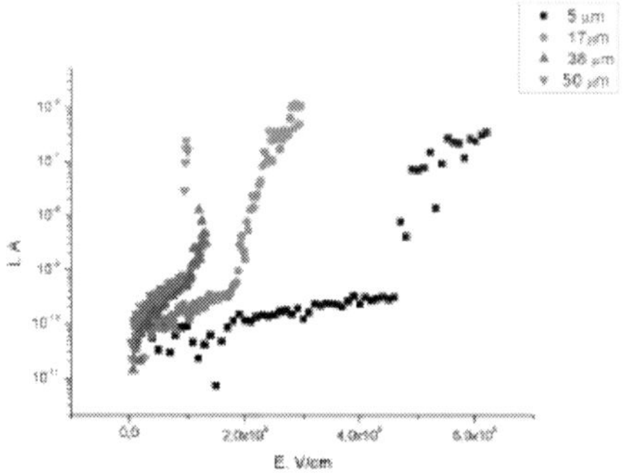

Figure 5. Field emission current as a function of electric field and anode-cathode distance.

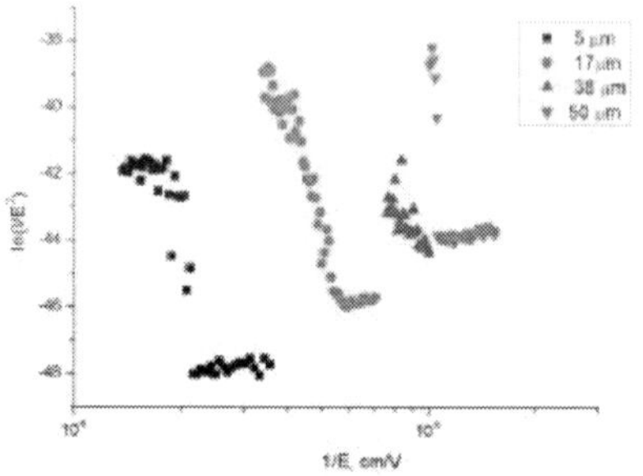

Figure 6. Fowler-Nordheim plot.

Table1. The dependence of β on the anode-cathode distance.

d, μm	K, V*m	β	maximal current, μA
5	7137900	116	1.3
17	3100000	264	1.0
38	589718	1413	0.012
50	557191	1500	0.2

3. Conclusion

A technique of preparation of point emitters from rod-like ZnO crystals using the FIB Quanta 200 3D device has been developed. The threshold values of the electric field, at which the emitters break down, is measured as 250 V/μm. Rod-like ZnO crystals revealed substantial amplification coefficient β, which strongly depends on d. We believe that rod-like ZnO crystals are encouraging candidates for field emission materials. Low emission current stability is associated with deficient vacuum in the SEM chamber. In the future experiments the characteristics will be measured in ultra high vacuum.

Acknowledgement

This work was supported by the Russian Foundation for Basic Science Project # 11-08-00790-a.
We are grateful to Dr A L Vasiliev for HREM of ZnO rods.

References

[1] Ramgir NS, Mulla IS, Vijayamohanan K, Late DJ, Bhise AB, More MA and Joag DS 2006 *Appl. Phys. Lett.* **88** 042107
[2] Huang YH, Bai XD, Zhang Y, Qi JJ, Gu YS and Liao QL 2007 *J. Phys.: Condens. Matter* **19** 176001
[3] She J, Xiao Z, Yang Y, Deng S, Chen J, Yang G and Xu N. *2008 ACS NANO* **2** no. 10, 2015
[4] Edgecombe CJ and Valdre U 2002 *Philos Mag B* **82** 987

GaAsBi atomic surface order and interfacial roughness observed by STM and TEM

F Bastiman, Y Qiu and T Walther

Department of Electronic & Electrical Engineering, University of Sheffield, Sir Frederick Mappin Building, Mappin Street, Sheffield S1 3JD, UK

f.bastiman@sheffield.ac.uk

Abstract. GaAsBi is an interesting ternary material for opto-electronic applications. Bi fractions of 11-13% allow 1.55 μm emissions from a range of bulk and QW structures. GaAsBi has shown strong room temperature photoluminescence. The temperature insensitive band gap and large spin orbit splitting are attractive optoelectronic features, however the typical full width half maximum is 2.5 times greater than GaAs. In solid source molecular beam epitaxy (MBE), near stoichiometric fluxes and low growth temperatures are necessary to achieve the desired Bi content. In order to explore photoluminescence linewidth broadening and the accommodation of strain a joint scanning tunnelling microscopy (STM) and scanning transmission electron microscopy (STEM) study of quasi-bulk and QW structures has been undertaken.

1. Introduction

The bismuth alloy $GaAs_{1-x}Bi_x$ has a number of interesting properties, namely band gap reductions of 88 meV/% and large spin orbital splitting [1, 2]. The narrow, temperature insensitive band gap offered by such an alloy is particularly well suited to the fabrication of infra red emitters and photodetectors [2], however certain complications in the growth process have limited the impact of this alloy.

The strong tendency towards surface segregation and droplet formation of Bi during epitaxy has led to unconventional low temperatures growth parameters. Dilute bismuthides can be grown at around 430 °C up to ~1% Bi [3], however for high compositions lower growth temperatures are necessary. It is common to grow at temperatures as low as 280 °C in order to ensure high quality, smooth epilayers are produced with >10% Bi fractions [4]. X-ray Diffraction (XRD) and photoluminescence (PL) data show high quality crystalline material with strong room temperature PL can be achieved. There are two restrictions to molecular beam epitaxy (MBE) growth of this ternary alloy. Firstly, in the 335 - 400 °C growth temperature range the film thickness/composition is restricted via strain related undulations and eventually dislocations. Secondly, below 335 °C undulations and dislocations are reduced due to thermal limitations, however point defects strongly affect material quality.

In this paper, scanning tunnelling microscopy (STM) and scanning transmission electron microscopy (STEM) are performed on samples grown at each of these limits in order to investigate observed XRD and PL phenomena.

2. Experiment

This work was performed on an Omicron MBE-STM system fitted with As_4, Ga, In, Bi and Mn effusion sources and RHEED for dynamic surface analysis. Full details of the system are discussed elsewhere [5]. An undoped on axis (± 0.1°) GaAs(001) epi-ready substrate was cleaved to 11 x 3.5 mm^2 in order to accommodate the sample mounting.

The samples were degassed with no further *ex situ* processing. The degas procedure entailed initial heating to a substrate temperature, T_s = 400 °C for 30 minutes, followed with heating to T_s = 600 °C under an As_4 flux for conventional oxide removal. The structure of sample LE2 featured an 80 nm buffer layer grown at 590 °C followed by cooling to 400 °C after which a 15 nm GaAsBi layer was deposited and capped with 30 nm of GaAs at the same temperature. To provide a sharper lower interface, the structure of sample B029 featured a thicker 160 nm buffer layer which was then annealed for 30 minutes at 600 °C to produce wide, flat terraces. After cooling to 320 °C, a 17 nm GaAsBi layer was deposited, followed by a 17 nm spacer, a further 4 nm QW and finally a 40 nm cap of GaAs. In each sample the RHEED showed a strong 2x1 reconstruction throughout the GaAsBi layer and an *n*x3 reconstruction during the cap. With the exception of the buffer layer, all growth was performed at the initial GaAsBi growth temperature and the samples were not subject to high temperature anneals.

T_s values of 295 °C and 400 °C [6] were identified from As cap removal and *c*(4×4)/(2×4) transition in the absence of external flux, respectively. Samples were quenched for *in vacuo* scanning tunnelling microscopy (STM) imaging purposes before reheating and capped with GaAs for *ex situ* analysis. The (004) high resolution X-ray diffraction (HRXRD) $\theta/2\theta$ scans of the as-grown samples were measured to verify crystal quality. The epilayer thicknesses and Bi composition were determined by fitting the HRXRD scans using RADS Mercury software, assuming a GaBi lattice constant of 6.324 Å [7]. For the photoluminescence (PL) measurements, the sample was excited with a continuous-wave 532 nm wavelength diode-pumped solid state laser. The PL was dispersed using a monochromator and then detected by a liquid nitrogen-cooled germanium detector. Scanning transmission electron microscopy (STEM) was performed in a JEOL 2010F field-emission transmission electron microscope with scan unit and annular dark-field (ADF) and bright-field (BF) detectors, which was operated at 197 kV, with 0.25nm probe size, semi-angle of beam convergence of 9.5 mrad and an inner detection angle for ADF of ~50 mrads.

3. PL and XRD characterisation

The large lattice mismatch (~12%) between binary GaAs and theoretical GaBi (6.324 Å) [7] leads to compositional/thickness epitaxial limitations. Strong room temperature PL is observed from samples with 0-9% Bi, covering the PL spectrum from 872 to ~1400nm. Whilst relatively thick bulk layers can be grown for <4% Bi, the strain causes undulations and eventually dislocations that limit the thickness of higher composition alloys. Furthermore, Bi segregation and Bi-Bi adatom interaction lead to the formation of Bi droplets for high Bi:As_4 fluxes [8].

Undulations and surface roughness seem to have limited effects on GaAsBi PL intensity, however the full width at half maximum (FWHM) rapidly increases upon the formation of Bi droplets. Figure 1 shows the PL spectra for LE2 grown at 400 °C with a nominal 5.6 % Bi composition determined from the PL peak shift. The intensity is only half that of a similar prepared sample where care was taken to avoid droplet formation, however the FWHM has increased from 75 meV (not shown) to 109 meV. An XRD spectrum for the same sample is shown in figure 2. It is evident from the broad peaks and lack of interfacial fringes that the layer has poorly defined interfaces. Best fit simulation data further implies that the Bi composition is not homogeneous throughout the layer, which probably contains 6.0% and 2.2% domains.

Sample B029 was in turn grown to investigate the effect of low temperature growth on interface roughening and FWHM. The PL spectrum is shown in Figure 1, and displays two distinct peaks, one at 1555nm from the 17nm QW and one at 1458 nm from the 4nm QW. The thicker QW exhibits a

FWHM of 106 meV and the thinner QW of 94 meV. It is unclear whether the linewidth narrowing is due to quantum effects or is a function of compositional variation. The XRD spectrum for B029 exhibits strong interfacial fringes, indicating the presence of abrupt interfaces, however attempts to fit a single compositional simulation peak to the experimental data proved non trivial. The difficulty arises from the different thickness of the QW, with the diffuse scattering from the thinner well and apparent compositional variation between the two GaAsBi layers. The 1555nm PL peak from the quasi-bulk 17nm layer indicates a Bi composition of ~11% from empirical fitting [9].

It is clear from Figure 1 that LE2 exhibits much strong RT-PL despite the interfacial roughness. The cause of the PL intensity reduction in B029 and the origin of the linewidth broadening will be discussed in the following sections.

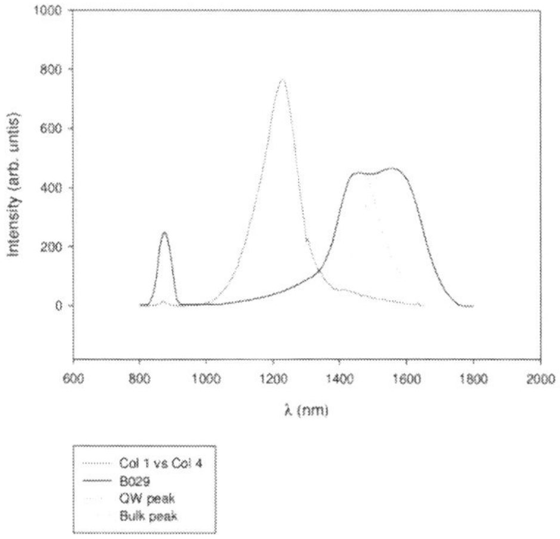

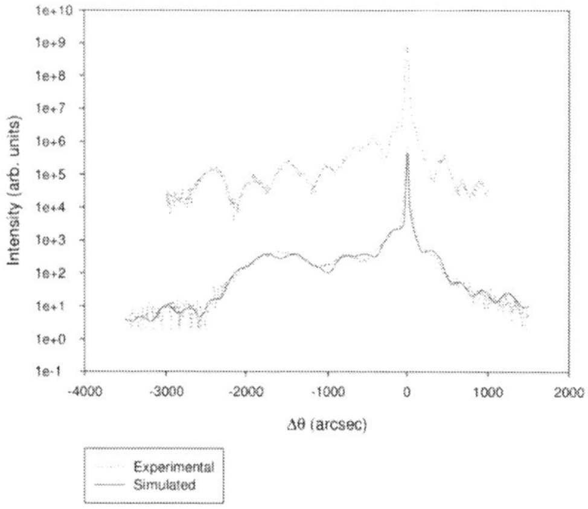

Figure 1. Room temperature PL intensity spectra for LE2 (red) and B029 (black, with dotted fits for bulk and quantum well peaks)

Figure 2. Experimental and simulated XRD spectra for samples LE2 (lower profile) and B029 (upper profile)

4. STM images

Undulations are commonly observed in ternary alloys with significant lattice mismatch. Figure 3 shows the upper surface of the QW in LE2 immediately prior to capping. The undulations have an average period of ~400 nm and a peak-to-trough amplitude variation of 2 – 3 nm. Each undulation is formed from dense stacks of 2D islands that are anisotropic, displaying elongation in the $[\bar{1}10]$ direction. The rough interfaces are certainly non-deal, especially for narrow QWs where thickness variation can affect the local band gap and quantum confinement. The undulation period protracts for both increasing thickness and Bi composition, however the amplitude of the undulations remains unchanged. The strain driven undulation clearly limits the thickness of smooth GaAsBi films at this composition, however the PL intensity of LE2 (Figure 1) is strong despite the rough interfaces.

Lowering the sample temperature enables both an increase of Bi incorporation and, for temperatures less than ~335 °C, the mechanism for undulation generation is thermally limited [4]. Figure 4 shows the upper surface of the 17 nm QW of sample B029. The larger pits are monolayer (ML) trenches and islands that are a feature of the buffer layer growth and anneal. The small, dense anisotropic islands are GaAsBi 2x1 islands. The limited thermal energy limits the adatom mobility leading to a larger number of smaller islands when compared to typical GaAs growth temperatures, however the surfactant nature of Bi enables the growth of flat epilayers even at these reduced temperatures.

Before capping the lower temperature grown sample exhibits superior interfacial quality, as supported by the XRD spectra in Figure 2. However, despite this apparent improvement in the epilayer interface the PL spectrum is weaker than the higher temperature sample and the FWHM is of a comparable width.

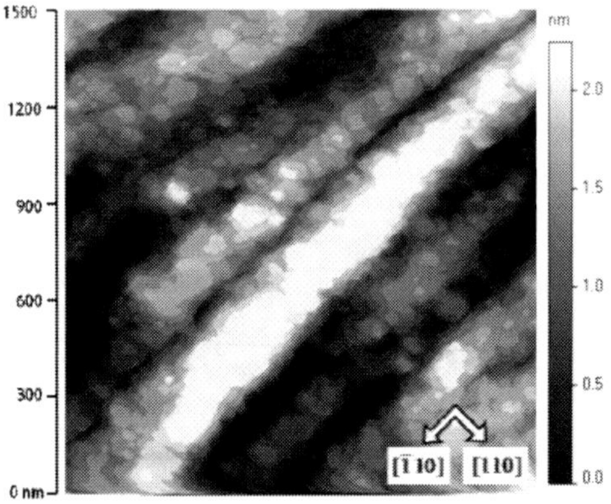

Figure 3. 1500 x 1500 nm^2 STM image of upper surface of the QW in LE2 prior to the capping layer

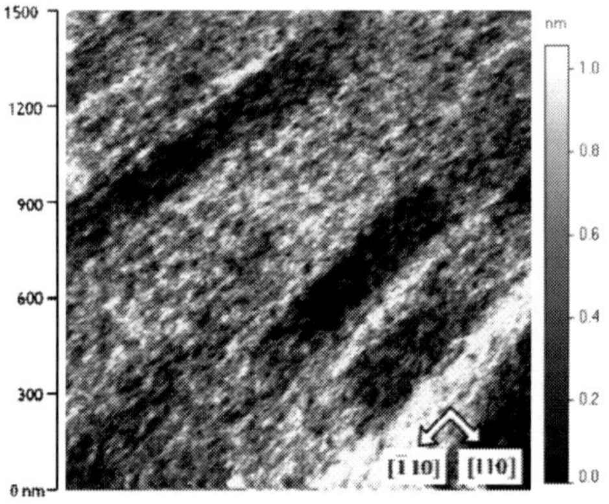

Figure 4. 1500 x 1500 nm^2 STM image of upper surface of the 17nm QW in B029 prior to the capping layer

5. STEM observations

Annular dark field (ADF) scanning transmission electron microscopy (STEM) images of the buried layers in cross-section are shown for both samples in Figure 5a. The [001] growth direction is indicated by the arrow in each case. Imaging of the upper sample (LE2) revealed a number of discontinuities similar to the one shown in the centre of the image. These discontinuities are accompanied by a rough interface which is particularly pronounced in the stark contrast at the upper surface. Whilst the STM image of LE2 revealed the presence of this undulation, the presence of such discontinuities was not derivable from the images.

Figure 5b shows a higher magnification ADF image of a portion of LE2, highlighting the poor contrast at the non-abrupt interfaces. The tendency of Bi to surface segregate at the higher growth temperature appears to have affected the interface abruptness and presumably lead to a fractional depletion of the Bi content of the upper monolayers (ML).

In contrast, the lower sample in Figure 5a (B029) displays continuous layers across the entire width of the sample. The layer thickness also displays excellent uniformity and the interfaces are abrupt. It is clear that the growth is homogeneous across the sample and that the lower growth temperature hinders the surface segregation of Bi. The higher image contrast of sample B029 compared to LE2 indicates a higher Bi content of the GaAsBi layers.

Figure 6a shows a higher resolution ADF-STEM image of B029 emphasising the strong contrast between the GaAs and GaAsBi layers. The profile data of Figure 6b reveals the upper interface of each QW is around 1nm wider than the lower interface. Indeed, the reduction of roughness from the GaAsBi-GaAs interface of the 17 nm QW to the GaAs-GaAsBi interface of the 4 nm QW reveals that the GaAs recovers toward a smoother state even at this lower growth temperature. This is probably due to the surfactant nature of the small amount of Bi that is always present during the spacer and cap layer growth and causes the nx3 reconstruction to be observed during growth. The tail in the profile of Figure 6b is probably due to Bi segregation.

The results of interfacial roughness from the two samples anti-correlates with observed PL intensity reduction, implying that the interface is not a dominant factor in establishing the PL intensity from GaAsBi QWs. Further, the FWHM is comparable in the PL data and so it is not clear that the interfacial roughness is contributing to the linewidth broadening. The different structure of the two samples prevents a direct comparison, but it can be assumed that the roughening and discontinuities in LE2 may contribute to the observed degradation in the PL linewidth, whilst the same type of degradation in B029 could be caused by anti-site and other low temperature growth defects. In order to explore the atomic structure of the buried layers, higher resolution STEM imaging of the atomic structure is clearly necessary.

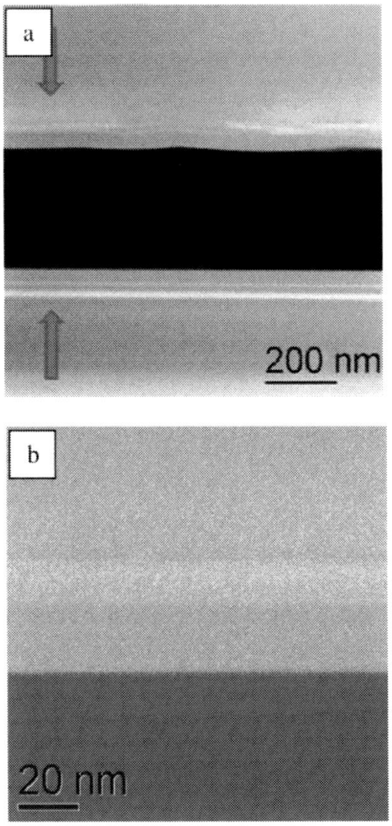

Figure 5. (a) ADF-STEM image of LE2 (top) and B029 (bottom). Arrows indicate growth direction. (b) ADF-STEM image at 250kX magnification of continuous section of LE2

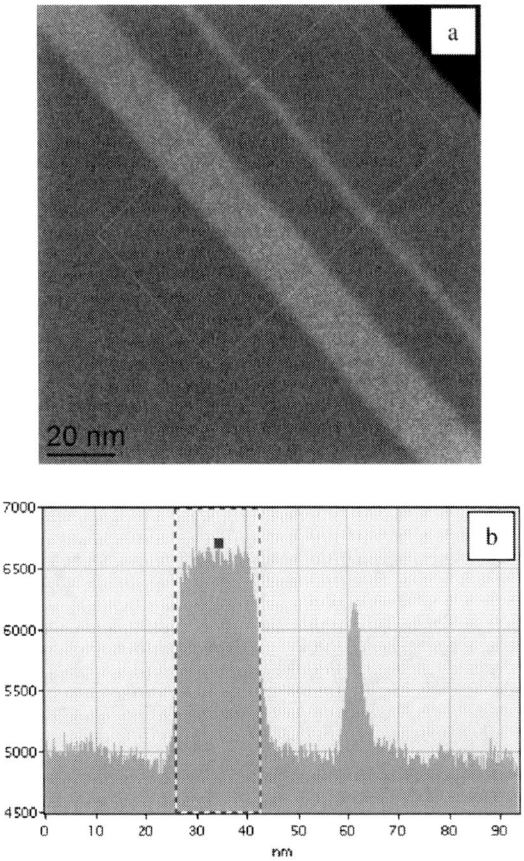

Figure 6. (a) ADF-STEM image of B029 (b) profile of rectangle in (a) exhibiting interfacial tail at the upper interface of the GaAsBi QW. The reduced contrast of the 4nm QW compared to the 17nm wide layer indicates a lower Bi content, e.g. around ~8% instead of ~11%.

The PL and XRD simulation provide two independent means of quantifying the Bi content of the structures. In all instances the data correlates to within 0.1%. Energy-dispersive X-ray spectrum (EXDS) analysis and quantification of Bi and Ga X-ray line intensities yields apparent Bi concentrations that vary with the lines used for quantification. EDXS data from the centre of the 17nm

QW of B029 is displayed in figure 7. Ga_L and Ga_K are clearly present around 1.1 and 10 kV, together with a number of smaller peaks that represent the Bi_M and Bi_L lines. Each line pair can provide a value for a composition estimate, along with a statistical fit error: 7.7±0.7% for Bi_M/Ga_K,10.7±1.8% for Bi_L/Ga_L, 14.6±1.7% for Bi_M/Ga_L. No calibrated *k*-factors for any Bi X-ray lines are currently available (this is under investigation). All values are therefore based on the nominal *k*-factors of Oxford Instrument's ISIS300 software. If we use pure GaAs as a reference, then the As_K/As_L ratio of 1.45±0.05 measured would imply a specimen thickness of t=81±12nm [10], which is realistic and can be fed into an absorption correction. A quantification of Bi_L or Bi_M lines relative to As_K and Ga_K then yields x_{Bi}=8.9±1.0%, while the ratio quantification of Bi_L or Bi_M relative to As_L and Ga_L lines yields 13.2±1.4%. Again, the difference is systematically bigger than simple counting and fitting errors can explain. Coincidentally, the nominal composition of x_{Bi}=11% lies right between these two alternatives, indicating the need for *k*-factor calibration using a thickness series for GaAsBi of known composition.

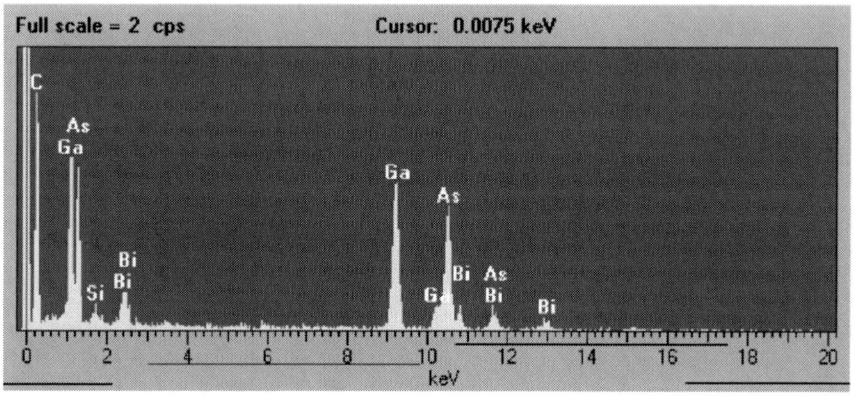

Figure 7. Energy-dispersive X-ray spectrum (EDXS) from centre of 17nm thick layer of GaAsSb in above sample B029

6. Summary

PL linewidth broadening is not caused by interfacial roughness, especially in the case of the lower growth temperatures where abrupt interfaces are observed. Strong RT-PL is present in discontinuous, undulating layers. Sharp abrupt interfaces are possible at low temperature due to reduced segregation, however, the PL quality is significantly reduced, probably due to LT-growth which could lead to anti-site and point defects. High resolution investigation will be necessary to identify the cause of linewidth broadening as MBE grown GaAsBi reaches strain and temperature driven compositional limits.

References

[1] Francoeur S, Seong MJ, Mascarenhas A, Tixier S, Adamcyk M and Tiedje T 2003 *Appl. Phys. Lett.* **82**, 3874
[2] Fluegel B, Francoeur S, Mascarenhas A, Tixier S, Young EC and Tiedje T 2006 *Phys. Rev. Lett.* **97**, 067205
[3] Bastiman F and David JPR 2011 *J. Cryst. Growth*, to be submitted
[4] Lu X, Beaton DA, Lewis RB, Tiedje T and Whitwick MB 2008 *Appl. Phys. Lett.* **92**, 192110
[5] Bastiman F, Cullis AG, Hopkinson M and Green M 2008 *Microscopy of Semiconducting Materials 2007, Springer Proc. in Physics* **120**, 471
[6] Reginski K, Muszalski J, Preobrazhenskii VV and Lubyshev DI 1995 *Thin Solid Films* **267**, 54
[7] Janotti A, Wei S-H and Zhang SB 2002 *Phys. Rev. B* **65**, 115203
[8] Bastiman F and David JPR 2011 *J. Cryst. Growth*, accepted
[9] Xianfeng L, Beaton DA, Lewis RB, Tiedje T and Yong Z 2009 *Appl. Phys. Lett.* **95**, 041903
[10] Walther T 2010 Proc. 16th MSM conference, Oxford, *J. Phys. Conf. Ser.* **209**, 012029

Microstructure of nanoscratched semiconductors

J Y Huang[1], F A Ponce[1], P G Caldas[2], C M Almeida[3] and R Prioli[2]

[1] Department of Physics, Arizona State University, Tempe, AZ 85287-1504, USA
[2] Departamento de Física, Universidade Pontifícia Católica do Rio de Janeiro, Brazil
[3] Divisão de Metrologia de Materiais, INMETRO, Rio de Janeiro, 25250-020, Brazil

E-mail: jingyi.huang@asu.edu

Abstract: The crystal defect structure of InP following scratches with an atomic force microscope (AFM) tip has been studied by transmission electron microscopy. The (100) surface of InP single crystals were scratched along <100> and <110> directions with an AFM diamond tip under applied normal forces ranging from 7 to 120 μN. For a given applied force, deeper plastic flow has been observed in the <110> case than in the <100> case. Under increasing applied forces, plastic flow was first observed at 15 μN for <110>, and at 30 μN for <100>. The crystal deformation during a scratch event has been observed to consist of three stages: (a) surface conformation to the shape of the AFM tip during the early stages of the process, and it is associated with slip at the nanoscale with the formation of an inverted triangular region with high dislocation density; followed by (b), a sudden displacement (*pop-in*) that signals the beginning of downward slip at the bulk scale; and (c), for sufficiently large forces, recovery of the bulk elastic strain by upward slip (*pop-up*). The <110> scratch direction is aligned along the {111} slip planes, and yields at lower applied forces than in the <100> case. Dislocations introduced by scratching are mostly screw type, as determined by series of **g**·**b** analysis.

Key words: InP, AFM, TEM, nanoscratching, dislocation, strain

1. Introduction

During past few decades there has been much interest in the study of mechanical deformation in semiconductors.[1-3] Recent advances in instrumentation have produced indenters and tips with sizes at the nanometer scale,[4-6] such as atomic force microscopes (AFM) with a diamond tip with diameter of ~80nm.[7] Studies have shown that contact damage at the nanoscale can be used to grow nanostructured patterns.[8,9] Dislocation arrays produced by nano-indentation and nano-scratching have been proved to be useful as nucleation sites for growth of nanocrystals.[10] Therefore, it is important to understand the deformation mechanism at the nanometer scale in semiconducting materials. In this paper, we report the results of a study of the microstructure resulting from nano-scratching (100) InP surfaces along different crystalline direction using various normal forces.

2. Experimental procedures

The material used in this study is an InP thin film with a thickness of 500 nm grown by metal organic vapor phase epitaxy in an AIX200 reactor, at 630 °C and 50 mbar, on epi-ready wafers of semi-insulating (100) InP. This layer was grown to minimize surface damage resulting from wafer cutting and polishing. Then the crystal was cleaved into 1cm×1cm square pieces to fit the stage of an AFM (model Nanoscope IIIa Multimode), equipped with a diamond crystal mounted at the end of a stainless steel cantilever, which was used to scratch the crystal surface. The diamond tip is cono-spherical in shape, with facets that make an angle of 25° with the scratched surface, and end with a spherical radius of ~40nm. A scratch pattern was obtained by the forward movement of the tip with a constant tip-to-sample relative velocity. We produced two sets of scratches, along <100> and along <110> directions, using normal forces of 7.5, 15, 30, 60, and 120 μN. Each scratch line was separated by 700nm evenly with the purpose of eliminating the interaction between adjacent scratches. Transmission electron microscopy (TEM) foils were prepared in cross section, for view normal to the scratch lines, using a focused ion beam with the surface protected by a Pt layer.

3. Results

The TEM images in figure 1 are of the scratches on the (001) InP surfaces along <100> and <110> directions. These images were taken with $\boldsymbol{g}$ = (002) diffraction conditions. No plastic deformation was observed for a normal force of 7μN. The onset of plastic deformation is 15μN for scratches along the [110], and 30μN for the [100] case.

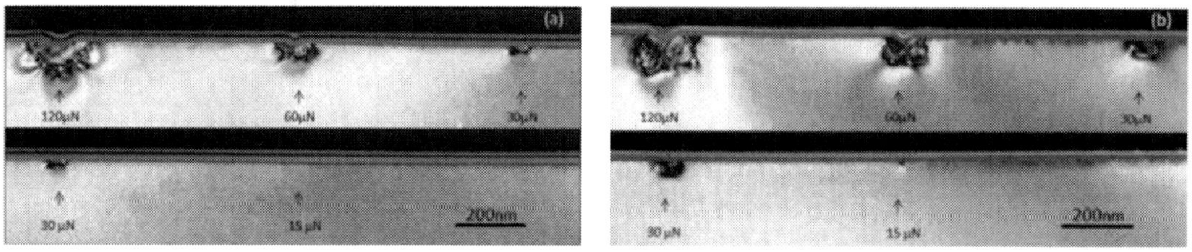

Figure 1. Bright field TEM images of scratches along (a) <100> and (b) <110> directions.

Figure 2(a) is a magnified TEM image of the [110] scratch, produced with a force of 120μN. Figure 2(c) is a further magnification corresponding to the dashed box in (a), showing a *pop-up*, defined as a volume of material that is displaced up towards the surface. No such *pop-up* is observed for the [100] scratch in figure 2(b). This can be explained in terms of the <110>{111} slip system of the zinc-blende structure. A scratch along the [110] direction will result in opposite shear stresses for the two {111} slip planes that are parallel to the scratching direction, leading to a butterfly shaped region with dislocations on both sides of the scratch; while sequential slipping on the other two {111} planes that are not parallel to the scratch direction forms the triangular shaped, highly defected region just below the scratch line, which could be popped up by the recovery of the elastic strain upon removal of the tip. However, this does not happen in the [100] case, because the four {111} slip planes experience exactly the same shear stress, leading to rotation on the {111} planes, and causing a lock of dislocations that prevents further defect propagation. This can also explain the onset difference of plastic deformation for different scratch direction. As shown in figure 2(c), surface modulation to the nanometer scale by the AFM tip has been also observed. When the tip is first pressed into material, the surface deforms to agree with the shape to the tip, but since the radius of the tip is only around 40 nm, after pressing to a certain depth, sequential slip will happen and result in an inclined 25° surface that matches the facets of the tip, as shown schematically in figure 3(d). Displacement due to slip on {111} by $\boldsymbol{b}$ = ½ <110> is represented by the arrow, and the plane made by sequential slip is shown as the bold line.

Journal of Physics: Conference Series **326** (2011) 012061 doi:10.1088/1742-6596/326/1/012061

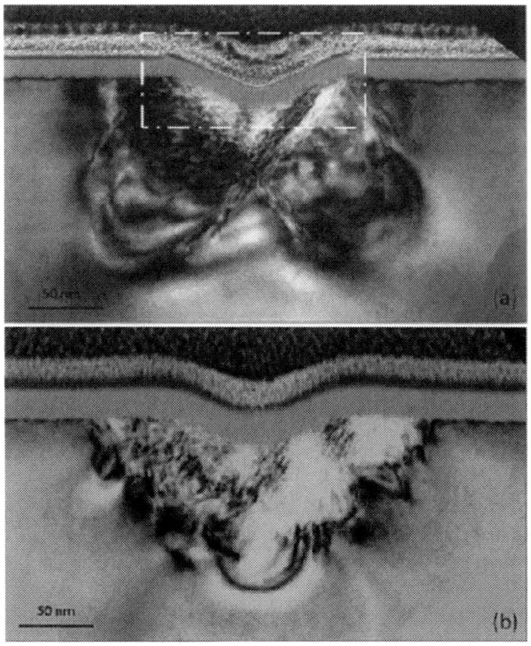

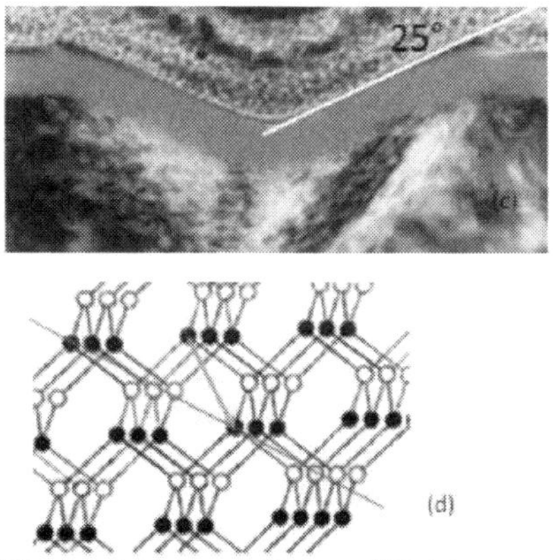

Figure 2. (a) and (b) are TEM images of [110] and [100] scratch under F = 120μN, diffraction condition is $\boldsymbol{g}$ = (002). (c) is magnified image of the dashed box in (a). (d) is atomic arrangement of FCC structure showing slip on {111} by $\boldsymbol{b}$ = 1/2<110> sequentially will result in an inclined 25° surface modulation.

After confirmation of the slip system, we took a series of TEM images on the 120 μN scratch by using different diffraction conditions as shown in figure 3 and figure 4 to determine the dislocation type.

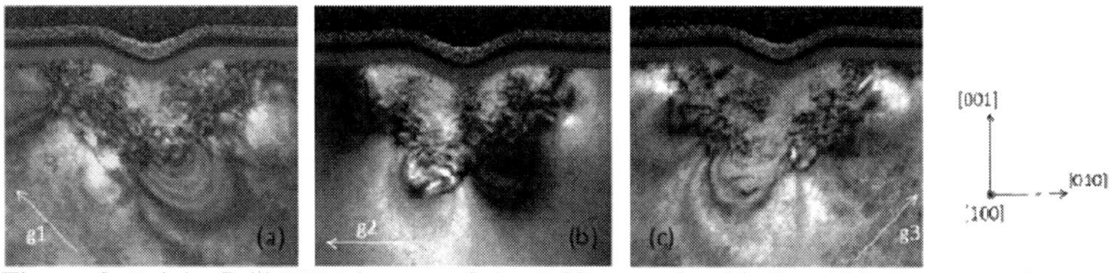

Figure 3. Bright field TEM images of the [100] scratch with F = 120μN: (a) $\boldsymbol{g}_1$=(0-11); (b) $\boldsymbol{g}_2$=(0-10); (c) $\boldsymbol{g}_3$=(011).

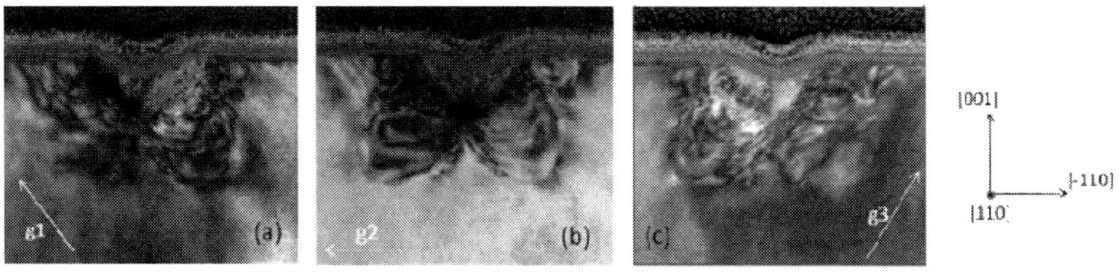

Figure 4. Bright field TEM images of the [110] scratch with F = 120μN: (a) $\boldsymbol{g}_1$=(1-11); (b) $\boldsymbol{g}_2$=(1-10); (c) $\boldsymbol{g}_3$=(-111).

Table 1. $g \cdot b$ analysis on possible Burgers vectors in [100] scratched zinc-blende structure.

	b_1=(01-1)/2	b_2=(0-1-1)/2	b_3=(101)/2	b_4=(10-1)/2	b_5=(110)/2	b_6=(1-10)/2
g_1=(0-11)	√	×	√	√	√	√
g_2=(0-10)	√	√	×	×	√	√
g_3=(011)	×	√	√	√	√	√

Table 2. $g \cdot b$ analysis on possible Burgers vectors in [110] scratched zinc-blende structure.

	b_1=(01-1)/2	b_2=(101)/2	b_3=(10-1)/2	b_4=(011)/2	b_5=(110)/2
g_1=(1-11)	√	√	×	×	×
g_2=(1-10)	√	√	√	√	×
g_3=(-111)	×	×	√	√	×

Since the dislocations introduced by scratching are due to slip on the <110>{111} system, possible Burgers vectors of these dislocations are limited to those in the two tables above. Table 1 and table 2 give the $g \cdot b$ analysis results which will be combined with figure 3 and figure 4 to explain the visibility and invisibility of those dislocations and to determine those dislocation type. We know that dislocations with Burgers vector b can be visible in TEM images taken with g condition only when $g \cdot b \neq 0$, as summarized in table 1 and table 2. A $g \cdot b$ analysis of figures 3 and 4 using tables 1 and 2 leads us to the understanding that these dislocations are of screw type, or at least have a screw component.

4. Conclusion

In conclusion, the microstructure of nano-scratched InP has been studied. We find that the threshold of plastic deformation starts at different normal forces depending on the scratch direction. The [110] case has onset for a normal force of 15 μN, while the onset of [100] case is around 30 μN. A *pop-up* event, defined as a volume of material that is displaced up towards the surface, has been observed only in the [110] scratched sample for an applied normal force of 120 μN. Specific surface morphologies caused by nano-scale scratching have been observed. Dislocations generated by nano-scratching have been proved to be related to <110>{111} slip system in zinc-blende crystal structure and these scratch-generated dislocations are mostly of the screw type.

References

[1] Jasinevicius R G, Pizani P S, and Duduch J G 2000 *J. Mater. Res.* **15**, 1688.
[2] Weppelmann E R, Field J S, and Swain M V 1995 *J. Mater. Sci.* **30**, 2455.
[3] Kucheyev S O, Bradby J E, Williams J S, Jagadish C, Toth M, Phillips M R, and Swain M V 2000 *Appl. Phys. Lett.* **77**, 3373-5.
[4] Wasmer K, Parlinska-Wojtan M, Gassilloud R, Pouvreau C, Tharian J, and Micher J 2007 *Appl. Phys. Lett.***90**, 031902.
[5] Le Bourhis E and Patriarche G 2003 *Prog. Cryst. Growth Charact.* **47**, 1.
[6] Ponce F A, Wei Q Y, Wu Z H, Fonseca-Filho H D, Almeida C M, Prioli R, and D. Cherns 2009 *J. Appl. Phys.* **106**, 076106.
[7] Caldas P J, Prioli R, Almeida C M, Huang J Y, Ponce F A 2011 *J. Appl. Phys.* **109**, 013502.
[8] Fonseca-Filho H D, Prioli R, Pires M P, Lopes A S, Souza P L, and Ponce F A 2007 *Appl. Phys. Lett.* **90**, 013117.
[9] Taylor C, Marega E, Stach E A, Salamo G, Hussey L, Muñoz M, and Malshe A 2008 *Nanotechnology* **19**, 015301.
[10] Fonseca-Filho H D, Almeida C M , Prioli R, Pires M P, Souza P L, Wu Z H, Wei Q Y, and Ponce F A 2010 *J. Appl. Phys.* **107**, 054313.

AUTHOR INDEX

AUTHOR INDEX

AUTHOR INDEX

Institute of Physics
Dirac House, Temple Back
Bristol BS1 6BE UK

CURRAN ASSOCIATES INC
proceedings
.com